# Babylon

# Topoi
# Berlin Studies of the Ancient World

Edited by
Excellence Cluster Topoi

Volume 1

De Gruyter

# Babylon

## Wissenskultur in Orient und Okzident

Herausgegeben von
Eva Cancik-Kirschbaum
Margarete van Ess
Joachim Marzahn

De Gruyter

ISBN 978-3-11-022211-1
e-ISBN 978-3-11-022212-8
ISSN 2191-5806

*Library of Congress Cataloging-in-Publication Data*

Babylon: Wissenskultur in Orient und Okzident / herausgegeben von Eva Cancik-Kirschbaum, Margarete van Ess und Joachim Marzahn.
p. cm. – (Topoi. Berlin studies of the ancient world; Bd. 1)
Papers given at a conference that took place on the occasion of the exhibition "Babylon: myth and truth" held in Berlin in 2008.
Includes bibliographical references and index.
ISBN 978-3-11-022211-1 (acid-free paper) – ISBN 978-3-11-022212-8 (electronic)
1. Babylon (Extinct city) – Intellectual life – Congresses. 2. Learning and scholarship – Iraq – Babylon (Extinct city) – History – Congresses. 3. Science – Iraq – Babylon (Extinct city) – History – Congresses. 4. East and West – History – Congresses. 5. Babylon (Extinct city) – Antiquities – Congresses.
I. Cancik-Kirschbaum, Eva Christiane. II. Ess, Margarete van. III. Marzahn, Joachim.
DS70.5.B3B34 2011
935–dc22

2010050456

*Bibliografische Information der Deutschen Nationalbibliothek*

Die Deutsche Nationalbibliothek verzeichnet diese Publikation in der Deutschen Nationalbibliografie; detaillierte bibliografische Daten sind im Internet über http://dnb.d-nb.de abrufbar.



Einbandbild: Schreitender Löwe – Detail der Fassade des Thronsaals von Nebukadnezar II. im Vorderasiatischen Museum, Staatliche Museen zu Berlin. Tonziegel, farbig glasiert und gebrannt, hergestellt in Babylon (6. Jh. v. Chr.).
Foto: © bpk / Vorderasiatisches Museum, SMB / Olaf M. Teßmer
Satz: Dörlemann Satz GmbH & Co. KG, Lemförde
Druck: Hubert & Co. GmbH & Co. KG, Göttingen
∞ Gedruckt auf säurefreiem Papier

Printed in Germany

www.degruyter.com

## Berlin Studies of the Ancient World

Mit dem vorliegenden Band „Babylon – Wissenskultur in Orient und Okzident" starten die „Berlin Studies of the Ancient World". Diese altertumswissenschaftliche Publikationsreihe – herausgegeben zunächst vom Exzellenzcluster Topoi und, nach der Gründung im Jahr 2011, vom Berliner Antike-Kolleg (BAK) – versammelt die ganze Breite der in beiden Einrichtungen vertretenen Disziplinen: die Prähistorische, Vorderasiatische und Klassische Archäologie, die Ägyptologie und Altorientalistik, die antiken Sprach- und Literaturwissenschaften, die Philosophie und Theologie, und schließlich die Fächer, die sich mit der Wirkungsgeschichte der antiken Kulturen und der Geschichte des Wissens und der Wissenschaften beschäftigen. Der Titel „Topoi", den diese Reihe mit dem Exzellenzcluster gemeinsam hat, bezeichnet den weit gespannten inhaltlichen Ansatz. Als „Räume" untersucht werden die natürliche oder vom Menschen veränderte Umwelt, der verwaltete Herrschaftsbereich, das sprachlich, bildlich oder gedanklich formulierte räumliche Konzept und die kosmologische oder metaphysische Vorstellung. Wie haben sich solche räumlichen Ordnungen und Wissenssysteme im Altertum gemeinsam und in wechselseitiger Abhängigkeit voneinander entwickelt? Der Exzellenzcluster und das BAK bearbeiten diese Frage aus einer Vielzahl an Perspektiven, die sich in ihren Fächern ebenso widerspiegelt wie in den sie tragenden Berliner Institutionen: den beiden großen Universitäten, der Freien Universität Berlin und der Humboldt-Universität zu Berlin, dem Deutschen Archäologischen Institut, der Stiftung Preußischer Kulturbesitz mit ihren Museen und der Staatsbibliothek, der Berlin-Brandenburgischen Akademie der Wissenschaften und dem Max-Planck-Institut für Wissenschaftsgeschichte.

Als Herausgeber treten wir dafür ein, dass die „Berlin Studies of the Ancient World" hohen Standards in Text, Bild und Gestaltung entsprechen, dass sie zugleich aber für die wissenschaftliche Gemeinschaft weithin zugänglich sein müssen. Im Verlag de Gruyter verfolgt die neue Reihe beide Ansprüche mit einem für die Geisteswissenschaften in Deutschland bislang einzigartigen Publikationsmodell. Alle Bände werden sowohl in gedruckter Form als auch als eBook erscheinen. Ausgewählte Titel sind darüber hinaus kostenfrei zugänglich – der vorliegende Band bezeichnet auch in dieser Hinsicht den Anfang. Mit diesem Modell folgen wir in Topoi und BAK den berechtigten Forderungen nationaler und internationaler Institutionen der Wissenschaftsförderung nach der Erreichbarkeit öffentlich finanzierter Forschungsergebnisse und nutzen die Möglichkeiten des Internets als Forum weltweiter Wissensverbreitung. Wir wünschen uns, dass die Veröffentlichungen der Reihe „Topoi. Berlin Studies of the Ancient World" auf diesem Wege zum Ausgangspunkt einer internationalen, unbeschränkten Fachdiskussion werden mögen.

Friederike Fless – Gerd Graßhoff – Michael Meyer

## Berlin Studies of the Ancient World

This volume Babylon – *Wissenskultur in Orient und Okzident* (Babylon: Cultures of Knowlegde in Orient and Occident) inaugurates the "Berlin Studies of the Ancient World," a publication series whose focus is the classical and ancient world. Editorship will pass from the Excellence Cluster TOPOI to the Berliner Antike-Kolleg (BAK) following its planned establishment in 2011. The series unites the spectrum of disciplines represented by these institutions: from prehistoric, ancient Near Eastern, and Classical archaeology to Egyptology, Assyriology and Classics, from the linguistics of ancient civilizations and their literatures to history, philosophy, theology, and the historical impact of ancient cultures on the history of knowledge and the sciences. The series title "TOPOI," shared by the Excellence Cluster of the same name, indicates the full breadth of the topics addressed and the methods associated with them. The series will explore, under the category of space, the natural environment and its transformation by human societies; the administered spaces of governmental authority; linguistic, pictorial, and spatial concepts as expressed in different media, and finally cosmological and metaphysical models of space as well. How did such spatial orders and knowledge systems evolve jointly and interdependently in ancient societies? The Excellence Cluster TOPOI and the BAK are addressing these questions from a number of perspectives, as reflected in the subjects represented as well as in the network of Berlin-based institutions, which are supporting the BAK: the two major universities, Freie Universität Berlin and Humboldt-Universität zu Berlin, the German Archaeological Institute (DAI), the Stiftung Preußischer Kulturbesitz (SPK), the Berlin Brandenburg Academy of Sciences (BBAW), and the Max Planck Institute for the History of Science (MPIWG).

As the editors of the series, we shall strive to ensure that "Berlin Studies of the Ancient World" conforms to the highest standards of quality with regard to text, image, and design, and that it is at the same time widely accessible to the research community as a whole. Published by de Gruyter, the series will pursue both objectives on the basis of a model that remains unique to date for the humanities and social sciences in Germany: all volumes will appear simultaneously in printed form and as e-books. Moreover, selected titles will be available free of charge – in this regard as well, the present volume represents a first. This model allows TOPOI and the BAK to conform to the legitimate expectations of a national and international institution devoted to the promotion of scholarship, namely that the results of publicly-financed research should be made as accessible as possible, and we are committed to exploiting the resources of the Internet as a forum for the worldwide dissemination of knowledge. It is our hope that the publications of the series "TOPOI. Berlin

Studies of the Ancient World" will serve as points of departure for a wide-ranging and international scholarly discussion.

Friederike Fless – Gerd Graßhoff – Michael Meyer

# Inhaltsverzeichnis

**Babylon – Wirkungsgeschichten**

# Einführung

„Babylon – Wissenskultur in Orient und Okzident" lautete das Thema einer internationalen Tagung, die vom 26. bis 28. Juni 2008 im Theodor-Wiegand-Saal des Pergamon-Museums in Berlin stattfand. Die Tagung begleitete die Eröffnung der Ausstellung „Babylon – Mythos und Wahrheit", in der versucht wurde, die heutigen Wissens- und Überlieferungshorizonte zum Begriff „Babylon" und seiner Umwelt nachzuzeichnen.[1] Der Schwerpunkt der Tagung lag auf der Bedeutung der Stadt Babylon als Ort von und als Metapher für Gelehrsamkeit, Weisheit und Wissenschaft.

Die Anfänge Babylons liegen weitgehend im Dunkeln. Im zweiten Jahrtausend entwickelt sich aus einem altorientalischen Stadtstaat mittlerer Dimension in der nördlichen Alluvialebene des Zweistromlandes die Hauptstadt eines Territorialreiches. Die Geschichte dieses babylonischen Reiches kennt Phasen der politischen Vormachtstellung und ebenso der Fremdherrschaft und des Niedergangs. Die Einnahme Babylons durch die Truppen des Perserkönigs Kyros im Jahre 539 v. Chr. beendet die politische Selbständigkeit Babyloniens – nicht aber die Geschichte der Stadt. Denn die politische Bedeutung Babylons war eng verknüpft mit seiner Rolle als Heimat und Kultort des Stadt- und späteren Reichsgottes Marduk, wie Babylon auch als Verkehrs- und Handelsmetropole überregionale Bedeutung besaß. Die Wechselbeziehungen zwischen Religion und Politik werden im 'Aufstieg' des Marduk vom Stadtgott an die Spitze des traditionsreichen mesopotamischen Pantheons besonders auffällig. Auch wenn die Mechanismen dieses Prozesses und die Gewichtung weiterer Faktoren individueller und institutioneller Art mit politischen, wirtschaftlichen und kulturellen Rahmenbedingungen noch wenig klar sind, liegt vermutlich in diesem Zusammenspiel der Schlüssel für die Entwicklung der Stadt zu einem Zentrum mesopotamischer Gelehrsamkeit, und ihre Bedeutung als Hort tradierten Wissens und höchster theologischer und wissenschaftlicher Autorität, die den Untergang der altorientalischen Reiche überdauern sollte. In dem dichten Strom aus Legenden, Fiktionen und Mythen, in dem diese Stadt fortlebt, sind unerschöpfliches Wissen und eine tiefe Kenntnis von den innersten Zusammenhängen des Kosmos zentrale Topoi. Chaldäische Weisheit auf Gebieten wie Himmelskunde, Vorzeichenwissenschaft (Divination), Medizin, Mathematik und Theologie, aber auch in Methoden der Schriftauslegung und Kommentierung verbindet nicht nur die zeitgenössischen Kulturen der Alten Welt mit dem Zentrum der Gelehr-

1 Die Berliner Ausstellung ist in zwei Begleitbänden dokumentiert „Babylon – Wahrheit", hrsg. v. Joachim Marzahn und Günther Schauerte; „Babylon – Mythos", hrsg. v. Moritz Wullen und Günther Schauerte – jew. München 2008.

samkeit im Zweistromland. Transformiert oder deklassiert zu Esoterik, Magie, Okkultismus und praewissenschaftlicher Spekulation ziehen sie sich durch viele Wirkungsgeschichten in Orient und Okzident.

Die 'Wiederentdeckung' nicht nur der Stadt Babylon, sondern des Kulturraumes *Alter Orient* eröffnete im Verlaufe des 19. Jh. n.Chr. den Zugang zu Originalquellen, die lange verschüttet waren. Monumente, Objekte, Bilder, vor allem aber die große Zahl der keilschriftlichen Texte bieten seither Einblick in mehr als drei Jahrtausende mesopotamischer Wissenskultur. Und sie zeigen: In der Hauptstadt des babylonischen Reiches hatte sich eine Gelehrtenkultur entwickelt, die bis in die Spätzeit großes Prestige in der gesamten Alten Welt genoss. In den Keilschrifttexten der Bibliotheken Babylons, aber ebenso in den Bauwerken der Stadt, in den Objekten und Monumenten werden Technologie und technisches Wissen, Logistik und anwendungsorientierte Wissensbereiche, systematisch organisierte Wissenschaften, Langzeitbeobachtungen sichtbar. Die mächtigen Institutionen dieser Wissenskultur, zuständig für Produktion, Vermittlung und Speicherung, treten hervor, ihre Interdependenz mit Politik und Wirtschaft, ihre Einbindung in religiöse Vorstellungswelten werden sichtbar. Die Wege des Wissens in der Alten Welt, nah und fern, die Personen, ihre Biographien, ihre Fragen werden in Texten und materieller Kultur greifbar. Mehr und mehr wird an diesem wiedererstehenden Babylon verständlich, weshalb dieser Ort immer *auch* Inbegriff des antiken Weltwissens war. In der Konfrontation mit den Originalquellen ergibt sich die Chance, die tatsächlichen Inhalte, Formen und Überformungen zu präzisieren und damit den Einfluss Babylons, des Alten Orients auf die griechische Philosophie und Wissenschaft sowie auf die jüdische und arabische Wissenschaftstradition – und damit letztendlich auch auf die Wissenstraditionen der Moderne – näher zu bestimmen.

Gleiches gilt auch für die archäologischen Hinterlassenschaften Babylons. Zwischen 1899 und 1917, bzw. erneut zwischen 1962 und 1990 von deutschen, italienischen und irakischen Archäologen mittels wissenschaftlicher Ausgrabungen untersucht, konnten nicht nur die beeindruckende Gesamtfläche der Stadt festgestellt, sondern auch die äußerst monumentalen Paläste und Tempelbereiche der neubabylonischen Zeit nachgewiesen und wieder erfahrbar gemacht werden. Die Kenntnis der Topographie Babylons, auch seines Heiligtums für Marduk mit dem „Turm von Babylon" trägt wesentlich zum Verständnis der Texte und der theologischen Deutung der Stadt bei und erlaubt uns gleichzeitig, ein materiell greifbares Verständnis der damaligen Architektur zu entwickeln. Auch die Hausarchitektur mit ihren vielfältigen Objektfunden gibt einen tiefen Einblick in die Alltagskultur und die erstaunlich entwickelte Wohnqualität der Zeit. Große Mengen an dort aufgefundenen beschrifteten Objekten und Textdokumenten gehören neben sonstigen archäologischen Funden zu den Fundamenten der Babylon-Forschung. Noch allerdings sind nur vergleichsweise kleine Bereiche der Stadt ergraben. Limitiert durch den hohen Grundwasserspiegel, die schiere Größe der Stadt und die seit Jahrzehnten erschwerenden politischen Umstände stagniert die archäologische Forschung. Gerade die Anfänge Babylons, aber auch die Besiedlung aus der Zeit nach dem Ende der mesopotamischen Reiche bedürfen daher weiterer Untersuchung.

Die Tagung versuchte den Wissensort Babylon als Teilphänomen der antiken und nachantiken Kultur- und Wissensgeschichte zu fassen. Von besonderem Interesse ist dabei der 'normierende' Charakter, der von diesem Ort ausgeht. Er ist nur zum Teil an die schriftliche Autorität der babylonischen Texttradition gebunden, und er resultiert offenbar auch aus der Autorität des 'Vor-Ort-Studiums', d.h. der räumlichen Anbindung an die existierende altorientalische Wissenstradition, wie sie in Babylon ihre spezifische Ausprägung findet. Die Tagungsbeiträge befassten sich – ausschnitthaft und paradigmatisch – mit Aspekten wie der Struktur der Gelehrten-Gruppen, der Soziologie des Wissens, mit Archiven und Bibliotheken als Wissens-Speichern, mit den an den Ort Babylon und seine mythotheologische Bedeutung gebundenen Hermeneutiken. Mit Blick auf die Wirkungsgeschichte traten die Rolle Babylons als Referenzort für griechische Philosophie und Wissenschaft ebenso in den Blick wie die Rolle babylonischer Wissenskultur in der jüdischen und arabisch-islamischen Tradition. Mindestens ebenso wichtig wie der Blick in die Vergangenheit und die Resümierung des bereits 'Gewussten' war jedoch die Frage nach den Perspektiven der Forschung an und über Babylon sowie nach dem Umgang mit den nicht oder nur teilweise publizierten Funden und Befunden. Ebenso wie man in diesem Band den expliziten Blick nach Osten, also den Austausch mit den östlich angrenzenden Kulturregionen und darüber hinaus vermisst, wird man mit Recht anmerken, dass Wissen und Wissenskultur nicht auf textgebundene Überlieferungen beschränkt sind, sondern Artefakte, Architektur und Befunde Teil dieser Phänomene sind. Allerdings kann gegenwärtig die weitere Erforschung der archäologischen Denkmäler und Befunde nur sehr eingeschränkt fortgeführt werden. Die Zukunft wird zeigen, wie weit es gelingt, sowohl das Studium der direkten materiellen Zeugen als auch jenes der geistigen Hinterlassenschaften wieder enger miteinander zu verzahnen. Die Konferenz konnte hierzu lediglich einen Ausblick bieten und auf die lohnende Notwendigkeit weiterer Feldforschungen verweisen sowie dies die politischen Umstände zulassen.

Dieser Band versammelt die schriftlichen Ausführungen von 22 Vorträgen[2] und wir möchten an dieser Stelle Referenten und Autoren für ihre Beiträge – mündlich wie schriftlich – sehr herzlich danken. Den Auftakt bildet unter dem Rubrum **Babylon – Heute** ein Blick auf die Situation der Ruinenstätte des antiken Babylons, wie sie sich zu Beginn des 21. Jahrhunderts darbietet. John Curtis analysiert in einer Retrospektive *The Present Condition of Babylon* auf die Ruinen von Babylon zwischen 1914 und 2003 Eingriffe, Veränderungen und Zerstörungen der archäologischen Monumente. Die aktuellen Schadensbilder, aber auch ein Ausblick auf die seit 2004 in Kooperation mit der UNESCO entwickelten und seit 2009 in bilateraler Kooperation zwischen der irakischen Antikenverwaltung

2 Das vollständige Programm findet sich im Anschluss.

und dem World Monuments Fund tatsächlich begonnenen Stabilisierungs- und Konservierungsarbeiten sind Gegenstand des Beitrags von Maryam U. Musa *The Situation of the Babylon Archaeological Site until 2006*.

Aus der systematischen Erforschung der Ruine Babylon seit 1899 stammen stratifiziert freigelegte Archive und Bibliotheken, die die Rolle von systematischen Akkumulationen von Wissen in unterschiedlichen institutionellen wie privaten Kontexten über zwei Jahrtausende dokumentieren. Olof Pedersén gibt in *Excavated and Unexcavated Libraries in Babylon* einen Überblick über den Stand der Forschung. Eindrücklich wird deutlich, welche Perspektiven in der intensiven Erforschung gerade dieser Überlieferung liegen.

Unter **Babylon – Hauptstadt** steht das Wissen um die politische und historische Verortung der Stadt Babylon im Mittelpunkt. Wilfred G. Lambert fragt in seinem Beitrag *Babylon: Origins* nach den Indizien, die sich für die Ursprünge Babylons aus dem Namen der Stadt und dem Namen des Stadtgottes, Marduk, ergeben. Dabei wird deutlich, dass die Stadt Sippar eine besondere Rolle für die Frühzeit Babylons gespielt haben muss. Der Weg Babylons zu der *einen* „Hauptstadt“ im Sinne des Wortes ist Gegenstand des Beitrags von Dominique Charpin, *Babylon in der altbabylonischen Zeit: eine Hauptstadt von vielen ... die als einzige übrig blieb*: Neben der Erringung der politischen Vormachtstellung spielt für das spätere Bild von Babylon die Rolle der Stadt in der Bewahrung sumerisch-akkadischer Traditionen eine wichtige Rolle. Diesem Blick auf die Formation des Zentrums setzt Frederick M. Fales in *Moving Babylon: on the Aramean and Chaldean presence in Southern Mesopotamia* die Perspektive der chaldäischen und aramäischen Stämme entgegen, die im 1. Jahrtausend einen wichtigen Faktor der Lokal- und Regionalgeschichte darstellen. Francis Joannès beschäftigt sich mit der Rolle öffentlicher monumentaler Schriftlichkeit in spätbabylonischer Zeit. Sein Beitrag *L'écriture du pouvoir à Babylone sous Nabuchodonosor II* zeigt, in welchem Umfang sakrales und profanes Bauen als Teil einer Erinnerungskultur in tradierten Topoi und ihren Transformationen thematisiert wird.

**Babylon – Wissensort** ist der Frage nach den systematischen und epistemischen Ausprägungen von Wissen gewidmet. Dina Katz untersucht in ihrem Beitrag *Recycling Mythological Traditions Toward a New Theology* das Weltschöpfungsepos Enuma Eliš mit Blick auf die intentionale Verarbeitung verschiedener Traditionen. Als Auftragsarbeit der politischen Elite schafft die literarische Komposition eine mythologisch-historische Begründung für die Bedeutung Babylons und illustriert in besonderer Weise die Aktualisierung tradierten in Mythen gefassten theologischen Wissens und der darin begründeten Weltsichten und Weltordnungen in zeitgenössischen Kontexten. Systematische Beschreibung und Reflexion solch universaler Ordnungsstrukturen bilden das Thema des Beitrages *Die Wissenschaft von der Zukunft. Überlegungen zur Bedeutung der Divination im Alten Orient* von Stefan M. Maul. Aus der Zusammenschau von Methodik, zugrunde liegendem Weltbild und konkreter Organisation dieses extrem einflussreichen Wissenssektors wird seine Funktion als Korrektiv zu den hierarchischen Strukturen der gesellschaftlichen Ordnung deutlich. Mark J. Geller beschäftigt sich in seinem Aufsatz über *Die theoretischen Grundlagen der* *babylonischen Heilkunde* mit dem Problem der scheinbar fehlenden theoretischen Hinterlegung praktischen Wissens. Mit Imagination, Logik, Beobachtung und Erfahrung verfügt die babylonische Heilkunde über epistemologische Verfahren, die der vor-hippokratischen Medizin nicht unähnlich sind. Hans Neumann wendet sich mit seinen *Bemerkungen zu einigen Aspekten babylonischen Rechtsdenkens im Spannungsfeld von Theorie und Praxis* einer vergleichbaren Problematik für das Gebiet der Rechtskunde zu. Während die Gesetzessammlungen sowie Rechts-, Gerichts- und Prozessurkunden Zeugnisse der juristischen Praxis sind, belegen literarische Prozessurkunden, dass die Schreiber neben einer theoretischen juristischen Ausbildung auch Fähigkeiten im Bereich juristischer Diskussion erhielten, womit Theorie und Praxis in einem produktiven Verhältnis zueinander standen. Damit rückt die Frage nach der Rolle normierender und normierter Texte für die Tradierung und Vermittlung von Wissen in den Blick. Nils P. Heeßel geht in *„Sieben Tafeln aus sieben Städten". Überlegungen zum Prozeß der Serialisierung von Texten in Babylonien in der zweiten Hälfte des zweiten Jahrtausends v. Chr.* der Frage nach der Genese kanonischer Textserien nach. An einer Reihe von Texten wie z. B. der Serie *„šumma ālu"* oder der vierten Tafel der *bārûtu*-Serie kann gezeigt werden, dass die Standardisierung von Texten im 2. Jahrtausend bereits sehr fortgeschritten war, die Serienbildung aber erst später, nach der Kassitenzeit eingesetzt haben dürfte. Wilfried van Soldt beschreibt die Rolle Babylons für die Weitergabe von systematischem Wissen in Nachbar- und Randgebiete der Keilschriftkulturen anhand von Schultexten aus Anatolien, Syrien, der Levante, Ägypten und Assur. In seinem Beitrag *The Role of Babylon in Western Peripheral Education* kann er anhand der babylonischen lexikalischen Serie ḪAR-ra = *ḫubullu* belegen, dass in den westlichen Randgebieten Schulmaterial aus Babylonien, vielleicht aus Babylon selbst, verwendet wurde, welches möglicherweise von den aus Babylon stammenden Lehrern mitgebracht wurde. Mathieu Ossendrijver schließlich lenkt mit *Science in Action: Networks in Babylonian Astronomy* den Blick auf die sozialen Strukturen und Arbeitszusammenhänge von Wissenschaft. Er untersucht, wie die babylonischen Astronomen der Seleukidenzeit als Teilnehmer eines Netzwerkes von Tempelgelehrten gemeinsam an astronomischen Beobachtungen und Berechnungen zusammengearbeitet haben. Als Hauptquelle für die Rekonstruktion des Netzwerkes dienen die Kolophone der gelehrten Texte aus der Bibliothek des Rēš-Tempels in Uruk.

Dem Blick der Zeitgenossen, Nachbarn und Epigonen ist der vierte Abschnitt des Bandes **Babylon – Wirkungsgeschichten** gewidmet. Er beleuchtet ausschnitthaft die synchronen und diachronen Bewegungen, in denen Wissen aus Babylon und über Babylon als Ort des Wissens in der Alten Welt zirkulierte.

Den Griechen war bewusst, vieles von den Völkern des Orients übernommen und perfektioniert zu haben. Hierbei kann neben Ägypten besonders auf Babylon als Teil des literarischen Diskurses, Ort von allgemeinem Interesse und Stadt der Weisheit verwiesen werden. Michael Erlers Beitrag *Chaldäer im Platonismus* zeigt mögliche Einwirkungen babylonischer Vorstellungen und die Rolle der barbaros philosophia für das griechische

Denken. Umgekehrt fragt Daniel T. Potts in seinem Aufsatz *The politai and the bīt tāmartu: The Seleucid and Parthian Theatres of the Greek Citizens of Babylon* nach institutionalisierten Orten des Wissens griechischer Prägung in Babylon. Die Existenz eines griechischen Theaters und griechischer Immigranten in Babylon kann anhand der Ausgrabungsbefunde und eindeutiger Aussagen in den „Astronomischen Tagebüchern" des 2.–1. Jh. v. Chr. belegt werden.

Exemplarisch für die Rezeption Babylons in den Qumran-Schriften analysiert Reinhard G. Kratz mit *Nabonid in Qumran* das fragmentarisch erhaltene „Gebet des Nabonid" und das biblische „Buch Daniel". Die Untersuchung dieser und weiterer Texte aus Qumran zeigt den Weg des letzten Chaldäerherrschers in die aramäische Überlieferung und seine Bedeutung für Qumran. Dass die Perspektiven auf Babylon sehr unterschiedlich sind, macht der Beitrag von Willem F. Smelik, *Myth and Truth: Babylon in the Babylonian Talmud* deutlich, wenn die Rezeption Babylons unter drei verschiedenen Gesichtspunkten untersucht wird: Wie sieht das babylonische Judentum das antike Babylon? Wie konstruiert das babylonische Judentum seine eigene Geschichte in Babylonien? Und schließlich, wie werden Akkulturationsprozesse des babylonischen Judentums im sassanidischen Babylonien bewertet? Die Frage nach den Kenntnissen der Kirchenväter über die antike Stadt Babylon und die Untersuchung ihrer geschichtlichen und landeskundlichen Bemerkungen über die Stadt sowie eine Synchronisierung mit verschiedenen Ausgrabungsfunden ergibt ein ambivalentes Bild, wie Christoph Markschies in *Babylon bei den Kirchenvätern* darstellt. Die Kirchenväter verfügten offenbar über relativ unspezifische Kenntnisse über Babylon. Demgegenüber stehen antike christliche Texte, in denen Babylon besonders negativ konnotiert dargestellt wird. Insgesamt scheint dies der Tendenz der Zeit zu entsprechen, Städte enthistorisiert und theologisch überhöht darzustellen. Die hebräische Bibel und die Schriften der Kirchenväter zählten – neben griechischen und lateinischen Autoren – zu den wichtigsten Quellen der Renaissancegelehrten über den Alten Orient. In der wissenschaftlichen Diskussion über die Ursprache der Menschheit spielte die Erzählung vom Turmbau zu Babel eine wichtige Rolle. Jean-Jacques Glassner referiert in *Noch einmal: Der Turm zu Babel und die vielen Sprachen* ein Stück früher Forschungsgeschichte, in der die Suche nach Herkunft und Identität der Kelten in der Wiederentdeckung Assurs und Babylon als Wiege der Menschheit mündete. Eine ganz andere Form des Nachlebens dokumentiert der Beitrag von Gotthard Strohmaier, *Ḫarrān – die Stadt des Sin in islamischer Zeit*. Dem von Nabonid von Babylon geförderten Kult des Mondgottes Sîn war in der Stadt Ḫarrān ein langes Nachleben beschieden. Untersuchungsgegenstand sind die ḫarrānischen Sabier, die sich weder zum Christentum noch zum Islam bekannten, und durch den am Hof tätigen Astronomen Ṯābit ibn Qurra (um 826–901) eine Vorrangstellung in Bagdad erworben hatten. Als eine wichtige Quelle zur Rekonstruktion babylonischer Traditionen dienen sabische Metallarbeiten mit astronomischen Tierkreiszeichen, deren Motive bis in die abendländische Kunst des Mittelalters fortbestanden. Der Beitrag Andrzej Reiches über *Die altorientalischen Denkmäler in den Museen in Polen* beschließt den

Band mit einem ausführlichen Überblick über den Bestand altvorderasiatischer Sammlungen in polnischen Museen seit dem 19. Jahrhundert einschließlich Ankauf-, Schenkungs- bzw. Ausgrabungsgeschichte. Er steht exemplarisch für die Geschichte von Artefakten aus Babylon in Museen und Sammlungen auf der ganzen Welt und führt auf diese Weise gewissermaßen in die Gegenwart und den Anlass der Tagung zurück.

In Vorbereitung und Durchführung der Tagung haben mit dem Deutschen Archäologischen Institut, dem Institut für Altorientalistik der Freien Universität und dem Vorderasiatischen Museum zu Berlin drei Institutionen kooperiert, denen sowohl die Erforschung der altorientalischen Kulturen als auch die Vermittlung der Forschungsergebnisse zentrales Anliegen ist. Eine öffentliche wissenschaftliche Tagung im Herzen der Ausstellung „Babylon – Mythos und Wahrheit" und damit *medias in res* im Sinne des Wortes symbolisiert diese beiden Anliegen in besonderer Weise. So gilt unser Dank dem Präsidenten der Stiftung Preußischer Kulturbesitz Prof. Dr. Dr. h.c. mult. Hermann Parzinger, dem damaligen Generaldirektor der Staatlichen Museen Prof. Dr. Peter-Klaus Schuster und der Direktorin des Vorderasiatischen Museums Berlin Prof. Dr. Beate Salje, die den Theodor-Wiegand-Saal des Pergamon-Museums als Tagungsort zur Verfügung stellten und der Tagung großzügige Gastgeber waren.

Stellvertretend für all jene, die durch ihren tatkräftigen Einsatz zum guten Gelingen dieser Tagung beigetragen haben, möchten wir an dieser Stelle Dr. Nicole Brisch, Dr. Grégory Chambon, Christine Kainert, William Pethe und Marco Ruhlig sehr herzlich danken.

Für die finanzielle Förderung der Tagung danken wir dem Exzellenzcluster 264 *Topoi – The Formation and Transformation of Space and Knowledge in Ancient Civilizations*, dem Außenamt der Freien Universität Berlin und dem Deutschen Archäologischen Institut.

Besonderer Dank gebührt der Deutschen Orient-Gesellschaft, der die Förderung der wissenschaftlichen Beschäftigung mit Babylon seit den Anfängen der Grabungen im Jahre 1899 ein besonderes Anliegen ist.

Für die Aufnahme der Tagungsbeiträge in die Reihe „TOPOI – Berlin Studies of the Ancient World" danken wir den Sprechern Prof. Dr. Friederike Fless und Prof. Dr. Christoph Rapp. Das Verlagshaus W. de Gruyter, namentlich die Lektoren Dr. Sabine Vogt und Dr. Mirko Vonderstein haben die Drucklegung sachkundig begleitet. Zugleich mit der gedruckten Ausgabe des Bandes erscheint eine e-Version, die online über das Verlagsportal zugänglich sein wird. Leider können aus urheberrechtlichen Gründen einige Abbildungen nicht gezeigt werden, doch abgesehen davon entspricht die e-Version dem gedruckten Band.

Im Sinne wissenschaftlicher Mehrsprachigkeit und unterschiedlicher Fachgepflogenheiten haben wir auf eine orthographische Vereinheitlichung in der Schreibung von Eigennamen verzichtet. Auch Siglen und Zitierweise antiker Texte in den einzelnen Beiträgen richten sich nach den in der jeweiligen Disziplin gebräuchlichen Formen. Die englische Fassung des arabischen Beitrags von Maryam U. Musa überarbeitete Margarete van Ess auf Basis einer Übersetzung von Naima Ghanma. Für die Übersetzung der Kurzfassungen ins

Arabische danken wir Dr. Laith Hussein, für die Durchsicht Hussein Bakkor M.A. Für die englischen Kurzfassungen danken wir Dr. Justin C. Johnson. Und – last but not least – möchten wir Kerstin Maiwald M.A. und Sandra Feix M.A. danken, die mit großem Engagement am Zustandekommen des Bandes mitgewirkt und u.a. die Herstellung der Druckvorlagen und der Indices besorgt haben.

Eva Cancik-Kirschbaum – Margarete van Ess – Joachim Marzahn

## "Babylon – Wissenskulturen zwischen Orient und Okzident" – Programm, Referenten und Podiumsgäste

**Donnerstag, 26.6.2008**

**Begrüßung und Eröffnung**

**J. Curtis (London)**, *The Condition of Babylon in December 2004*

**E. Stone (New York)**, *The View from Above: Damage to Babylon and other Iraqi Sites as Seen from Space*

**M. U. Musa (Bagdad)**, *Damages sustained in the Ancient City of Babylon*

**M. Erler (Würzburg)**, *Chaldäer im Platonismus*

**W. Smelik (London)**, *Babylon and the Babylonian Talmud*

**D. T. Potts (Sydney)**, *The politai & the bīt tāmartu: The Seleucid & Parthian Theatres of the Greek Citizens*

**D. Katz (Leiden)**, *Reconstructing Babylon: Recycling Old Traditions in the Service of a New Theology*

**Freitag, 27.6.2008**

**W. G. Lambert (Birmingham)**, *The Background of Babylon and its Rise to Eminence*

**S. M. Maul (Heidelberg)**, *Zukunftswissen und die Wissenschaft von der Zukunft*

**M. J. Geller (London)**, *Where is "Theory" in Babylonian Medicine?*

**H. Neumann (Münster)**, *Babylonisches Rechtsdenken im Spannungsfeld von Theorie und Praxis*

**O. Pedersén (Uppsala)**, *Excavated and Unexcavated Libraries in Babylon*

**S. Shukri (Bagdad)**, *Babylonian Sites on the Euphrates*

**N. P. Heeßel (Heidelberg)**, *Sieben Tafeln aus sieben Städten: Babylons Rolle als Zentrum der Wissenschaft*

**W. van Soldt (Leiden)**, *Wissensexporte: Babylon und die Curricula der Peripherie*

**M. Ossendrijver (Tübingen)**, *Exzellente Netzwerke in der babylonischen Astronomie*

**D. Charpin (Paris)**, *Babylone à l'époque paléo-babylonienne: une capitale parmi des autres ... qui est restée la seule*

**Sonnabend, 28.6.2008**

**Chr. Markschies (Berlin)**, *Babylon bei den Kirchenvätern*
**C. Wunsch (Jena/Leipzig)**, *Das babylonische Recht und seine Rezeption durch exilierte Judäer*
**I. Finkel (London)**, *The Cyrus-Cylinder*
**F. Joannes (Paris)**, *L'écriture publique du pouvoir dans la Babylonie du Ier millénaire*
**M. F. Fales (Udine)**, *Babylon Surrounded: Arameans and Chaldeans*

**Sonnabend, 28.6.2008**

**R. Kratz (Göttingen)**, *Nabonid in Qumran*
**G. Strohmaier (Berlin)**, *Ḫarrān – die Stadt des Sin in arabischer Zeit*
**A. Reiche (Warschau)**, *Altvorderasiatische Sammlungen und die Rezeption Babylons in Polen*
**J.-J. Glassner (Paris)**, *Noch einmal: Der Turm zu Babel und die vielen Sprachen*

Als Podiumsgäste nahmen teil:
Vertreter des College of Arts der Universität Bagdad, der Generaldirektion der irakischen Antikenverwaltung sowie des Kulturausschusses des irakischen Parlaments.

# Babylon – Heute

*John Curtis*

## The Present Condition of Babylon

When I was asked to speak at this conference, the title I was given was "The Condition of Babylon in December 2004", reflecting the fact that in December 2004 a meeting was held in Babylon to assess the condition of the site prior to its transfer from military control back to the care of the Iraq State Board for Antiquities and Heritage (hereafter SBAH). However, while it would be perfectly possible to speak about the condition of Babylon at that particular moment, it would be of greater interest and more meaningful to put this subject into a wider context and consider what happened at Babylon both before and after the period of military occupation. Therefore I am proposing to describe the situation at Babylon from the end of the German excavations in 1917 up to the present day, and accordingly I have changed the title of the lecture to "The Present Condition of Babylon"[1].

In the course of the German excavations directed by Robert Koldewey between 1899 and 1917, much was discovered about the topography of ancient Babylon and many of the buildings in the centre of the ancient city were wholly or partially excavated.[2]

However, the remains of mudbrick buildings are notoriously difficult to preserve and once they are exposed to the elements they rapidly deteriorate. The disintegrating mudbrick buildings, plus the trenches and spoil tips that are a feature of all major excavations, combined to present a site that was difficult to interpret and visually unappealing. This is reflected in early views of the ruins (e.g. Fig. 1).

There were, it is true, some redeeming features. The foundations of the Ishtar Gate, with unglazed molded bricks showing bulls, lions and dragons, were still visible, and the Lion of Babylon was a well-known and popular attraction for visitors to the site. Although the site of the Hanging Gardens, one of the Seven Wonders of the World, has never been convincingly identified, they figured largely in the public imagination (e.g. Fig. 2).

All these features, combined with the fame and notoriety of ancient Babylon, were enough to ensure that after the state of Iraq came into existence in 1920, Babylon became one of the iconic symbols of the new state. Thus, the Lion of Babylon was shown on various Iraqi stamps from 1941 onwards (Fig. 3).

In an effort to improve the quality of the visitor experience at Babylon in the 1960s, the Ninmakh Temple was reconstructed in is entirety and in the centre of the ancient site a half-size replica of the above-ground part of the Ishtar gate was built, showing the brilliantly co-

1 See also J.E. Curtis, The Site of Babylon Today, in: I.L. Finkel/M.J. Seymour (eds.), Babylon. Myth and Reality, London 2008, 213–220.

2 R. Koldewey, Das wieder erstehende Babylon, hg. von B. Hrouda, München 1990.

Fig. 1 | Postcard showing the 'Ruins of Babylon', 1932 (J. E. Curtis).

Fig. 2 | Colored postcard featuring 'The Hanging Gardens of Babylon', c. 1935 (J. E. Curtis).

Fig. 3 | 8-fils stamp showing the Lion of Babylon, first issued 1943 (J. E. Curtis).

lored glazed bricks with representations of bulls, lion and dragons (Fig. 4). The upper part of the gate had in fact been found in thousands of fragments which were taken to Berlin and painstakingly restored to produce the monument that is now visible in the Vorderasiatisches Museum. In spite of these welcome innovations at the site itself, however, Babylon still remained a disappointment for most visitors.

It was not altogether surprising, therefore, that Saddam Hussein should have attempted to turn Babylon into a symbol in keeping with his nationalistic ambitions for the republic of Iraq. Accordingly, the 'Archaeological Restoration of Babylon Project', managed by the SBAH, commenced on 14$^{th}$ February 1978, and had as its objective an ambitious and extensive reconstruction of the centre of the ancient city. Even the Iraq-Iran war, which began in September 1980, was not allowed to interfere with or hold up the plans for the reconstruction, which continued unabated. As early as 1982, a set of seven coins was issued commemorating the restoration project and bearing images of Babylon. These show the Ishtar Gate, the Lion of Babylon, the Stele of Hammurabi (now in the Louvre in Paris) and a reconstruction of the Babylon ziggurat. These coins were distributed in a presentation case together with a small metal tablet with a cuneiform inscription recording the fact that Saddam Hussein was the restorer of Babylon (Fig. 5). Hand in hand with the 'restoration' project was a large amount of excavation work undertaken by the SBAH, much of it reported on in *Sumer* volume 41.[3]

3 M. S. Demirji (ed.), Researches on Babylon in the Two International Symposiums, 2$^{nd}$ and 3$^{rd}$, held in 1979 and 1981, Sumer XLI, Baghdad 1983.

Fig. 4 | Half-sized replica of the Ishtar Gate in the centre of Babylon (J. E. Curtis).

The 'restoration' project was on a vast scale, and involved the creation of three artificial lakes on the site and the formation of three gigantic artificial mounds. On the top of one of these mounds, overlooking the site, a palace was built for Saddam Hussein. Also in the course of the 'restoration' project, the Southern Palace of Nebuchadnezzar II (604–562 BC), containing five major courtyards and about 250 rooms, was largely rebuilt. One of the grandest rooms, the throne room of Nebuchadnezzar II (the Al-'Arsh Hall), was adapted for use for concerts and receptions. The main entrance to the palace was rebuilt to a height of 30 m and many of the walls were rebuilt to a height of almost 20 m. For the reconstruction, new bricks were laid on top of bricks surviving from the original structure. Many of the original bricks were stamped with an inscription in Babylonian cuneiform of Nebuchadnezzar II, and Saddam Hussein continued this ancient practice by having many of the new bricks stamped with his own inscription in Arabic, reading in translation 'In the era of Saddam Hussein, protector of Iraq, who rebuilt the Royal Palace'. There was further restoration work on buildings including the Ishtar Gate and the Processional Way, and on the Hellenistic period theatre, where seating was restored so that 2500 people could be accommodated for performances. Also, the Lion of Babylon was given pride of place in open ground adjacent to the Southern Palace. Clearly, much of the 'restoration' work undertaken at Babylon during the time of Saddam Hussein went far beyond what is normally acceptable and has been roundly condemned by conservation groups.

Fig. 5 | Set of 7 coins bearing images of Babylon issued in 1982 to commemorate the 'Archaeological Restoration of Babylon Project'(J. E. Curtis).

Fig. 6 | Medal showing Saddam Hussein with Nebuchadnezzar issued to commemorate the First Babylon International Festival in 1987 (Trustees of the British Museum).

Fig. 7 | 25-dinar banknote issued in 2001 showing the Ishtar Gate and the Lion of Babylon (J. E. Curtis).

The first phase of the reconstruction project was mostly finished in time for the First Babylon International Festival in September 1987. This was a lavish event lasting for a month and featured music and dance by performers from about 30 different countries. In the promotional literature issued by the Festival Committee, Saddam Hussein was compared with great figures from Babylonian history like Hammurabi and Nebuchadnezzar II, and a special medal was issued showing the profiled portrait of Saddam Hussein overlapping that of Nebuchadnezzar II (Fig. 6).

The 1988 festival was even grander, but lasted for only 10 days. There were processions of figures dressed in Babylonian costumes and playing music on Babylonian style musical instruments, and at the end of the Processional Way there were gigantic portraits of Saddam Hussein and Nebuchadnezzar II. After that, festivals were held nearly every year up until 2002, sometimes accompanied by an archaeological conference, and as late as 2001 a new 25 dinar banknote was issued showing on the reverse the Ishtar gate and the Lion of Babylon (Fig. 7).

In the time of Saddam Hussein, then, the treatment of Babylon was cavalier, to say the least, but it was treated with scarcely any more respect by coalition troops after 2003. To start with, there were few problems. Thus, during the invasion and conquest of Iraq by the USA and its allies in March-April 2003 there was relatively little damage at Babylon. The small museum near the entrance to the site had been cleared out and the entrance bricked up before the war began, which was a sensible precaution, but the nearby gift shop was looted and burnt. The only other damage noted by a British Museum group that visited the site on 18th June was to one of the *mušḫuššu* (dragon) figures in the foundations of the Ishtar Gate. Although Babylon had been established on 21st April 2003 as a 'Forward Combat Operations Centre' for American forces, who were soon joined by Polish troops, the military presence in June 2003 appeared to be still on a modest scale, even though there were American troops based in Saddam Hussein's palace and Polish military vehicles were parked quite close to the Ishtar Gate. However, this all changed after 3rd September 2003 when Babylon became Camp Alpha for the Multinational Division Central South. From this point onwards the camp escalated rapidly in size, so that at its greatest extent it covered 150 hectares in size and was home to 2000 troops. The camp was established right in the heart of ancient Babylon (Fig. 8), straddling the north wall of the Inner City and including many of the famous landmarks of Ancient Babylon such as the Ishtar Gate and the Lion of Babylon, the rebuilt Southern Palace of Nebuchadnezzar II and the Greek Theatre, the restored Temples of Ninmakh, Ishtar and Nabu-ša-hare, and the Babylonian Houses. Most of the infrastructure works for the camp were contracted out to the firm KBR (Kellogg, Brown and Root), and entrances to the camp were given names such as Warsaw Gate, Reno Gate and Ronson Gate. These designations caused particular offence to the Iraqi staff at Babylon when existing Babylonian names could have been used.

During the summer of 2004, as news spread about the scale of military activity at Babylon, there was growing unease amongst the international community and an increasing number of complaints on the worldwide web. Foremost amongst those protesting about the desecration of Babylon was Professor Zeinab al-Bahrani of the University of Columbia. Such was the depth of feeling around the world, that at the end of December 2004 the coalition authorities took the decision to close down the camp and hand control of Babylon back to the Iraqi side (Fig. 9). In preparation for this handover, Polish archaeologists attached to the Polish forces prepared a lengthy document entitled *Report Concerning the Condition of the Preservation of the Babylon Archaeological Site,* which is a thorough survey describing, as its title implies, the condition of the site and the surviving and restored monuments.[4] It is an extremely useful piece of work, and an important source of reference, but what the report does not do is detail the damage caused between March 2003 and December 2004. Consequently, I was asked by Dr Mufid al-Jazairi, then Minister of Culture, to

4 M. Olbrys (ed.), Report Concerning the Condition of the Preservation of the Babylon Archaeological Site, privately distributed typescript and DVD, 2004.

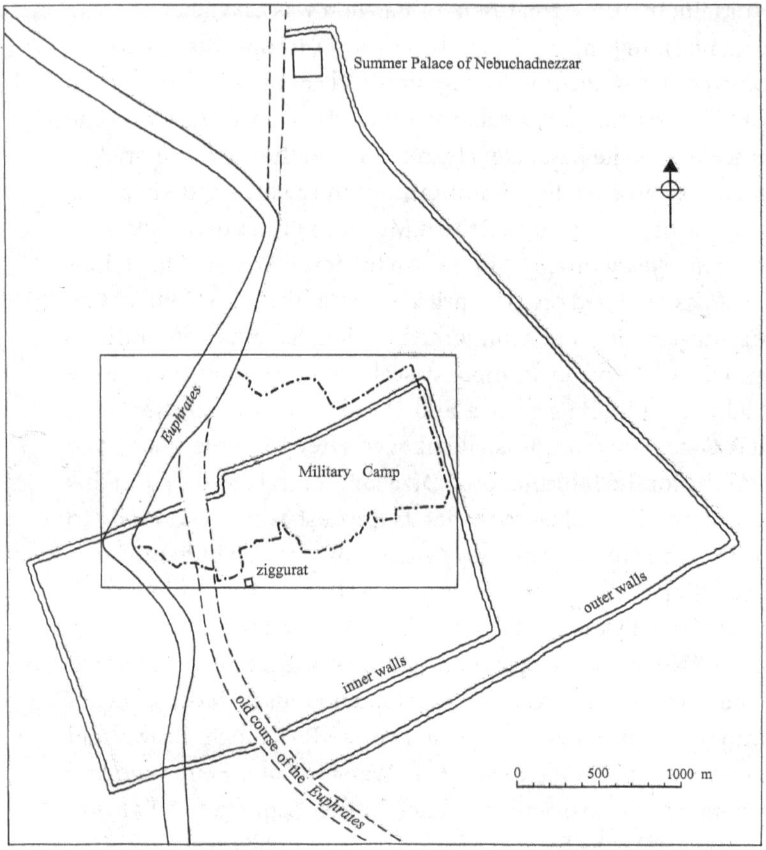


Fig. 8 | Greater Babylon showing the position of the military camp (J. E. Curtis).

attend the handover ceremony on 11th–13th December 2004 and prepare an independent report on damage caused during the military occupation. This report was subsequently published on the British Museum website.[5] At the handover ceremony, participants in the meeting, including Mr Borhan Shaker of the SBAH, Dr Rene Teijgeler, then senior advisor of the US Embassy to the Iraqi Ministry of Culture, the Polish Ambassador, Stanislaw Smolen, the Polish archaeologists, and several senior US army officers, were guided around the site by Dr Maryam Umran Musa and her two assistants, Mr Haidar Abdul Wahid and Mr Raed Hamed.

It quickly became clear that during the period of its occupation as a military camp substantial damage had been caused to ancient Babylon. We were shown 11 trenches dug by the military, often through previously undisturbed archaeological deposits. The longest of these was about 170 m long, 2 m deep and 1.0–1.5 m wide (Fig. 10).

5 J.E. Curtis, Report on meeting at Babylon 11th–13th December 2004, http:/www.britishmuseum.org/iraqproject.

Fig. 9 | The military camp at Babylon being dismantled in December 2004 (J. E. Curtis).

Fig. 10 | A military trench just outside the southern boundary of the Babylon camp (J. E. Curtis).

Fig. 11 | A military "cutting" at Babylon involving the removal of topsoil (J. E. Curtis).

It was just beyond the line of barbed wire marking the southern boundary of the camp, quite close to the site of the ancient ziggurat Etemenanki. The purpose of this ditch was apparently defensive, to prevent vehicles from driving right up to the wire. Thrown up on the sides of the trench were piles of earth containing pottery (potsherds and at least one complete pottery vessel), bones and fragments of brick with inscriptions of Nebuchadnezzar II Then, there were about 14 'cuttings', areas where topsoil had been removed, probably by a mechanical shovel, sometimes to a depth of 6 m (Fig. 11).

The largest of these was an area measuring 30 m × 20 m. Again, the digging of these 'cuttings' was indiscriminate, in that while in some instances the earth that was removed may have been from old spoil-tips sometimes dating back to the German excavations, in other instances the earth probably came from previously undisturbed archaeological deposits. The next problem was the so-called 'fuel farm', an area where there had been gigantic tanks of fuel, separated by earthen berms (banks). It was here that the military vehicles had come to refuel (Fig. 12), and inevitably there had been considerable spillage and leakage, with the result that fuel will have worked its way down into the ground contaminating the archaeological deposits beneath.

Before the war, some areas of the site had been flattened and covered with gravel, including an area for helicopters to land, but during the construction of the camp this helipad was greatly enlarged and there was a vast increase in the amount of flattened area elsewhere, to provide a base for all the temporary structures needed for a military camp as well as parking lots for the military vehicles. Sometimes, the gravel laid on these flattened areas was chemically treated to keep down dust. In many places around the camp, but particularly in

Fig. 12 | The 'fuel farm' at Babylon (J. E. Curtis).

Fig. 13 | Sandbags, HESCO containers and concrete T-walls at Babylon (J. E. Curtis).

the vicinity of the gates, there were still in December 2004 large numbers of HESCO containers (Fig. 13).

These are large wire-mesh cages lined with fabric which are filled with earth. They serve the same purpose as sand-bags, but are very much bigger. The earth in many of these HESCO containers is mixed with potsherds, bones, and even fragments of inscribed brick (Fig. 14), showing that the earth comes from either undisturbed or redeposited archaeological contexts.

At some point in the life of the camp, it was pointed out that it was bad practice to fill these HESCO containers with earth from Babylon, so earth was then brought in from an undisclosed location outside the ancient city. The problem with this is that Iraq is effectively a vast archaeological site, and wherever one digs there is a high chance of finding archae-

Fig. 14 | HESCO containers at Babylon filled with archaeological deposits (J. E. Curtis).

ological remains. So, some of the HESCO containers will have been filled with archaeological deposits from outside Babylon, and when the containers disintegrate as they are deigned to do (they are biodegradable) the contents will spill out contaminating the archaeological record at Babylon. Another problem is related to the driving of heavy military vehicles around the site. In many places wheel-marks and signs of surface disturbance were visible, but it is not usually clear how much damage this might have caused to the fragile archaeological deposits beneath. In one instance, however, it is painfully obvious. This is in the north part of the famous Processional Way, where original paving slabs have been broken by military vehicles which should never have been driven along this street (Fig. 15).

Arguably the most serious damage of all occurred in the Ishtar Gate. We mentioned above that in June 2003 damage was noted to one of the dragon (*mušḫuššu*) figures in the foundations of the gate, but by December 2004 a further eight figures had been damaged, making a total of nine in all (Fig. 16). In every case the damage seems to have been caused by a souvenir hunter or hunters trying to remove molded bricks as mementoes. As this monument was in a secure military camp throughout this period, the implication must be that the damage was caused by a person or persons attached to the coalition forces. Lastly, it was pointed out to the inspection party that one of the roofs of the reconstructed Ninmakh Temple had collapsed, allegedly as a result of vibration caused by helicopter movements.

This violation of the site of Babylon caused anger and outrage around the world, and many people turned to UNESCO as the most appropriate body to assess and redress the situation. Consequently, UNESCO convened a meeting of experts for a special session on Babylon at its Paris headquarters on 24th June 2005, and this was followed by a second special session in Berlin on 22nd November 2005, to coincide with the 150th Birthday of Ro-

Fig. 15 | Broken paving-slabs along the Processional Way at Babylon (J. E. Curtis).

bert Koldewey. A third special session took place in Paris on 12th November 2007. This lengthy delay between the second and third sessions was occasioned by the deterioration of the security situation in Iraq. At all three meetings, the participants called for the establishment of an ICC[6] sub-committee 'for the protection, conservation and management of the archaeological site of Babylon', and they requested that a full assessment of the damage caused to Babylon in the period March 2003 to December 2004 should be drawn up, based on reports by the Iraq State Board of Antiquities and Heritage (SBAH), the report by the Polish archaeologists referred to above, my own report, and information supplied by individuals including Professor Elizabeth Stone, Professor R. Parapetti, and Professor John Russell. The preparation of this assessment should be coordinated by Dr Margarete van Ess. The first meeting of the ICC Babylon sub-committee finally took place in Berlin on 25th June 2008. A draft assessment was produced based principally on the SBAH report, but drawing on the other sources referred to above. At the time of writing (March 2009), it is hoped that the UNESCO report will be published shortly. After the Berlin meeting, Tamar Teneishvili[7] and I visited Babylon 25th–27th February 2008 at the request of UNESCO to check some final details for the report and also to ascertain whether there had been further damage at the site since December 2004[8]. In fact, the inspection team found no evidence of signifi-

6 International Coordination Committee for the Safeguarding of the Cultural Heritage of Iraq.

7 Tamar Teneishvili is attached to the UNESCO office for Iraq in Amman.

8 This visit was facilitated by Diane Siebrandt (Cultural Heritage Liaison Officer in the US Embassy, Baghdad) and Dr Ismail Hijara (Cultural Advisor to the Babil Provincial Reconstruction Team). Thanks are also due to Dr Maryam Umran Musa who again guided the inspection party around the site.

Fig. 16 | Dr Maryam Umran Musah pointing out damage to a *mušḫuššu* (dragon) figure in the Ishtar Gate at Babylon (J. E. Curtis).

cant damage, either malicious or accidental, since December 2004. However, there is much evidence of deterioration resulting from neglect and lack of maintenance, particularly with the reconstructed buildings.

The ICC sub-committee also recommended that UNESCO should be involved with the SBAH in drawing up a management plan for the site of Babylon. At the same time, the US government has awarded a grant to the World Monuments Fund to enable that body to draw up a conservation plan together with the SBAH and the Getty Conservation Institute. It is to be hoped that all these bodies can work effectively together under the auspices of the SBAH to produce a conservation management plan that is worthy of the site of Babylon. It is also to be hoped that Babylon will soon become a World Heritage Site. It was included by the Iraq government on their tentative list on 29th October 2003, but it is yet to be accorded World Heritage Site status.

## ADDENDUM

The UNESCO 'Final Report on Damage Assessment in Babylon', edited by Margarete van Ess and John Curtis, accompanied by a CD containing reports by the State Board of Antiquities and Heritage (Iraq), the Bureau of Defence Matters (Poland), Dr John Curtis, Prof Elizabeth Stone, and Prof Roberto Parapetti, was published in Paris on 10th July 2009.

For a comprehensive survey of recent events at the site of Babylon see now Roberto Parapetti 'Babylon 1978-2008'[9]. There is also some discussion of the stratigraphy of Babylon in the areas of the Ninmakh Temple and the Ishtar Gate in W. Allinger-Csollich, S. Heinsch and W. Kuntner[10].

9 R. Parapetti, Babylon 1978-2008, Mesopotamia XLIII (2008), 129–166.

10 W. Allinger-Csollich/S. Heinsch/W. Kuntner, Babylon. Past, present, future. The project "comparative studies Babylon-Borsippa": a synopsis, in: Proceedings of the 6th International Congress of the Ancient Near East, vol. 1, Wiesbaden 2010, 29–38.

*Maryam U. Musa*

## The Situation of the Babylon Archaeological Site until 2006[1]

The city of Babylon is located 90 km south of Baghdad, within the boundaries of the city of al-Hillah. Geographically, it sits in the centre of a group of well known ancient cities like Kish, Borsippa, Kutha, Sippar, Jemdat Nasr and al-Uqayr as well as not far from the site of the tomb of the Prophet Dhu al-Kifl (Ezekiel).

Between 1899 and 1917 scientific excavations were conducted in the city by a German expedition headed by Robert Koldewey. This expedition revealed important parts of the city. Other parts were uncovered during the excavations of the Babylon Archaeological Restoration Project that began in 1977. Many parts of Babylon, however, have yet to be excavated (Fig. 1).

The city of Babylon underwent preservation works within the context of the Babylon Archaeological Restoration Project as well as reconstruction works by the Babylon International Festival Authority. A festival was started in 1987 and held annually, mainly in the Greek Theatre, the Throne Room of the Southern Palace and the Temple of Ninmakh. The preservation works covered the vast majority of archaeological structures unearthed during the excavations conducted both by the German expedition and the Babylon Archaeological Restoration Project. However, the preservation works were neither of respective international standard nor suited to the city. The city's archaeological value has been compromised by some of the modifications and additions that were made and by modern buildings and installations that have been built in scattered areas of the city, as will be mentioned below.

During the war in 2003, the archaeological city was subjected to vandalism and destruction. The Nebuchadnezzar and Hammurabi museums suffered severe damage, vandalism and plunder. Everything in the two museums was stolen. Fortunately, the objects exhibited in the two museums were gypsum replicas. They, nevertheless, were partly stolen and partly found scattered and destroyed on the grounds of the two museums. The offices of the Babylon Archaeological Restoration Project (including those of the Study Centre) and the Museum office did not escape the plunderers and robbers. Whatever could not be stolen was burned. Everything in the Babylon Library and archive was destroyed, including im-

1 This report was carried out in collaboration with Wahbi Abd-al-Razzaq, Jawad Muhammad Jawad, Rabi' Ja'far Hashim, Husayn Ali Muhammad, Ra'd Hamid Abdallah, Adil Kazim Amin, Ali 'Abid Shalgham, Maytham Hamzah Shahid, Haydar Abd-al-Wahid Uraybi.

Parts of this report were made public in M. U. Moussa, The Damages Sustained to the Ancient City of Babel as a Consequence of the Military Presence of Coalition Forces in 2003, in: P. G. Stone – J. Farchakh Bajjaly (eds.), The Destruction of Cultural Heritage in Iraq, Woodbrigde 2008, 143–150; a more detailed report has been published in Arabic as annex to the "Report on Damage Assessment in Babylon" of the "International Coordination Committee for the Saveguarding of the Cultural Heritage of Iraq, Sub-committee on Babylon (UNESCO 2009).

Translation of the report: Naima Ghanma; some stylistic alterations and explication of content: Margarete van Ess.

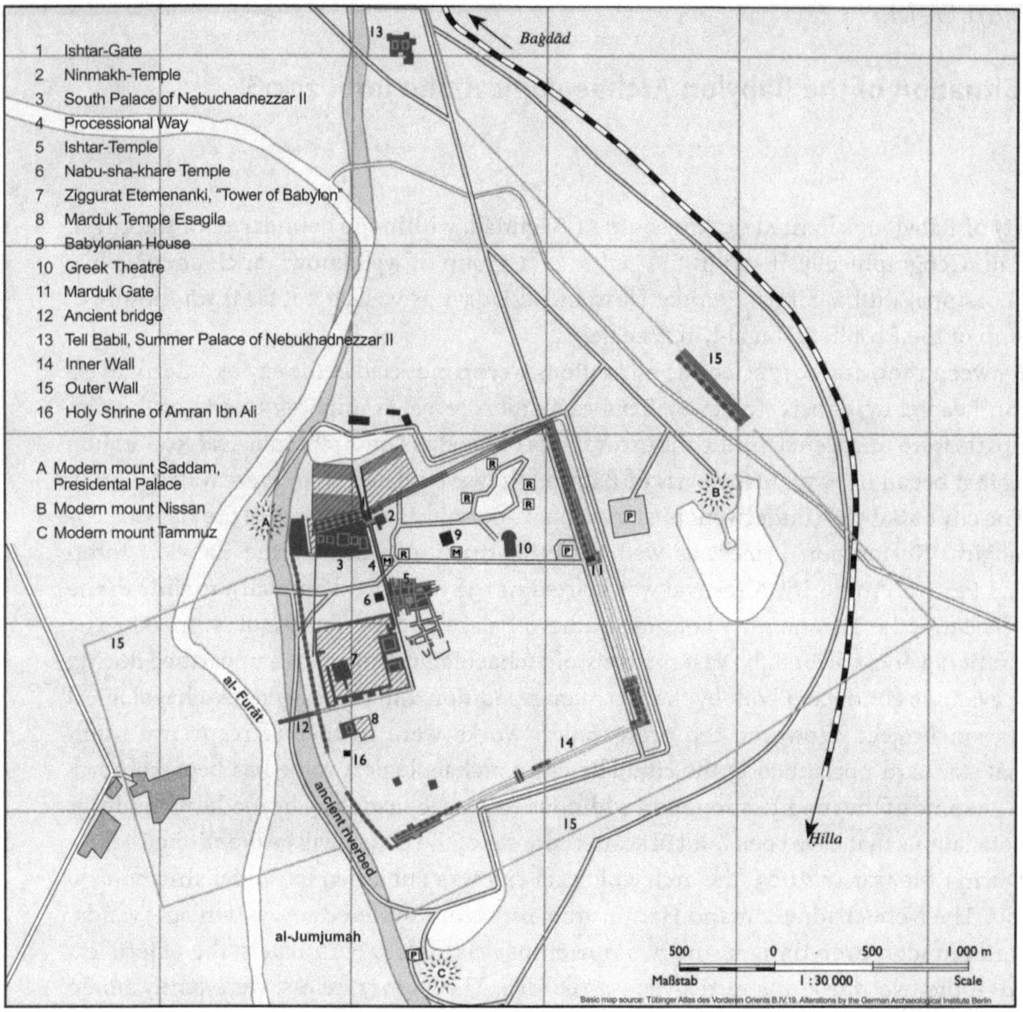


Fig. 1 | Map of Babylon (DAI-Orientabteilung).

portant reports, maps and studies on the results of excavation and preservation works performed under the Babylon Archaeological Restoration Project.

A key archaeological structure subjected to destruction is the reconstructed Greek Theatre. The Greek Theatre was previously prepared for the annual International Festival and equipped with everything needed for this purpose. While destruction mainly focused on the modern structures and equipment, they endangered the original part of the theatre as well. Being unique in type, the theatre counts among the important archaeological structures in Babylon, in particular, and Iraq, in general. The visitor can perceive its archaeological features upon entering the city on the right hand side of the main street leading to the city centre.

Fig. 2 | al-Hawliyah canal along the main street (State Board of Antiquities and Heritage, Iraq).

However, we were not able to inventory damage done to the Greek Theatre due to the entry of the Multinational Forces at the time.

Upon their entry into the archaeological city of Babylon in April 2003, the Multinational Forces used the city as a camp. They evacuated it on December 22, 2004.

The use of Babylon as a camp was a grave encroachment on this internationally known archaeological site. During their presence in Babylon, the Multinational Forces caused major direct damage to the city by digging, cutting, scraping, leveling, and establishing earthen barriers as well as indirect damage, again, resulting from their presence. Key structures that were damaged include the "Inner Wall", the "Ishtar Gate", the "Processional Way", the "Temple of Ninmakh", the "Temple of Ishtar", the "Nabu-sha-khare Temple", the "Greek Theatre", the "Royal Palaces" and the "Babylonian Houses".

## Pre-2003 Damage

Before detailing the damage sustained by the city in 2003 after the Multinational Forces entered the city, one should mention damage to the city that occurred before 2003:

1. Al-Hawliyah moat or canal

It surrounds the city of Babylon on the east and partially on the north and south. It is 4.5 km long and 15 m wide. The trench is lined on the bottom and sides with cement. This trench runs along the main street from the north. It links the city of Babylon with the outer perimeter (Fig. 2).

Fig. 3 | Nissan Lake and mount Nissan (State Board of Antiquities and Heritage, Iraq).

Fig. 4 | Tammuz Lake (State Board of Antiquities and Heritage, Iraq).

2. The new lakes

The moat is linked to three lakes on the north, south and east. A fourth lake north of the Greek Theatre is connected to the Babylon Euphrates channel to the north.

- Lake No. 1: Known as Nissan Lake, is located in the southeast part of the city between the Inner and Outer Walls. With an area of 116 dunam[2], it is the largest of the lakes (Fig. 3).
- Lake No. 2: Known as Saddam Lake, is located in the northern part of the city and has an area of 16 dunam.

2 In Iraq the dunam covers 2,500 $m^2$ (26,910 sq ft).

Fig. 5 | Modern Presidential Palace on top of Mt. Saddam (State Board of Antiquities and Heritage, Iraq).

- Lake No. 3: Known as Tammuz Lake, is located to the south of and within the sacred precinct of the city, near the village of Jumjumah. It has an area of 17 dunam (Fig. 4).
- Lake No. 4: Located in the northeast part of the city, inside the Inner Wall north of the Greek Theatre. It has an area of 10 dunam.

3. Earthen mounds

Three artificial mounds were erected under the Babylon International Festival plan. Each has a height of 30 m, an area of 30 dunam, and a slope of 100 m. These mounds include:

- Mt. Tammuz: It is in the south of the city, within the Sacred Precinct.
- Mt. Nissan: It is southeast of the city.
- Mt. Saddam. It is north of the city, overlooking the Euphrates River from the east. One of the Modern Presidential Palaces was built on it (Fig. 5).

4. Parking lots

Parking lots were built in scattered areas of the city, in varying dimensions:

- 1st lot: Southeast of the city. It has an approximate area of 58 m × 116 m, a part of which (26 m × 16 m) is unpaved.
- 2nd lot: Along the eastern part of the Inner Wall on the east and the moat. It has an area of 260 m × 271 m.
- 3rd lot: Located opposite to the inner side of the Marduk Gate, at the head of the street leading to the archaeological city. It has an area of 100 m × 126 m.
- 4th lot: West of the Greek Theatre. It has an area of 53 m × 98 m.
- 5th lot: North of the Babylonian House (area of the airfield built by the Multinational Forces). It has an area of 98 m × 54 m.

Fig. 6 | Parking lot near the Modern Ishtar Gate (State Board of Antiquities and Heritage, Iraq).

- 6th lot: Across from the rest house north of the Ishtar Temple. It has an area of 18 m × 67 m.
- 7th lot: Along the road leading to the Modern Presidential Palace on the north, near the new Ishtar Gate. It has an area of 16 m × 100 m (Fig. 6).
- 8th lot: It is located south of the Modern Presidential Palace, across from three modern buildings. It has an area of 52 m × 127 m.

5. Modern buildings

They include:

- Restaurants and service buildings in the centre of the city, around the lake north of the Greek Theatre and in the pavilions area.
- The Modern Presidential Palace and buildings surrounding it.
- The Nebuchadnezzar and the Hammurabi Museums, the Study Centre, the Museum Office and the rest house that were built in the centre of the city (Fig. 7).

6. Old encroachments

They consist of: the main street, which cuts through the city from north to south, from the Summer Palace to the moat in the southern part; the street that branches off from it, which formally led to the village of Kuwayrish (currently the area of the Modern Presidential Palace); the street running along the outer side of the moat to the north, east, and south of the city; the road linking the village of Jumjumah and the archaeological city of Babylon from the south to the city of al-Hillah; and the road built on the southern and eastern side, along the Outer Wall of the city (the road cuts through the wall); other streets in the north of the city that link the Modern Presidential Palace with contemporary buildings.

Fig. 7 | Modern buildings in the centre of Babylon (State Board of Antiquities and Heritage, Iraq).

Fig. 8 | Modern Parks were created within the archaeological site of Babylon (State Board of Antiquities and Heritage, Iraq).

7. Parks scattered throughout the city
These include the so-called "Pavilions Park" located in the centre of the city, near the Inner Wall and in the vicinity of the Temple of Ninmakh, in addition to parks located in the area of the Modern Presidential Palace (Fig. 8).

8. Encroachments caused by agriculture and construction
These encroachments, caused by citizen landowners, have resulted in the removal of parts of the Walls. Much of the land – specifically the western part of the city, the village of Sinjar, and other parts of the village of Jumjumah on the eastern side – remains private property.

Fig. 9 | Defensive trench (State Board of Antiquities and Heritage, Iraq).

The features of the western side of the city of Babylon disappeared many years ago due to encroachment by agriculture and development on the archaeological area, excluding a small portion of the remains of the city wall in the southwestern part of it.

On the eastern side of the city, one can observe the remains of the Outer Wall in the form of extended mounds. Between the mounds, there are areas from which the remains of the wall have been removed in the eastern and southern portions thereof. Some houses incorporate remnants of the Outer Wall. Encroachments have also been caused by traffic on the Hillah–Baghdad road and the secondary road parallel to the wall, which was built during the Babylon International Festival in 1987.

9. Unscientific preservation

Additions and modifications have been made to numerous archaeological structures in the city, particularly the Southern Palace, the northern portion of the Processional Way, a number of temples, the eastern portion of the Inner Wall, the Babylonian Houses and the Greek Theatre.

10. Modern excavation works

These consist of a series of defensive trenches dug during the recent war in 2003 in scattered locations of the city near the Main Palace, the Southern Palace and the Greek Theatre. The trenches were approximately similar in surface area (Fig. 9).

11. The German excavations

The German excavations covered important parts of the city, e.g., the temples, palaces, streets, gates etc. These excavations yielded valuable information. However, they also caused major damage, because the excavated areas were left exposed to the elements – e.g.

rain, wind, ground water, salt, plants and human activity – without providing for the barest conservation or protection requirements.

These factors, in the aggregate, have directly or indirectly affected the site and exposed all the structures excavated by the Germans to comprehensive deterioration and destruction.

## Post-2003 Damage

As aforementioned, the military presence in the city generated direct damage due to the fortification and defense requirements of a military zone. Furthermore, that presence has also caused indirect damage. Both types shall be separately detailed as follows:

## Direct Damage

### *Excavation Works*

The Multinational Forces dug many trenches of varying sizes in different parts of the city, including important areas thereof. The terrain of the Babylon site in its entirety – be it mounds or level ground – is historical ground and contains diverse antiquities. Any intervention into the ground therefore destroys archaeological features.

Three trenches were excavated by heavy machinery by the Multinational Forces along the Inner Wall of the city, to the east and in the so-called Homera area northeast of the Greek Theatre. The trenches are of various dimensions, running along the approximate length of the Inner Wall as follows:

1. First trench
It has a size of 30 m × 2–3 m × 1.15–2 m (Fig. 10).

2. Second trench
It has a size of 13 m × 2–3 m × 1.15 m.

3. Third trench
It has a size of 14 m × 2 m × 2 m and runs to the south of and along the second trench.

The archaeological deposits removed from these trenches were used to create a barrier. They contain pottery fragments of different shapes and sizes as well as ancient baked bricks, some of which bear cuneiform brick stamps of the Babylonian King Nebuchadnezzar II.

Fig. 10 | Trench 1 along the Inner Wall (State Board of Antiquities and Heritage, Iraq).

Fig. 11 | Trench 5 north of the modern Presidential Palace (State Board of Antiquities and Heritage, Iraq).

4. Fourth trench:

It is situated west of the Gate of Marduk, on the left hand side of the main entrance to the city, along the Marduk Street according to the Babylon map based on the Babylon topographic texts. This fourth trench is 30 m in length, has a width of 2 m, and runs between 25 cm to 1 m in depth.

5. Fifth trench

It is located north of the Modern Presidential Palace, to the left of the Shatt al-Hillah. This area constitutes the northern access to the city. The fifth trench has a length of 18 m, a width of 2 m and a depth of 1.15 m (Fig. 11).

Fig. 12 | Trench 7 next to the Ziggurat area (State Board of Antiquities and Heritage, Iraq).

6. Sixth trench

It is located on the same expanse, on the right hand side of the entrance. It has a length of 25 m, a width of 2–3 m and a depth of 1.15 m. The archaeological deposits removed from both trenches contain fragments of ancient baked brick and various pottery fragments.

7. Seventh trench

It is considered to be the one that has caused most damage, because it was dug at the head of the northeast corner of the Ziggurat, the Tower of Babylon, known in cuneiform texts as E-temen-an-ki – a blatant encroachment on the city and the tower itself. The trench has a length of 62 m, a width of 2–3 m and a depth of 2.5 m. The archaeological deposits removed from the trench were placed on both sides thereof. They contain various pottery and baked brick fragments (Fig. 12). An archaeological cross section on the western side of the trench, i.e. near the Tower of Babylon revealed six layers. Of special importance is layer 2 which contains baked brick fragments of different sizes mixed with bitumen.

8. Eighth trench

This is the longest trench dug in the city. It is located in the religious area of the city, north of the seventh trench. It runs 162 m in length, 1–1.5 m in width and 2 m in depth (Fig. 13). A trench with a length of 10 m branches off from its eastern side. The archaeological deposits removed from the trench contain pottery and baked brick fragments of various sizes. Some of the fragments bear brick stamps of the Babylonian King Nebuchadnezzar II. The most prominent find from the dumps is a partly broken small cup. In order to evaluate the damage, an area of 3 m × 2 m in the middle-most portion of the trench was cleaned and documented (Fig. 14). Four layers could be distinguished. The first consists of light

Fig. 13 | Trench 8, north of trench 7 (State Board of Antiquities and Heritage, Iraq).

Fig. 14 | Trench 8, cleaned section with archaeological layers (State Board of Antiquities and Heritage, Iraq).

yellow clay with a thickness of 45 cm. Interspersed in it is a layer of pottery fragments with a thickness of 5 cm. The second layer is light in color and has a thickness of 35 cm. The third is dark in color and has a thickness of 40 cm. The fourth layer comprises clay and pottery fragments. It has a thickness of 30–50 cm. In the lowest part of the trench, we found a row of sandy bricks, each measuring 32 cm × 32 cm × 10 cm, a typical brick size of the neo-babylonian period. This row of unbaked brick points to the existence of an unbaked brick wall that may be part of a structure that extends to a deeper level and a greater distance. Pottery fragments were also found here. This trench has begun to collapse and cause additional damage.

9. Ninth trench

This trench is located approximately 50 m north of the aforementioned one, southwest of the Nabu-sha-khare Temple, near the area of the excavations conducted by the Babylon Archaeological Restoration Project in 1982. It has a length of 15 m, a width of 3 m, and a depth of 2 m. This trench is surrounded by archaeological deposits that have been removed from it. They contain various archaeological fragments, including pottery, baked bricks and bones. The salient pieces include a fragmented glazed pottery vessel.

These trenches are distributed approximately on the edges of the extended area of the military camp. The trenches are surrounded by barbed wire from the east, south and north.

10. First pit

In the southern part of the city, at the crossing of the roads that lead to the village of Jumjumah to the west, and the road that links the archaeological city of Babylon to the city of al-Hillah, there is a pit with an almost circular shape, a diameter of 7 m, and a depth of 5.2 m. The archaeological deposits removed from the pit were packed into bags or used to fill containers.

11. Second pit

Located nearby the former pit, another pit has been dug. It is larger in area and is approximately circular in shape, with a diameter of 10 m and a depth of 2 m.

In addition to these trenches, another one dug by the Multinational Forces was observed in photographs taken by the latter. It is located in the northern part of the area north of the Modern Presidential Palace. The Multinational Forces filled in the pit at the time.

### *Cutting Works*

Cutting works, i.e. non-archaeological removal of parts of mounds, were conducted in scattered areas of the city on archaeological mounds located in the south, centre, north and northeast parts of the city.

1. First cut

Its dimensions are 8 m × 7 m × 6 m. It is located northeast of the Greek Theatre in a chain of mounds that extends along the length of the former main road between Baghdad and al-Hillah (Homera area). The cut begins approximately in the northeastern corner of the Inner Wall. The results of the German excavations in Babylon indicate that the larger part of the archaeological deposits in the area was originally transported from the area of the Ziggurat after Alexander attempted to rebuild it. They cleaned the Ziggurat and transported the rubble to this area. It is almost a certainty that the soil in the area is archaeologically rich; however, it has not been scientifically excavated yet.

Fig. 15 | Cut 2, northwestern part of Homera mound (State Board of Antiquities and Heritage, Iraq).

2. Second cut
It is situated to the south of, and bigger than, the first cut. It has a volume of 30 m × 19 m × 7 m (Fig. 15).

3. Third cut
It is south of the second cut and has the dimension of 23 m × 24 m × 5 m.

4. Fourth cut:
It has the dimension of 5 m × 16 m × 12 m.

5. Fifth cut
This cut is in the southern part of the city, west of the street that links the city to the village of Jumjumah. The objective of the cutting works in this area was to create earthen barriers. Most of the earthen barriers and obstacles located in this area contain fragments of baked brick and pottery. The fifth cut has a length of 38 m, a width of 6 m, and a height of 3 m. The cross-section resulting from the cutting works contains various fragments.

6. Sixth cut
It is on the eastern side. Similar to the former cut, it contains ancient pottery fragments, etc. It has a length of 30 m, a width of 1–8 m and a height of 6 m.

7. Seventh cut
It is north of the cutting works mentioned in paragraphs 5 and 6 above, on the western side of the road, near the al-Kurah area (brick kilns), south of the southern airfield built by the Multi-

Fig. 16 | Cut 7, southwestern part of the city, west of the street Jumjumah – Presidential Palace (State Board of Antiquities and Heritage, Iraq).

national Forces. The cut was made to build a road for vehicles transporting supplies and food to residential locations. The cut is 60 m in length, 3 m in width and 3 m height (Fig. 16).

8. Eighth cut

It is situated south of al-Kurah. This cut covers a large area estimated at 600 sq m. The archaeological deposits removed from the cut contain fragments of baked brick and diverse pottery fragments.

9. Ninth cut

It is located on the eastern side of al-Kurah and has a length of 100 m. The archaeological deposits removed from this cut were used to fill HESCO[3] containers and to create earthen obstacles. Parts of the mounds located on both sides of the road were removed, as evidenced by the pottery fragments in the HESCO containers (Fig. 17).

10. Tenth cut

The area south of the Southern Palace was subjected to an encroachment consisting of a cut with a length of 30 m. The archaeological deposits removed were mixed with sand brought from outside Babylon and used to create earthen barriers.

11. Eleventh cut

It is east of the Temple of Ishtar, near the observation post. The cut has a length of 3 m, a width of 2 m, and a height of 3 m.

3 Sandbags made by the US-owned Handling Equipment Specialty Company.

Fig. 17 | Cut 9, southwestern part of the city, west of the street Jumjumah – Presidential Palace, south of cut 7 (State Board of Antiquities and Heritage, Iraq).

12. Twelfth cut
It is located north of the Babylonian House. A 150 m long cut was made in the mounds.

13. Thirteenth cut
This cut is located in the northern part of the airfield. It has a length of 180 m.

14. Fourteenth cut
This cut is located east of the Babylonian House and west of the Greek Theatre. The cutting works resulted in the removal of an area measuring 12 m × 13 m with a height of 1–2 m. Fragments of pottery and baked bricks are evident. The cut also extends to walls built of baked brick, clearly perceived on the surface (Fig. 18). To shed light on the damage, an area of 5 m × 5 m was cleaned (Fig. 19). Immediately below the surface, at a depth of no more than 10 cm, a wall was exposed being part of a structural unit built with baked bricks, with a wall thickness of 40–60 cm. All structures are endangered by erosion.

Pottery found in connection with the structures date back to the Parthian period. In addition, a tomb was unearthed built with fragments of baked brick and unbaked bricks with a length of 2 m and a width of 70 cm, coated on the inside with plaster with a thickness of 2 cm. This tomb goes back to a more recent time than the wall, as some of the wall was removed to build the tomb. The southern part of the tomb was completely destroyed, perhaps because of the cut. In the southwest corner of the square, a large jar was found. Its entire upper part is missing due to the cut made by the Multinational Forces in the area as mentioned above. The archaeological deposits were very hard due to compaction by the heavy machinery that made the cut. The archaeological deposits removed by these cutting works were transported to another location.

Fig. 18 | Cut 14, between Babylonian House and Greek Theatre (State Board of Antiquities and Heritage, Iraq).

Fig. 19 | Cut 14, archaeological structures unearthed by cleaning of an 5 × 5 m area (State Board of Antiquities and Heritage, Iraq).

### *Scraping and Leveling Works*

The encroachments caused by the Multinational Forces in Babylon include the scraping and leveling of several archaeological areas and mounds, which were then covered with sand and gravel. This operation covered large areas of Babylon and entailed the use of heavy equipment to compact the archaeological deposits, which destroyed all antiquities beneath the surface.

Fig. 20 | Scraping and leveling of the area south of the Greek Theatre (State Board of Antiquities and Heritage, Iraq).

1. The Area South of the Greek Theatre and the Babylonian House

Previously, there was a small parking lot in this area. The lot, which was not paved, was recently expanded by the Multinational Forces. A 2075 sq m area, including the old parking lot, was leveled, covered with sand and gravel, and surrounded with HESCO sandbags. The Multinational Forces used this parking lot to park large trucks and heavy machinery. This area has not been excavated. It is located within the boundaries of the centre of the city. In addition, a road was built to link the lot to the temples area (Fig. 20).

The area surrounding the Greek Theatre to its south and west, of 2000 sq m, has been covered with large stones, as has the area located east of the Greek Theatre, which was used as a parking lot for different vehicles.

2. The Area Surrounding the Babylonian House and the Hammurabi Museum

This 10,000 sq m area, and particularly to the south and east of the Babylonian House, is served as the headquarters of the American KBR Company. It was leveled, covered with gravel and sand and occupied by various pieces of military equipment of varying shapes and sizes and caravans.

Furthermore, the western part of the Babylonian House, in the close vicinity of an excavation location carried out by the Babylon Archaeological Restoration Project in 1999, has undergone leveling works in preparation for further use, albeit pottery fragments having been found scattered on the surface of the area (Fig. 21).

Fig. 21 | Scraping and leveling of the area surrounding the Babylonian House and the Hammurabi Museum (State Board of Antiquities and Heritage, Iraq).

Fig. 22 | Leveling of mounds north of the Babylonian House (State Board of Antiquities and Heritage, Iraq).

3. The Area North of the Babylonian House

The Multinational Forces have transformed this area into an airport tarmac. The old parking lot has been expanded and an area of 30,000 sq m has been leveled and flattened (excluding the old parking lot).

South of the asphalt-paved street, and north of the Babylonian House, several mounds were cut, and the area was leveled, covered with large stones and used as a residential area (Fig. 22).

4. The Area North of the Modern Presidential Palace

This 80 m × 70 m area, which is located in the northwestern part of the eastern part of the city, has been leveled and covered with large stones.

Fig. 23 | Leveling of the area near the gate to Jumjumah (State Board of Antiquities and Heritage, Iraq).

Fig. 24 | Leveling of the area south of the Temple of Ninmakh (State Board of Antiquities and Heritage, Iraq).

5. The Area South of the Archaeological City near the Gate of Jumjumah
This encroachment involves the leveling of a portion of this area, which was then covered with gravel and made into a landing area for aircraft. In addition, residences consisting of caravans were set up (Fig. 23).

6. The Area South of the Temple of Ninmakh
An approximate area of 3300 sq m, located between the Temple of Ninmakh to the north and the Study Centre to the south was covered with sand and used as a residential location and a vehicle parking lot. It has, furthermore, been surrounded by barbed wire (Fig. 24).

Fig. 25 | Earthen berms northeast of the Greek Theatre (State Board of Antiquities and Heritage, Iraq).

7. The Area North of the First Airfield
A road with a length of 130 m and a width of 8–10 m was built leading to the observation tower north of the airfield. The tower was built on a mound in this area. The road was leveled, covered with gravel, and surrounded by concrete blocks. The area covered totals 1300 sq m. In addition, another road with a length of about 300 m and a width of 8–10 m was built east of the tower. It leads to another tower near the lake, north of the Greek Theatre. The encroached area surrounding the tower is 400 sq m.

### *Earthen Barriers*

The Multinational Forces created earthen barriers at the northern, southern, and eastern entrances to the city and in scattered locations within the city. These barriers consist of either linearly arranged heaps or containers (baskets) with a length of 1.5 m and a width of 1.10 m.

1. Berms Consisting of Heaps
- To the northeast of the Greek Theatre, there are three linear earthen berms of varying lengths, running parallel to each other. The archaeological deposits removed from the first trench were used to make one of these berms. The length of the first berm is 30 m (Fig. 25). The length of the second berm is 29 m. The length of the third berm is 21 m.
- East of the Greek Theatre is an earthen berm with a length of 35 m. It contains pottery fragments that can be seen to the right of the road leading to the centre of the city. This berm is made from surrounding earth that was cut into.

Fig. 26 | Earthen berms west of the Greek Theatre (State Board of Antiquities and Heritage, Iraq).

- West of the Greek Theatre, there is a 1200 sq m area surrounded by earthen berm divided into seven squares. These berms were intended to create areas for the storage of fuel. The archaeological deposits used to make these berms contain pottery and ancient baked brick fragments. Some of the baked brick fragments bear cuneiform inscriptions. This indicates that the archaeological deposits were indeed taken from Babylon itself. They have lost their archaeological value as they have been removed from their context. To the south, there is a 1300 sq m area that has been covered with gravel. Traces of heavy vehicles are clear in this area (Fig. 26).
- Northwest of the Modern Presidential Palace are long berms with a total length of 138 m. Most of these berms contain ancient baked brick. Many of these bricks contain cuneiform inscriptions in the form of an impression of Nebuchadnezzar II's stamp.

2. Barriers Consisting of HESCO Bags

HESCO bags are present along both sides of the Jumjumah road, within the boundaries of the city of Babylon, covering a length of 380 m on the left and 250 m on the right hand side of the road. The bags are all filled with archaeological deposits from the city of Babylon containing pottery and baked brick fragments.

Along the northern entrance to the city, north of the Modern Presidential Palace, on both sides of the entrance, are HESCO bags filled with archaeological deposits covering a length of 200 m. Some of the bags contain fragments of pottery and baked brick. Each HESCO container has a height of 1.5 m and a width of 1.10 m. Simple tapping by hand can provide an estimate of the archaeological deposits' volume used in the filling (Fig. 27).

Fig. 27 | HESCO container in the northern part of the city (State Board of Antiquities and Heritage, Iraq).

Fig. 28 | Barbed wire implanted into archaeological structures (State Board of Antiquities and Heritage, Iraq).

### *Barbed Wire and Steel Stakes*

The Multinational Forces surrounded the city with barbed wire, secured with steel stakes. The barbed wire is not limited to a specific location or number. Rather, it was set up in different locations on the perimeter of the central area established as a camp for the Multinational Forces and in the city centre, as needed. The damage in this case was caused by the implantation of stakes in archaeological ground and on several mounds and walls, as happened to a wall of the central area and a wall of the sacred precinct (Fig. 28).

Fig. 29 | Ishtar Gate, destroyed *mušḫuššu* (State Board of Antiquities and Heritage, Iraq).

*Concrete Blocks*

Concrete blocks of various sizes and shapes were found placed at the entrances to the city, on its periphery and within it.

## Direct Damage to Archaeological Structures

**Gate of Ishtar:**
The gate of Ishtar is one of the most prominent, towering archaeological structures in the city, serving as its northern access. The damage to the Gate includes cuts on nine of the bodies of the animals adorning the gate. These animals depict the legendary dragon-snake, the symbol of Marduk, the god of the city of Babylon (Fig. 29).

**Southern part of the Processional Way:**
The city's main street is the Processional Way, named according to the religious processions passed thereon between the Temple of Esagila and the Akiti House (New Year's Festival House). The southern part of the Processional Way was uncovered during the Babylon Archaeological Restoration Project excavations in 1979. Major damage is observable on the Processional Way, starting from the Nabu-sha-khare Temple to the south caused by movement of heavy vehicle. The vehicles crushed the paving of the street, causing it to crumble (Fig. 30). Three rows of concrete blocks were placed in the middle of the Processional Way, directly atop the paving. These approximately 2-ton concrete blocks were transported in the middle of the Processional Way by heavy vehicles.

Fig. 30 | Processional Way, crushed burned bricks of the paving (State Board of Antiquities and Heritage, Iraq).

Fig. 31 | Processional Way, cut into the wall protecting the Processional Way (State Board of Antiquities and Heritage, Iraq).

Removal of these heavy blocks, therefore, was carried out by helicopter to prevent further damage to the Processional Way. This was done on November 29, 2004.

In addition, a row of HESCO bags containing archaeological deposits taken from the eastern Wall of the sacred precinct were placed on the way, and barbed wire was attached by steel stakes to the wall itself and in the middle of the Processional Way. There is also a cut in the wall itself with a length of 2.5 m, a depth of 50 cm and a height of 1.5 m (Fig. 31).

## Indirect Damage to Archaeological Structures

Most of the archaeological structures were significantly damaged, due to military movements of the Multinational Forces in Babylon. This caused cracks, fractures and openings to the walls and roofs of these structures. The presence of the Multinational Forces in Babylon made these structures inaccessible for preservation works and prevented us from monitoring them regularly as before was conducted annually or whenever necessary. The State Board of Antiquities and Heritage also monitored these structures carefully and continuously so as to protect them, preserve them and ensure that they never get to the state in which they are in at the present time.

## Direct Damage to Modern Buildings

The modern buildings in Babylon were damaged significantly during and after the war. Due to the Multinational Forces's entry of the city we were not able to inventory the damage at the time. Later the Multinational Forces used materials from those buildings, leaving frames devoid of doors, windows and electrical fixtures and by this open to further destruction by erosion. These buildings include: the Modern Presidential Palace and buildings surrounding it, other buildings located throughout the city, e.g. the restaurants located around the lake north of the Greek Theatre, the rest house, the buildings located in the pavilions area, the Office of the Babylon Archaeological Restoration Project, the Administration and Study Centre and the museums.

## Post-2004 Activities

In February 2005 the Minister of Culture of Iraq asked UNESCO to support international evaluation and activities for immediate conservation of structures in Babylon in order to avoid further damage and to come up with a full documentation of the alterations and damages caused to the city of Babylon. (Fig. 32) In addition, a catalogue of urgent interventions was set up by the staff of the State Board of Antiquities and Heritage, including the following requests:

1. To treat the damage caused to the city before 2003, e.g., the moat, lakes and modern buildings. It is believed that the lakes and the moat surrounding the city should be drained after the necessary studies are conducted.

2. To relocate the modern buildings in the centre of the city, e.g., the restaurants located around the lake and north of the Greek Theatre, the service buildings in the pavilions area, the Nebuchadnezzar Museum, the Hammurabi Museum, the museum office and the Study Centre. These buildings should be relocated around the Modern Presidential Palace.

Fig. 32 | Map of Babylon Showing Pre-2003 and Post-2003 Damage (Satellite Imagery aquired by US State Department; interpretation by Elizabeth Stone.)

3. To treat the damage caused directly by the military presence, including the digging, cutting and leveling works. Cleaning of cross-sections should be carried for all destructive interventions to ascertain the extent of the damage. The damage should then be documented, sketched and photographed.

4. To remove the refuse left over from the military presence and remove military fortifications, such as concrete blocks, HESCO sandbags, barbed wire and other refuse.

5. To restore imposing archaeological structures subjected to deterioration and destruction during the past three years due to the military presence.

6. To establish a museum commensurate with the importance of the city. The two museums currently in the city do not fulfill this purpose and are not well located. It is recommended to use the Modern Presidential Palace in the city as a museum. The State Board of Antiquities and Heritage has granted its authorization for this purpose.

7. To fence the city entirely.

8. To use the structures attached to the Modern Presidential Palace for a Babylon research centre.

9. To construct a visitors centre outside the boundaries of the archaeological area.

10. To conduct the necessary studies on groundwater, a key problem affecting the city.

11. To increase the number of souvenir shops selling archaeology and heritage related books outside the boundaries of the archaeological area, similar to the practice in other countries.

12. To supply electricity, water, telephone and Internet services.

13. To construct a service division in the city.

14. To construct a tourist hotel and residential complex for employees outside the boundaries of the city including a medical service point.

15. To provide appropriate houses and tracts of unoccupied agricultural land; this benefits those citizens who own land within Babylon.

In spring 2005 an international committee under the auspices of UNESCO was created that several times met in Paris and Berlin[4]. Activities in the city of Babylon, however, were only started end of the year 2009, due primarily to the deterioration of the security situation in Iraq that dramatically slowed down the possibility to follow up activities there, but in addition to difficulties to create an international project for Babylon as suggested by the Ministry of Culture of Iraq. In July 2009, a "Report on Damage Assessments in Babylon" by the Sub-Committee on Babylon of the International Coordination Committee for the Safeguarding of the Cultural Heritage of Iraq was handed over to the public that lists the damages caused to Babylon as agreed upon by representatives of all former evaluation teams and authors of different damage reports[5].

In 2009, representatives of the State Board of Antiquities and Heritage, Iraq and the World Monuments Fund agreed upon the project "The Future for Babylon" that assesses the conditions of the site and develops a list of priority conservation projects to stabilize the archaeological ruins[6]. Work started in 2009 and is continued in 2010. Therefore, little progress has been made for the preservation of Babylon so far.

Concrete blocks and parts of the barbed wire that surrounded and protected the military camp have been removed and the cleaning and painting of modern structures has been carried out until February 2005 by SBAH in cooperation with the Polish Military Contingent in Iraq. In 2010, within the project "The Future for Babylon" cleaning work of formerly restorated archaeological buildings started as well as the evaluation and mapping of the conditions of preservation of monuments.

4 Paragraph added by Margarete van Ess.

5 http://unesdoc.unesco.org/images/0018/001831/183134e.pdf (24. May 2010).

6 http://www.wmf.org/project/future-babylon (24. May 2010).

*Olof Pedersén*

# Excavated and Unexcavated Libraries in Babylon

Ancient libraries from Babylon itself can be considered a main source, perhaps even the most important one, about ancient science, religion, literature, etc. The libraries were the places where "knowledge" written mostly with cuneiform script on clay tablets was collected. Babylon, like many other cities, has preserved remains of ancient archives and libraries sometimes in a good archaeological context, sometimes without. It will be given here an overview of what has been excavated in Babylon with more or less proper archaeological methods, to what extent it has been published, and what may still await publication and excavation. The main focus will be on libraries from the Neo-Babylonian period, but other periods and finds will also be discussed as far as the material allows.

The libraries in Babylon and Babylonia were famous among the Assyrians and many of the texts in the Middle and Neo-Assyrian libraries in Assur and Nineveh stem originally from this area. King Assurbanipal is well known in different ways for acquiring literary texts from libraries in Babylonia for his library in Nineveh.[1] What is known of the city of Babylon and the libraries found there?

## Babylon in the geographical setting

The ruins of Babylon are situated on the fertile flood plain some 85 km south of the modern Iraqi capital Baghdad. On the immediate south side of Babylon are the outskirts of the modern city of Hilla, the capital of the modern Iraqi province that has also been given the name of Babylon.

The nearest ancient towns at a distance of 15 to 30 km were Kish, Borsippa, Dilbat, and Kutha. Downstream and upstream along the Euphrates, Marad and Sippar were situated 60 km away. In all these towns, remains of libraries have either been found or their existence is probably. The landscapes between were full of smaller settlements, not so well studied, and we know nothing about possible libraries in such places.

As is well known, today the remains of Babylon consist of a number of sites with the modern names Kasr, Merkes, Amran, Ishin-Aswad, Sahn, and Homera within the Inner City, and Babil in the northern corner of the Outer City. The Western City has received only minor attention (*Figure 1*).

1 S. Parpola, Assyrian Library Records, Journal of Near Eastern Studies 42 (1983), 1–30.

## Modern history of Babylon

Our knowledge about Babylon has increased enormously during the last 200 years (*Table 1, Figure 1*). Before this time, most of the knowledge about Babylon came from different literary traditions, e. g. the Bible and classical authors. During the last centuries archaeological activities have changed our understanding.

During the Middle Ages, Babylon was in Arabic literature referred to as the place where baked bricks could be mined. The brick-mining activity continued during the 19th century on a large scale. The bricks originally used for building activities during the reign of Nebuchadnezzar II were after mining reused for construction of modern buildings in Hilla and other Iraqi cities.[2]

Hormuz Rassam and other early excavators had to face the problem that they had official permission to excavate for antiquities whereas at the same time brick miners had official permission to take bricks away in order to construct modern buildings. The first attempt was to try collaboration between the two groups. Koldewey as an architect interested in ancient buildings succeeded in stopping the brick miners first in Kasr, only later on in Babil. But at that time large sections of the old buildings had gone.

The early excavations had a focus on finding cuneiform clay tablets; other aspects of findings were in the early years considered to be less promising. In fact the first major excavation period started after large amounts of cuneiform clay tablets came on the antiquity market in 1876. The British Museum therefore sent Hormuz Rassam 1878–1882 to excavate for clay tablets in Babylon. The early excavations were concentrated to a large extent in the Amran and western Ishin-Aswad areas of Babylon. Here great amounts of clay tablets from archives and libraries were found but unfortunately proper recording of the find spots has not been made.

The Royal Museums (later called The State Museums) in Berlin and the Deutsche Orient-Gesellschaft sent Robert Koldewey and some other German architects 1899–1917 for large-scale excavations in Babylon. Most of the information we now have about Babylon is the result of these enterprises. Much of the activities were spent on the palaces in Kasr, on temples in Kasr and Merkes, and on private houses in Merkes, but there was also a lot of excavation activity in Ishin-Aswad, Amran, Sahn, Homera, and Babil. Large sections of the city walls were excavated around the inner city as well as parts of whose of the outer city.

The German archaeologists had quickly to run away when allied troops approached Babylon in 1917 during World War I. In the following years no German was allowed to visit Babylon and some of the best find objects were stolen from the not always well protected excavation house during this period. After the establishment of the Iraq Museum in 1923, a proper division of all objects that remained in the excavation house was arranged between

2 C. Janssen, Bābil, the City of Witchcraft and Wine, Mesopotamian History and Environment, Series I, Memoirs III, Ghent 1995; R. Koldewey, Das wiedererstehende Babylon, 5th ed., ed. B. Hrouda, Berlin/München 1990.

Table 1. | Modern history of Babylon with main excavations and other information.

| | | |
|---|---|---|
| 2000 | | Allied military base 2003-2005 |
| | (Italian excavations 1987-1989) | |
| | **Iraqi excavations and reconstructions 1978-** | |
| | German excavations DAI 1962-1972 | |
| 1950 | | |
| | | Division of finds 1926<br>Iraq Museum 1923<br>Objects stolen 1917-1926 |
| 1900 | **German excavations Koldewey 1899-1917** | |
| | British excavations Rassam 1878-1882 | |
| | | Tablets on antiquity market 1876 |
| | (French excavations Fresnel & Oppert 1852-1854) | |
| 1850 | (British excavations Layard 1850-1851) | |
| | (Mignan 1827) | |
| 1800 | | |

the museums in Baghdad and Berlin in 1926. Supplementary German excavations by DAI in 1962–1972 clarified some of the remaining archaeological problems.

Iraqi excavations and reconstructions have been conducted from 1978 onwards. The monumental reconstructions of the South Palace of Nebuchadnezzar, the Street of procession, three or four temples, and several private houses have increased the tourist aspect of the city. Several main findings of clay tablets have been done in excavated buildings.

A much needed geomagnetic exploration of the city area of Babylon has never occurred. Such an exploration could have solved several of the major questions concerning the size of the inhabited city, and the use of the different sections of the large city. Instead Babylon, due to the 2003 war, was for a period used as an allied military base. Today, we know where some of the main town walls were situated, but we do not know the extent of the inhabited areas inside and outside these walls, and these and many other questions remain to be solved.

The results of the early British excavations can now be found in the British Museum in London. Other early excavations generated clay tablets and other find objects to several more museums. The finds from the German excavations can now be found in the Vorderasiatisches Museum in Berlin, in the Iraq Museum in Baghdad, and in the Archaeological Museums in Istanbul. Objects stolen from the German excavation house during World War I or after could be in any collection in the world. Find objects from the later Iraqi excavations are preserved in the Iraq Museum in Baghdad.

Publications about the architecture of the palaces, temples, town walls, and of several private houses appeared rather quickly. Most of the excavated clay figurines have recently

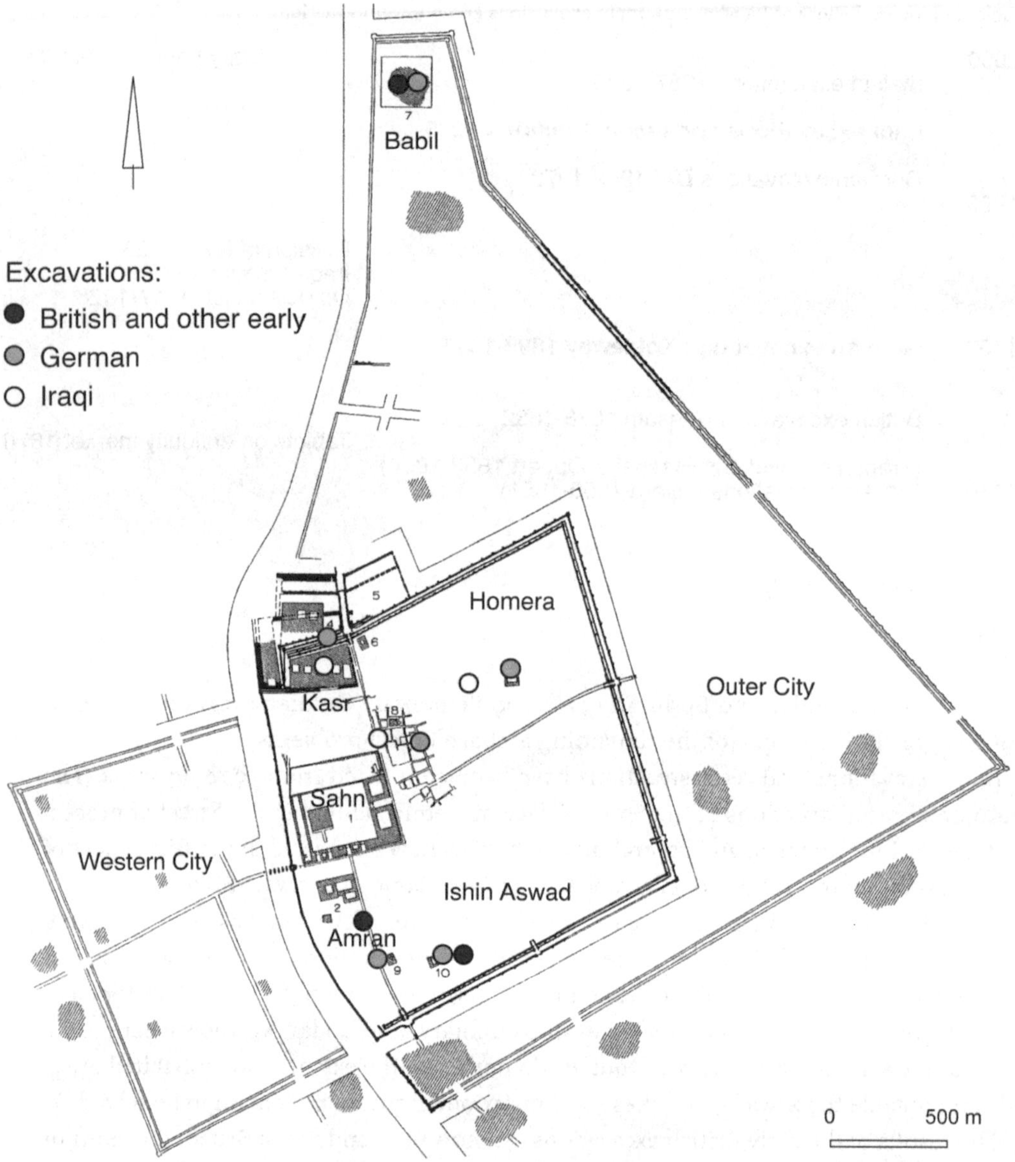


Fig. 1. | Babylon: Plan of the Inner City, Outer City, and the Western City. (Plan adopted from O. Pedersén, Archive und Bibliotheken in Babylon: Die Tontafeln der Grabung Robert Koldeweys 1899–1917 [= Abhandlungen der Deutschen Orient-Gesellschaft 25], Berlin/Saarbrücken 2005, Fig. 1, 23, and 50.)

Table 2. | Ancient history of Babylon with main historical periods, political subdivision, and some rulers.

| | | | |
|---|---|---|---|
| 1 BC | | Parthian | |
| | | Hellenistic | Alexander |
| | | Achaemenid | Darius I |
| 500 BC | | | |
| | Neo-Babylonian 900-100 BC | Neo-Babylonian | Nebuchadnezzar II |
| | | | Assurbanipal |
| | | Neo-Assyrian | |
| 1000 BC | | | |
| | | Middle Babylonian | |
| | Middle Babylonian 1600-900 BC | | |
| | | Kassite | Kurigalzu I |
| 1500 BC | | | |
| | | | Samsuditana |
| | Old Babylonian 2000-1600 BC | Old Babylonian | Hammurapi |
| 2000 BC | | | |
| | | Ur III | |
| | | Old Akkadian | |
| 2500 BC | | Early Dynastic III | |

been published,[3] but most of the other find objects, including the majority of the cuneiform clay tablets remain unpublished. Only about 7% of the interesting and important clay tablets from the German excavations have been published, most from Berlin, but only one single tablet from Baghdad.[4]

## Ancient history of Babylon

The historical terminology to be used in this presentation divides the Babylonian history in three main cultural phases: Old Babylonian, Middle Babylonian, and Neo-Babylonian. A more detailed political subdivision and a focus on selected kings may of course also have been possible (*Table 2*).

3 E. Klengel/N. Cholidis, Die Terrakotten von Babylon im Vorderasiatischen Museum in Berlin, Teil I: Die anthropomorphen Figuren (= WVDOG 115), Saarwellingen 2006.

4 For details see O. Pedersén, Archive und Bibliotheken in Babylon: Die Tontafeln der Grabung Robert Koldeweys 1899–1917 (= Abhandlungen der Deutschen Orient-Gesellschaft 25), Berlin/Saarbrücken 2005.

The oldest evidence for dating Babylon is the Early Dynastic III potsherds found on the surface of Amran and west of Homera.[5] Whereas the high ground water level has made excavations in the earliest levels so far impossible, textual evidence refers to Babylon in a date formula from the reign of the Old Akkadian king Šar-kali-šarri. From the Ur III period there are several references to Babylon, especially to the Ensi of Babylon (RGTC 1 and 2, s. v.). Limited excavations were conducted on late Old Babylonian and Middle Babylonian levels. Extensive excavations could only be performed on the Neo-Babylonian level. From all these three periods, remains of several archives and libraries consisting of cuneiform clay tablets have been unearthed.

Archives are defined here as collections of documents, existing in one, possibly a few exemplars, e.g. letters, legal, economic, or administrative documents. From the point of their contents, libraries have been defined as collections of literary, religious, and scientific texts; what we in modern time print as books. In the following there will be a concentration on libraries and archives from the German excavations referred to as A (old), M (middle), and N (new) followed by a number refer to the treatment of them in my German study of the archives and libraries in Babylon.[6]

## Old Babylonian Period

The Old Babylonian Period covers the years 2000–1600 BC. Due to the modern high ground water level, excavations could only provide evidence for the second half of this period, and especially its last 100 years. The excavation of the Old Babylonian levels was limited to a few years in the middle of the German excavation, and was only possible due to a broken dam lowering the ground water level.[7] Other excavations before or after the German do not seem to have been able to reach Old Babylonian levels.

Babylon during the Old Babylonian period may have consisted of the Inner and Western City with a total city area of some 400 ha or 4,000,000 m$^2$ inside the town walls. The New Town in the northeast of the Inner City and the Ištar Gate at the north city wall are attested in late Old Babylonian cuneiform texts from Babylon.[8] We consider here the inner town wall around the Inner and Western City to represent the size of the late Old Babylonian city (*Figure 2*).

5 M. Gibson, The City and Area of Kish, Coconut Grove, Miami 1972, 149.

6 O. Pedersén, Archive und Bibliotheken in Babylon: Die Tontafeln der Grabung Robert Koldeweys 1899–1917 (= Abhandlungen der Deutschen Orient-Gesellschaft 25), Berlin/Saarbrücken 2005.

7 Ibid. 17–68.

8 H. Klengel, Die östliche Neustadt Babylons in Texten altbabylonischer Zeit, in: Societies and Languages of the Ancient Near East (= Fs. I. M. Diakonoff), Warminster 1982, 169–173; O. Pedersén, Archive und Bibliotheken in Babylon: Die Tontafeln der Grabung Robert Koldeweys 1899–1917 (= Abhandlungen der Deutschen Orient-Gesellschaft 25), Berlin/Saarbrücken 2005, A2; R. Pientka, Ein spätaltbabylonischer Kaufvertrag aus Babylon, in: C. Wunsch [ed.], Mining the Archives. Festschrift for Christopher Walker on the Occasion of His 60th Birthday (= Babylonische Archive Band 1), Dresden 2002, 201–214.

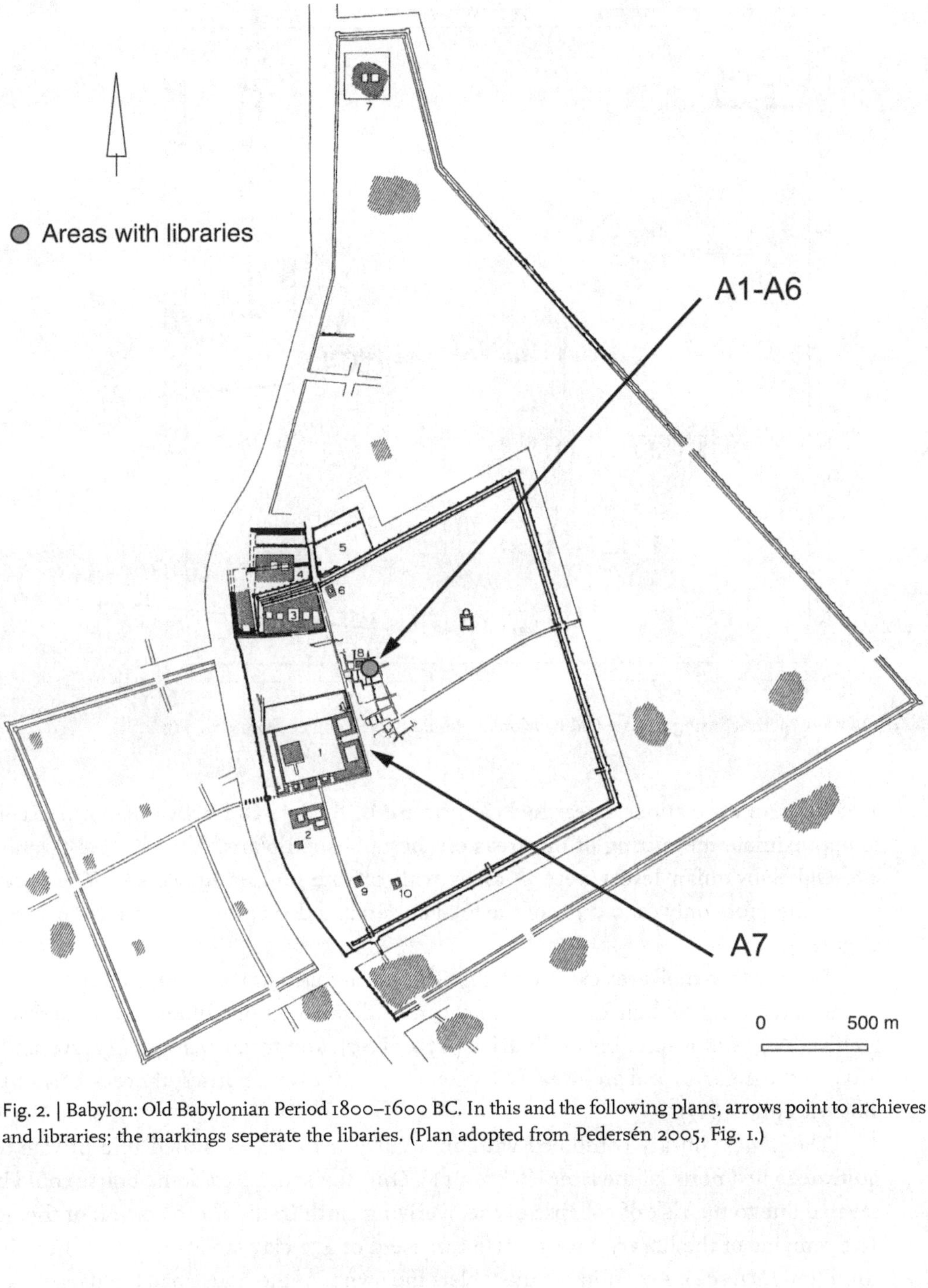


Fig. 2. | Babylon: Old Babylonian Period 1800–1600 BC. In this and the following plans, arrows point to archieves and libraries; the markings seperate the libaries. (Plan adopted from Pedersén 2005, Fig. 1.)

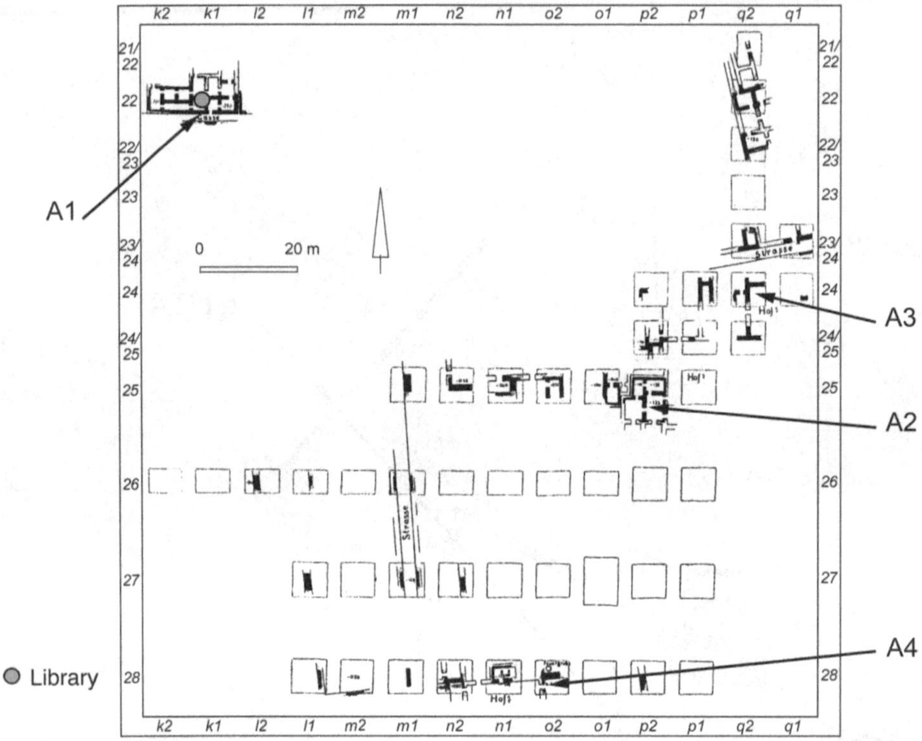


Fig. 3. | Babylon: Merkes during the late Old Babylonian Period. (Plan adopted from ebd., Fig. 2.)

The German excavations unearthed ca 1700 m² of the late Old Babylonian city according to approximate measuring of the areas on the published plans. Almost all excavations of late Old Babylonian levels were in areas with private houses in Merkes. The excavated area is therefore only ca. 0.04 % of the total city area, indeed a very small section of the city (*Figure 3*).

Despite the small area excavated, the German archaeologists unearthed some 967 clay tablets. according to their find spots they can be divided on one library combined with an archive (**A1**), minor remains of libraries (**A5** and **A7**), and three archives (**A2**, **A3**, and **A4**). Most of the libraries and archives have been found in private houses; there is only one possible exception (**A7**).

The largest library combined with an archive (**A1**) was excavated in a private house quite deep just north of the later Ištar temple. Only the south part of the house could be excavated due to the risk of collapse of the overlying earth in the north section of the house. The remains of the library with archive consisted of 452 clay tablets many of them in bad condition (*Figure 4*). According to the tablets the owner of the house had the name Marduk-nāṣir. He had the professional title ugula dumu.meš é.dub.ba.a, i.e. head of a school. In the

Fig. 4. | Old Babylonian library A1 from private house in Merkes. The excavated remains of the library consist of 452 tablets belonging to Marduk-nāṣir, ugula dumu.meš é.dub.ba.a, head of a school. (Photo with permission of VAM Berlin.)

library, there were several epics often written in Sumerian, incantations, older royal inscriptions, etc. A group of 38 school tablets cite e. g. selected literary works. The documents in the archival section concern often agricultural matters and date mostly 1742–1711 BC.

The other remains of libraries are smaller and less completely excavated. In one only partially excavated private house, the archaeologists unearthed a library combined with archive (**A5**) consisting of 46 tablets, mostly literary, lexical and documentary tablets. Under the later street of procession an older building was only partially cut by the excavators. The 33 tablets, literary and lexical texts as well as school tablets, are probably part of a larger only partially excavated library (A7).

The excavated Old Babylonian archives were all from private houses. The largest archive (A2) with some 240 tablets belonged to the business man (*tamkārum*) Kurû and his companions. Most tablets date from 1616–1599 BC, but some are older. Another somewhat older archive (A3) belonging to Ilšu-bani and his father has been unearthed in a private house.

Some 14% of the excavated tablets from the German excavations of the Old Babylonian period in Babylon have been published. This is far more than for any other period, and can essentially be contributed to the publication[9] of half of the number of tablets in archive A2. However, it should not be forgotten that the great majority of the excavated cuneiform texts from the Old Babylonian period still await their publication.

Despite the fact that only 0.04 % of the city has been excavated for the Old Babylonian level, several libraries and archives have been unearthed. Much more are of course still hidden in the earth, under later levels and partly under the ground water.

## Middle Babylonian Period

The Middle Babylonian Period falls within the years 1600–900 BC. Due to the modern high ground water level, excavations of the Middle Babylonian levels was limited to a few years in the middle of the German excavation, and was only possible due to a broken dam lowering the ground water level. In fact the excavations of the Old and Middle Babylonian levels were conducted in the same trenches in the Merkes area, only on different levels. Other excavations before or after the German do not seem to have reached Middle Babylonian levels.

Babylon during the Middle Babylonian period may have consisted of the Inner and Western City – like in the Old Babylonian period – with a total city area of some 400 ha or 4,000,000 $m^2$ inside the town walls. We consider the inner town wall around the Inner and Western City to represent the size of the late Middle Babylonian city (*Figure 5*).

The German excavations unearthed ca 2160 $m^2$ of the late Middle Babylonian city according to approximate measuring of the areas on the published plans. Almost all excavations of Middle Babylonian levels were in areas with private houses in Merkes. The excavated area is therefore only ca. 0.05 % of the total city area, indeed a very small section of the city (*Figure 6*).

Despite the small area excavated, the German archaeologists unearthed some 564 clay tablets. They can according to their find spots be divided on three text collections that may be called libraries (M3, M4, and M6)as well as six archives (M1, M2, M5, M7, M8, and M9). Most of both libraries and archives were found in private houses (one possible exception is M12).

The remains of the main Middle Babylonian library M4 were found hidden in a private house. The 61, mostly large and well-preserved clay tablets had been carefully placed in sand under a floor (*Figure 7*). Ṭāb-ṣilli-Marduk, *bārû*, "diviner", seems to have been the owner of the tablets. Among them, there were 35 often large tablets with cuneiform texts describing omens often from intestines, and another 12 tablets show the same types of omens by means of pictures drawn on the tablets. In another private house, the remains of a library M3 consisted of 25 clay tablets. One larger clay tablet has a picture of an animal, other tablets were with omens and lexical material, and a few were school tablets. In a room of a pri-

9 H. Klengel, Altbabylonische Texte aus Babylon (= Vorderasiatische Schriftdenkmäler 22), Berlin 1983.

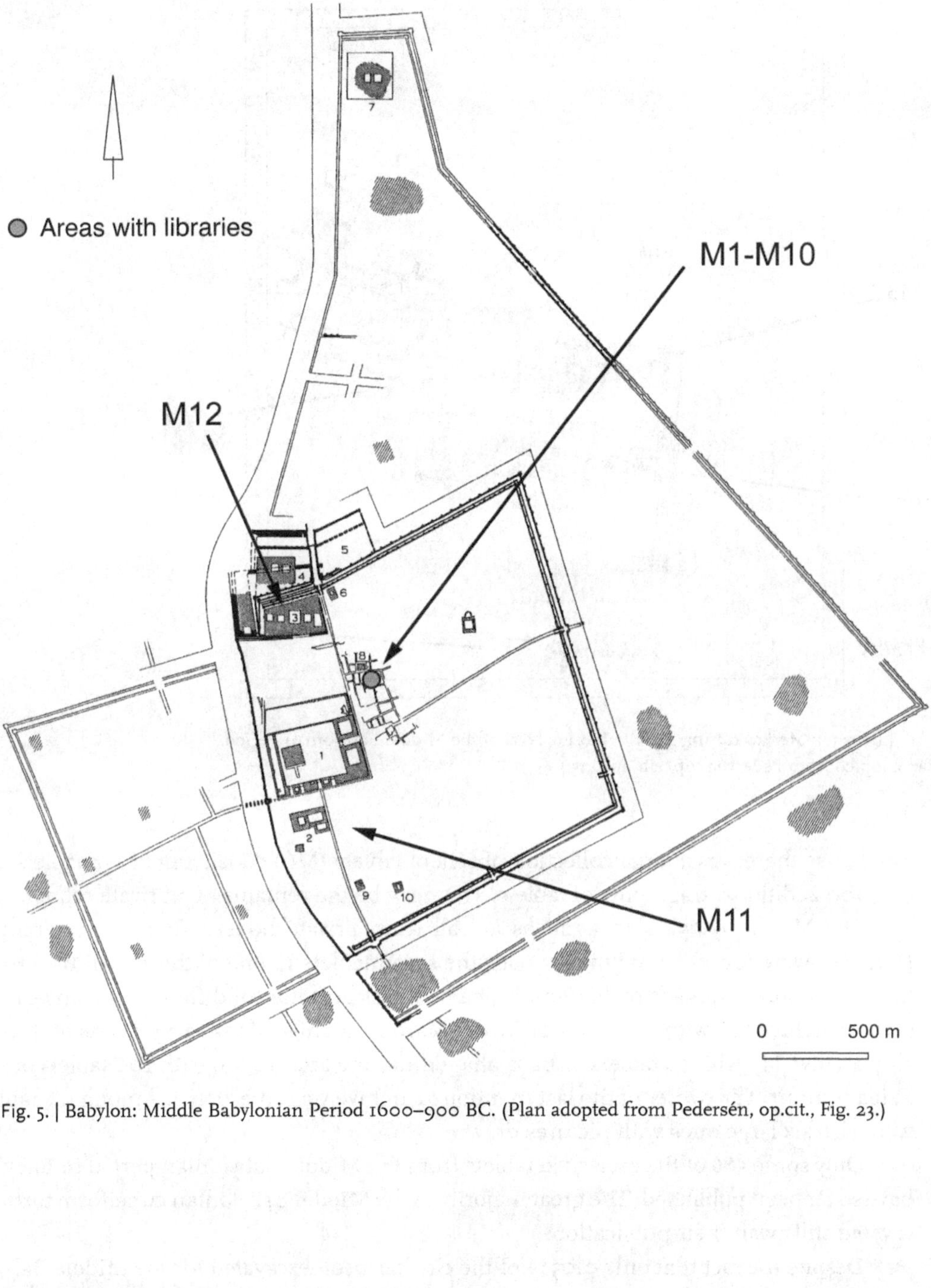


Fig. 5. | Babylon: Middle Babylonian Period 1600–900 BC. (Plan adopted from Pedersén, op.cit., Fig. 23.)

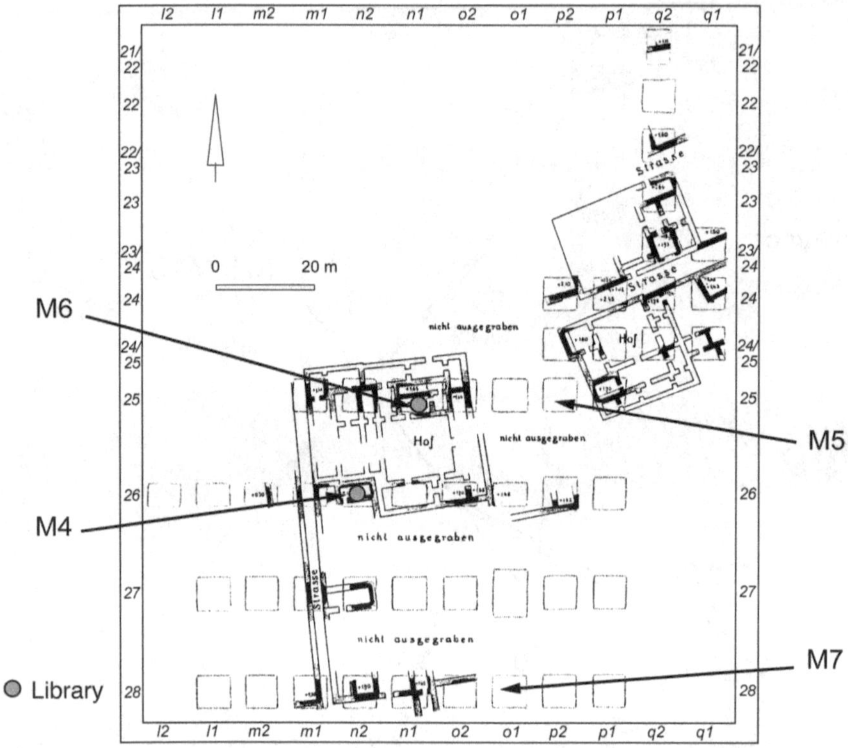


Fig. 6. | Babylon: Merkes during the late Kassite level of the Middle Babylonian Period. (Plan adopted from Pedersén, op. cit., Fig. 25.)

vate house there was a large collection of school tablets (M6) consisting of 154 tablets and ca. 1000 additional fragments of tablets. This may be the remains of a private school.

The Middle Babylonian archives are all from private houses. Two of the archives (M1, M9) were found stored in clay pots; the seven tablets in one of these (M1) document the acquisition of houses by Ešrû-rabi. The business activities of different persons can be seen in archive M2 with 39 tablets of Kiribti-Marduk, archive M5 with 34 tablets of Sîn-uballiṭ, archive M7 with 10 tablets of Baba-aha-iddina, and archive M8 with 100 tablets of Itti-Ezida-lummir. The owner of the last mentioned archive was an exorcist. Among his tablets were also six large ones with pictures drawn.

Only some 5 % of the excavated tablets from the Middle Babylonian period in Babylon have so far been published. The great majority of the Middle Babylonian cuneiform texts excavated still await their publication.

Despite the fact that only 0.05 % of the city has been excavated for the Middle Babylonian level, several libraries and archives have been unearthed. Much more are of course still hidden in the earth, under later levels and partly underground water.

Fig. 7. | Middle Babylonian library M4 from private house in Merkes. The excavated remains of the library consist of 61 tablets belonging to Ṭâb-ṣilli-Marduk, bārû, "diviner," and some colleagues. (Photo with permission of VAM Berlin.)

### Neo-Babylonian Period

The Neo-Babylonian Period discussed here covers approximatively the period 900–100 BC. The Neo-Babylonian levels are near the surface and this fact made it possible to conduct much more extensive excavations than for earlier periods.

Babylon during the Neo-Babylonian period consisted of the Inner City, the Western City, and the Outer City with a total area of ca. 8 $km^2$ or 800 ha inside Nebuchadnezzar's city walls. Most archaeological activity has been concentrated to sections in the Inner City (*Figure 8*).

The German excavations unearthed some 120,000 $m^2$ of the Neo-Babylonian city according to approximate measuring of the areas on the published plans. Excavations of Neo-Babylonian levels were conducted in several areas of the city. Palaces, temples, and private house were unearthed in Kasr, Merkes, Sahn, Amran, Ishin-Aswad, and Homera, all inside the east Inner City. Inner and partly outer city walls were also excavated. Hardly anything has been excavated in the Western and Outer Cities, except part of the palace in Babil. The

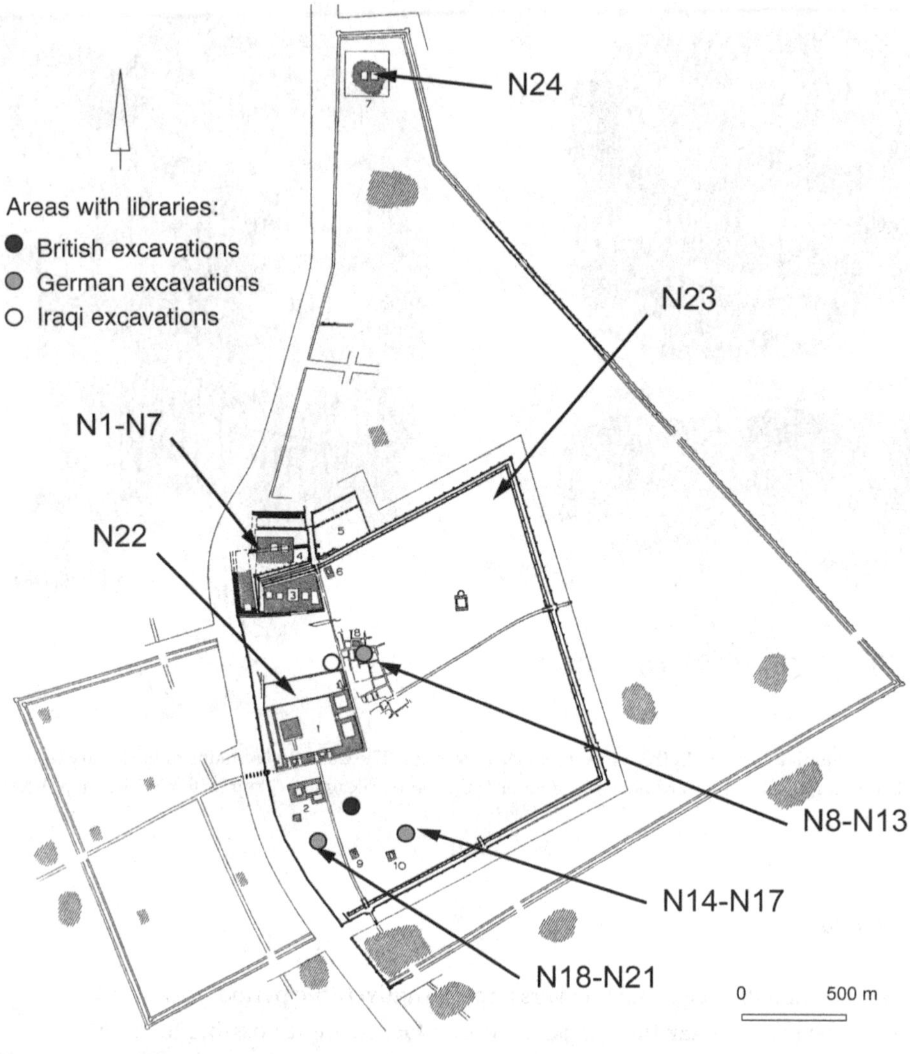


Fig. 8. | Babylon: Neo-Babylonian Period 900–100 BC. Plan adopted from Pedersén, op. cit., Fig. 50.)

excavated area can be calculated as some 3% of the Inner and Western City or ca. 1.5% of the total city area (*Figures 9, 10, and 12*).

Despite the rather small area excavated, the German archaeologists unearthed some 3498 clay tablets. They can according to their find spots be divided on some libraries (N8, N15, N18, N19) and minor remains of libraries (N10, N14) as well as several archives (N1–3, N5, N6, N8, N9, N10, N11, N12, N14, N15, N16, N18, N19) and minor remains of archives (N20, N23). The libraries were found in private houses, and in one or two cases in temples. The archives were from private houses, temples, and palaces.

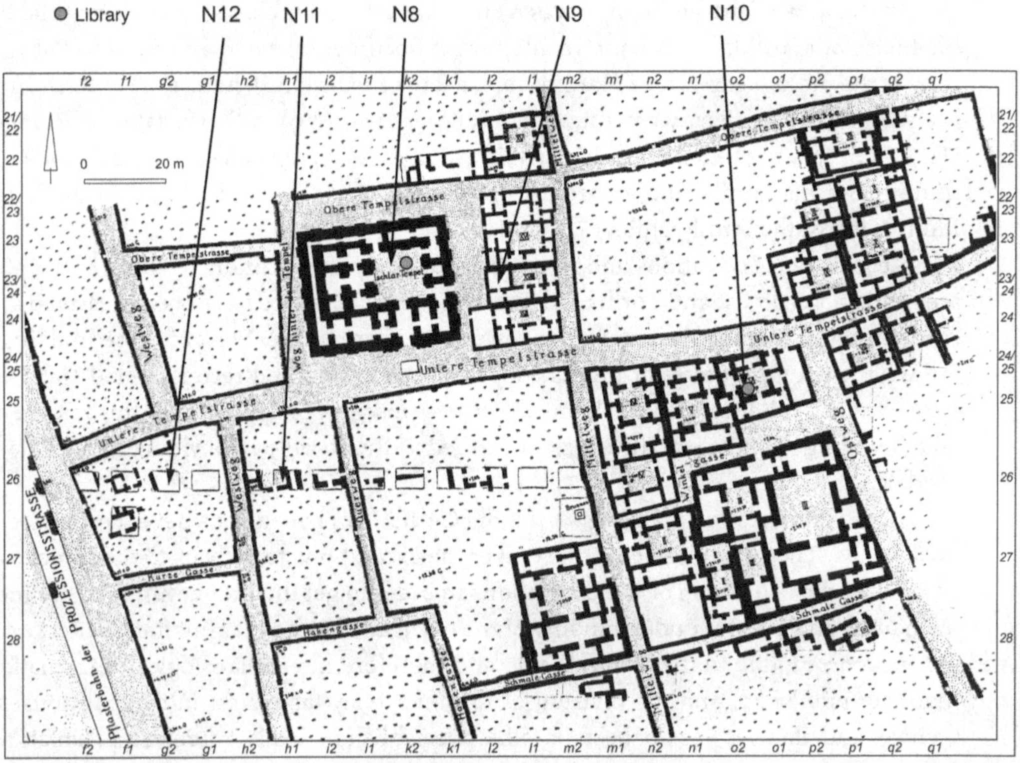


Fig. 9. | Merkes area of Babylon with Ištar temple surrounded by private houses. Neo Babylonian libraries N8 found in the temple and N10 in a private house. The other marked collections of clay tablets are archives with private documents. (Plan adopted from Pedersén, op. cit., Fig. 73.)

Before the German excavations a large number of sometimes large archives and libraries had been excavated during British and other early excavations in most cases without proper recording of the find spots. More recently, Iraqi excavations have unearthed some important collections of clay tablets in private houses, in a temple, and in the palace.
In the Merkes area (*Figure 9*), the German excavations opened up sections of the city consisting of a number of streets, several building blocks with private houses and the Ištar temple. The archaeologists unearthed libraries combined with archives in the temple (N8) and in one of the private houses (N10). Archives were found in other private houses.

In the Ištar temple, the excavators found the remains of a library combined with an archive (N8) consisting of 44 tablets. Among the tablets are descriptions of the city of Babylon with its temples and gods, as well as some omens, prayers, and lexical lists in addition to some archival documents. This may just have been a small handbook library for the temple.

In the area of private houses was a small library and archive (N10) with 36 tablets in the house of Marduk-šuma-uṣur. In addition to documents, it contained a Neo-Babylonian copy of the prologue of Codex Hammurapi, a large medical text, and some omens. In other houses private archives were unearthed, one archive (N9) with 113 tablets belonged to Nabû-ittannu and others, an archive (N11) with 49 tablets in two clay pots belonged to Bēl-ušallim and his family, and an archive (N12) with 163 tablets was found in the house of Iddin-Nabû of the family Egibi.

Recent Iraqi excavations and reconstructions have been conducted both in the Merkes area and in areas west and north of it. Two large private houses have been excavated and reconstructed in Merkes west of the reconstructed Ištar temple. West of these buildings, on the west side of the Street of procession, in the excavated and reconstructed Nabû of the *ḫarû* temple large amounts of school tablets were discovered deposited in this temple of the god of writing. A private archive was unearthed in the so-called Babylonian house north of Merkes.

In the Ishin-Aswad area (*Figure 10*), the German archaeologists never completed their excavations. They had planned at the end of the excavations to go back to the trenches they had dug at the beginning of the excavations in order to get the proper limits of the individual houses and a better understanding of the city plan like the examination in Merkes. That never occurred, due to the World War I, and therefore the archaeological information remains unpublished, and the division of the many clay tablets on individual houses and archives and libraries has not been possible to establish in detail. Therefore all tablets found in private houses within this area are here treated together as one library combined with archive (N15), even if a proper finishing of the excavations would not only have given much more tablets but also would have allowed a division of the tablets on more than one library with archive.

The archaeologists unearthed 457 tablets from libraries and archives (N15) in the private houses cut by excavation trenches in Ishin-Aswad. Among the owner of the tablets are the two brothers Līširu and Nabû-kāṣir. There is a great variety of the contents of the tablets in the libraries here, but unfortunately many tablets are not in a good condition (*Figure 11*). Among others there are omens, medical texts, hymns, temple texts, and sign lists, as well as a number of documents in the archival sections.

Some 315 tablets were unearthed in the Ninurta temple during the German excavations. A group of 190 tablets from the antiquity market excavated before the German dig belong according to content together with the tablets from the German excavations and form altogether an archive and library (N14). The core of the tablet is a large private archive of Ṭābija and his family. The library consists of only 32 of the tablets with omens and other religious content. The Išḫara temple showed traces of previous excavations and did only contained a small rest of an archive (N16) with a handful of tablets.

Koldewey was shown the place where the many tablets of the large Egibi archive are said to have been found in several clay pots in the southeast part of Ishin-Aswad. The re-

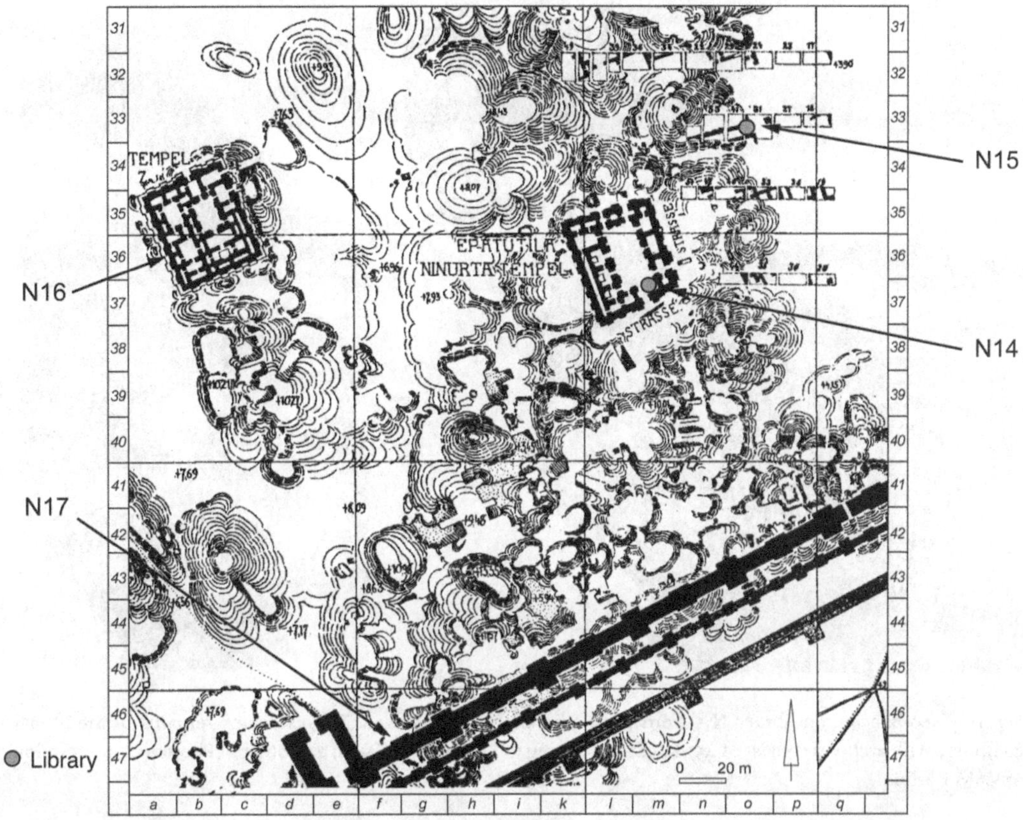


Fig. 10. | Ishin-Aswad area of Babylon with Išḫara and Ninurta temples surrounded by private houses. Neo-Babylonian combined libraries and archives N14 found in the Ninurta temple and N15 in an area with private houses. The other marked collections of clay tablets are archives with private documents. (Plan adopted from Pedersén, op. cit., Fig. 98.)

construction of what may have been the largest Neo-Babylonian family archive will still take several years.[10]

In the Amran area, especially in its south part (*Figure 12*), the German excavators found the latest dated remains of libraries combined with archives in private houses. The tablets are from the late Neo-Babylonian, often the Hellenistic period. In a private house, a library and archive (N19) with some 60 tablets were unearthed, perhaps the remains of a larger collection. Several of them were astronomical; another large group consisted of hymns, incantations, and other types of religious texts. Among the texts were also an archive section

10 C. Wunsch, Die Urkunden des babylonischen Geschäftsmannes Iddin-Marduk: Zum Handel mit Naturalien im 6. Jahrhundert v. Chr. (= Cuneiform Monographs 3A & 3B), Groningen 1993; dies., Das Egibi-Archiv: I. Die Felder und Gärten (= Cuneiform Monographs 20A & 20B), Groningen 2000.

Fig. 11. | Neo-Babylonian library N15 from private house in Ishin-Aswad. The excavated remains of the library combined with archives consist of 457 tablets belonging to Līširu, Nabû-kāṣir, and others. (Photo with permission of VAM Berlin.)

with some Achaemenid and Hellenistic cuneiform documents. A similar but smaller group of 28 tablets (N18) was found in two excavation trenches cutting private houses.

It was also in the same south section of Amran, south of the Marduk temple, that the early British excavators unearthed large remains of cuneiform clay tablet from libraries or archives from Neo-Babylonian, especially Hellenistic times. Many of the tablets were astronomical and literary. Due to the lack of excavation documentation in early times, there is no detailed evidence from where the tablets were unearthed.

It is possible, but not certain that the archive of the Nappāḫu family[11] from early clandestine excavations came from a private house in Amran or perhaps Ishin-Aswad.

The main sanctuary in Babylon, the Marduk temple was never properly excavated, except to a very small part of the building. The almost 20 m later material above the floor of the temple could only allow the Germans to make some tunnels in order get an idea of the plan of the construction. Only a few rooms with 8 tablets (N20) were excavated, and we do not know what was in the most of the other rooms. Koldewey noticed that there was no

11 H. D. Baker, The Archive of the Nappāḫu Family (= Archiv für Orientforschung. Beiheft 30), Wien 2004.

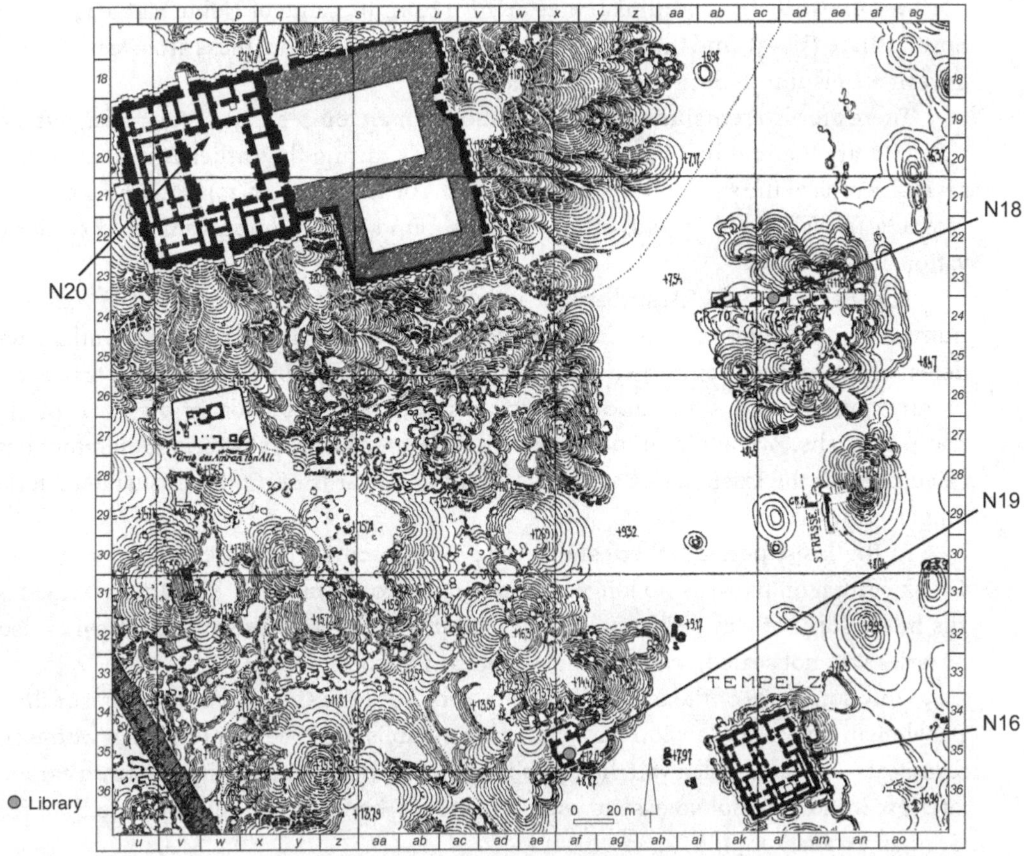


Fig. 12. | Amran area of Babylon with Marduk temple surrounded by private houses. Also Išḫara temple in Ishin-Aswad area. Neo-Babylonian libraries N18 and N19 found in private houses. The other marked collections of clay tablets are archives with documents. Early British excavations brought many astronomical and literary texts from libraries in the southwest section of the plan. (Plan adopted from Pedersén, op. cit., Fig. 112.)

traces of earlier excavation down to the temple level so earlier findings of tablets were not from the temple.

In the Sahn area with the large buildings around the large courtyard with the zikkurat in the middle, the excavation activity was limited due to lack of resources. Most of the walls were only followed in order to get a plan. Only a very limited number of rooms were excavated and during these circumstances only 12 tablets (N22) could be found. In Sippar and Uruk, excavations of such buildings revealed large archives, but in Babylon we do not know if the situation would be the same.

In the Kasr area with the Neo-Babylonian palaces, no remains of libraries were found. They may have existed, but if so they have been moved in antiquity. Two main archives were however unearthed, an administrative archive from the time of Nebuchadnezzar II in the South Palace (N1–3), and the two hundred years later private business archive of the Persian governor Bēlšunu (N6) in the Main Palace.

There are also remains of archives in the Ninmaḫ temple (N5) in Kasr area, but if there was a library there, it had been taken away already during the earlier British excavation of several rooms of the same building. The remains of an archive (N23) from private houses in Homera fit well together with other remains found during unprovenienced earlier excavation activity.

The Assyrian king Assurbanipal is well known for acquiring literary texts from a number of private libraries in Babylonia for his library in Nineveh some months after the defeat of his brother Šamash-šum-iškun in Babylon in 648 BC. The three lists of acquired literary texts include some 2000 clay tablets and writing-boards in the form of almost 100 polyptychs.[12] In addition to the clay tablets excavated in Babylon, we therefore have to calculate with the existence of a number of writing-boards of more perishable material no longer preserved.

In the latest periods, there should also have been an increased use of Aramaic and Greek written on papyrus no longer preserved due to the climate. Remains of sealed papyrus in the form of clay sealings have been found in the Kasr excavation. Papyrus used in libraries was not sealed, so we cannot expect any such remains.

Only some 5 % of the excavated clay tablets from the German excavations of the Neo-Babylonian period in Babylon have so far been published. The great majority of the cuneiform texts excavated still await their publication. The degree of publication from earlier and later excavations is not so easy to establish, it may be more or less complete than for the German excavation.

Despite the fact that only 1.5 % of the city inside the outer city walls (or 3 % inside the walls around the Inner and Western City) has been properly excavated for the Neo-Babylonian level, several libraries and archives have been unearthed. Much more than what has so far been found is of course still hidden in the earth.

## Perspectives

Much has been done so far with Babylon, but much more is there for the future. Therefore our understanding of the libraries in Babylon as places where ancient knowledge was stored is far from being comprehensive and probably not even representative. A lack of publications of the already excavated material and the relatively small excavated area make our comprehension of the ancient knowledge in Babylon rather limited. There are a number of publications and new studies of old material still to be prepared, and in addition

12 S. Parpola, Assyrian Library Records, Journal of Near Eastern Studies 42 (1983), 1–30.

some important examinations in the ruins of Babylon itself that were planned but never could be realised. We now and in the future have better methods and theories, and we have the ability to base it on a century of already detailed research.

Basic for future work with Babylon would be a more complete publication of the find objects including all the clay tablets. Only 7 % of the clay tablets from the German excavations have so far been published. Other find objects may often not be better studied. Excavations before and after the German dig also have given huge amount of not yet well-studied tablets and other find objects.

Due to the small area excavated compared with the complete city area, the realisation of a geomagnetic exploration of the city area is important and would result in a much better understanding of the size and use of the inhabited area of the city. Further excavations are needed both in order to clarify problems known from the German and other older excavations, and to answer new questions to be put to this large city due to new theories and methods. The discovery of libraries from other groups of professionals would enlarge the diversity of our understanding of Babylon.

With only 0.05 % of the Old and Middle Babylonian city and 1.5 % or 3 % of the Neo-Babylonian city properly excavated, most libraries, archives and other find objects must still remain unexcavated in Babylon. With an even distribution of the finding of clay tablets in Babylon, there would be several hundred once existing libraries and archives still in the ruins, and even if we calculate with an uneven distribution only a minor part of the intellectual or economic evidence have been unearthed. Babylon was a much more diversified and complex intellectual and economic power than the small selection of material now available show. We are only at the beginning of an understanding of the richness and diversity of the knowledge in Babylon.

# Babylon – Hauptstadt

*Wilfred G. Lambert*

## Babylon: Origins

It is a fact that Babylon is very little known in the Third Millenium B.C. It only came to prominence when Hammurabi made it the dominant power in southern Iraq in the earlier second millennium B.C. The earliest unambigous mention occurs in a year-name of the Akkadian king Šar-kali-šarri, RTC 118 rev., restored from parallel phrases in other year-names of the same king, RTC 86, 87 and 124, also found in part on Adab 405 in the Istanbul museum, cited from an unpublished transliteration of I. J. Gelb[1]:

| | |
|---|---|
| *[i]n 1 šanat (mu)* | In the year |
| *šar-kà-lí-šàr-ri* | when Šar-kali-šarri |
| *[uš-ši bī]t an-nu-ni-tim* | laid [the foundations] |
| *[ù bī]t il-a-ba*$_4$ | [of the] temple of Annūnītum |
| *in* KÁ.DINGIR$^{ki}$ | [and of the] temple of Ilaba |
| *iš-ku-nu* | in Babylon. |
| *ù* $^{m}$*šar-la-ak* | and when he defeated |
| *šarri gu*$_5$*-ti-im*$^{ki}$ | Šarlak, |
| *ik-mi-ù* | king of Gutium. |

The wording here implies the foundation of temples which did not previously exist. Sumerian and Babylonian kings often write of "building" temples, when certainly repairs, or at the most rebuilding, is meant. Here it is different: the new temples imply that Babylon was not a trivial place to be so honoured, and it also implies that at least one temple already existed there, namely that of the city patron deity, Marduk. Annūnītum is an expanded form of the name Anūna, probably of Semitic origin.[2] Ilaba is a little known god in Babylon, but also known from Ugarit and certainly Semitic.[3]

Babylon is well attested in administrative documents of the Third Dynasty of Ur, where it is recorded as being under a governor (ensi) and as paying taxes to the central government.[4] Babylon was of course in the Semitic speaking Akkad, not in the Sumerian speaking Sumer of the Sumer and Akkad divide.

1 D. Frayne, RIME II p. 183 iii k.

2 See the present writer, A Babylonian Prayer to Annūna, in: H. Behrens/D. Loding/M. Roth (eds.), Dumu-e$_2$-dub-ba-a, Philadelphia 1989, 321–325.

3 See the present writer in: Ugarit Forschungen 13 (1981), 299–301.

4 W. W. Hallo, A Sumerian Amphictyony, Journal of Cuneiform Studies 14 (1960), 88ff. and RGTC II, 21–22.

The beginning of the Second Millennium was marked by a migration of Amorites down the Euphrates valley into Sumer and Akkad, where their tribal leaders took over and ruled the cities and were gradually assimilated into Babylonian civilization, as Sumerian culture declined outside the scribal schools. Hammurabi's ancestors took over Babylon and formed a long-lasting dynasty, which Hammurabi propelled into headship of the whole area, and though his successors often lacked his skills, the city retained its prestige until the end of the Persian Empire. These well-known facts can be found in all the standard histories. Hammurabi's success was owed to a combination of circumstances. First, he inherited the rule over Babylon at a young age and so had the time to develop plans and bring them to fruition. Secondly, there is no evidence that he was a great general. It appears that he succeeded by making moves slowly and only at the right moment. Thus his military successes had lasting impact. Thirdly, he was an extremely shrewd and utterly unscrupulous diplomat. Fourthly, he was perhaps unique in his age in keeping a firm control on all his officials, which no doubt endeared him to his subjects, so that on his death no rebellion against his successor took place.

Against this background we may probe the origins and cultural status of the city more deeply. Ur III writings of the name of the city are ká-dingir$^{ki}$, often with a resuming -ma proving an Akkadian name *Bāb-ilim*, "Gate of the God". Place names of this kind are rare, but note Ka-Geštinanna (RGTC II, 88: Ur III, RGTC III, 128: OB), and Ka-Enlilla (RGTC III, 127: OB). Ur III readings with a resumptive -ra (RGTC II, 21) perhaps need collation, but in the Second and First Millennia this is a normal writing, and a Sumerian version of the name Ka-Dingirra, is surely proved. I. J. Gelb in Journal of the Institute of Asian Studies I (1955), 1–4, argued that this standard reading was only achieved by folk etymology. He cited a place near Lagash in Ur III documents (now RGTC II, 194–195) ($^{giš}$) tir-ba-bíl/bìl/bil$_4$-la. This place has nothing to do with Babylon, but Gelb's idea is entirely conceivable. There is evidence, but this "Forest of Babilla" is entirely unrelated. In the Cassite period the name of Babylon can be written with the traditional logograms, but, when not, phonetic writings attest an unexpected A vowel in the second syllable: *pan/pa-an/pa-am-ba-li*; *bà-bà-lam/lim*[5]. This is inexplicable from the traditional common writing and would remain a mere curiosity did it not drawn attention to what may well be the earliest royal inscriptions from Babylon itself, hitherto unnoticed. It is part of a limestone votive inscription published by Ferris J. Stephens[6]:

5 B. Landsberger, Journal of Cuneiform Studies 8 (1954), 67–68[172]; RGTC V, 47–49; Archiv für Orientforschung 32 (1985), 1 4.

6 F. J. Stephens, in: Yale Oriental Series. Babylonian Texts IX (1937), 2, see fig. 1.

| | |
|---|---|
| .... | (Beginning lost) |
| ⸢en$_5$⸣-[si] | governor of |
| BAR.KI.BAR | BAR.KI.BAR |
| dumu a-ḫu-ì-lum | son of Aḫu-ilum |
| lú ì-lum-be-l[í] | man of Ilum-beli |
| lú ur-kù-bí | man of Ur-Kubi |
| dím é | builder of the temple of |
| $^d$amar-utu | Marduk |
| mu-gub-am$_6$ | set up (this votive) |
| .... | (End lost) |

The ruler responsible for this votive is unknown, but what remains is largely clear, though "man of" is not a known term in such contexts. Aḫu-ilum and Ilum-beli are well known Old Akkadian name types. Ur-Kube belongs to an extremely common Sumerian name type ("Man (literally "dog") of (divine name)"), but Kubi (a demon) seems not otherwise to occur in Sumerian personal names, though it is not rare in Akkadian personal names[7], either declined (Kubum, etc.) or not (Kubi, note i-din-ku-bi)[8]. The writing dím é (=bāni bīt) suggests that the text is to be read as Akkadian, not Sumerian. The date was given as "Zeit Mesilims" by D. O. Edzard[9], while T. Jacobsen assigned it on palaeography to Early Dynastic II[10]. A recent examination of the original by B. R. Foster (private communication) supports its being genuinely archaic and not a copy from a later scribal school such as that of Sippar.

A key problem is in the second preserved line: what is BAR.KI.BAR. The present writer would take the KI as the determinative for places, kept in the middle from a very archaic traditional writing when sign-order was not rigorous, so that the place name is written BAR.BAR, which could be Babbar (note that older Sumerian writes UD.UD for babbar, but later only UD) or Baba(r/l): ba$_{15}$-ba$_{15}$[11]. In any case the writing can explain the Cassite writings of the name Babylon just given. If this is correct, it supports the idea that "gate of the God" results from folk etymology, perhaps a deliberate attempt to boost the place.

The name of the god of the place remains to be investigated: $^d$amar-utu. Amar-utu is in fact a male personal name of a well known type: "bull-calf of Utu". Lack of the divine determinative is no problem. Names of this type at Fara normally, but not invariably, take the divine determinative, while the common Ur-Utu regularly lacks it[12]. There are other cases of gods bearing human names: Dumuzi and $^d$inim-ma-ni-zi (Ninurta's vizier), but so far no expla-

7 J. J. Stamm, Die akkadische Namengebung, 1939 (= Mitteilungen der Vorderasiatischen Gesellschaft 44), 306.
8 Textes cunéiformes 1 238 48.
9 D. O. Edzard, in: H. W. Haussig (ed.), Wörterbuch der Mythologie I/1, Stuttgart 1965, 96.
10 T. Jacobsen, in: Journal of the American Oriental Society 88 (1968), 105[2].
11 R. Borger, Mesopotamisches Zeichenlexikon, Münster 2003 [= Alter Orient und Altes Testament 305], 70. 121.
12 F. Pomponio, La Prosopografia dei Testi Presargonici di Fara, Rome 1987.

nation of this phenomenon is certain. The spelling of the name Amar-utu is given in a variety of sources, not all of which are given here. The Old Babylonian period can be represented by Diri-Nippur:

ma-ru-tu-uk/ma-ru-tu-u$_4$[13]

The Middle Assyrian copies of the god list AN=Anum can be used for a Middle Babylonian example:

$^{d}$ma-ru-tu-uk-ka[14]

For the First Millenium the acrostic prayer of Ashurbanipal is significant since the spelling is part of the acrostic:

ma-ru-du-uk[15]

The almost universal modern "Marduk" is extremely hard to find in ancient sources: It occurs in a Late Assyrian copy of a literary text: $^{d}$mar-duk (KAR 166 rev. 8). In the so-called "Canon of Ptolemy", conveniently cited in Greek in RlA VI, 101, two Babylonian kings' names include: Mardok- and mordakou, but no one will put too much weight on texts that have reached us by more than a millennium of Greek copyists. The modern usage of "Marduk" may have been influenced by a man's name which first appears in Middle Babylonian times: Mar/šd/tuku/a/i, etc., of unknown origin and meaning. The resemblance of this to Marūtuk is of course of no consequence, and it is unfortunate that T. Jacobsen[16] quoted "$^{d}$mar-duk" as occurring twice in F. E. Peiser's Urkunden aus der Zeit des dritten babylonischen Dynastie[17], when in fact it is the name of a man involved in the business transactions with name written $^{m}$mar-tuk. Ancient Hebrew copying traditons were better informed. The Ben Asher family that supplied the most respected pointing system to the Hebrew Bible in the later First Millennium A.D. regularly distinguished between the man Mord$^{o}$kay and the god M$^{e}$rōdak, a distinction already attested in the LXX version of the relevant Biblical books.

The emesal form of the name is not hard to find, though modern scholars have not searched for it. An Emesal litany K 9270+9289 BA V 659 (collated) in obv. 3 reads:

umun $^{d}$a-ma-ru-uk-kám

13 MSL XV 36 10 44 and 11 09.

14 R. L. Litke, A Reconstruction of the Assyro-Babylonian God-Lists, New Haven 1998, 91 (Z. 195).

15 J. Craig, Assyrian and Babylonian Religious Texts I, 1895, 31:1–8.

16 T. Jacobsen, in: Journal of the American Oriental Society 88 (1968), 105$^{4}$.

17 Berlin 1905.

which allows the second writing of the name in Enūma Eliš VI 138 to be seen as such:

$^{d}$ma-ru-uk-ka/ku

The intervocalic T has dropped, as in adda/a' a "father", and the endings are the product of late or middle-period scribes.

The meaning of the name is transparent: "Bull-calf of (the god) Utu". No doubt amar was originally taken as a metaphor for "son", but the short form of amar, mar (a common phenomenon in words of that phonetic pattern) allowed the ancient scholars to identify it with the Akkadian māru "son" (actually from the root mr'), as plainly asserted in Enūma Eliš I 101–102 and VI 127–128.

What has been presented so far may seem of purely philological relevance, but that is not the case. It allows us to probe into the cultural and religious prehistory of Babylon. Sumer and Akkad were originally, it appears, a land of city states, independent both politically and religiously. Each city state had its own patron deity (very rarely two such gods), but in Sumer the totality of city gods added up to a comprehensive pantheon with no duplication. Did an ecumenical conference in early times work out such a scheme for their mutual benefit? The larger cities were of course the core of such city states, the smaller places had their deities worked into the pantheons of the larger. Babylon, with its god "Bull-calf of Utu" (the sun-god) is marked as being under the influence of Sippar, where Šamaš was the patron. The Sumerian sun-god Utu was at home far away in the south. But Šamaš/Utu was second- or even third-ranking in the Babylonian pantheon as it developed at the beginning of the Second Millennium B.C. In Semitic circles he had been much higher. Yaḫdun-Lim of Mari calls him "King of heaven and earth" (šar šamê u erṣetim: RIME IV, 605: 1–2; see the whole context), which amounts to supreme power in the universe. And Hammurabi himself in one inscription calls Šamaš "great lord of heaven and earth, king of the gods" (en gal an-ki-bi-da lugal dingir-re-e-ne-ke$_4$ = bēlum rabi'um ša šamā'i u erṣetim šarru ša ilāni: RIME IV, 334: 1–4). On an entirely different level a conventional greeting in Old Babylonian letters is "May Šamaš and Marduk keep you in good health" (Šamaš u Marūtuk liballiṭūka). The sequence of the two gods is normal in one sense: fathers had precedence over sons. That Marduk's status in Babylon was never challenged is proof of its antiquity. No doubt in the Third Millennium Babylon was within the religious orbit of Sippar.

However, as time passed the status of Šamaš did not improve, indeed it declined and we may suspect that the priests of Babylon were not too happy. Another ancestry of their city god was proclaimed. Marduk was identified with Asalluḫi, son of Enki/Ea of Eridu in the far south of the country. By the end of the Third Millennium a generally accepted pantheon had been worked out for Sumer and Akkad, with An, Enlil and Enki as the top trinity, occasionally joined by the Mother Goddess.

The Prologue to Hammurabi's laws begins with a statement of the promotion of Marduk, son of Ea, to having power over the peoples and being one of the Igigi gods, a group of top gods.

This reflected Hammurabi's having made Babylon politically supreme in Sumer and Akkad, with a theological consequence that Marduk had be to a top god. One might think that Hammurabi himself was responsible for this, asserting that Marduk was son of Ea by virtue of having also the name Asalluḫi. But a literary letter of Sin-iddinam of Larsa to the goddess Nin-isina published by W.W. Hallo[18], mentions:

> [$^{d}$asa]l-lú-ḫi lugal Ká-dingir-ra$^{ki}$ dumu $^{d}$íd-lú-ru-gú Asalluḫi, king of Babylon, son of Idlurugu.

Sin-iddinam ruled well before Hammurabi, so it appears that persons in Babylon before Hammurabi wanted to exalt their god by changing his paternity from Utu to Ea. The move was so successful that from Hammurabi's time no known source explicity alludes to Šamaš as Marduk's father. The problem of the name was overcome by ignoring the final -k. By normal Sumerian grammar the genitive -ak loses the k at the end of a word or name, so "Son of the Sun-god" became "The son, the sun-god": this sun-god not simply giving humans light and heat, but Marduk giving light to the gods, the higher calling, as plainly stated in Enuma Eliš I, 101–102 and VI, 127–128.

18 W. W. Hallo, in: B.L. Eichler (ed.), Kramer Anniversary Volume, Neukirchen-Vluyn 1976 (= Alter Orient und Altes Testament 25), 209–224, line 16.

*Dominique Charpin*

# Babylon in der altbabylonischen Zeit: eine Hauptstadt von vielen ... die als einzige übrig blieb*

Sehr häufig kann der Historiker den Beginn von Ereignissen nicht fassen. Im Falle Babylons aber können auf die Frage nach dem Ursprung seiner kulturellen Ausstrahlung sogar Elemente der Antwort hierauf aus seiner Vorgeschichte gewonnen werden. Man ist sich darüber einig, dass der Platz Babylons in Mesopotamien im 3. Jahrtausend wenig bedeutsam war. Es wird in den meisten Fällen vermutet, dass Babylon erst ab Hammurabi eine vorherrschende Rolle gespielt hat, und dies nicht nur in politischer, sondern auch in kultureller Hinsicht. Ein Beispiel von vielen für diese Sichtweise finden wir in den Schriften von H.W.F. Saggs:

> „Hammurabi (...) has a claim to greatness, for it was he who gave Babylon, a formerly insignificant city, a status it never lost. He raised it to be for a millennium and a half the national capital, and the city after which the whole south Mesopotamia received the name *mat Babili*, the Land of Babylon. And although it was superseded politically towards the end of the first millennium BC, Babylon remains significant as a cultural symbol even to the present day."[1]

In diesem Beitrag soll gezeigt werden, dass diese Sichtweise historisch nicht exakt ist.

Es ist bekannt, dass unsere Schwierigkeit hinsichtlich der älteren Geschichte Babylons auf dem Umstand beruht, dass die altbabylonischen Schichten der Stadt während der Grabungen der DOG kaum angeschnitten werden konnten. Als 1906/7 durch ein Unglück ein Damm im Norden von Hillah brach und sich der Euphratlauf so zeitweilig verlagerte, konnte zwar bis auf 12 Meter Tiefe hinunter gegraben werden, die Ausgrabungen in dieser Tiefe hatten aber nur eine sehr begrenzte Ausdehnung.[2] Aus diesem Grund muss die kulturelle Ausstrahlung Babylons in der ersten Hälfte des zweiten Jahrtausends aufgrund von Quellen untersucht werden, die nicht aus Babylon selbst stammen.

In meinem Beitrag will ich folgende These vertreten: Babylon wurde erst *nach* der Regierung Hammurabis, vor allem aber erst gegen Ende der altbabylonischen Zeit zum

* Man findet hier den Text des Vortrags, wie er im Rahmen des Kolloquiums gehalten wurde, ergänzt um einige Belege und Zusätze. Mein Dank geht an N. Ziegler und J. Rothkamm für die deutsche Übersetzung meines Textes.

1 H.W.F. Saggs, Babylonians, London 1995, 107f.

2 Siehe die Abbildung des Plans in O. Pedersén, Archive und Bibliotheken in Babylon. Die Tontafeln der Grabung Robert Koldeweys 1899–1917 (= Abhandlungen der Deutschen Orient-Gesellschaft 25), Berlin 2005, 18 (nach WVDOG 47, 1926, Taf. 8).

Zentrum, besser gesagt: zur Hüterin der sumero-akkadischen Tradition. Meine Argumentation ist im Folgenden in drei Abschnitte gegliedert: Zuerst wollen wir den sehr bescheidenen Platz Babylons zum Zeitpunkt der großen Eroberungen Hammurabis ab 1763 v. Chr. betrachten. Dann beabsichtigen wir, die Konsequenzen der Eroberungen für Babylon innerhalb des kulturellen Gefüges des Vorderen Orients zu untersuchen. Schließlich soll gezeigt werden, wie Babylon unter den Nachfolgern Hammurabis zur Hüterin sumerischer Traditionen wurde.

## 1. Babylon bis zu den Eroberungen Hammurabis

Vor 1763[3] v. Chr. war der politische und kulturelle Einfluss der Stadt Babylon sehr beschränkt. Ihre Nachbarin Ešnunna war hier in beiderlei Hinsicht weitaus bedeutsamer.

### 1.1. Akkad, eine Region mit zwei Zentren

Es erscheint mir wichtig, sich der Differenz bewusst zu sein, die zu Beginn des zweiten Jahrtausends zwischen Sumer und Akkad immer noch bestand. Akkad fiel im Verlauf des 19. Jahrhunderts v. Chr. in das Einflussgebiet zweier Mächte, nämlich Ešnunna im Nordosten und Babylon im Südwesten. Wir können gut beobachten, wie die Vielzahl kleiner Stadtstaaten, die sich gegen 1900 gebildet hatten, allmählich von den Herrschern Babylons und Ešnunnas annektiert wurden.[4] Noch zur Zeit Zimrī-Lîms von Mari konnte man mit der geographischen Bezeichnung Akkad beide Königreiche ansprechen, wie die folgende Textpassage zeigt, in der ein Beamter aus Mari die Frage stellt:

> „Der Herr von Akkad, der mit ihnen den Eid geleistet hat, ist es der Herr von Ešnunna oder der Herr von Babylon? Ich konnte ihn nicht identifizieren."[5]

Dieses Zitat bezeugt also, dass die Ortsbezeichnung Akkad sowohl auf Babylon als auch auf Ešnunna verweisen konnte.[6] Diese beiden Städte sind nur 110 km voneinander entfernt, das heißt ungefähr ebenso weit, wie die beiden großen Rivalen in Sumer, Isin und Larsa, die 90 km auseinander liegen.

3 1763 v. Chr. entspricht dem Zeitpunkt der Eroberung Larsas durch Hammurabi, gemäß der Mittleren Chronologie.

4 Siehe D. Charpin, Histoire politique du Proche-Orient amorrite (2002–1595), in: D. Charpin/D. O. Edzard/M. Stol, Mesopotamien: Die altbabylonische Zeit, Fribourg/Göttingen 2004 (= Orbis Biblicus et Orientalis 160/4), 96 Amn. 359 (nachfolgend zitiert als OBO 160/4).

5 Archives royales de Mari 27 135: (31) *ù* l[ú] *ak-ka-du-um ša it-ti-šu-nu* (32) *ni-iš* dingir-*lim* [ì]*z-ku-ru lu-ú* lú èš-nun-na$^{ki}$ (33) *lu-ú* lú ká-din[gir-ra]$^{ki}$ *pa-ga-ar-šu ú-ul ú*$^{!}$*-w*[*e*]*-di*.

6 Für Akkad = Ešnunna noch unter Samsu-iluna, siehe Orbis Biblicus Orientalis 160/4, 341 und Anm. 1781.

Es ist in dieser Hinsicht aufschlussreich zu sehen, dass die einzige allgemeinverständliche Einführung zur Geschichte Ešnunnas den Titel „La rivale di Babilonia" trägt, und nur als Untertitel „Storia di Ešnunna ai tempi di Ḫammurapi".[7] Die Geschichte der Folgezeit jener beiden Städte führt dazu, dass wir heute die Bedeutung Ešnunnas in Relation zu der von Babylon setzen. Für das 19. Jahrhundert v. Chr. gilt aber, sowohl vom kulturellen als auch vom politischen Standpunkt aus, das Gegenteil.

### 1.2. Die politische und kulturelle Rolle Ešnunnas

Es ist klar, dass die Unterschätzung der Rolle Ešnunnas auf die Quellenlage zurückzuführen ist. Ohne ins Detail gehen zu wollen, soll hier an einige Schlüsselmomente der politischen Geschichte Ešnunnas erinnert werden. Nach der politischen Vereinigung des gesamten Diyalagebiets im 19. Jh. v. Chr. setzte sich die Expansionspolitik Ešnunnas in zwei Richtungen fort: nach Nordwesten in Richtung Habur-Dreieck und nach Westen in Richtung Mittlerer Euphrat.

– Unter Ipiq-Adad II. und Narām-Sîn gelangten Truppen bis nach Ašnakkum (das heutige Chagar-Bazar)[8] und Tarnip (wohl Tall Tamir)[9] einerseits und bis nach Puzurrān im Süden Maris andererseits;

– der Wettstreit begann erneut einige Jahrzehnte später zwischen Samsî-Addu und Dâduša; und schließlich versuchte Ibâl-pî-El erneut, fünf Jahre nach dem Tod Samsî-Addus diese Expansionspolitik in die beiden selben Richtungen zu führen.

Die Arbeiten an den Archiven von Mari der letzten 25 Jahre haben zeigen können, wie bedeutsam der kulturelle Einfluss Ešnunnas war. Diese Stadt hatte die Verwendung der altbabylonischen Schrift in Obermesopotamien und am Mittleren Euphrat veranlasst – nicht Babylon! Dies ist besonders deutlich in Mari sichtbar. Die Babylonisierung gegen Ende der Šakkanakku-Zeit fand unter dem direkten politischen Einfluss Ešnunnas statt.[10] Wir besitzen eine Tontafel im sogenannten „Šakkanakku-Schreibstil" und denselben Text in modernisierter Schreibweise. Das Syllabar, die Graphie, und einige Ideogramme zeigen an, dass

7 C. Saporetti, La rivale di Babilonia. Storia di Ešnunna ai tempi di Ḫammurapi, Rom 2002.

8 D. Lacambre/A. Millet Albà, Le nom ancien de Chagar Bazar, in: Ö. Tunca/Abd el-Massih Bagdo (Hgg.), Chagar Bazar (Syrie) III. Trouvailles épigraphiques et sigillographiques (2000–2002). Publications de la Mission archéologique de l'Université de Liège en Syrie, Louvain/Paris/Dudley 2008, 143–154.

9 Für die Identifikation, siehe meine Studie über „Un itinéraire paléo-babylonien le long du Habur", in: E. Cancik-Kirschbaum/N. Ziegler (Hgg.), Entre les fleuves. Untersuchungen zur historischen Geographie Obermesopotamiens im 2. Jahrtausend v. Chr., Gladbeck 2009 (= Berliner Beiträge zum Vorderen Orient 20), 59–74; die englische Fassung erschien unter dem Titel „An Old Babylonian Itinerary along the Habur" (= Dūr-Katlimmu Studies 1), in: H. Kuhne, Dūr-Katlimmu and Beyond, Studia Chaburensia 1, Berlin 2010, 33–47.

10 Siehe die Bibliographie zusammengestellt in: D. Charpin/N. Ziegler, Mari et le Proche-Orient à l'époque amorrite: essai d'histoire politique, Florilegium marianum V, Mémoires de NABU 6 (2003), 40.

diese „Babylonisierung der Schrift“ durch die Schule von Ešnunna veranlasst worden war. So wird sowohl in Ešnunna als auch in Mari der Laut /QA/ mit dem QA-Zeichen geschrieben, und nicht mit GA. Der Titel des Generals ist GAL MAR-TU und nicht UGULA MAR.TU wie in Babylon oder Larsa usw.

Einer der Hauptunterschiede zwischen Sumer und Akkad in altbabylonischer Zeit war der Gebrauch der akkadischen Sprache in den kommemorativen Inschriften. Es kann angenommen werden, dass auch hier die Inschriften aus Ešnunna jene aus Mari, vor allem jene Yahdun-Lîms, beeinflussten. Die Stele des Königs Dāduša, die jünger ist, bestätigt dieselbe Tendenz.[11]

Das hohe literarische Niveau, das die Schreiber in Ešnunna erlangt hatten, kann durch den akkadisch-sprachigen Text bezeugt werden, der die Taten des Königs Narām-Sîn feiert und der auf den Toren eines Heiligtums für den Gott Erra angebracht war.[12]

Die Kunstfertigkeit der Schreiber Ešnunnas war aber nicht allein auf die schöngeistige Literatur beschränkt. Die Ausgrabungen in Tall Harmal oder Tall Haddad beweisen ihre Kompetenzen auch in der Mathematik.[13] Die Grabungen im Tall Yelkhi oder Tall Haddad zeigen außerdem ihre Fähigkeiten in Bezug auf die Orakelschau;[14] meine Aufzählung ist an dieser Stelle nicht vollständig ...

### 1.3. Ešnunna nach dem Sieg Hammurabis

Die Geschichte Ešnunnas wurde in der Folgezeit durch mehrere Niederlagen geprägt. Wir wissen, dass das Königreich 1765 v. Chr. durch einen Angriff Elams verwüstet worden war: Wegen dieser systematisch erfolgten Zerstörungen haben wir so viele Texte aus Tall Harmal/Šaduppûm.[15] Die Bedeutung des Sieges, den Hammurabi über Ešnunna 1761 v. Chr. errang, kann auch anhand des Namens des 32. Jahres Hammurabis erkannt werden: Der Königsname wird hier von einem Epitheton UR-SAG („Held“) begleitet, was ungewöhnlich ist. Dieses Epitheton wird außerdem durch eine besondere Schriftsetzung hervorgehoben. Das Epithet UR-SAG befindet sich am Ende der Zeile nach dem LUGAL und nicht, wie sonst üblich, am Anfang der folgenden Zeile. Und bei diesem Format bleiben alle Ta-

11 Siehe D. Charpin, Données nouvelles sur la région du Petit Zab au XVIII$^{e}$ siècle av. J.-C., RA 98 (2004), 151–178.

12 Siehe zuletzt B.R. Foster, Before the Muses. An Anthology of Akkadian Literature. Third edition, Bethesda 2005, 124f.

13 T. Baqir, Mathematical Texts, Sumer 6 (1950), 5–28; ders., Mathematical Texts, Sumer 7 (1951), 129–169; F. al-Rawi/M. Roaf, Ten Old Babylonian Mathematical Problems from Tell Haddad, Himrin, Sumer 43 (1984), 175–218.

14 O. Rouault/C. Saporetti, Old Babylonian Texts from Tell Yelkhi (Hamrīn Project, Iraq), Mesopotamia 20 (1985), 23–52; F.N.H. Al-Rawi, Texts from Tell Haddad and elsewhere, Iraq 56 (1994), 35–43.

15 D. Charpin, A propos du site de Tell Harmal, NABU 1987/117.

feln, die nicht die abgekürzte Version des Jahresnamens enthalten.[16] Dieses Detail zeigt an, wie wichtig den Zeitgenossen der Sieg Babylons über Ešnunna erschien.

Der Name des 38. Regierungsjahrs Hammurabis berichtet darüber, dass Ešnunna durch eine Überschwemmung zerstört worden sei. Die Standard-Übersetzung dieses Jahresnamens macht Hammurabi für diese Überschwemmung verantwortlich. Wie dem auch gewesen sein mag, Hammurabi berichtet im Prolog seines Kodex, dass er die Flüchtlinge aus Ešnunna in Babylon aufgenommen habe:

> „Der fromme Fürst, der das Antlitz des Tišpak erglänzen ließ, der reine Mahlzeiten bereitete für Ninazu, der seine Einwohner rettete in der Not, der ihre Grundfesten festsetzte innerhalb Babylons in Wohlbefinden."[17]

Wir haben keinerlei andere Zeugnisse für die Anwesenheit dieser Bewohner Ešnunnas in Babylon. Es könnte aber sein, dass sich unter ihnen Schreiber befanden, die für die literarische Qualität der späteren Texte mitverantwortlich waren. Es ist signifikant, dass die Texte des Narām-Sîn, die ich gerade erwähnt habe, von ihrem Herausgeber Wilfred George Lambert mit den Inschriften B und C des Samsu-iluna verglichen worden sind.[18] Man kann daran die Art und Weise erkennen, in der Babylon von der zeitlich früheren Erfahrung Ešnunnas mit Königsinschriften profitiert hat.

Auch später noch kam es zum Wettstreit zwischen Ešnunna und Babylon: Samsu-iluna führte mehrere Feldzüge gegen Ešnunna, wie die Namen seines 20. und 24. Jahres beweisen; deren Ausgang war die Unterwerfung des Gebiets und der Bau der Festung Dūr-Samsu-iluna in Hafadji. Unter Abī-ešuh wurde Ešnunna, das zu dieser Zeit von Ahušina regiert wurde, erneut unabhängig. Danach haben wir nur noch sporadische Erwähnungen Ešnunnas bis in die Zeit Samsu-ditanas,[19] woraufhin die Diyalaregion für mehrere Jahrhunderte weitgehend unbesiedelt bleiben sollte.

## 1.4. Zusammenfassung

Wir sehen also, dass die sowohl in kultureller als auch in politischer Hinsicht wichtigste Stadt im Zentrum des heutigen Irak bis 1763 v. Chr. Ešnunna war und nicht Babylon. Erst nach diesem Datum veränderte sich die Lage, und dieser Punkt sollte unterstrichen werden.

16 Siehe zum Beispiel M. Birot, Tablettes économiques et administratives d'époque babylonienne ancienne, conservées au Musée d'art et d'histoire de Genève 8.

17 Prolog des Codex iv 33–44 (vgl. R. Borger, Der Codex Hammurapi [= Texte aus der Umwelt des Alten Testaments I,1], Gütersloh 1982, 43).

18 W.G. Lambert, Studies in Nergal, Bibliotheca Orientalia 30 (1973), 357–363 (358).

19 Siehe zuletzt D. Charpin, Économie et société à Sippar et en Babylonie du nord à l'époque paléo-babylonienne, RA 99 (2005), 133–176 (150 f.).

## 2. Die Folgen der Eroberungen Hammurabis

Der Aufstieg Babylons erfolgte dann sehr schnell. Der Anstoß dazu kam von außen, denn der siegreiche Widerstand Babylons gegen den elamischen Eindringling im Jahr 1765 v. Chr. verdeutlichte allen die militärische Stärke Babylons. Das darauf folgende wichtigste Datum ist 1763 v. Chr., d.h. das 30. Jahr Hammurabis, in dem der König von Babylon Rīm-Sîn von Larsa besiegte, und Babylons Truppen das gesamte Land Sumer einnehmen konnten.

### 2.1. Die kulturellen Folgen der Eroberung

Diese militärische Eroberung und die Annexion des Südens hatten weitreichende Folgen. Von einem kulturellen Standpunkt ist das wohl markanteste Phänomen die Entwicklung einer sumerischen Kultur in Babylon. Im Gegensatz zur bisher gängigen Meinung gibt es in der Tat keine einzige sumerisch-sprachige Königsinschrift Hammurabis vor der Eroberung Larsas.

– Es soll gleich eingangs angemerkt werden, dass die sumerische Ziegel-Inschrift Hammurabis, RIME 4.3.6.1 falsch datiert worden ist. D. Frayne hat sie mit dem Namen des vierten Jahrs Hammurabis in Zusammenhang gebracht, in der die Errichtung einer Mauer für das Kloster in Sippar kommemoriert wird. Die Inschrift RIME 4.3.6.1 des Hammurabi behandelt aber den Bau eines Deichs, um Überschwemmungen zu vermeiden.[20]

– Die einzige „zweisprachige" Inschrift Hammurabis vor der Eroberung des Südens wäre die Inschrift Nr. 2, die den Bau einer Umfassungsmauer von Sippar behandelt. Wie alle diese zweisprachigen Texte handelt es sich dabei nicht um eine echte Bilingue, sondern um Inschriften, die in zwei Varianten, einer sumerischen und einer akkadischen Version, existierten. D. Frayne hat diese Inschrift mit den Namen der Jahre 23 und 25 in Verbindung gebracht. Es besteht aber eine größere Verwandtschaft zum Namen des 43. Jahres. Es erscheint mir daher viel wahrscheinlicher, dass die Inschrift zu diesem späteren Zeitpunkt verfasst wurde.[21]

Wir sehen damit, dass wir bis jetzt keine einzige Inschrift eines Herrschers von Babylon vor der Eroberung des Südens Mesopotamiens besitzen. Dies scheint mir kein Zufall zu sein. Es ist klar, dass die Anwesenheit von Schreibern aus dem südlichen Mesopotamien die kulturelle Situation in Babylon verändert hat. S. Seminara hat in seiner Arbeit über altbabylonische Königsinschriften die *typologische* Verwandtschaft von Königsinschriften aus Babylon und jenen aus Larsa bemerkt, er hat sich aber nicht für die histo-

20 RA 100 (2006), 144, s. S. 273f. zu „Hammu-rapi 1".

21 Vgl die Z. 38–40, die fast wortgetreue Entsprechungen sind und den Vergleich Hammurabis mit einem Gott. Für weitere Einzelheiten, siehe D. Charpin, L'exercice du pouvoir par les rois de la I<sup>ère</sup> dynastie de Babylone: problèmes de méthode, in: G. Wilhelm (Hg.), Organization, Representation and Symbols of Power in the Ancient Near East, Winona Lake, im Druck.

rischen Umstände dieser Kontinuität interessiert.[22] Diese scheinen mir aber eindeutig zu sein. Ein weiteres Beispiel kann hier angeführt werden. Die Hymnologie in Larsa erlebt unter Rîm-Sîn eine Neuerung: die Königshymnen enden nun mit der Doxologie Rîm-Sîn lugal-mu. Solcherlei Hymnen gibt es nun auch für Hammurabi und Samsu-iluna, und es besteht für mich kein Zweifel, dass wir hier einen Einfluss der gebildeten Schreiber aus Larsa haben.

Die Arbeiten von N. Ziegler über die Konsequenzen von militärischen Eroberungen erlauben es uns, die Situation zu verstehen:[23] Nach der Einnahme eines Königreichs wurde das Palastpersonal (natürlich, aber keineswegs ausschließlich, die weibliche Palastbevölkerung) vom Sieger deportiert, nicht aber die gewöhnliche Bevölkerung der Hauptstadt: die Priester aus Larsa sind so in ihrer Heimat geblieben, die Palastschreiber Rîm-Sîns hingegen wurden wohl zum Großteil nach Babylon geführt.

## 2.2. Die historische Bedeutung Hammurabis

Die historische Bedeutung Hammurabis einzuschätzen, ist eine delikate Angelegenheit. Ich habe das Gefühl, dass man hier vielleicht einen Hang zur Übertreibung hat. Wenn Marc Van De Mieroop schreibt:

> "The king's accomplishments were considered exceptional by his contemporaries, and the honors his courtiers bestowed on him surpassed those allotted to other kings of his dynasty",[24]

wüsste man gerne, auf welche Quellen er sich beruft. Ich für meinen Teil kann dem nicht folgen.[25] Auch wenn er hervorhebt, wie einzigartig der Personenname Hammurabi-ilî sei,[26]

22 S. Seminara, Le iscrizioni reali sumero-accadiche d'età paleobabilonese. Un'analisti tipologica e storico-letteraria, Atti della Accademia Nazionale dei Lincei Anno CDI – 2004 8/3, Roma 2004; siehe meine Bemerkungen zu diesem Thema in: Les inscriptions royales suméro-akkadiennes d'époque paléo-babyloniennes, RA 100 (2006), 131–160 (157).

23 N. Ziegler, Le harem du vaincu, RA 93 (1999), 1–26; ebd., Kriege und ihre Folgen: Frauenschicksale anhand der Archive aus Mari, H. Neumann/R. Dittmann/A. Schuster Brandis/C. Eder (Hgg.), Krieg und Frieden im Alten Vorderasien, Wiesbaden (= SANTAG. Arbeiten und Untersuchungen zur Keilschriftkunde), im Druck.

24 M. Van De Mieroop, King Hammurabi of Babylon. A Biography, Malden/Oxford 2005, 122.

25 Ich würde ihm noch weniger folgen, wenn er bemerkt: „A medical text of the first millennium includes the entry: ‚A salve for the eyes from Hammurabi, a proven [medication].' Hammurabi became thus the example of the erudite ruler, whose skills included scholarship and medicine" (ebd., 131). Hier führt ihn der Wunsch nach Sensationen zu einer Überinterpretation (in Widerspruch zur Bemerkung auf S. 140, wonach Hammurabi ungebildet gewesen sei): der betreffende Text sagt einfach *tēqītu ša Hammurabi*, „Salbe des Hammurabi", d.h. „aus der Zeit des Hammurabi" oder „die dazu diente, Hammurabi zu heilen" (BAM 159 IV 16 sq. = von Weiher SpTU II 50: 12 und Kommentar auf S. 196).

26 „The name Hammurabili, meaning ‚Hammurabi is my god', appeared, something unparalleled in his dynasty" (S. 127).

täuscht er sich, denn es gab ebenfalls Leute, die Ammiditana-iluni oder Ammiṣaduqa-iluni hießen.[27] Dies für ein Charakteristikum des südlichen Mesopotamien zu halten ist ebenfalls falsch: Wir kennen auch einen Samsî-Addu-ilî ... im obermesopotamischen Raum.[28]

Eines der Argumente, die vorgebracht werden, um die Bedeutung der Regierungszeit Hammurabis zu unterstreichen, ist, zu zeigen, welche Erinnerung er hinterlassen hat. Victor Hurowitz hat jüngstens diesbezüglich eine Arbeit verfasst.[29] Damit die Darstellung an Signifikanz gewinnt, müsste man dieselbe Arbeit für die anderen babylonischen Herrscher machen: man könnte dann feststellen, dass auch diese Spuren hinterlassen haben. So wird die Erinnerung an Sîn-muballiṭ nicht nur durch seinen Sohn Hammurabi wachgehalten (Inschrift *Hammurabi* 7), sondern auch, was außergewöhnlich ist, durch seinen Enkel Samsu-iluna (Inschrift *Samsu-iluna* 2). Den Königen Sabium und Apil-Sîn wurden Chroniken gewidmet, die noch im 1. Jahrtausend abgeschrieben wurden.[30] Sicher, der Herausgeber der medizinischen Liste SA.GIG, Esagil-kîn-apli, wird als Abkomme des Asalluhi-mansum vorgestellt, bezeichnet als Weiser-*apkallu* des Hammurabi. Aber die seleukidenzeitliche Liste der Herrscher mit ihren *ummânu* (Gelehrten), die von van Dijk publiziert worden ist,[31] enthält auch zwei *ummânu* des Abī-ešuh,[32] nicht aber Gelehrte der Zeit Hammurabis ... Man könnte auch erörtern, dass eine der Orakelanfragen Hammurabis während seines Feldzugs gegen Kazallu durch die Jahrhunderte hindurch kopiert worden ist, um dann in der siebenten Tafel der *tamītu*-Anfragen-Reihe in neuassyrischer Zeit enthalten zu sein.[33] Dem kann entgegengesetzt werden, dass dieselbe Anfragenreihe auch Orakelanfragen Abî-ešuhs[34] oder Samsu-ditanas[35] enthält ...

27 H. Klengel, Ḫammurapi und seine Nachfolger im altbabylonischen Onomastikon, Journal of Cuneiform Studies (1976), 156–160 (159 4a und 5a).

28 In den Texten aus Chagar-Bazar; vgl. Philippe Talon, Old Babylonian Texts from Chagar Bazar. Akkadica Supplementum, vol. 10, Brüssel 1997, 133 und meine Bemerkungen in RA 100 (2006), 155.

29 V.A. Hurowitz, Hammurabi in Mesopotamian Tradition, in: Y. Sefati et al. (Hgg.), An experienced scribe who neglects nothing. Ancient Near Eastern studies in honor of Jacob Klein, Bethesda 2005, 497–532. Siehe seither R. Pientka-Hinz, Midlifecrisis und Angst vor dem Vergessen? Zur Geschichtsüberlieferung Ḫammu-rapis von Babylon, in: K.-P. Adam (Hg.), Historiographie in der Antike, Berlin/New York 2007, 1–25 (diese Arbeit konnte ich erst nach Abfassung des vorliegenden Beitrages zur Kenntnis nehmen).

30 E. Leichty/C.B.F. Walker, Three Babylonian Chronicle and Scientific Texts, in: G. Frame/L. Wilding (Hgg.), From the Upper Sea to the Lower Sea. Studies on History of Assyria and Babylonia in Honor of A. K. Grayson, Leyden 2004 (= Publications de l'Institut historique et archéologique néerlandais de Stamboul 101), 203–212; siehe zu diesem Thema meine Bemerkungen in: RA 100 (2006), 181b–182a.

31 J.J. van Dijk, Die Inschriftenfunde, III.: Die Tontafeln aus dem Palast des Sinkašid, Berlin 1962 (= UVB 18), 39–62 (S. 44 W 20030,7 l. 15; Kommentar S. 51).

32 Siehe ferner W.W. Hallo, Toward a History of Sumerian Literature, in: S. J. Lieberman, Sumerological Studies in Honor of Thorkild Jacobsen on his Seventieth Birthday, Chicago 1975 (= Assyriological Studies 20), 181–203 (187) und D. Charpin, Immigrés, réfugiés et déportés en Babylonie sous Hammu-rabi et ses successeurs, in: D. Charpin/F. Joannès (Hgg.), La circulation des biens, des personnes et des idées dans le Proche-Orient ancien, Actes de la XXXVIIIe Rencontre Assyriologique Internationale (Paris, 8–10 juillet 1991), Paris 1992, 207–218 (212 Anm. 39).

33 Siehe die Editionen dieses Textes in: W.G. Lambert, Babylonian Oracle Questions (= MC 13), Winona Lake 2007, Nr. 1: 1–25 und meinen Kommentar in: RA 91 (1997), 188–190.

34 MC 13, Nr. 3c: 22–47.

Bleibt aber der Gesetzeskodex Hammurabis. Auch hier können mehrere Anmerkungen gemacht werden:

1) Der Fund ist einem Zufall zu verdanken, nämlich der Grabungstechnik Morgans, eines Ingenieurs, der Kubikmeterweise Erde freilegte. Die amerikanischen Ausgräber von Tall Asmar haben bei weitem nicht soviel Erdreich durchsucht und so sind die Gesetze aus Ešnunna nur dank Schulabschriften bekannt: es kann aber nicht ausgeschlossen werden, dass eine Stele ebenfalls existiert hatte.[36]

2) Es kann daran erinnert werden, dass zwei Stelenfragmente mit der Abschrift eines Teils des Kodex Lipit-Eštar erhalten sind:[37] Hammurabi war also nicht unbedingt innovativ in dieser Hinsicht. Es ist allerdings richtig, dass sein Gesetzeswerk in den folgenden Jahrhunderten kopiert worden ist, da es auch nach dem Ende der altbabylonischen Zeit noch immer in den Tempeln Nordbabyloniens sichtbar war – im Gegensatz zum Kodex Lipit-Eštar, der unter den Gemäuern Nippurs vergraben war.

## 2.3. Zusammenfassung

Die Eroberungen Hammurabis haben keine spektakulären Veränderungen mit sich gebracht. Nur die Ereignisse *nach* seiner Herrschaft können die bedeutsame Rolle erklären, die diese Stadt spielen sollte und die ihren Ruhm begründete.[38]

35 MC 13, Nr. 1: 26–95.

36 Man muss gleichwohl anmerken, dass in Susa zahlreiche Statuen von Königen Ešnunnas gefunden wurden (A. Spycket, La statuaire du Proche-Orient ancien, Leyden/Kölln 1981, 237–239), aber keine Stele eines Kodex Dâduša ...

37 Der Kodex Lipit-Eštar ist nicht nur durch Abschriften auf Tontafeln bezeugt, sondern auch durch Fragmente von Stelen aus Stein (F. R. Kraus, Königliche Verfügungen in altbabylonischer Zeit, Leiden 1984 [= Studia et documenta ad iura Orientis antiqui pertinentia 11], 20 § 1 A), vgl. M. Roth, Law Collections from Mesopotamia and Asia Minor, Atlanta 1995, 35 Amn. 1, die von „two fragments of a stone stela that could be Lipit-Ishtar's original monument" spricht.

38 Unter diesem Gesichtspunkt ist das Urteil von W.G. Lambert mehrdeutig: „The reign of Hammurabi was politically important, but equally significant for culture. He destroyed Mari and unified control which had previously rested with many cites. Whether for this reason only, or for other reasons also, the diversity of Mesopotamian culture ended. The details of this presumably gradual process are lost, but when the obscure end of the First Dynasty of Babylon and the equally obscure beginning of the Cassite period are over, we find many of the old Sumerian cities such as Nippur and Uruk back as leaders of culture, but with the significant addition of Babylon." (Babylonian Wisdom Literature, Oxford 1960, 13; vor fünfzig Jahren geschrieben!). Die Passage setzt den Akzent zunächst auf Hammurabi, aber anschließend soll der Einigungsprozess der babylonischen Kultur in der Folge stattgefunden haben.

## 3. Babylon, Bewahrerin sumerischer Traditionen

Seit der Veröffentlichung von Yale Oriental Series 13 durch J.J. Finkelstein im Jahr 1972 haben wir Quellen zur Immigration von Bewohnern des Südens Sumers nach Nordbabylonien während der Regierungszeit Samsu-ilunas. Ich möchte an dieser Stelle auf diesen Punkt zurückkommen, der von zentraler Bedeutung für die Kulturgeschichte Babylons ist, der aber bisher noch nicht genügend unterstrichen wurde.[39] Der Grund, weshalb dieses Phänomen bis heute nur ungenügende Beachtung gefunden hat, ist simpel: wir haben keinen narrativen Bericht über diese Ereignisse, keine Königsinschrift spricht davon. Nur durch das mühselige Sammeln von Informationen in den Archiven konnten diese Emigrationswellen rekonstruiert werden.

### 3.1. Die zwei aufeinander folgenden Auswanderungswellen

Die Flüchtlinge kamen in zwei sukzessiven Wellen an. Zuerst wurden die am südlichsten gelegenen Städte Mesopotamiens aufgegeben. Die historischen Umstände, die dazu geführt hatten, sind unbekannt. Es hatte kurz zuvor eine allgemeine Revolte südmesopotamischer Städte gegen Samsu-iluna gegeben. Unabhängige Könige waren in Larsa (Rîm-Sîn II.) und Uruk (Rîm-Anum) auf den Thron gestiegen. Samsu-iluna gelang es, diese Revolte in seinem zehnten Regierungsjahr niederzuschlagen. Aber bereits für das Jahr 12 fehlen Texte aus Ur, Larsa oder Uruk, ebenso wie aus kleineren Städten wie Kutalla. Wir wissen nicht, was passiert ist; gemäß einigen Autoren hatte ein elamischer Überfall zur Aufgabe der Gebiete geführt, gemäß anderen eine ökologische Katastrophe.[40]

Die zweite Auswanderungswelle fand um das 30. Jahr Samsu-ilunas statt: es war der Zeitpunkt, zu dem Isin und Nippur aufgegeben wurden. C. Wilcke hat vorgeschlagen, dass Samsu-iluna selbst die Überschwemmung des Südens verursacht hatte,[41] es handelt sich dabei aber um eine noch unverifizierte Hypothese. Wir wissen allerdings, dass einige Texte aus Nippur nach Regierungsjahren Ilumma-ilums datiert sind, dem Meerlandkönig, ehe die schriftliche Überlieferung dort gänzlich endet.

39 Man muss den Irrtum von J. Sasson bedauern, der bezüglich des Endes der Herrschaft Hammurabis anmerkte: „He also resettles people deported from conquered territory. In Babylon and in towns under his control, such as Kish and Ur, priestly refugees from diverse defeated cities, including Isin, Uruk, and Larsa, are allowed to build temples in which they could continue the worship of the gods of their ancestors" (King Hammurabi of Babylon, in: J. M. Sasson et al. (Hgg.), Civilizations of the Ancient Near East, New York 1995, 907a). Das Phänomen datiert in Wirklichkeit in die Zeit Samsu-ilunas. Der Fehler verbreitet sich unglücklicherweise in den Werken, die aus dritter Hand geschrieben sind, wie das von G. Leick, The Babylonians. An Introduction, Londres/New York 2003, 35.

40 Siehe Orbis Biblicus Orientalis 160/4, 342–346.

41 C. Wilcke, Weltuntergang als Anfang. Theologische, anthropologische, politisch-historische und ästhetische Ebenen der Interpretation der Sintflutgeschichte im babylonischen Atram-hasīs-Epos, in: A. Jones (Hg.), Weltende. Beiträge zur Kultur- und Religionswissenschaft, Wiesbaden 1999, 63–112.

Häufig ist unterstrichen worden, dass die Aufgabe und die brutale Zerstörung der Städte Südmesopotamiens für uns heute ein Glücksfall sind. Diesem Umstand haben wir zu verdanken, dass zahlreiche Manuskripte sumerischer literarischer Texte sowohl in Ur als auch in Nippur gefunden worden sind: Schul-Übungstexte, die nicht vernichtet worden sind, um erneut beschrieben zu werden.[42] Was aber noch wichtiger sein dürfte, ist der Umstand, dass die Bewohner all dieser Städte nach Babylon oder in benachbarte Städte geflohen waren.

### 3.2. Die Integration der Immigranten

Dank spätaltbabylonischer Archive aus Babylon, Sippar und Kiš konnte die Eingliederung der Immigranten in die Gesellschaft Babylons verfolgt werden.

– Ich habe den Fall des Klerus von Uruk untersucht, der nach Kiš ausgewandert war;[43] Händler aus Uruk sind ebenfalls in Dilbat bezeugt.

– Die Texte aus Babylon, die in Vorderasiatische Schriftdenkmäler 22 von Horst Klengel publiziert worden sind, dokumentieren die Anwesenheit von Familien aus Larsa in Babylon.[44]

– In ihrer Doktorarbeit hatte R. Pientka diese Beobachtungen aufgenommen und vervollständigt. Sie konnte zeigen, dass es Hinweise für die Anwesenheit von Leuten aus Lagaš in Kiš gibt, von ehemaligen Bewohnern Isins in Sippar und von Familien aus Nippur in Babylon.[45]

Einige Texte, die seit 1999 publiziert sind, erlauben, das Bild zu vervollständigen, und zeigen an, dass es auch Emigranten aus Larsa in Sippar gab.[46]

Es gibt dementsprechend keinen eindeutigen Zusammenhang zwischen dem Herkunftsort der Immigranten und ihrem Zielort. Leute aus Uruk landeten zum größten Teil in Kiš aber auch in Dilbat; ebenso konnten Exilanten aus Larsa in Sippar, aber auch in Babylon ansässig werden. In einigen konkreten Fällen kann der kulturelle Einfluss der Bewohner Südmesopotamiens klar nachvollzogen werden. So ist zum Beispiel ein Exemplar

42 Siehe z.B. D.O. Edzard, Geschichte Mesopotamiens. Von den Sumerern bis zu Alexander dem Großen, München 2004, 131.

43 D. Charpin, Le Clergé d'Ur, 403–415.

44 D. Charpin, Un quartier de Babylone et ses habitants, Bibliotheca Orientalia 42 (1985), 265–278.

45 R. Pientka, Die spätaltbabylonische Zeit: Abiešuḫ bis Samsuditana. Quellen, Jahresdaten, Geschichte (= Imgula 2), Münster 1998, 177–211 (Kap. 4 „Das Fortbestehen süd-babylonischer Traditionen"), insbesondere 188–195 (Lagaš – Kiš; Larsa – Babylon; Isin – Sippar; Nippur – Babylon).

46 Vgl. vor allem M. Jursa, „Als König Abi-ešuh gerechte Ordnung hergestellt hat": eine bemerkenswerte altbabylonische Prozessurkunde, RA 91 (1997), 135–145 die nachfolgende Bibliografie zu diesem Text: F. van Koppen, Abum-waqar Overseer of the Merchants at Sippar, NABU 1999/80; B. Lion, Divorces du nord et du sud, NABU 2001/97; D. Charpin, Übersetzung und Kommentar in „Lettres et procès paléo-babyloniens", in: F. Joannès (Hg.), Rendre la justice en Mésopotamie, Paris 2000, 93–95. Nr. 49.

einer Zylinderinschrift Warad-Sîns mit Erwähnung des Baus der Stadtmauer von Ur in einem Privathaus im Merkes in Babylon gefunden worden.[47]

Das Desaster der Raubgrabungen und die noch nicht genügend erarbeitete Paläographie erklären, warum es den Sumerologen bis heute nicht möglich war, die altbabylonischen Manuskripte, die sich in den Museen befinden, präzise zu datieren. Ein Großteil von ihnen ist gewiss die Frucht der Arbeit dieser Emigranten der spätaltbabylonischen Zeit.

### 3.3. Bilanz

Es wird gerne der Vergleich gezogen zwischen dem großen Hammurabi und seinem wenig erfolgreichen Sohn Samsu-iluna, der unfähig war, das Erbe seines Vaters aufrecht zu erhalten. Man muss sich aber verständlich machen, dass es vom kulturgeschichtlichen Standpunkt paradoxerweise gerade der Zufluss der Kultur des Südens Mesopotamiens nach Norden war, der am Anfang von Babylons Prestige steht.

Die Serie TIN-TIR$^{ki}$ bezeugt die Existenz eines Tempels für die Šarrat-Larsa, der gewiss unter den Nachfolgern Samsu-ilunas auf Veranlassung der Immigranten aus Larsa errichtet wurde. Wir wissen nicht, wann die Stadtviertel von Babylon den Namen Eridu oder Kullab erhielten: es erscheint mir aber äußerst wahrscheinlich, dass die Spekulationen über Babylon als Zusammenschau der bewohnten Welt im Umfeld der Exilanten aus Sumer gegen Ende der altbabylonischen Zeit ihren Ursprung nahmen, und nicht erst in der Kassitenzeit, wie man gerne vermutet.[48]

### Zusammenfassung

Einige Bemerkungen zum Abschluss: Zuerst einmal ist gesichert, dass Babylon zu Beginn und während der ersten Hälfte der sogenannten altbabylonischen Zeit keineswegs eine Hauptstadt von der Wichtigkeit Ešnunnas war, weder vom politischen noch vom kulturellen Standpunkt aus gesehen. Es ist eindeutig, dass unsere moderne Terminologie unzureichend ist. „Altbabylonisch" ist ein Ausdruck, der verwendet wird, um einen Zustand der Sprachgeschichte zu beschreiben. Seine erweiterte Verwendung als Bezeichnung für eine Epoche von vier Jahrhunderten im Anschluss an den Fall der Dritten Dynastie von Ur, ist

47 A. Falkenstein, Eine Inschrift Waradsins aus Babylon, Baghdader Mitteilungen 3 (1964), 25–40. Referenz in: O. Pedersén, Archive und Bibliotheken in Babylon: Die Tontafeln der Grabung Robert Koldeweys 1899–1917 (= Abhandlungen der Deutschen Orient-Gesellschaft 25), Berlin/Saarbrücken 2005, 23 BaB 46431; ein Foto dieses Zylinders ist leider nicht publiziert worden.

48 A. R. George, Babylonian Topographical Texts, Louvain 1992 (= Orientalische Literaturzeitung 40), 5: „The Kassite period, then, was a time when scholars began to investigate the theological and cosmological implications of Babylon's political pre-eminence".

unglücklich, denn damit wird, mehr oder weniger unbewusst, eine Bedeutung Babylons suggeriert, die übertrieben ist. Ich persönlich ziehe es vor, von der „amurritischen Zeit" zu sprechen, auch wenn mir hierbei bisher nur wenige Kollegen gefolgt sind.[49]

Für die Zeit nach Hammurabi stellt man nicht so sehr eine gesteigerte Ausstrahlung Babylons fest, sondern eine Ausstrahlung in Abwesenheit von konkurrenzfähigen Zentren. Mari, dann Ur, Larsa und Uruk, später Isin und Nippur, und letztlich auch Ešnunna: all diese großen Hauptstädte oder religiösen Zentren waren gegen Ende der altbabylonischen Zeit verschwunden. Manche von ihnen lebten in der Kassitenzeit wieder auf, nicht so Mari oder Ešnunna. Einige Städte Nordbabyloniens, wie Sippar oder Kiš, hatten seit der Regierungszeit Sumu-la-Els aufgehört, eine politische Rolle zu spielen, andere wie Borsippa oder Dilbat hatten nie eine solche gehabt: Babylon allein war übrig geblieben ... zu ihrem großen Ruhm ...

Es ist auch klar, dass das Phänomen des Aufstiegs Marduks innerhalb des Pantheons unter den Nachfolgern Hammurabis[50] nun in diesem Kontext verstanden werden muss. In den Jahren nach der Regierung Hammurabis sieht man seine steigende Bedeutung in der privaten Frömmigkeit der Bevölkerung; dann aber, in der kassitischen Zeit wurde er zum Haupt des Pantheons erhöht, ein langer Prozess, der wohl endgültig unter Nabukadnezar I. abgeschlossen wurde.

Babylon verdankte seinen Erfolg dem Zusammentreffen zweier aufeinanderfolgender historischer Phänomene. Es wurde im Gebiet von Akkad politisch dominant nach der Ausgrenzung und der nachfolgenden Vernichtung von Ešnunna. Dann hat es das Erbe Sumers antreten können, nachdem Hammurabi Rîm-Sîn besiegt hatte, vor allem aber, als die Einwohner des Südens nach Babylon und in die umgebenden Städte immigrierten. Dieses doppelte Erbe kann die führende Rolle Babylons ab der Kassitenzeit erklären; eine Rolle, die Babylon bis zum Ende der keilschriftlichen Tradition innehaben sollte.

49 Man hat mir entgegengehalten, dass die Amurriter schon vor der Zeit, die ich amurritisch zu nennen vorschlage, in Babylon präsent sind. Das ist wahr, aber ich habe niemals einen solchen Einwand gegen die Bezeichnung „kassitische Zeit" gesehen, bei der es genau dasselbe Phänomen gibt ...

50 So wie es sehr gut beschrieben wurde von W. Sommerfeld, Der Aufstieg Marduks (= Alter Orient und Altes Testament 213), Neukirchen-Vluyn 1982; zusammengefasst in: W. Sommerfeld, The rise of Marduk – Some Aspects of Divine Exaltation, Sumer 41 (1979–81), 97–100; letzte Aktualisierung durch den Autor in: Art. „Marduk A. Philologisch I. In Mesopotamien", RlA 7 (1987–90), 360–370.

*Frederick M. Fales*

# Moving around Babylon: On the Aramean and Chaldean Presence in Southern Mesopotamia

1. A glimpse at the human and social "landscape" of the Babylonian region in the early centuries of the 1st millennium BC, as depicted in the textual material of the age, is sufficient to suggest that the main cities of ancient tradition and fame – from Babylon itself to Borsippa to Sippar to Nippur to Isin to Uruk and Ur – would have only contributed to a limited extent to the overall population. Specifically, all around these well-established urban seats of political power or of religious authority or of learning – where a dwindling population of ancient Sumero-Akkadian stock survived – a massive, and often hovering, presence of Aramean and Chaldean tribally-based groups had been progressively gathering. These groups were in part settled in specific ecological niches of the alluvial plain between the lower reaches of the Tigris and Euphrates, and were in part (seasonally or permanently) circulating in the countryside among the named cities, with no particular scruples about going in and out from the urban precincts themselves.

This brief essay has the scope of touching upon the available evidence on these peoples of non-native origin who were, comprehensively, "moving around Babylon" between the 8th and the 7th century BC. But the notion, here propounded, of "moving around Babylon", should not be understood as a mere historical-geographical guideline; there is decidedly more to it than meets the eye. During this crucial historical period, in effect, Babylon functioned consistently as a political target or prize of excellence, *per se* and in antagonism with the invasive actions of the Assyrians. In a nutshell, Babylon – the city of cities of Mesopotamian tradition – needed to be taken and ruled, or at other times protected and defended to the hilt, against the powerful northern Mesopotamian neighbours, if it was ever to become the political and administrative "center of the world" for the originally non-Mesopotamian peoples of the southern alluvium, as it eventually would. In this particular light, therefore, a number of varying and diverse socio-cultural protagonists, going back to Aramean or Chaldean ethnicities, may be caught in the textual record as "moving around Babylon" at the time.

The historical origins of the Arameans on the lower Tigris and Euphrates alluvium still remain quite unclear.[1] As is well known, Aramean tribal groups are most prominently

1 J. A. Brinkman, A Political History of Post-Kassite Babylonia, 1158–722 B.C., Roma 1968; Id., Prelude to Empire. Babylonian Society and Politics, 747–626 B.C., Philadelphia 1984; E. Lipiński, The Aramaeans. Their Ancient History, Culture, Religion, Leuven 2000; F. M. Fales, Arameans and Chaldeans: Environment and Society, in: G. Leick (ed.), The Babylonian World, New York-London 2007, 288–298.

attested from the $11^{th}$ century B.C. onward as new occupants of strategic areas (and perhaps of pre-existing fortified settlements) throughout the northern Mesopotamia and Syrian steppe – the so-called Jezirah – a base from which they would oppose the Assyrian military thrust towards the Euphrates fords and thence westward into the Transeuphratene and the Levant until 850 BC.

In parallel, official Babylonian texts indicate that tribal groups variously labelled as "Arameans" or "Suteans" (a traditional designation for West Semitic nomads) carried out the looting of Sippar and other cities in the northern alluvial plain at different moments of the $11^{th}$ and $10^{th}$ centuries. At largely the same time, even the Assyrian main cities on the upper and middle Tigris were being menaced by Aramean marauders; and – as one reconstruction goes[2] – the strong Assyrian armed reaction which ensued during the late $10^{th}$–early $9^{th}$ century could have forced the tribal groups to migrate downstream, where they occupied land from the Tigris riverbank to the nearby Euphrates near Sippar, and especially in the vast southeastern plain between the Tigris and Elam. Certainly, some residual connections between these southern Arameans with the middle Euphrates area might act as evidence of this progressive descent downstream: as in the case of the *Haṭallu* tribe, which is mentioned in the annals of the philobabylonian rulers of Suhu around 770/760 BC, and reappears, with some of its sub-groups, in the long list[3], which summarizes the names of the Aramean tribes defeated by Tiglath-pileser III (745–722 BC). There are, however, other theoretical reconstructions of this scenario, directly linking the Aramean takeover lower reaches of the Tigris to the plundering actions of the $11^{th}$ century, i.e. implying that the Arameans had very early on already split between northern (Jezirah) and southern branches.[4]

It is difficult to decide on the matter, as things stand. As is well known, the characteristics and statistics of human presence in southern Mesopotamia have been the object of a number of regional or local surveys and analyses in an anthropological-archaeological perspective during the last half-century.[5] These surveys, while unfortunately not covering in full the territory of the southernmost Tigris-Euphrates alluvium, have provided a valid and detailed picture of human occupation from a diachronic perspective for a number of specifically observed areas, ranging from the Diyala basin around present-day Baghdad to the Euphrates basin to the city and area of Kiš and finally to the Uruk hinterland. The combined data of the surveys points to an overall long-term trend for the period between the $12^{th}$ and

2 E. Lipiński, The Aramaeans. *cit.*, 412ff.

3 H. Tadmor, The Inscriptions of Tiglath-pileser III, King of Assyria, Jerusalem 1994, 158–161; E. Lipiński, The Aramaeans. *cit.*, 426–428.

4 J. A. Brinkman, A Political History of Post-Kassite Babylonia, *cit.*, 281–283.

5 R. McC. Adams, Land behind Baghdad, Chicago-London, 1965; R. McC. Adams, Heartland of Cities, Chicago-London 1981; R. McC. Adams/H.-J. Nissen, The Uruk Countryside, Chicago-London 1972; McG. Gibson, The City and Area of Kish, Coconut Grove 1972; S. W. Cole/H. Gasche, Levees, Floods, and the River Network of Northern Babylonia: 2000–1500 and 1000–500 BC – A preliminary report, in: J. Renger (ed.), Babylon: Focus mesopotamischer Geschichte, Berlin 1999, 87–110.

the late 8th century BC in the lower Euphrates region and in that of the Diyala.[6] It is marked by a general decline in population levels and by a diminution of urbanism, with a corresponding increase of economic and social ruralization. Another feature which characterizes this phase is that of extensive abandonments of settlements in both surveyed areas, with limited compensation in the foundation of new sites. However, it is difficult to state whether, and to what extent, this trend should be viewed in connection with the shifting away and drying up of specific watercourses – which in its turn, according to some, represented the outcome of a perceptible climatic change towards aridity[7] – or rather due to social and political disruptions caused by internal causes.

Let us thus rest with what we actually know about the Aramean presence in southern Mesopotamia. In one of the earliest testimonials, from the time of Tukulti-Ninurta II (885 BC), the Utu' or Itu' tribe – which occupied the west bank of the Tigris around present-day Samarra – is shown as having mixed living quarters, perhaps in relation to seasonal transhumance, comprising "encampments" (*maškanāte*) consisting of tents as well as actual (agricultural) "villages".[8] A century and a half later, when the lower Tigris catchment area became the object of intense Assyrian military pressure, we can recognize nearly 40 distinct names of medium- to small-sized tribal entities in Tiglath-pileser III's list of "unsubmissive Arameans". These tribes would seem to have retained their basic West Semitic ethnolinguistic traditions (both in personal and group onomastics), and to have held fast to their kinship-based social structure with only minimal yield to the pressures of adjacent sedentary states.

The Arameans, in the main, also seem to have resisted the power of attraction of indigenous Babylonian culture with its prestigious network of beliefs and lore, ennobled by a great antiquity. This social and cultural "separateness" is all the more noteworthy in that many of the Aramean tribes were in close contact with the Babylonian settlements for everyday matters: thus, e.g. the vast group of the Puqudu was active for a time in the area surrounding the ages-old cultural and political center of Nippur, to the extent of frequenting the city *en masse* to participate in a festival (*isinnu*) during the month of Ululu.[9] Other textual attestations for this tribe point, on the other hand, to a variety of non-urban settings for its predominant economic activities; some of its main grazing grounds were in the general area of Lahiru, eastwards of the Tigris between the Diyala and Der; while a number of reports place the Puqudu in the marshy areas further south, along the Babylonian-Elamite border. It is in an even more southernly location, along the lower reaches of the Tigris and

6 J. A. Brinkman, Prelude to Empire. Babylonian Society and Politics, *cit.*, 8–11.

7 J. Neumann/S. Parpola, Climatic Change and the Eleventh-Tenth Century Eclipse of Assyria and Babylonia, Journal of Near Eastern Studies 46 (1987), 161–182; W. Kirleis/M. Herles, Climatic Changes as Reason for Assyro-Aramaean Conflicts? Pollen Evidence for Drought at the End of the 2nd Millennium BC, State Archives of Assyria Bulletin 16 (2007), 7–37.

8 Cf. J. N. Postgate, art. "Itu'", in: RlA V, 221–222.

9 S. W. Cole, Nippur in Late Assyrian Times, Helsinki 1996, 9–13.

of its inner branches, that we find the Puqudu in league with the Chaldean chieftain Marduk-apal-iddina (Merodach-baladan) II of Bit-Yakin against the Assyrians during the years 712–709 BC. In later phases, this tribe will be again associated with anti-Assyrian activities, but now operating from the southernmost sector of the alluvium, from where it sometimes reached out westward to constitute a menace for the philo-Assyrian governors of Uruk and Ur. In general, compared to the Chaldean groups, the family units of Arameans prove to have had a far great mobility, not only infiltrating themselves between one enclave and another, but even giving rise to interregional movements, between the Middle Euphrates and Lower Mesopotamia.[10]

A further characteristic of the Arameans lies in their on-going rejection of an ideology of unified leadership encompassing complexes wider than individual kinship-based groups. To the opposite, in fact, as the case of the geographically ubiquitous Puqudu might show, it is the kinship-based group itself that seems to have split up in various internal ramifications, albeit retaining its common tribal denomination. The social (and, when necessary, military/political) leadership of each Aramean ramification or clan unit went back to a specific *nasiku*, "sheikh", as indicated by written records from the reign of Sargon II onward. In point of fact, it has been noted that *nasikus* are attested in these documents in connection with a multiplicity of institutional or even purely geographical entities: i.e. not only tribes, but also lands, cities, even rivers.[11] This very feature might indicate a certain degree of segmentation and renewed identification as being in progress within the tribal units themselves. In particular, the fact that Tiglath-pileser III's list comprised, alongside many indisputably tribal groupings, entities elsewhere known only as toponyms (Rapiqu, Hiranu, Rabilu, Radê, Karma', etc.), might not point so much to an "Assyrian fabrication or simplification"[12], as to the reality of a process of social and territorial subdivision which was underway among the groups themselves.

To be sure, some tribal units had presumably attained early on an ideal balance between their demographical dimensions, their specific territorial quarters, and their distinctive ethnicity. This would seem to be the case of the Utu'/Itu', who, after their subjugation by Tiglath-pileser III, were integrated into the ranks of the Assyrian administration as a trustworthy corps of "military police" characterized by their original ethnonym (similarly to the "Swiss guards", the specific military force of the Vatican for the last five centuries). The auxiliary bowmen represented in the Assyrian palace bas-reliefs might refer to this group.[13]

On the other hand, some ethnosocial units of the same type are reported by the contemporary texts as still being in the process of internal accretion, as in the case of the

10 *Ibid.*, 23–28.

11 J. A. Brinkman, A Political History of Post-Kassite Babylonia, *cit.*, 274–275.

12 *Ibid.*, 271.

13 J. E. Reade, The Neo-Assyrian Court and Army: Evidence from the Sculptures, Iraq 34 (1972), 87–112; J. N. Postgate, The Assyrian Army in Zamua, Iraq 62 (2000), 89–108.

Rupu' who had incorporated the smaller group of Q/Gamu according to a letter from Nippur.[14] And finally, the vaster and geographically more dispersed tribal complexes, such as the Puqudu and the Gambulu, while still retaining their distinctive self-identification, prove to have developed a number of inner clan-internal subdivisions with reference to different "sheikhs", who united their military and political efforts or took individual courses of action, depending on the circumstances. In this case, the possibility that (periodical or random) "conventions" of all the clans could have been organized to decide common tribal policies, is realistic, but has not hitherto surfaced as a specific occurrence in the textual record.

Differently from the long-attested Arameans, the Chaldeans (*Kaldu*) are – quite surprisingly – not documented in the written sources before 878 BC. Their place names, and especially those of their vast territorial and political enclaves, were characterized by the noun *Bīt*, "household", followed by the linguistically West Semitic personal name of an eponymic ancestor figure, exactly as in the case of the contemporary Aramean states of the Jezirah and Transeuphratene.[15] This feature allows us to postulate a connection of the Chaldeans with the northern and western Arameans in the general perspective of a shared heritage of ethnicity; while some slight hints in the texts might more specifically point to political affiliations of long standing between the Chaldeans and the Aramean tribes of the Middle Euphrates area: e.g. as we shall see below, the rebel Chaldean leader Mukīn-zēri had political contacts with the Arameans of Hindanu, some 300 kms upstream on the Euphrates.

On the other hand, it must be admitted that no straightforward structural similarities between the two main non-indigenous groupings present in the Babylonian area may be traced on the social, economic, and cultural level; quite the contrary is in fact true. This contrast is particularly evident if one observes the settlement patterns of the Chaldeans, and their general socio-economic profile. To be sure, the social structure of the Chaldeans was rigidly centered upon the tribal unit of which all subjects were jointly "members" (*mār*, literally, "son" of the eponymic ancestor) – similarly to the Aramean tribal "households" of the northern Jezirah and inner Syria – but it would be more precise to state that such units represented in fact tribal confederations, which must have undergone a relatively long process of social coalescence, although the latter has left no trace in the written record.

The leader of each tribal confederation was indicated in the Assyrian texts as *ra'su*, "chieftain". The fact that all such chieftains mutually recognized their status within a wider territorial-political complex which ideally united the different Chaldean confederations, is evident from a Nimrud letter from the time of Tiglath-pileser III[16], in which the young Merodach-baladan is described as "one of the chieftains of the land of Chaldea" (*ina libbi*

14 S. W. Cole, Nippur in Late Assyrian Times, *cit.*, 83, 5–7.
15 For etymologies, cf. E. Lipiński, The Aramaeans, *cit.*, 418–420.
16 H. W. F. Saggs, The Nimrud Letters, 1952, London 2001, 25–26: 5'–6'; cf. § 2 below.

*re'asāni ša māt Kaldi*). On the other hand, infighting among the various Chaldean chiefs seems to have represented an endemic feature in the framework of the overall subdivision of the southern Mesopotamian area during the earliest periods, and did not cease entirely at the time of Assyrian intervention. Thus, e.g., the archive of the governor of Nippur[17] includes a letter (no. 34) relating a raid undertaken by a Chaldean army (probably including Bit-Amukkani and Bit-Dakkuri, allied at the time) against the Chaldean-dominated city of Larak in order to despoil the harvest. But certainly the Assyrian presence played a major part in determining the choices and alliances of each tribe.

While nominally retaining a social and political structure based on kinship ties, the Chaldeans appear to have taken on a basically sedentary way of life in their southern Euphrates enclaves, with occupations in agriculture, stock raising, and intra-regional trade.[18] The Chaldeans in the Assyrian record (texts and palace reliefs) are shown tending to large tracts of land within their well-watered territorial niches, where they practised agriculture (including date-palm cultivation) and breeding of horses and cattle; however, looking at things on a seasonal basis, they may well have carried out specific forms of "social dimorphism", entailing intensive agricultural exploitation in the area alongside the rivers, alternated with periods of massified transhumance in steppe sectors.

The territories of the three main Chaldean groupings (Bit-Dakkuri, Bit-Amukkani, Bit-Yakin) extended in a sort of arc along the "living" Euphrates branches from the Borsippa region to the Uruk countryside to the southernmost reaches of the Euphrates around Ur and into the marshlands to the east. Specifically, Bit-Dakkuri probably occupied the cultivated areas along the river from the area of Borsippa (present Birs Nimrud) to that of Marad (present Diwaniyah), and thus definitely to the northwest of Bit-Amukkani[19], which controlled a territory in the central sector of southern Mesopotamia, more or less between Nippur and Uruk. Bit-Yakin occupied the more southern sector of the alluvial plane, including ample marshy areas – hence the definition of Merodach-baladan as "king of the Sealand" by Sennacherib.

The structures for communal living within these enclaves comprised not only rural villages and small townships, but also a fair percentage of walled cities. Thus Sennacherib, describing his first campaign into Babylonia (703 BC), boasted of having besieged and conquered 33 walled cities and 250 townships of Bit-Dakkuri; 8 walled cities and 120 townships of Bit- Ša'alli; 39 walled cities and 350 townships of Bit-Amukkani; 8 walled cities and 100 townships of Bit-Yakin – a grand total of 88 major urban sites with defensive structures and 820 smaller settlements of mainly rural character in their environs.[20]

17 S. W. Cole, The Early Neo-Babylonian Governor's Archive from Nippur, Chicago 1996.

18 J. A. Brinkman, Prelude to Empire. Babylonian Society and Politics, *cit.*; S. W. Cole, The Early Neo-Babylonian Governor's Archive from Nippur, *cit.*

19 S. W. Cole, Nippur in Late Assyrian Times, *cit.*, 31.

20 D. D. Luckenbill, The Annals of Sennacherib, Chicago 1924, 36–50 and 54–56; E. Frahm, Einleitung in die Sanherib-Inschriften, Wien 1997, 9.

But the strategic position of the main Chaldean enclaves along the westernmost and southern axes of the alluvium also had crucial implications for commerce. The lists of precious goods offered already in the 9th century, and then again under Tiglath-pileser III, by the Chaldean chiefs as tribute to the Assyrians, which included elephant hides and tusks, ebony and sissoo wood, prove that the Chaldean tribes had gained full control of the trade routes cutting through the Babylonian region[21], and were thus on the receiving end of a vast commercial network which reached Mesopotamia from the Levant, Northern Arabia, and Egypt by land. Various pieces of information are in agreement in pointing to the fact that a specific "southern Mesopotamian axis" of trade, based on seamanship and the recently introduced large-scale exploitation of the camel as pack-animal, was progressively constituted, such as to antagonize the northern Mesopotamian routes dominated by the Assyrian empire – and which would eventually replace the latter. If we range together a number of specific factors, such as [1] the presence of Arabian allies in the military efforts by the Chaldeans[22], [2] a strong thrust for political and military cooperation offered by the Elamite state on the basis of economic advantages[23], and [3] evidence for direct contacts with the Levant[24], we may begin to form a coherent picture of the political substructure behind this new commercial axis.

And finally, from the cultural point of view, the Chaldeans embraced Babylonian ways quite soon after their arrival. Both leaders and commoners of the Chaldeans mentioned in the texts bore fully Babylonian personal names, with devotional reference to the traditional Sumero-Akkadian pantheon of the region. This aspect – so obvious in the written record as to be a given – does not seem to have been investigated in depth as regards its possible causes. Much clearer, on the other hand, are the political implications of this clear cultural stance, viz. the capacity of the Chaldeans to enter the arena of military appropriation and territorial supremacy in the Southern Mesopotamian region boasting exactly the same rights as the local population of ancient stock. And the consequences of this capacity would not be long in bearing fruit, under the specific stimulus provided by continuous Assyrian interference in Babylonian affairs.

2. Let us at this point attempt to gain a closer view of indigenous and non-indigenous components of the southern Mesopotamian region as "moving around Babylon", especially

21 J. A. Brinkman, A Political History of Post-Kassite Babylonia, *cit.*, 198–199; G. Frame, Babylonia 689–627 B.C. A Political History, Leiden 1992, 37.

22 E.g. Basqanu, the brother of Iati'e, queen of the Arabs, who was captured by Sennacherib in 703 BC: I. Eph'al, "Arabs" in Babylonia in the 8th Century B.C., Journal of the American Oriental Society 94 (1974), 108–115; S. Parpola (ed.), The Prosopography of the Neo-Assyrian Empire: I/1–2, Helsinki 1998–1999, 276a.

23 Beginning with Merodach-baladan's time; cf. J. A. Brinkman, Elamite Aid to Merodach-Baladan, Journal of Near Eastern Studies 24 (1965), 161–166.

24 E.g. the offer of an anti-Assyrian alliance by Merodach-baladan to Hezekiah of Judah in *2 Kings*, 20,12.

in light of the Assyrian effort to subjugate and control the entire southern Mesopotamian region. In consideration of the space of this contribution, but also in view of the fact that the written sources are not evenly distributed over the full century they span, I would on the present occasion limit myself to one specific case-study, relevant to the earliest phases of Assyrian occupation: the revolt of Mukīn-zēri during Tiglath-pileser III's reign as described in the Nimrud letters.[25]

Nabû-mukīn-zēri, as his full name went, was the chief of the tribal political unit of Bit-Amukkani[26]; specifically, he was also a personality of great prestige and a crucial reference-point for the neighbouring urban communities such as Nippur, as we know from the local "governor's archive".[27] As the standard reconstruction of events goes[28], we have it that, in a period of strong dynastic instability in Babylon, Mukīn-zēri removed another rebellious candidate to the throne, Nabû-šuma-ukīn (II), and assumed kingship in 732. Tiglath-pileser III, being occupied by the siege of Damascus, could not react immediately; but the following year he marched to the south against Mukīn-zēri in his residence, the fortified city of Šapiya or Sapiya. This simplified reconstruction is essentially based on the only two passages of Assyrian royal inscriptions that deal with the revolt and its punishment:

> "I enclosed Mukīn-zēri of Bit-Amukkani in Sapiya, his royal city. I inflicted a heavy defeat upon him before his city gates. I cut down the orchards and the *mušukkannu*-trees around the city walls, and did not leave a single one. I cut down the date palms throughout his land. I ripped off their fruit and filled the meadows. All of his cities I utterly destroyed and burnt down.
> I destroyed Bit-Šilani, Bit-Amukkani, and Bit-Ša'alli completely, like mounds after the flood, and reduced them to mounds of ruins."[29]

> "I smashed [Bit-Amukkani] and brought all its people to Assyria. I enclosed Mukīn-zēri their king in Sapiya, his city."[30]

25 Cf. F. M. Fales, Tiglat-Pileser III tra annalistica reale ed epistolografia quotidiana, in: F. Pecchioli Daddi/ M. C. Guidotti (eds.), Narrare gli Eventi, Roma 2005, 163–191. I have benefited from a preliminary English translation of this article by Dr. Greta van Buylaere, Cambridge, whom I thank wholeheartedly.
26 S. Parpola (ed.), The Prosopography of the Neo-Assyrian Empire: II/1–2, Helsinki 2000, 764a–b.
27 S. W. Cole, The Early Neo-Babylonian Governor's Archive from Nippur, *cit.*, texts no. 6, 17–18, 21–22, 97; cf. Id., Nippur in Late Assyrian Times, *cit.*, 31–32, for the political relationships between Mukīn-zēri and the *šandabakku* of Nippur.
28 J. A. Brinkman, A Political History of Post-Kassite Babylonia, *cit.*, 236. See also Id., art. "Mukīn-zēri", RlA VIII, 410–411.
29 Summary Inscription 7:23–25; H. Tadmor, The Inscriptions of Tiglath-pileser III, King of Assyria, Jerusalem 1994, 162–163.
30 Summary Inscription 11:16; H. Tadmor, The Inscriptions of Tiglath-pileser III, *cit.*, 196–197.

In both inscriptions, the episode of the siege is squeezed between two other narrative pieces, which include more details, concerning the Chaldean polities adjacent to Bit-Amukkani. After the attack against Šapiya, Tiglath-pileser's official text recalls the dire defeat and capture of the Chaldean chieftains Nabû-ušabši of Bit-Šilani (in the city of Sarrabanu) and Zak/qiru of Bit-Ša'alli (in the capital of Dur(-ša)-Balihayyu).

The fact that this phase could have been, however, partly simultaneous to that of the fight against Bit-Amukkani, might be deduced from the common recapitulation of the destruction with the *topos* on the "post-diluvial *tell*" which ends the narration about Mukīn-zēri in the more extensive inscription. Bit-Šilani should have been adjacent to Bit-Amukkani, considering that its major city, Sarrabanu, joined the latter at the time of Sennacherib. Bit-Ša'alli should, on the other hand, have been located slightly to the northwest of Bit-Amukkani, on the Euphrates side.[31]

Immediately after the description of the siege, then, Tiglath-pileser's larger text has a report on rich tribute brought by other Chaldean chieftains, who voluntarily came into the presence of the king, viz. Balassu of Bit-Dakkuri, Nadinu of Larak, and – especially – Marduk-apla-iddina of Bit-Yakin. Merodach-baladan makes here his debut in Assyrian royal inscriptions, of which – as is well-known – he will become a major protagonist as adversary to the Assyrians until the time of Sennacherib. He is explicitly said by Tiglath-pileser to have brought his own enormous tribute (of gold, pearls, beams of ebony, medicinal plants, multicoloured garments, cattle and sheep) "to Sapiya, in my presence": thus, we may again assume a basic contemporaneity of this episode with the Assyrian takeover of Mukīn-zēri's stronghold immediately after his defeat.

As shall be pointed out below, the Nimrud letters bear out both these hypotheses in full, since all the protagonists of these passages may be shown to have operated at the same time on one or the other side of the military and political divide between the Assyrians and Mukīn-zēri. But how long did the actual engagement between the Assyrian king and this precursor of all later Chaldean rebels actually last? On p. 98 I give a synoptic chart for the main events of the years 731–729[32] according to information drawn from the official sources: the Eponym Chronicle, the royal inscriptions (annals and summary inscriptions, stele and rock reliefs) or other sources (in italics).

On the basis of the indication in the "Eponym Chronicle" for the 15$^{th}$ *palû* ("against Šapiya"), it would be possible to date the siege of the city to the first year of this new involvement of Tiglath-pileser III in the Babylonian region, that is, in 731, and likewise to link to it the voluntary submission of Merodach-baladan – following in this Tadmor's reconstruction.[33] On the other hand, a further chronicle text (the "Babylonian Chronicle") provides three additional pieces of information: (a) that Mukīn-zēri ruled for three full

31 S. W. Cole, Nippur in Late Assyrian Times, *cit.*, 31.
32 Based on H. Tadmor, The Inscriptions of Tiglath-pileser III, *cit.* 232–237 and passim.
33 H. Tadmor, The Inscriptions of Tiglath-pileser III, *cit.*, 272.

| **Legend**: TP III = Tiglath-pileser III; ARI = Assyrian royal inscriptions (annals and summary inscriptions); √ = existing narration; / = non-preserved or non-existing narration. | | | | |
|---|---|---|---|---|
| Year / Regnal year | Eponym Chronicle (quotation or summary) | ARI | Stele, relief, other official source | Summary of the information in the royal inscriptions and/or in other available sources |
| 731 (15th *palû*) | "Against Šapiya" | √ | / | Defeat of the Chaldean chieftains of central and southern Babylonia. Siege of Šapiya, seat of Nabû-mukīn-zēri of Bit-Amukkani. |
| 730 (16th *palû*) | "(The king stayed) in the Land" | / | / | / |
| 729 (17th *palû*) | "The king grasped the hands of the god Bel" | / | (*Babylonian Chronicle*) | *Defeat of Nabû-mukīn-zēri, after 3 years of rulership.* TP III ascends the Babylonian throne and participates in the New Year's Festival (in Nisannu of the Babylonian year 728). |

years over Babylonia, that (b) in his third year, Tiglath-pileser launched an attack on the region, devastating Bit-Amukkani and defeating the rebel, and that (c) soon after, the Assyrian king ascended the Babylonian throne.[34] And since the "Eponym Chronicle" has the notation "the king stayed in the country" for the 16th *palû* whereas the 17th *palû* is characterized as (the year in which) "the king grasped the hands of the god Bēl" – referring thus to the enthronement of Tiglath-pileser in Babylonia – it seems plausible to date the final demise of Mukīn-zēri to 729 BC, in agreement with most specialists.

The essential chronology of the political-military events under examination is thus sufficiently clear, as is also their basic plot (revolt of Mukīn-zēri → Assyrian military intervention → other Chaldean chiefs pay homage to TP III → three years of overall hostilities → final victory and dethronement of the rebel → the Assyrian ruler assumes kingship of Babylonia). However, as noted long ago[35], a number of questions pertaining to the political and social context in which the rebellion of Mukīn-zēri took place still remain open. In this light, it is worth pursuing research on the matter by looking at the contemporaneous letters which describe, or touch upon, the rebellion, which come from Nimrud, the ancient capital city of Kalhu.

34 J.-J. Glassner, Chroniques mésopotamiennes, Paris 1993, 180.
35 J. A. Brinkman, A Political History of Post-Kassite Babylonia, *cit.*, 238.

It barely needs recalling that Nimrud/Kalhu, discovered by Austen Henry Layard and initially believed to be Nineveh, was again excavated by a British expedition led by Max Mallowan about one century later. During the campaign of 1952, ca 240 epistolary texts re-emerged in the northeastern area of the site, and specifically in the area known as the "*ziggurat* terrace." These documents were entrusted to the then young philologist Henry William Saggs for a complete publication: however, only 105 of these letters were published in a series of articles appearing from 1955 to 1974 in the journal *Iraq*, after which a long silence ensued. These 105 letters, identified by the *siglum* ND and a number progressively applied to all "small finds" of the Nimrud excavation, epigraphic or not (e.g. ND 2700), received from Saggs an editorial ordering, with the abbreviation N(imrud) L(etters) and a Roman number. Finally, in late 2001, Saggs, by now retired, produced the edition of all the 240 letters in the series *Cuneiform Texts from Nimrud*, complete with copies, transliterations and translations. Despite this vast effort, this edition does not, unfortunately, meet the standards which would have made a half a century's wait justified, since it largely fails to take into due account the relevant philological-linguistical progress made in the meanwhile, and it is characterized by the absence of all historical comment or chronological specification concerning these epistolary texts. Thus, while awaiting a further ("definitive") edition, there is much detailed groundwork to be done on such materials.

The sole group of letters from Nimrud/Kalhu for which historical attention of a certain detail was provided in the 2001 edition is the one relating to the Mukīn-zēri revolt.[36] Saggs connected a total of 40 texts to this historical episode, of which – in his opinion – 32 were liable to be ordered chronologically or at least on the basis of internal clues.[37] However, some doubts about the identity of senders and addressed kings, and, more generally, about the exact chronological phase of several of the letters advise a more cautious reduction of this number to a bit more than one half.

In this light, a twofold division of these sources is in order: a first category (A), in which the Chaldean usurper is directly mentioned (or at least his people, and toponyms with a well-known connection to the sedition), and a second (B), in which the same Mukīn-zēri does not appear, and the links between names/toponyms and the political and military scene of the years 731–729 are more tenuous. The texts of category A (1–12) are presented in the synoptic chart below; while the most important of category B will be merely quoted (by ND number) in the ensuing discussion.

36 See already H. W. F. Saggs, The Nimrud Letters, 1952 – Part I: The Ukin-zer Rebellion and Other Texts, Iraq 17 (1955), 21–50.

37 H. W. F. Saggs, The Nimrud Letters, 1952, London 2001, 9.

| No. | Text | Sender and addressee ([] = fragmentary context) | Contextual data, when present | Summary of the context |
|---|---|---|---|---|
| 1 | ND 2674 (= H.W.F. Saggs, The Nimrud Letters, 1952, London 2001, 14–17 = NL VI)[38] | [], to "my brother" | written by Merodach-balad an (?) | Answer to a request of news by the addressee: Mukīn-zēri has arrived in Babylon with cavalry, blocking the inhabitants of the city within the walls. Fear also in Borsippa, from where the inhabitants evacuate the horses. |
| 2 | ND 2695 (= Saggs 2001, 18–19) | Dummuqu to the king | | Report: Mukīn-zēri has entered Babylon with ten horses, offering peace to the people. Allusion to 5–6000 horses and mules (fragmentary). He has confirmed: "Do not fear!" |
| 3 | ND 2632 (= Saggs 2001, 19–21 = NL I) | Šamaš-bunay and Nabû-nammir to the king | | Report of a mission: the two Assyrian officials go to Babylon with proposals of tax exemptions on behalf of TP III, but the citizens, allies of Mukīn-zēri – and in the presence of one of his envoys – do not let them enter. Animated discussion takes place before the city gate: the two Assyrians announce the factual arrival of the king, but the locals maintain that they do not believe the king will come. Threatening to punish the allies and followers of Mukīn-zēri, the two announce their will to remain in Kar-Nergal until the arrival of TP III.<br>Report on the Li'tamu tribe, which sends a message to proclaim its true loyalty to the king.<br>Report on the city of Dilbat and the actions of Mukīn-zēri over there: fragmentary. |
| 4 | ND 2494 (= Saggs 2001, 21–22 = NL LXVI) | [], to the king | | Answer to a request of the king to record how many Babylonians "have come over to my side": no one has come. |

38 Henceforth abbreviated as Saggs 2001.

| No. | Text | Sender and addressee ([] = fragmentary context) | Contextual data, when present | Summary of the context |
|---|---|---|---|---|
| 5 | ND 2717 (= Saggs 2001, 22–25 = NL II) | [], to the king | area between Marad and Dilbat | Report: the Assyrian forces and those of Mukīn-zēri are opposite each other near the territory of a tribe, maybe of Aramaeans (the name is lost), with the common purpose of obtaining an alliance with the latter. The tribesmen seem reluctant to take sides and to continue towards Marad, as the Assyrians – who send a cavalry corps to negotiate – would like.<br>Report of a raid of Mukīn-zēri against the inhabitants of Larak "when he left Sapiya," and of a raid in the opposite direction by the people of Larak, who took away 20,000 sheep of the Chaldean chieftain, while they were grazing.<br>Report about Babylon: Mukīn-zēri has urged the Babylonians to destroy the date palms of Dilbat, however, only some servants of the temple of Bel have accepted, the others were unwilling. |
| 6 | ND 2603 (= Saggs 2001, 25–26 = NL V) | [], to the king | | Report: Mukīn-zēri has written to Merodach-baladan, defining him as one of the main Chaldean chieftains, urging him for an alliance, and accusing Balassu (of Bit-Dakkuri) of being one of his rivals, who, moreover, would have ruined the Chaldean land. His message is intercepted and brought to the author of this letter, who is with Balassu. The latter shows signs of deep anxiety, since Mukīn-zēri is the son of his sister. The Assyrians tend to soothe him, encouraging him to go to the aid of the people of Larak, as previously agreed. |

| No. | Text | Sender and addressee ([] = fragmentary context) | Contextual data, when present | Summary of the context |
|---|---|---|---|---|
| 7 | ND 2363 (= Saggs 2001, 35–36, previously unpublished) | [], to the king | | Very fragmentary report: Balassu, the city Dur-ladini (of Bit-Dakkuri) and Nadinu (Chaldean chieftain of Larak) are mentioned in connection with movements of people and means of transportation. The sender says to have returned from Dur-ša-Balihayyu (the capital of Bit-Ša'alli). |
| 8 | ND 2360 (= Saggs 2001, 56–57 = NL IV) | [], to the king | | Very fragmentary report: Mukīn-zēri and his troops are mentioned, perhaps coming out from the city of Hamete. |
| 9 | ND 2636 (= Saggs 2001, 61–63 = NL VII) | [], to the king | | Very fragmentary report: Mukīn-zēri is mentioned, and – perhaps – the fact that Bit-Amukkani has abandoned him. |
| 10 | ND 2700 (= Saggs 2001, 64–66 = NL III). | [], to the king | | Report on several events. A messenger of Mukīn-zēri, named Yadib-ilu', native to a family of Hindana, had been captured by Assyrian auxiliary forces to be sent to the king, but then he fled and went back to Mukīn-zēri. |
| 11 | ND 2388 (= Saggs 2001, 57–58, prev. unp.) | [], to [] | | Message between two officials (?): fragmentary. The fact that the king has spoken with a messenger of the "sons of Mukīn-zēri" is mentioned. |
| 12 | ND 2385 (= Saggs 2001, 45–46 = NL LXV) | Aššur-šallimanni to the king | | Report: the forces of the sender, together with the commander in chief have opened the "fortress" (= Šapiya). They reached the urban gates, and defeated the enemy: "Mukīn-zēri is killed/defeated, and his son Šum-ukīn is killed/defeated."<br>As to the rations of the grain ordered by the king for Merodach-baladan, the sender has now – after the battle – provided them. |

NOTES TO THE CHART. **No. 1.** As noticed by Saggs (2001, 16), neither of the two correspondents can be the Assyrian king as he is several times mentioned, not only without the habitual formula of courtesy, but even in a passage with a stinging tone (rev. ll. 15–17): "Until you have brought a written reply to my tablet, I shall not agree to send a second message to the king of Assyria". Following Saggs, Merodach-baladan could have been the author. **No. 2.** Dummuqu was perhaps the founder of a noble Babylonian family, as might be made out from later sources, which recall his descent.[38] In any case, the precise reason for the loyalty and dependability of Dummuqu vis-à-vis the Assyrian king is unknown. **No. 3.** In the letters of Nabû-nammir[39] he is defined as an "official active in Babylonia." **No. 5.** This letter may be from Šamaš-bunaya, as it deals with Babylon (see letter no. 3). It is certainly not from Aššur-šallimanni, as "the governor of Arrapha" is mentioned. **No. 6.** On Balassu, see now S. Parpola (gen. ed.), The Prosopography of the Neo-Assyrian Empire: I/1–2 (ed. K. Radner), Helsinki 1998–1999, 256–258. **No. 12.** The sender is known as the governor of Arrapha, and eponym for the year 735 BC.[40] Brinkman[41] suggested that Šapiya might be an abbreviated form of the toponym *Ša-pî-Bēl*, the future capital of the province of Gambulu. Note, however, that the derivation from an Akkadian toponym contrasts with the double spelling Sapiya/Šapiya, which rather refers back to the various ways (resp. in Neo-Assyrian and Neo-Babylonian) to transcribe the Aramaic phoneme š. On the other hand, in *ibid.* 764b, the name is (erroneously) read as Nabû-eṭir.

We may now attempt to provide these epistolary sources with a general chronological and/or logical order, in accordance with the information supplied by the official inscriptions of Tiglath-pileser III. **Nos. 1–2** seem to refer to the early stages of the revolt of Mukīn-zēri: the first relates *tout court* of the arrival of cavalry of the Chaldean chieftain near the city of Babylon, blocking the inhabitants within the walls, interrupting the agricultural works. The alarm of the attack extends immediately to nearby Borsippa, from where the horses are evacuated – whether in view of defence, or out of terror for raids, we do not know. In the second letter (no. 2) the picture is more detailed: in fact, we are given to see the rebel with ten of his cavalrymen arriving at the regional capital, with a peace offering for the Babylonians, if they should side with him; a fragmentary allusion to a contingent of 5,000–6,000 horses and mules might refer to forces of Mukīn-zēri awaiting outside the city.

In **no. 3**, the alliance with the rebel on the part of the Babylonians – perhaps willy-nilly – seems to have been already sealed, and thus the mission of two officials of the Assyrian king who offer fiscal privileges to the Babylonian citizens in front of the city gates fails miserably, under the attentive gaze of a man of Mukīn-zēri. The two make attempts – with a setting which in the negative recalls the biblical episode of the *rab šaqē* before Jerusalem[42] – to convince the locals that the arrival of the Assyrian king is impending. However, the Babylonians appear very doubtful on this count, and they definitely refuse the Assyrians entrance

38 Cf. S. Parpola (ed.), The Prosopography of the Neo-Assyrian Empire: I/1–2, Helsinki 1998–1999, 387b, where however, this text is not cited, although it is in Id., The Prosopography of the Neo-Assyrian Empire: II/2, Helsinki 2000, 765a, where it is suggested that the sender wrote from Kutâ/Cutha.

39 Cf. S. Parpola (ed.), The Prosopography of the Neo-Assyrian Empire: II/2, Helsinki 2000, 855a.

40 Cf. F. M. Fales, Cento lettere neo-assire, Venice 1983, 34–37 and 66–67; S. Parpola (ed.), The Prosopography of the Neo-Assyrian Empire: I/1–2, Helsinki 1998–1999, 217b.

41 J. A. Brinkman, Merodach-baladan II, in: Studies presented to A. Leo Oppenheim (Chicago 1964), 11[28].

42 *2 Kings* 18:17–37; cf. G. Frame, Babylonia 689–627 B.C. *cit.*, 140.

within the walls; thus the officials are forced to return to the encampment at Kar-Nergal (near Kutâ), while threatening the Babylonians with heavy punishments in the future. In any case, it may be noted that there is no hint in this letter (or elsewhere in the epistolary material under examination) concerning the fact that Mukīn-zēri had seized the Babylonian crown for himself, as is instead indicated in the Babylonian chronicle.

The scenario drawn by this letter is, however, very rich regarding other developments, both military and diplomatic: Mukīn-zēri is said to be moving towards the nearby city of Dilbat, which was probably pro-Assyrian[43], whereas the Aramean tribe of Li'tamu, already subjugated by Tiglath-pileser III (as we know from official texts) has sent a message of loyalty to the Assyrians. Also the brief message **no. 4** seems to refer to this phase, with the sender communicating to the king that he cannot record – at least for the moment – any pro-Assyrian defection in the Babylonian ranks.

From letter **no. 5**, which gathers together reports from different areas, we gain a picture of the troops of Mukīn-zēri in action over the entire northern sector of lower Mesopotamia, from the flood basin of the Euphrates to that of the Tigris. In a first report, we find the Assyrian and Chaldean troops encamped opposite each other in an unspecified location, while the respective representatives attempt to obtain an alliance with an Aramean tribe (the name is lost); the latter is holed up in a settlement, instead of moving off towards Marad, as had been requested by the Assyrians. The second report must contain a flash-back, since it recalls a plundering action on the part of Mukīn-zēri, after leaving his stronghold of Šapiya, against the eastern city of Larak (of which the Chaldean chieftain Nadinu, an ally of the Assyrians according to the official texts, was in charge). Subsequently – so the report goes – the inhabitants of Larak took revenge by raiding 20,000 sheep of Mukīn-zēri at pasture. Finally, the sender's focus shifts back to the west, to Babylon, with the news that only some Babylonians had followed Mukīn-zēri in his foray to pillage the palm trees of Dilbat, whereas most of them were reluctant – which may be taken as good news for the Assyrian king.

In text **no. 6** Merodach-baladan enters the scene, albeit indirectly. Mukīn-zēri had made a concrete offer of alliance to the chief of Bit-Yakin, blandishing him as one of the great leaders of the Chaldeans, and at the same time accusing Balassu of Bit-Dakkuri of being a traitor, who brought the Chaldeans to ruin (the two had in fact been military allies in the past, as recalled above). Mukīn-zēri's message, having been intercepted and confiscated by the Assyrians, is read to Balassu, who shows some anxiety about continuing his policy of loyalty towards Tiglath-pileser III, due to his family ties with Mukīn-zēri (who is his nephew) and for fear of concrete reprisals on the part of the rebel. The Assyrians cannot but blandish him in their turn, urging him to proceed to Larak – as already agreed – thus providing some relief to that city. And in point of fact, a letter of group B, ND 2365, confirms the presence of Balassu at the side of the Assyrians in military operations.

43 J. A. Brinkman, A Political History of Post-Kassite Babylonia, *cit.*, 237.

Further letters of group B, i.e.: ND 2389 (by Merodach-baladan, who declares his loyalty to Tiglath-pileser III), ND 2663 (movement of Assyrian troops in connection with the road to Larak) and ND 2779 (mentioning Merodach-baladan together with the Aramean tribes of Li'tamu and Hagaranu) should refer to this same phase – in which the Assyrians kept fast or even strengthened their alliance with Bit-Dakkuri, Larak, and Bit-Yakin.

Subsequently, we might range the very fragmentary letter **no. 7**: here Balassu and Nadinu are mentioned together, in connection with war activities, and there is a hint to an Assyrian presence in Dur-ša-Balihayyu, the capital of Bit-Ša'alli. This mention might imply that – at the time of the rebellion of Mukīn-zēri – the war operations against this other centre of anti-Assyrian sedition had been finished, in some agreement with the order of events to be made out from the royal inscriptions of Tiglath-pileser III. Another fragmentary letter (ND 2444), by an unknown author, mentions one Nabû-ušebši and people passing from his jurisdiction to the Assyrian one: according to Saggs[44] this might be a graphical variant of the name of Nabû-ušabši, the Chaldean chief of Bit-Šilani.[45]

The subsequent development of the war, both on the field, and in its political-diplomatic aspects, is relatively obscure. Two letters (**nos. 10, 11**) show the capture of messengers sent to Mukīn-zēri (or to his "sons") by the Assyrians: the fact that one of these messengers was an Aramean of Hindana/u – an area on the middle-lower Euphrates known for its long-standing cultural and political connections with Babylonia – allows us to surmise that Mukīn-zēri had attempted to extend his alliances to territories subjected to Assyrian political dominion. From text **no. 8** it is clear that Mukīn-zēri's activity in battle entailed the capture of centres hostile to him; but on the other hand, he might have progressively lost the support of his own people (letter **no. 9** seems to hint to desertions in Bit-Amukkani).

Some further letters of group B might also refer to the conclusive part of the hostility against Bit-Amukkani. Thus, of the three letters signed by Šamaš-bunaya, a high-ranking official (cf. below), ND 2663 and ND 2628 relate movements of Assyrian auxiliary troops and of civilian refugees in the central-eastern zone of the alluvium between Nippur and Larak, and are perhaps to be placed in the intermediate phases of the hostility; while ND 2403, very fragmentary, might refer to the political reconciliation of Tiglath-pileser III with the city of Babylon, after the hostilities against Mukīn-zēri were concluded.

We arrive thus to letter **no. 12**, which is the most significant of this group of epistolary documents from Kalhu. This text bears a first-hand account of the capture of Šapiya, the starting place of the hostilities, where Mukīn-zēri seems to have been forced to return due to the progressive narrowing of the circle of alliances which initially supported him. The report on the battle sent to the king is authored by the governor of Arrapha, Aššur-šallimanni,

44 H. W. F. Saggs, The Nimrud Letters, 1952, London 2001, 40.

45 Cf. S. Parpola (ed.), The Prosopography of the Neo-Assyrian Empire: II/2, Helsinki 2000, 901a.

a eunuch, who is mentioned in the official texts of Tiglath-pileser III in connection with another victory in southern Mesopotamia:

> "That Puqudu (tribe) and the city Lahiru of Idibirina, and the cities of Hilimmu (and) Pillutu, which are on the border of Elam, I annexed to Assyria, and I placed them under the control of my eunuch, the governor of Arrapha."[46]

The content of letter no. 12 lends itself to some detailed observations. First of all, it may be noted that the exact place of the battle is not specified, since it was obviously well-known to both the sender and the addressee, and is described merely as "the fortress" (*bi*[*r-ta*], l. 4): this led Saggs to believe that the location of the battle was Babylon.[47] However, as observed by Brinkman[48], there is no proof of Chaldean resistance in that city, and – as said above – the letters hitherto examined rather leave the impression of an alliance with the rebel *nolens volens* on the part of the Babylonians. Thus, the most likely location for the battle remains Šapiya; especially because of the references to "the city gates" (*abullāte*) and the consequent "defeat" (*duāku/dâku*) on ll. 7–9 of the obverse,

> *ina libbi abullāte / niqṭirib diktu / niddu'ak*
> "we have reached within the city gates, and we have inflicted a defeat"

which recalls of course, in the same terms and in the same order, the description of the battle of Šapiya as provided by the official inscriptions:

> *ina Sapie āl šarrūtišu ērišu diktašu ma'attu ina pān abullātešu adūk*
> "in Sapiya, his royal city, I enclosed him; I inflicted a heavy defeat upon him before his city gates".[49]

On the other hand, a basic inconsistency between the two sources should be underscored here: the official text speaks of a pitched battle *before* the gates, while the epistolary text indicates a hand-to-hand combat taking place *inside* the enemy city itself. But this is not all: the author of the letter goes on to describe how he and the other Assyrian commander, the *turtānu* – who might have been the Šamaš-bunaya mentioned above – were forced to advance inch by inch against a stiff resistance by the Chaldeans:

> *issurri šarru bēlī / iqabbi mā ina libbi abullāte / idukū ilāni ša šarri bēlīya / šumma 1 ammatu 1 rūtu ina qabli āli / la erabūni / šarru bēlī liš'al*
> "Perhaps the king, my lord, will say: '(already) within the city gates they defeated them' – (but) I swear by the gods of the king, my lord, that they (=the Assyrian soldiers) penetrated cubit after cubit, span after span, within the city! May the king, my lord, request information (on this count)!"

46 H. Tadmor, The Inscriptions of Tiglath-pileser III, *cit.*, 160–161, Summary Inscription 7, ll. 13–14.
47 H. W. F. Saggs, The Nimrud Letters, 1952, London 2001, 11.
48 J. A. Brinkman, A Political History of Post-Kassite Babylonia, *cit.*, 238[1529].
49 H. Tadmor, The Inscriptions of Tiglath-pileser III, *cit*, 160–161, Summary Inscription 7:23.

And finally: what happened to Mukīn-zēri? Here we have to deal with a further point of inconsistency between the sources. Against the total silence of the royal inscriptions on the personal fate of the defeated Mukīn-zēri, we may note the precise description supplied by the governor of Arrapha in his letter (ll. 9–14):

> *Mukīn-zēri / dēki Šum-ukīn mārušu / dēki ālu kašdu šû / šarru bēli lū hadi / šarru bēlī / pān mār šiprini liddin*
> "Mukīn-zēri has been defeated/killed; his son Šum-ukīn has been defeated/killed; the whole city has been taken; may the king, my lord, rejoice. May the king, my lord, give audience to our messenger."

This passage has given rise to a small *crux interpretum*, due to the fact that the verb *duāku/dâku* means "to defeat" (especially with the internal object *diktu*, as in the passages quoted above), but also "to kill," with a certain indifference of use, both in Neo-Assyrian and in Standard Babylonian. Thus, Brinkman[50] opted for the hypothesis of the defeat, indicating that "Mukīn-zēri remained at large", while Saggs espoused since 1955 the hypothesis of a killing, on the basis of the repetition of the verbal form applied to both father and son, which would have been superfluous in the case of a mere defeat. To be sure, partially in favour of Saggs, it may be observed that if the two had survived the clash – as prisoners or as fugitives – the question would certainly have deserved some extra words by Aššur-šallimanni. On the other hand, a Babylonian deed[51] is dated to the fourth year of Mukīn-zēri's reign: and, barring the case of an error or mere misinformation on the part of the scribe[52], it might somehow support the notion that Mukīn-zēri was still alive and active in at least part of the territory of southern Mesopotamia in 728 BC. Thus, at the end of the day, the question of Mukīn-zēri's exact fate remains open.

Summing up, the letter of Aššur-šallimanni contains the account of a fierce battle which took place at Šapiya/Sapiya, and which ended with the Assyrians entering the stronghold through the gates, conquering the city through hand-to-hand combat, and defeating – or perhaps even killing – the two Chaldean chiefs at the head the defence, Mukīn-zēri and his son Šum-ukīn. This letter thus provides us with a rare account of a "classic" Assyrian victory, presumably based on a high number of attacking troops (since two army corps might have been joined for the purpose), on the availability of a superior technology for all forms of armed encounter, and – no doubt – on the unconditional devotion to the imperial cause shown by the Assyrian commanders. In short, this letter describes one of those battles of siege and subsequent conquest of enemy cities that the Assyrian bas-reliefs illustrate

50 J. A. Brinkman, A Political History of Post-Kassite Babylonia, *cit.*, 238[1529].
51 A. T. Clay, Babylonian Business Transactions of the First Millennium B.C. (= Babylonian Records in the Library of J. Pierpont Morgan, I), New York 1912, note 22; cf. J. A. Brinkman, A Political History of Post-Kassite Babylonia, *cit.*, 239, 284.
52 As, e.g., suggested by J.-J. Glassner, Chroniques mésopotamiennes, Paris 1993, 262.

in great detail, against the largely stereotyped accounts of the contemporaneous royal inscriptions.

Further historiographic results are the following: (A) the siege of Šapiya certainly did not take three years; while it was probably undertaken in 731 BC (= 15th *palû*) by Tiglath-pileser III himself, it must have been lifted subsequently for unknown reasons. The Nimrud letters speak of a siege/attack on Šapiya only in letter no. 12, presumably to be dated to 729 BC (= the third year of Mukīn-zēri's revolt), with the commanders writing to Tiglath-pileser who was residing in Assyria (as he had already done during the previous year, his 16th *palû*). (B) The conquest of Šapiya must have taken place at a later date of 729 BC than the editing of Summary Inscription no. 7, the purpose of which was to celebrate the restoration of the palace at Kalhu. This would explain the basic differences between the official inscription, which has the Assyrians engaging the enemy *before* the city gates, and Aššur-šallimanni's letter, which shows the troops operating fully *inside* the same gates. The obvious conclusion is, therefore, that the official texts of Tiglath-pileser were not subsequently "updated" with more recent information regarding the final outcome of the war against the Chaldean rebel – or, alternatively, that such an updated version has not come down to us.

But let us return once more to the main theme of this contribution, which is that of a view of non-indigenous groups "moving around Babylon", with alternatively peaceful and hostile stances vis-à-vis Assyrian interference. The Mukīn-zēri epistolary dossier shows once more the full cast of characters which was introduced in section § 1, above, albeit now endowed with movement and with complex reciprocal relations.

In the first place, we are given to observe the inhabitants of the ancient cities of the southern alluvium – and especially the people of Babylon itself – as floundering in the political and military net that has been cast around them: deeply distrustful of the Assyrians, but no less terrified of the Chaldeans, they practice forms of passive resistance which seem to lead nowhere, and often become the helpless prey of raids and other forms of violence. The second group which stands out is that of the Aramean tribesmen: they seem to resent the military struggle taking place around them as much as the city-dwellers, but possibly only insofar as it impinges upon their freedom of movement and their socio-economic autarchy. Accordingly, they choose to side with one or the other opponent, but do not seem particularly bent on participating to the action in person – at least, not for the time being. And finally, we have the Chaldeans, whom the revolt of Mukīn-zēri catches in the process of (re)defining their mutual political relations: in other words, this first major interference of the Assyrians in Babylonian affairs seems to represent a unique opportunity to measure the respective strength of the individual confederations and of their smaller camp-followers – but not without some personal and political qualms reflecting the tight gentilic structure among the groups, as in the case of Balassu of Bit-Dakkuri, who was the uncle of Mukīn-zēri of Bit-Amukkani.

As is well known, all these actions and interrelations of the main protagonists of the southern Mesopotamian scenario were destined to become even more complex quite soon thereafter, with Merodach-baladan's rise to power. It may thus be useful to recall that Aššur-šallimanni's letter bears a further passage, relevant to Tiglath-pileser's order to send grain rations to the chief of Bit-Yakin, obviously as reward for his allegiance to the Assyrians (reverse, ll. 7–10):

> "Regarding the barley rations which are destined to Merodach-baladan, of which the king, my lord, had spoken – after I saw the defeat, I wrote that they should transport (them)"

No doubt, Merodach-baladan's commitment to Tiglath-pileser's cause should be dated as early as 731 BC – in agreement with Tadmor's reconstruction – when the chief of Bit-Yakin brought rich tribute to Šapiya, subjecting himself to the Assyrian ruler together with Bit-Dakkuri and Larak. Subsequently, as we have seen, he was courted by Mukīn-zēri, but – with a certain foresight – chose to decline the offer by Bit-Amukkani and to declare anew his fealty to the Assyrian king, perhaps even participating to some extent to military operations. Thus, as soon as Mukīn-zēri's revolt had been definitely quenched, he was awarded by the Assyrians the benefit of a large consignment of barley – which probably came out from the grain stores of conquered Šapiya.

However, Tiglath-pileser III was destined to die a mere two years after taking over the Babylonian throne; and – after the brief reign of Shalmaneser V – it was his son Sargon II[53] who would be called to deal, for a very long time and often on the very same theatre of operations, with the chief of Bit-Yakin. Merodach-baladan had by that time prepared the political and military scenario so as to be able to seize anew the throne of Babylon, and to rule on much more efficacious terms than his predecessor Mukīn-zēri. So adroit were his moves, in fact, that he would survive his new and powerful adversary even through defeat and exile, and yield in the end only to the overwhelming might of a third Assyrian king. Thus, when his final demise against Sennacherib occurred, no less than thirty years had gone by after his first appearance on the scene at S/Šapiya – when he was forced to bind himself to the cause of Tiglath-pileser, but was surely already "moving around Babylon" in his plans.

53 F. Thomas, Sargon II., der Sohn Tiglath-pilesers III, in: Festschrift für Kurt Bergerhof, Neukirchen-Vluyn 1993, 465–470.

*Francis Joannès*

## L'écriture publique du pouvoir à Babylone sous Nabuchodonosor II

Au Ier millénaire av. J.-C., le statut de la ville de Babylone comme capitale religieuse de la basse Mésopotamie est bien connu, de même que sa transformation en «centre du monde» par les lettrés liés au culte de Marduk: c'est à cet endroit que le roi des dieux, chef du panthéon, avait installé sa résidence, c'est là que l'assemblée des dieux prenait les décisions affectant l'ordre même de l'univers. À ce titre, Babylone était la première des villes-saintes de Babylonie et elle jouissait d'un statut tout à fait particulier.

Parallèlement, tout au long du Ier millénaire, Babylone s'affirme comme la capitale politique du pays, principale résidence royale, même si quelques attestations sont connues pour d'autres lieux de séjour du souverain[1].

La coexistence en un même lieu de la sphère d'autorité du pouvoir central et de la sphère d'autorité du pouvoir divin, avec les problèmes que peut générer ce recouvrement, est un phénomène qui a été étudié pour l'Assyrie et qui explique en partie la construction de certains ensembles palatiaux (Kār–Tukultī-Ninurta, Kalhu, Ninive, Dūr–Šarrukīn). Il paraît logique de supposer que la même situation génératrice de conflits de compétence était également présente à Babylone. Le passage des lignes 20 et suivantes de la IVème tablette de l'Épopée d'Erra distingue d'ailleurs nettement, en les opposants, les gens de l'intérieur de la ville de ceux qui habitent le palais du roi[2].

L'un des moments les plus intéressants pour étudier ce phénomène est celui où la ville s'est trouvée capitale de l'empire néo-babylonien et où elle a bénéficié d'un programme de grands travaux qui ont concerné les murailles de la cité, de nombreux temples – dont, au premier chef, celui de Marduk – et le palais du roi, donc la période qui va de 626 à 539 av. J.-C. L'un des points de vue que l'on peut adopter pour évaluer ce rapport roi/dieu, est celui que nous procurent certaines inscriptions royales rédigées pendant cette période, qui situent effectivement le dieu et le roi l'un par rapport à l'autre. C'est ce à quoi se réfère le titre même de cette communication: «l'écriture publique du pouvoir», ce qui ne signifie pas forcément qu'elle soit visible par tous, mais que la rédaction de cette écriture du pouvoir, que

1 Cf. M. Jursa, «Palast. A. IVb. Neubabylonisch», in RlA 10, 3/4, 208–212.

2 Cf. J. Bottéro, Mythes et rites de Babylone, Paris 1985, 243: «*Car, délaissant la ville, et sorti-hors, tu as revêtu l'apparence d'un lion et tu es entré au Palais: dès qu'elles t'ont aperçu, les troupes ont pris les armes, et le cœur du Gouverneur, revanchard contre Babylone, s'est déchaîné; il a expédié ses soldats, comme pour spolier un ennemi, poussant au pire le capitaine de l'armée et lui disant: "Cette ville à laquelle je t'envoie, homme, n'y respecte nul dieu, n'y redoute personne: mets-y à mort petits comme grands, sans épargner un seul bébé, encore à la mamelle! Après quoi, pille tous les trésors accumulés de Babylone!" L'armée du roi, rassemblée, s'est donc introduite en la ville, l'arc enflammé, le glaive dégaîné.*»

ce soit pouvoir du roi ou pouvoir de la divinité tutélaire, le situe dans son rapport politique et religieux à la population de Babylone.

Par ailleurs, quand on parle d'"écriture", ce terme est à prendre au sens large: il désigne des textes écrits, (principalement en cunéiforme), mais aussi une certaine iconographie, associée au texte, fonctionnant même parfois comme un texte.

On cherchera donc ici, à partir des inscriptions laissées principalement par Nabuchodonosor II et à partir de certains de ses programmes architecturaux, à évaluer le rapport entre la Babylone de Marduk et la Babylone du roi.

L'une des manifestations de cette écriture du pouvoir est la présence de briques inscrites ou estampillées, portant la titulature royale. C'est un usage particulièrement ancien en Mésopotamie, mais il a connu sous le règne de Nabuchodonosor II une mise en œuvre particulièrement abondante, si l'on en juge par le nombre d'exemplaires qui en ont été retrouvés.

Quand on parle des briques de la ville de Babylone, en dehors des tablettes, prismes et cylindres de fondation, on doit d'ailleurs faire la distinction entre des briques portant des inscriptions en rapport avec un lieu précis (le rempart, le Palais Sud, la voie processionnelle) et celles qui comportent une simple titulature, que l'on a appelée «inscription standard» de Nabuchodonosor II, qui a été estampillée sur des briques retrouvées aussi en d'autres lieux que Babylone (Kiš, Borsippa, Habl aṣ-Ṣahr[3]).

S'agit-il seulement d'une manifestation du prestige royal? En fait, le souverain y est assez clairement présenté dans un rôle d'auxiliaire des dieux et sa titulature y est réduite à l'affirmation de son domaine territorial de compétence (*šar Bābili*), de ses obligations religieuses comme pourvoyeur des temples (*zānin Esagil u Ezida*) et de sa légitimité familiale (*māru/aplu ašarēdu ša Nabû-apla-uṣur*). On sait par ailleurs que ces inscriptions ne sont normalement pas visibles, puisque très souvent placées sur la partie inférieure des briques. Même s'il s'agit d'une écriture «publique», ce n'est donc pas une écriture lisible ni visible. De plus, les inscriptions estampillées sont souvent composées en caractères archaïques, identifiables seulement par les lettrés pourvus d'une véritable «culture cunéiforme». On peut donc y voir non seulement une manifestation de piété (comme pour les inscriptions de fondation), mais aussi une indication pour les générations futures du nom des intervenants royaux successifs[4], voire même, de manière très pragmatique, un procédé administratif de gestion au moment de la fabrication des briques. Des briques portant la même inscription standard se trouvent par exemple sur le site de Habl aṣ-Ṣahr, une partie du mur de

3 H. Gasche *et alii*, Ḥabl aṣ-Ṣahr 1983–1985: Nebuchadnezzar II's Cross-Country Wall North of Sippar (= Northern Akkad Project Reports 1), Ghent 1987, 3–46; H. Gasche *et alii*, Ḥabl aṣ-Ṣahr 1986, nouvelles fouilles. L'ouvrage défensif de Nabuchodonosor II au nord de Sippar (= Northern Akkad Project Reports 2), Ghent 1989, 23–70.

4 La réutilisation, à la Porte d'Ištar, de briques estampillées au nom de Sennachérib n'est pas seulement l'indice d'une mobilisation massive de tous les matériaux disponibles, y compris ce qui aurait pu être mis au rebut. C'est aussi une manière de rappeler le passé de la ville et de l'insérer dans une oeuvre postérieure.

Médie fouillée dans les années 80. Dans son étude des très nombreuses briques inscrites retrouvées aussi sur le site de Habl aṣ-Ṣahr, J. Black[5] avait procédé à un classement typologique des briques ayant reçu une inscription estampillée et il avait émis l'hypothèse que la différenciation en inscriptions de 3 à 7 lignes, de même que les variantes d'écriture dans le nom du roi, dans celui de la ville, les alternances entre idéogrammes et écriture phonétique, renvoyaient à des estampilles dépendant chacune d'un atelier différent de fabrication des briques, et que ce pouvait être aussi un moyen de différenciation, et donc de comptabilisation de la production de chacun de ces ateliers. Il ne retenait donc pas le critère élaboré par P. R. Berger[6] de répartition chronologique des inscriptions sur briques d'après le nombre de lignes.

Que conclure de la présence en si grand nombre de ces briques inscrites à l'intérieur de Babylone? C'est une forme indirecte d'écriture du pouvoir et de manifestation de l'action royale: la plupart du temps non visible, destinée comme les documents de fondation à une lecture divine symbolique, elle ne fournit que peu de renseignements sur les actions du souverain. Nabuchodonosor II a donc compensé par le nombre ce qui ne pouvait se dire par les mots, mais il y a peut-être aussi dans ce système d'estampillage un pur et simple procédé de comptabilité et de réservation des matériaux rassemblés en grande quantité pour les constructions officielles.

Ainsi, Nabuchodonosor II marque par ce procédé hérité de ses devanciers, sa place d'initiateur des grands travaux dans Babylone, mais ne s'en réserve pas réellement la gloire. Ce n'est pas là qu'il faut chercher les monuments qui exaltent directement son prestige royal. Et l'on peut dire que s'il en fut le constructeur, il n'est pas l'occupant principal des lieux. C'est donc au dieu Marduk, et, vu les conceptions théologiques de l'époque, à son fils Nabû, qu'appartiennent en propre ces édifices avec le territoire sur lequel ils se dressent. Cela conduit à examiner un second aspect, qui est de déterminer où se fait la séparation entre le domaine de Marduk et celui du roi de Babylone. On peut l'étudier au travers du programme iconographique de la Porte d'Ištar, qui renvoie principalement à un prototype nippuréen.

On a dit, à plusieurs reprises, que Babylone était une ville-sainte, réplique de l'Apsû, mais en tant que résidence de Marduk et, surtout, une réplique de Nippur, prise en tant que capitale religieuse de la basse Mésopotamie dont Babylone avait repris plusieurs particularités.

5 Cf. note 3. J. Black cite Babylone, Borsippa, Kiš et Sippar comme lieu où l'on trouve ces briques inscrites avec la *Standard-Inscription* de Nabuchodonosor II. Certaines ont été réutilisées à Séleucie. Mais à Habl aṣ-Ṣahr, c'est toute la structure méridionale du mur qui a été construite en briques cuites estampillées. On a au total dix variations possibles de la Standard-Inscription, avec une répartition d'estampillages entre 3 et 7 lignes.

6 P. R. Berger, Die neubabylonischen Königsinschriften (= Alter Orient und Altes Testament 4/1), Neukirchen-Vluyn 1973.

Il suffira de citer ici les noms de la double muraille, Imgur-Enlil et Nemettī-Enlil pour rappeler ce rapport. On rappellera également la proposition d'A. George[7] de faire de ce double ensemble une ceinture-*nebēhu* qui renforçait la séparation entre extérieur et intérieur de Babylone, où l'autorité suprême était exclusivement celle de Marduk, même si le palais royal traditionnel, le *Südburg* selon la terminologie des fouilles allemandes, y était implanté. L'annonce que l'on pénétrait dans le domaine de Marduk, lorsqu'on s'engageait, à l'extérieur de la ville, sur la voie processionnelle qui conduisait vers la Porte d'Ištar, était concrétisée, de manière publique et particulièrement visible, par la présence d'animaux en briques, soit en relief, soit colorés, soit en relief et colorés, selon les étapes de la construction de la porte d'Ištar.

Selon les propres inscriptions de Nabuchodonosor II, les *mušhuššu* dont il a fait une ornementation, se trouvaient en deux endroits principalement à Babylone: sur le bateau cultuel de Marduk, et sur les seuils des différentes portes de Babylone, avec un traitement tout particulier pour la Porte d'Ištar[8]. Parmi ses successeurs, Nergal-šar-uṣur en signale la présence aux portes de l'Esagil, et Nabonide en a installé aux portes de l'Emah. Il s'agit d'animaux prophylactiques, et c'est le même rôle que jouent les taureaux-*rīmu* qui alternent sur la porte d'Ištar avec les dragons de Marduk. R. Koldewey signale par ailleurs la présence, sur l'esplanade située devant la porte d'Ištar, d'au moins une base qui a dû servir à porter une statue[9].

De fait, une autre inscription de Nabuchodonosor II nous apprend qu'il avait fait placer à cet endroit des statues de dragons-*mušhuššu* «féroces» (*šēzuzūtu*) et des taureaux-*rīmu* «furieux» (*ekdūtu*) en bronze, qui formaient comme une haie à ceux qui se préparaient à pénétrer à l'intérieur de la ville[10].

Le couple dragon-*mušhuššu* et taureau-*rīmu*, avec les qualificatifs qui leur sont associés, rappelle la très antique tradition des animaux dont Narām-Sîn d'Akkad avait orné le parvis de l'Ekur à Nippur[11]. L'accès à l'intérieur de Babylone était donc ornementé comme l'était l'accès de l'Ekur dans les temps anciens, ou bien *à la manière de* l'Ekur, si l'on admet qu'il s'agit sans doute de la reprise d'une tradition, plus que d'une copie d'un modèle dis-

7 A. R. George, The city wall of Babylon – a belt-cord (= Nouvelles Assyriologiques Brèves et Utilitaires), Paris/Rouen 1991, 101.

8 Cf. A.L. Oppenheim/E. Reiner et al. (eds.), The Assyrian Dictionary of the University of Chicago, M2, Chicago 1956, 270b.

9 R. Koldewey, Ausgrabungen der Deutschen Orient-Gesellschaft in Babylon, II: Das Ischtar-Tor, Leipzig 1918, 10: «Auf dem Vorplatz, etwa 6 m vor der westlichen Verbindungsmauer, steht ein rundes, säulenähnliches Postament von 1,40 m Durchmesser».

10 Cf. pour Babylone en général, VAB 4, 72 i 19–21: *a-na si-ip-pi* ká-gal-meš-*šú* am-meš urudu *e-ik-du-ú-tì ù* muš-huš *še-zu-zu-ú-tì ú-uš-zi-iz*; *ibid.*, 86 ii 9, 8–10 (// 90 i 45, 4 //106 i 59, 8 // 162B v 11): *ri-i-mu e-ri-i e-ik-du-ú-tì ù* muš-huš *še-zu-zu-ú-tì i-na* ká-gal-meš-*šú ú-uš-zi-iz-ma*; et pour la Porte d'Ištar en particulier: *ibid.*, 132 vi 5(!) et 17: *i-na a-gur-ri* na$_4$-za-gìn kù-*tì šá* am-meš *ù* muš-huš *ba-nu-ú qer-bu-us-sa na-ak-li-iš ú-še-piš*.

11 Cf. A. Westenholz, Old Sumerian and Old Akkadian Texts in Philadelphia Chiefly from Nippur, part 1: Literary and Lexical Texts and the Earliest Administrative Documents from Nippur, Malibu 1975.

paru depuis longtemps[12]. On en retient le principe que l'accès aux portes de la ville de Babylone était disposé comme l'accès à un sanctuaire, et que celui-ci était visiblement celui de Marduk.

On note aussi la valeur symbolique de la représentation du dragon-*mušhuššu*, si on le considère non comme une simple représentation, mais aussi en tant que pictogramme, c'est-à-dire ici un élément iconographique ayant valeur de signe et qui pouvait, comme certains signes de l'écriture cunéiforme, bénéficier d'un double niveau de lecture: d'une part c'était le gardien d'une zone sacrée, au même titre que les *šedu* et les *lamassu* des palais assyriens, mais d'autre part, il intervenait comme animal attribut de Marduk, une sorte de «marque de propriété».

Si toute cette zone intérieure de la ville était réservée au dieu, où le souverain a-t-il alors pu édifier un ensemble sur lequel il pouvait imprimer sa propre marque, et qu'il pouvait doter des signes de sa seule gloire? Il s'agit, très vraisemblablement, de ce que l'on appelle le Palais Nord, ou *Hauptburg*, situé à l'extérieur de la ville, au-delà de la double muraille.

On notera d'abord que Nabuchodonosor II a lui-même indiqué qu'il ne pouvait, à l'intérieur de Babylone, agrandir le Palais Sud au-delà des compléments qu'il avait apportés au palais de Nabopolassar. L'inscription dite «de la East-India Company» oppose le Palais Sud, résidence favorite de la royauté, mais qui fait plaisir aux dieux et qui n'empiète sur aucun espace supplémentaire, qui sont soigneusement énumérés par Nabuchodonosor II, et le Palais Nord, au-delà de la muraille, décrit comme l'endroit où se manifeste le mieux la puissance royale, et où ont été rassemblés les témoignages de la puissance de Nabuchodonosor II:

*Dans Babylone, l'espace de ma résidence ne convenait plus à la qualité de ma royauté; comme mon cœur était plein de respect envers mon Seigneur Marduk, dans Babylone, sa place-forte, pour agrandir ma résidence royale, je n'ai pas modifié la moindre rue, je n'ai pas déplacé le moindre de ses sanctuaires, je n'ai pas bouché le moindre de ses canaux. J'ai cherché tout alentour (un espace disponible pour) ma résidence*[13].

Bien que le Palais Nord n'ait été fouillé que de manière partielle, son allure générale telle qu'elle a pu être reconstituée, était particulièrement majestueuse et imposante, puisqu'il était édifié sur une haute terrasse et qu'un vaste espace vide de construction semble avoir séparé le corps de bâtiment principal de la muraille qui bordait la voie processionnelle. Celui qui accédait à ce palais, venant de l'est ou du nord-est, le voyait donc, de loin, se dresser dans toute sa majesté, réplique, en plus modeste, et à l'extérieur de la ville proprement dite, de l'Etemenanki installé au cœur de Babylone et culminant à plus de 90 m.

12 Cette tradition s'appuie également sur le parallélisme entre les monstres vaincus par Ninurta et par Marduk selon la tradition mythico-épique.

13 *i-na* ká-dingir-ra$^{ki}$ *ku-um-mu mu-ša-bi-ia a-na si-ma-at šar-ru-ti-ia la šu-um-ṣa aš-šum pu-lu-úh-ti* $^{d}$amar-utu en-*ia ba-šu-ù lìb-bu-ú-a i-na* ká-dingir-ra$^{ki}$ uru *ni-ṣi-ir-ti-šu a-na šu-un-du-lam šu-ba-at šar-ru-ti-ia su-uq-šu la e-ni*$_7$ bara$_2$-*šu la ú-ni-iš pa-la-ga-šu la e-es-ke-er ku-um-mu ra-ap-ši-iš aš-te-e'-e-ma.*

Mais le visiteur voyait aussi, en pénétrant dans l'espace palatial, les éléments de «l'écriture publique» du pouvoir royal qui ne se trouvaient pas à l'intérieur de la ville. Un certain nombre d'objets en pierre, certains inscrits, d'autres anépigraphes, ont en effet été retrouvés dans le vaste espace entre la voie processionnelle et la terrasse supportant le palais, ou à proximité immédiate. C'est ce qui, pour l'essentiel, avait été interprété par E. Unger, comme les éléments d'un *Schloßmuseum*[14], et dont on sait maintenant que ce n'était pas le cas[15]. Ces monuments remplissaient une fonction à la fois d'information et de propagande.

On y trouve, entre autres, des objets bien connus, que l'on peut mentionner pour mémoire:

– des trophées, tels l'inscription d'Adad-nīrārī II ou le relief néo-hittite représentant le dieu de l'Orage, ou encore la stèle du gouverneur du Suhū, Šamaš-rēš-uṣur.

– des inscriptions royales d'un type particulier, comme la stèle de Nabonide rapportant son accès au trône et légitimant ce souverain, ou le relief représentant une partie de la scène générale figurée à Behistun, dans la restitution qui en a été récemment proposée[16].

On peut y adjoindre le grand lion de Babylone en basalte, retrouvé dans l'angle nord-est de l'espace vide situé entre le bâtiment principal et la voie processionnelle. Il ne s'agit pas d'un monument inscrit, mais, vu sa localisation, il peut être considéré comme une manifestation concrète du prestige et de la gloire du souverain. Peut-être est-ce à cette décoration constituée de trophées divers que fait allusion l'inscription de Nabuchodonosor II qui indique à propos de ce Palais Nord:

*bal-ti uz-zu pu-luh-ti me-lam-me šar-ru-ti i-ta-ti-šu sa-ah-ra-a-ma*[17],

«Son extérieur était enveloppé de fierté, de fureur, de la terreur causée par l'éclat de la royauté ...»

Il reste certes impossible de déterminer combien de monuments inscrits étaient disposés dans cet espace conduisant de la voie processionnelle à la terrasse du Palais Nord, mais il est évident que c'est là qu'était célébrée la puissance royale, depuis l'époque des campagnes de Nabopolassar dans le pays de Suhū, jusqu'à la période troublée au cours de la quelle Darius Ier monta sur le trône royal perse et triompha de ses rivaux, dont Bardiya et les deux Nabuchodonosor (III et IV) en Babylonie.

Trois points majeurs se dégagent de cette présentation de l'écriture publique du pouvoir à Babylone, sous Nabuchodonosor II principalement:

1. Le roi a fait disposer en très grandes quantités des briques estampées ou incisées à son nom, mais d'une manière assez neutre pour ne pas concurrencer l'autorité souveraine

14 E. Unger, Babylon. Die heilige Stadt nach der Beschreibung der Babylonier (rééd.), Berlin 1970.

15 E. Klengel-Brandt, Gab es ein Museum in der Hauptburg Nebukadnezars II. in Babylon? (= Forschungen und Berichte, Staatliche Museen zu Berlin 28), Berlin 1991, 41–46.

16 U. Seidl, Ein Monument Darius' I. aus Babylon, Zeitschrift für Assyriologie und Vorderasiatische Archäologie 89 (1999), 101–114.

17 Ibid., *VAB* 4 138 ix 33–35.

de Marduk sur le territoire de la ville de Babylone. Il s'agit ici d'une écriture du pouvoir présente sous une forme détournée, et à usage non seulement de propagande, mais peut-être aussi simplement de comptabilité administrative.

2. L'autorité souveraine de Marduk sur l'intérieur de la ville de Babylone s'étendait jusqu'aux murailles. Les lieux névralgiques de communication avec l'extérieur, là où sont situées les grand' portes de la ville, sont ornés et protégés par des animaux prophylactiques, à l'image des grands sanctuaires dont l'Ekur de Nippur fut le prototype[18].

3. Le prestige royal ne pouvait donc se manifester que dans la zone palatiale édifiée à l'extérieur des murailles. Il y a donc dans la construction du Palais Nord à la fois des raisons sans doute purement matérielles[19], mais aussi idéologiques: il s'agissait de trouver un endroit où le roi fût hors les murs, donc hors de la sphère d'autorité directe de Marduk. La situation générale de l'ensemble palatial à cheval sur la muraille renvoie d'ailleurs aussi à la disposition du palais de Dur-Šarrukīn et le parallèle a pu être volontaire. Mais le Palais Nord était aussi l'endroit où le pouvoir royal n'était plus sous l'emprise directe de l'influence de Marduk et où le souverain pouvait donner libre cours aux manifestations publiques de sa puissance; par le biais de monuments inscrits ou non et constituant son «écriture publique».

On peut ajouter, pour finir, deux remarques: d'abord, si la répartition chronologique des groupes de tablettes retrouvées sur l'ensemble Palais Sud/Palais Nord a un sens, le précieux inventaire dressé par O. Pedersen[20] montre que la majorité des attestations postérieurs à la période néo-babylonienne *stricto sensu* proviennent du Palais Nord et que c'est là probablement que se trouvaient le structures administratives génératrices d'écrit. En d'autres termes, la fonction officielle et politique du palais était plus présente dans le Palais Nord, et c'est un élément qu'il convient de prendre en compte quand on parle du «palais royal» de Babylone. Le siège du pouvoir était donc plutôt à cet endroit que dans la partie méridionale (Palais Sud). Celui-ci comportait pourtant une structure monumentale, que l'on considère comme la salle du trône. Il reste pour le moment impossible à déterminer si elle a constamment joué ce rôle.

C'est donc dans le Palais Nord, au travers d'une certaine forme d'écriture publique du pouvoir royal, qu'était le plus visiblement concentrée l'essence du pouvoir royal, et cela veut dire qu'il faut faire de cette partie de l'ensemble palatial la plus importante en termes de siège du pouvoir. Le Palais Sud – malgré la présence de la salle du Trône – aurait alors eu une fonction plus résidentielle ou économique. Ceci n'est pour l'heure qu'une hypothèse de travail.

18 Il ne s'agit d'ailleurs pas d'un cas isolé, comme semble en témoigner la décoration des portes de l'Eninnu de Girsu restauré par Gudea, telle que la reconstitue W. Heimpel, The gates of the Eninnu, Journal of Cuneiform Studies 48 (1996), 17–29.

19 On pense à des problèmes d'infiltration d'eau dans le Palais Sud signalés par Nabuchodonosor II lui-même, ainsi qu'à problèmes de place pour la nombreuse population du palais, y compris les familles des rois vaincus.

20 O. Pedersén, Archive and Bibliotheken in Babylon: Die Tontafeln der Grabung Robert Koldeweys 1899–1917. Abhandlungen der Deutschen Orient-Gesellschaft 25, Berlin 2005.

Une seconde remarque porte sur la durée pendant laquelle l'exaltation des actions royales a été mise en évidence à Babylone: la présentation des monuments des vaincus ne date sans doute pas de la dynastie néo-babylonienne: le groupe de statues des Šakkannakku de Mari qui a été retrouvé dans un emplacement particulier de la voie processionnelle avant l'entrée de la Porte d'Ištar a pu avoir le même statut. La stratigraphie empêche de situer la présentation de cet ensemble à l'époque néo-babylonienne, mais il n'est pas exclu qu'il soit l'héritage d'une tradition faisant précéder l'entrée dans la ville par un passage entre des trophées. Savoir jusqu'à quand il faut faire remonter dans le temps la présentation de cet ensemble éloignerait du sujet présent, mais on retiendra qu'avant d'être réinterprété comme un ensemble cultuel, ce groupe monumental appartenait lui aussi à cette célébration symbolique du prestige royal dont on vient de voir de voir des illustrations au I$^{er}$ millénaire.

# Babylon – Wissensort

*Dina Katz*

## Reconstructing Babylon: Recycling Mythological Traditions Toward a New Theology

Many scholars agree that Enūma eliš, "When above," was formulated during the reign of Nebuchadnezzar I.,[1] to celebrate the renewal of Marduk's cult in Babylon after his statue was brought back from captivity.[2] The statue is the god himself, and a god in exile leaves his community divested of divine grace and humiliated, a traumatic religious and political event. The return of the statue signifies a return of divine grace, which in the life of a nation can be, or can be made to be, a constituting event. The return of Marduk to Babylon could have been such a constituting event, and Enūma eliš may have been an ideological testimony of it.[3] Enūma eliš is not merely a religious principle; it is also a political statement. Because Marduk and Babylon are inseparable the text institutes Babylon as the religious center of Babylonia instead of ancient Nippur.[4]

The historical background of the text, however, is not quite decided, and associating it with the return of Marduk from captivity, although likely, is uncertain.[5] For example,

1 The most comprehensive support for this date is advanced by W.G. Lambert, The Reign of Nebuchadnezzar I: A Turning point in the History of Ancient Mesopotamian Religion, in: W.S. McCullough (ed.), The Seed of Wisdom, Essays in Honour of T.J. Meek, Toronto 1964, 3–13. (Henceforth, Fs. Meek), Idem, "Studies in Marduk", Bulletin of the School of Oriental and African Studies 47 (1984), 1–9.

2 There were several statues of Marduk in Babylon, but that does not immediately mean that if one was captured another could simply replace it and that the cult could continue without interruption. If it were just a simple question of quantity and statues could conveniently replace one another then a text like "The Prophecy of Marduk" would be meaningless in every respect and would probably have never been composed. Texts referring to gods in exile, kings boasting about looting or retrieving divine statues and a city lamenting a god who has left his city, indicate that even when there are several statues of a god in the city each one had a life and significance of its own. It is not impossible that in cultic functions there was a hierarchy of statues which may have influenced their location. On the statues of Marduk in Babylon see now A.R. George, Marduk and the Cult of the Gods of Nippur at Babylon, OrNS 66 (1997), 65–70, with previous bibliography.

3 Tablet VII:145–end, but particularly ll. 145–148 suggest that the idea of a constituting event was not alien to the author and may have been in his purpose. Learning the events repeatedly and passing them on from generation to generation is a means to create a framework for a national identity. Compare: "And thou shalt relate to thy son on that day …," in the "Hagada of Passover", the celebration of Exodus in the Jewish tradition, performed literally every year for generations.

4 Enūma eliš is very clear about it. Unlike the Old Babylonian sources that indeed proclaim the superiority of Marduk over the Igigi and Anunaki (who here made Marduk their king) in Enūma eliš Enlil gave up his own name "Lord of the Lands" and thereby his position, and bestowed it on Marduk, because Marduk is the creator of the universe. He is not equal to Enlil but his successor and substitute; Enlil became Marduk. As for Nippur, it is not even mentioned (see below).

5 S. Dalley, Statues of Marduk and the Date of Enūma Eliš, Altorientalische Forschungen 24 (1997), 163–171, especially pp. 165–166.

Anšar's announcement that Marduk's name is Asalluhi[6] would fit best the Late Old Babylonian period. An early date associates the text with the first exaltation of Marduk by Hammurabi rather than with his return from captivity.[7] S. Dalley advocates the possibility that Enūma eliš originated in the Old Babylonian period, towards the end of Hammurabi's reign. She proposes that it underwent redactions and modifications until the Late Babylonian period.[8] The earliest known hymns praising gods go back to the third millennium, and one cannot deny the possibility that a hymn praising the glory of Marduk was composed in the Old Babylonian period. Also one cannot deny that scribal tradition was conserved from generation to generation. Following this line of reasoning one may conclude that either no original hymn was composed since the beginning of the second millennium, or that the early traditions were used with great skill so Enūma eliš is an original work of literature.[9] The end of the text strongly suggests that the latter is the case. The couplet VII:157–8 signifies that the author wrote down the text as it was disclosed to him personally, which submits an innovation and also sanctifies it.

The sack of Babylon by Muršili I. and the supposed return of Marduk (from Terqa) by Agum II. suggest a date in the Kassite period.[10] Considering literary properties, Oppen-

6 Tablet VI:101.

7 The greatness that was Babylon at the time of Hammurabi made the city the political power centre of Southern Mesopotamia and gave its chief god Marduk the same prominence, but not yet to the extent that he would replace Enlil. In that case, the political and religious horizons of Enūma eliš may have been merely illusory propaganda for internal purposes. More suitable to that period is the so called "Weidner Chronicle" (Grayson's chronicle 19) which focuses on Marduk and his cult in Esagil, but still leaves undiminished the status of Enlil, Anu and Ea. The date of the text, however, is unknown. The contents suggest the Old Babylonian period but literary features point to a later date, not before the Kassite period. The claim that Esagil, and therefore Marduk, existed since the dawn of history, and the concept that kingship moved around according to the god's will (the pattern of the Sumerian King List) suggest the time of Hammurabi in connection with the rise of Marduk. But the linkage between the exercise of kingship and the care for Esagil lends the text a sense of admonition which favours the time of Hammurabi's successors, when Babylon was still big but its political influence was gradually decreasing.

8 Dalley, op. cit.

9 A case in point is the text of Ištar's descent to the Netherworld, which is clearly dependent on the Sumerian tale of Inana's descent. However, against the common view, the two compositions differ on the most important issue, the subject: Inana's descent is a tale about Inana, while Ištar's descent is a story about the Netherworld. The long repetitions and symmetrical structure which are typical for later compositions create the impression that Enūma eliš is an expanded Old Babylonian narrative, particularly in combination with some historical evidence that can be dated to that period. However, since no author writes a text in a vacuum, Enūma eliš, whenever it was created, will naturally include elements of earlier materials. Therefore, the features that have roots in the Old Babylonian period do not necessarily indicate that an earlier version of Enūma eliš was written at that time. Indeed, in the Old Babylonian period Marduk was raised to kingship among the gods, a status that befitted the size and status of the Babylonian kingdom. However, then he was elevated next to Enlil, not as his substitute. Yet, the very essence of Enūma eliš was to establish Marduk as the substitute of Enlil, as the sole creator of the universe and as the founder of Babylon – the first and only holy city. Towards these goals the text was structured from the very first line. To date, this ideology has no precedent in the Old Babylonian material. Therefore, more compelling evidence is necessary to secure an Old Babylonian date for the origin of the text.

10 On the possible historicity of Agum-Kakrime see now A.H. Podany, The Land of Hana, Bethesda CDL 2002, 58–9, with previous literature.

heim detected the "stilted hymnic style of the Kassite period"[11] and Sommerfeld defends the Kassite period for historical reasons.[12] Jacobsen associated Tiāmat with *māt tâmtim* and saw in the text an allegorical interpretation of historical events. He suggested that the victory of Marduk represents the victory of the Kassite Ulamburiash over the Sealand dynasty, which made Babylonia a large territorial kingdom. According to Jacobsen this victory explains the choice of Tiāmat as the rival of Marduk and the silence about Nippur and Enlil.[13] The motif of the battle between the sea and the storm god and the elaborations on the nature of Tiāmat add an enhancing dimension to the text. However, that the status of Enlil was minimized to the limit and that Nippur was ignored altogether is a problem, since during the Kassite period Nippur and the temple of Enlil became notably prosperous and influential. In defense of a Kassite date one may speculate that Enūma eliš signifies the aspiration of the Babylonian king to trim down the power and influence of Nippur. But conversely the ideology of Enūma eliš would probably be more acceptable at a time when Nippur was economically and politically weak, so that its priesthood could not raise much resistance. Moreover, since the Kassite kings had their own patron deities, Šumalija and Šuqamuna, they had no reason to elevate Marduk above Enlil.

After the invasion of Tukulti Ninurta I. (1243–1207) Nippur underwent a long period of gradual decline, and when in the 12$^{th}$ century Marduk was brought back from Elam by Nebuchadnezzar I. (1125–1104) the conditions were ripe for a new ideology. This is the option advocated by Lambert which has more solid historical evidence and therefore seems more likely.[14] There are also literary reasons for a later date, motifs appropriated from the SB ver-

11 A.L. Oppenheim, Ancient Mesopotamia, Portrait of a Dead Civilization, Chicago 1964, 264.

12 W. Sommerfeld, Der Aufstieg Marduks, Neukirchen-Vluyn 1982, 174ff. But see the review in W.G. Lambert, Studies in Marduk, Bulletin of the School of Oriental and African Studies 47 (1984), 3–4.

13 Th. Jacobsen, The Treasures of Darkness, New Haven and London 1976, 189–90. In JAOS 88 (1968), 104–108 Jacobsen discussed the properties of Marduk and Tiāmat, and the West-Semitic origin of the motif of the battle between the sea and the storm god. In p. 107 Jacobsen dated Enūma eliš to "around the middle of the second millennium B.C." His assumption that the motif travelled eastwards into Mesopotamia in the Old Babylonian period is now confirmed by a text from Mari which suggests that in Mari it was integrated in a coronation ritual. See J.-M. Durand, Le mythologème du combat entre le Dieu de l'orage et la Mer en Mesopotamie, MARI 7 (1993), 41–61 and idem, Le Culte d'Addu d'Alep et l'affaire d'Alahtum, Florilegium marianum VII, Mémoires de NABU 8, Paris 2002, n°38. That the Mari text has Têmtim rather than the Ugaritic term Yam, suggests that the model for the duel with Tiāmat was borrowed from Mari, rather than directly from Ugarit. On the other hand, the attestation of A.AB.BA *ta-ma-tum* DAGAL-*tum* in a letter from Ugarit (Ras-Shamra – Ougarit XIV, 2001, p. 261, text no. 8), perhaps contemporary to Enūma eliš may indicate an Ugaritic influence. Whereas the model for the fight was conveniently borrowed from the west, the details of the description are well rooted in South Mesopotamian traditions, as is the other face of Tiamat, a motherly creation goddess. Note that a-ab-ba – *tâmtu* appears as a divine component in personal names as early as the Ur III period. The personal names suggest that this Tiāmat had a positive image, particularly the names A.AB.BA-*simti* and A.AB.BA-*bašti*. One Tiāmat-bašti was a lukur of King Šu-$^{d}$Sîn (RIME: 338, 3/2.1.4.29, with previous bibliography).

14 Lambert, Fs. Meek. The list of Marduk's names in An=Anum could have been the strongest argument for dating Enūma eliš to the Kassite period, but it is uncertain whether the extant sources of An=Anum tablet II reflect the earliest version of the text. The list is the subject of A. Seri, The Fifty Names of Marduk in Enūma eliš, Journal of the American Oriental Society 126 (2006), who advocates a later date.

sion of the Ninurta narrative, the Anzu myth *bīn šar dadmē*, rather than the Old Babylonian version. These motifs now function as crucial elements in the ideological purpose of the text. While the text could not have been earlier than the Kassite period a later date gives effective results and is therefore more likely. The way that the borrowed mythological materials were modified and interwoven into the texture of the story points to a well conceived construction planned in advance, rather than a text that developed gradually during transmission.

The political weakness of Babylonia at the end of the 2nd millennium, when Marduk was just a city god as compared to the eminence of the god Aššur, and the evident high scribal skills lead Michalowski to associate the composition mainly with the cultural tension between Babylonia and Assyria, one round in an ideological battle.[15] Such a view implies that the return of Marduk to Babylon was, at best, a marginal political and religious event. As Michalowski argues, the position of Marduk in Enūma eliš is indeed comparable to that of the god Aššur in Assyria. At the same time, however, Marduk's status is also comparable to that of Enlil in Nippur. But even if the literary architect of Enūma eliš followed the Assyrian model, it does not weaken the ideological tone of the text. Its political tendency advocates a religious and cultic reform rather than merely an intellectual competition with Assyria. How seriously the ideology of Enūma eliš was taken is shown by a related source, the introduction to a Late Babylonian incantation from Sippar, CT 13, 35–38, that describes the creation of the world by Marduk. The literary properties of the text clearly demonstrate superior skills, but that the good work was meant to impress the learned Assyrians is doubtful. Those who could read and copy the text already recognized the cultural superiority of Babylonia, so this text could not prove anything which they did not know already.

The return of Marduk to Babylon after years in captivity, as a result of a successful military campaign, signified the dawn of better times for the Babylonians.[16] The religious significance of this experience is attested by the 22nd of the 50 names of Marduk, "who brought all the refugee gods back into their sanctuary" (En. el. VII:53). This name echoes the return of Marduk from captivity and implies that his return to Babylon is relevant to the text. An additional source, the Prophecy of Marduk, points to the same effect. In historical terms, the liberation of a god from enemy hands offers a king a unique opportunity to consolidate his power. This would mean that the experience of Marduk's arrival in Babylon and the renewal of his cult may well have been a constituting event for the Babylonians. With it came the impetus for theological and political reform, namely centralizing the cult and thereby the power of the king in Babylon. Enūma eliš, which proclaims the superiority of Marduk over all the gods as the successor of Enlil, demonstrates the shift of authority, religious and political to Babylon.

15 P. Michalowski, Presence at the Creation, in: T. Abusch et al. (eds.), Lingering Over Words, Studies in Ancient Near Eastern Literature in Honor of William L. Moran, Atlanta 1990, 381–396. Henceforth Fs. Moran.
16 J. A. Brinkman, A Political History of Post-Kassite Babylonia, Roma 1968, 104–109.

Enūma eliš was recited to Marduk during the New Year festival. In keeping with the Mesopotamian concept that creation materializes through its pronouncement, that things come into existence by the force of the word, when they are given a name,[17] the reading of the text to or performing it before Marduk every year re-asserted that he is everything that the text describes and embodies all the names he is given. Marduk, however, was a young god, so the author of Enūma eliš faced a challenge of reconciling the gap between Marduk's actual status with his own objective for composing the text. He had to achieve his purpose to promote Marduk's rank as truly and unquestionably superior to Enlil in an acceptable manner. The millennia old theological principles that governed Mesopotamian religion could not be suddenly replaced by a totally new order. Elements in the text suggest that he achieved his aim by introducing innovations as a continuity of the older tradition. In order to institute the new order he used old traditions which he adjusted for his purpose. These recycled familiar traditions may have seemed a natural development from the past which would lend the text the authority of an old, well-established fact.

It has long been recognized that Enūma eliš depends to a large extent on older mythological narratives concerning Ea and Ninurta. The particular significance of these traditions is that the Ea mythology anchors Marduk in his traditional genealogy, and the association with Ninurta legitimizes Marduk's status as the successor of Enlil. Clearly the mythological motifs of these two gods best served the ideological purpose of the text and they must have been deliberately chosen by the author.[18] Some revisions were made to adapt the appropriated material to its new function. At first sight an impression is created of an eclectic, imitative text. But a closer look at the changes that were made confirms not only that the choice of these particular mythological traditions was well planned, but also that the adjustments were carefully executed and interwoven into the new text so that there is clear progress towards the final goal.

In line with the Mesopotamian literary convention, Enūma eliš begins with a cosmological introduction followed by theogony, giving the narrative a chronological and thematic framework. But this introduction is significantly different from the convention in three ways. It goes back earlier than any other known account, to the very beginning of time, before heaven and earth came into being. Enlil, the eternal head of the Mesopotamian pantheon, is not mentioned at all, as if he never existed. Anu, the ancient head of the Sumerian pantheon, is the third generation of creation, the father of Ea and the grandfather of Marduk.

The introduction is based on myths of Marduk's father Ea. All there was at the very beginning were two watery matters, Apsû and Tiāmat. The vision of primordial reality as sea

17 E.g. En. el. I:1–2: "When the sky above was not yet named, the earth below was not mentioned by name".

18 It appears, therefore, that the combat between the sea and the storm god, be it Adad or Tišpak, serves as a framework only. The Ninurta motifs are the subject of W.G. Lambert, Ninurta Mythology in the Babylonian Epic of Creation, in: Keilschriftliche Literaturen. Ausgewählte Vorträge der XXXII. Rencontre Assyriologique Internationale, Bonner Beiträge zum Vorderen Orient, Band 6, K. Hecker, W. Sommerfeld (eds.), Berlin 1986, 55–60 (Henceforth "Ninurta Mythology").

or, rather, bodies of water is not exclusive to the Sumerian cosmology,[19] but the union of Tiāmat with Apsû is matched only in the mythology of Ea. Apsû, the sweet underground water, is an important aspect of Enki's divinity, being his abode in Eridu. His involvement in the plot, the struggle and victory of Ea over Apsû, links Ea himself with the origin of time. Moreover, giving Apsû a function in parallel to Tiāmat to predict Marduk's victory over her suggests that this role was designed for the purpose of Enūma eliš.[20] The epithet of Tiāmat, "who bore them all" (l. 4. *muallidat gimrišun*), is very likely a late manifestation of Namma, Enki's mother, as in Enki and Ninmah line 17, "the first mother, the begetter of all the gods."[21] No male partner is mentioned with Namma, and therefore she seems to be a womb of the world which creates life by itself. Thus in essence, Namma is the embodiment of the material of creation. The concept of the sea as material of creation can also be traced in the Theogony of Dunnu: the plough penetrated Earth and thereby made furrows which created the sea, and the furrows (i. e. the sea) "gave birth by themselves to Sumuqan."[22] The motherly image of Tiāmat in the episode with Apsû,[23] combined with some descriptions implying that she was a closed body of water, suggest to me that Tiāmat was understood as a lake of amniotic water at the dawn of time. In a birth incantation the sea was used as a metaphor for the amniotic fluid.[24] Thus, Tiāmat mingling with Apsû suggests amniotic fluid mingling with semen, signifying bisexual reproduction.[25] In the hands of Marduk, Tiāmat, the sea of amniotic fluid, would become the building material of the whole universe, the

19 One may argue that also this image of Tiāmat was borrowed from the West Semitic mythology like the motif of the battle, i.e. that Tiāmat is a reflection of Aṯirat "lady of the sea, creatress of the gods", KTU 1.4 iii:25 (translation: D. Pardee, in W.W. Hallo et al. (eds.), The Context of Scripture, Leiden 1997, 258). Moreover, a letter from Daganba'ali to Urtenu, RS 92.2006 count among the guardian gods Ea *bēlu kabtu* and A.AB.BA *ta-ma-tum* DAGAL-*tum*, "Sea, the wide sea," see *Ras-Shamra – Ougarit* vol. XIV (2001) p. 261 text no. 8. However, this is also an old Sumerian concept, as in Namma, the ancient Sumerian goddess in Eridu and the mother of Enki, attested since the ED III period (Reallexikon der Assyriologie 9, 136–7). Her name is written with the sign ENGUR indicating a body of water, and in the ES vocabulary her name was interpreted as the female sexual organ (MSL IV, 4:24). The relation between the Ugaritic letter and the Sumerian sources is difficult to assess.

20 There does not appear to be any earlier representation of Apsû personified and acting as the creation force.

21 $^{d}$namma-ke$_4$ ama palil ù-tud diĝir šár-šár-ra-ke$_4$-ne.

22 J.J. Glassner kindly drew my attention to the Theogony of Dunnu, for which I thank him. I agree with his observation that the image of the plough is a metaphor for the male organ and penetrating the earth one for copulation, and that the sea, their daughter, has nothing to do with plain, salty sea-water. The Theogony of Dunnu is known from a single Late Babylonian copy of tablets from Babylon and Aššur, CT 46 no. 43. For text and translation, see Th. Jacobsen, The Harab Myth, SANE 2/3 (1984). Note that sea is written as Sumerian a-ab-ba rather than the expected Akkadian *tâmtu*. Presumably, the personal names with the a-ab-ba as divine component refer to that image of the sea, rather than to the monstrous Tiāmat.

23 Especially Tablet I:41–46.

24 See particularly YOS 11, 86; V. Dijk, OrNS 42 (1973), 503; M. Stol, Birth in Babylonia and the Bible, its Mediterranean Setting, Groningen/Styx 2000, 11, especially ll. 5 and 7: "In the water of intercourse the bone is created; in the flesh of muscles the baby was created. In the ocean waters (l. 5 me-e a.ab.ba), fearsome raging, in the far-off waters of the sea (l. 7 *i-na me-e ti-a-am-tim ru-qu-ú-tim*): where the little one is – his arms are bound ...".

25 Note that the Theogony of Dunnu preserves the old tradition that the sea is the material of creation, which procreates by itself. But it also explains how the sea came into being, in a bisexual act of copulation, a contradictory concept.

creation of which would establish his status above and instead of Enlil.[26] Therefore, I propose that the author deliberately chose Namma, Enki's mother and the primeval water of creation, as the model for Tiāmat. At the same time, water also embodies the sea, the monster famously vanquished by the storm god in mythology.[27] Thus, the sea embodies two opposing mythological motifs which can be used in one and the same narrative. And since the author of Enūma eliš adapted these mythological motifs to motifs of the Ninurta myths, the battle between the storm god and the sea could have been chosen deliberately for the purpose of Enūma eliš. Thus, the author cleverly laced the plot with an emotional drama of jealousy and guilt to turn the sea of primeval water, a traditionally motherly figure comparable to Namma, into a monster. In other words, the two opposing characteristics of Tiāmat can be seen to have been intentionally chosen and the change in the nature of Tiāmat to have been planned in advance rather than gradually developed in the plot during a process of transmission.

In the first four generations of gods, until Ea came to power, Enlil's name is noticeably missing.[28] His exclusion serves for the rise of Ea to supremacy which was necessary for the rise of Marduk. Ea's supremacy over the existing gods is achieved through the first theomachy, in which Ea killed Apsû, appropriated his royal symbols (crown and melammu) and used him as the basis for his own temple.[29] In this temple Marduk was created, a suitable heir for a victorious father and a direct descendant of Anu.

The reason for the conflict is noise disturbing the resting gods, a motif adapted from Atrahasīs and from Enki and Ninmah (in Sumerian).[30] The role of Enlil in Atrahasīs is accommodated by Apsû and the motherly role of Namma in Enki and Ninmah by Tiāmat. But in Enūma Eliš, at that phase the world was not yet created, the gods were not working and humanity was not yet conceived, and so the circumstances are different. Since the purpose of Enūma eliš is to raise young Marduk's status, the fundamental principle of the text is a generation gap. Hence, the source of the noise and the cause of the conflict are playing children rather than over-worked and rebellious gods or a crowd of noisy people.

The section that focused on Ea was the precursor to the account of Marduk's exploits. The two sections of the text are structurally and thematically parallel. Both descriptions

26 In Mesopotamian literature creation stories, such as the first lines of Enūma eliš, are usually used to introduce a chronological framework to the narrated events. But the creation of the universe by Marduk is another matter because it has everything to do with the ideological objective of the text. I can agree with Michalowski that the creation is not the focus of the text (Fs. Moran, 383). But the victory over Tiāmat is certainly not enough by itself to make Marduk superior to Enlil, which is the objective of the text. Therefore the creation story, which is indeed unique to Enūma eliš, is a central pillar in the argumentation of the author. That the gods put Marduk to a test for creation, Tablet IV:21–28, indicates how important this qualification was.

27 Adad/Addu in Mari letters, Tišpak in Ešnuna, Ba'al in Ugarit, but not yet Ninurta in Babylonia.

28 Already emphasized by Lambert in: Studies Marduk, Bulletin of the School of Oriental and African Studies 47 (1984), 5.

29 The classic example for parricide in Mesopotamian literature is still the Theogony of Dunnu.

30 Noise was treated extensively by Michalowki in: Fs. Moran, 385–389.

begin with an appeal to Tiāmat and end with the same formula, I:73 = IV:123: "After he had captured and slain his enemies". The motifs that governed the story of Ea are at work again in relation to Marduk, but the extent of the narrative is different. The progress of the conflict until Marduk killed Tiāmat takes hundreds of lines (from Tablet I:110 to IV:122), but until Ea killed Apsû (I:29–72) it takes only 43 lines. At the end Ea created his abode out of Apsû, but from Tiāmat Marduk created the whole universe.

In aspiring to raise Marduk above Enlil the author selects the most suitable model for his purpose, the exploits of Ninurta, particularly his fight against Anzu. But why was a west Semitic model, the far-off sea Tiāmat, chosen as the enemy rather than a feature characteristic of Babylonia? This arises from the mythological creative property of primordial water. This body of water is the material of creation, and by choosing Tiāmat the author can complete the victory of Marduk with the creation of the universe from the body of his dead enemy.[31] The creation of the world is the most important argument for the exaltation of Marduk.

The combat between Marduk and Tiāmat was formulated by themes that were taken mainly from the narratives about Ninurta: the slaying of Anzu and Lugal-e. Since the components that originally belonged to the Ninurta myths have already been identified and discussed by Lambert and more recently by Wilcke,[32] only a few examples will be mentioned here. The most obvious borrowings are the central elements that hardly needed a change, Ninurta's eleven enemies (Lugal-e:129–133) emerge as the eleven monsters that Tiāmat created against Marduk, the combat occurred in two phases and the news of the victory were carried by the wind.

The development of the conflict until the victory over Tiāmat depends on the heroic battles of Ninurta. Both Ninurta and Marduk were sent to crush the attack of evil forces and protect the interests of their group. Yet there are some rather significant differences. Marduk is not just powerful and reactive, as Ninurta was, but he initiated activity. They both fought against a threat to the gods, but Ninurta reacted to the attack while Marduk caused the conflict. Ninurta accepted the challenge obediently, but Marduk set conditions that would make him king of all the gods. Moreover, he created a long process in the assembly of the gods, which included a test to prove his competence, to legitimize beyond any opposition his status as a king. Ninurta protected the authority of his father Enlil and the already organized world from the powers of chaos, whereas Marduk fought the old sluggish generation to get the authority over the gods for himself, and then, he himself created and organized the world. The confrontation of Marduk with Tiāmat follows closely the example of Ninurta, with some adaptation, because as the plot develops it leads to Kingu rather than to

31 The monsters that Tiāmat created and Kingu are, actually, much at home in Babylonia. Thereby the author blended her into the mythological landscape of southern Mesopotamia.

32 Lambert, Ninurta Mythology; C. Wilke, Vom altorientalischen Blick, in: Anfang und Ursprung, Colloquium Raurricum Bd. 10, Berlin/New York 2007, 36ff.

Tiāmat as the rival of Marduk. Kingu initiated the war, and by holding the tablet of destinies he became the leader. Kingu is the actual parallel of Anzu, not Tiāmat. Yet the purpose of the text required that Marduk should kill Tiāmat, because above all Tiāmat is the manifestation of creation, the initial material of creation from which Marduk would subsequently have to create the world. From a literary point of view, Tiāmat should be the parallel of Apsû, whom Ea killed and used to create his abode, which is why the text is focused on Tiāmat and makes her the real power behind the opponents. From the beginning of the conflict she is portrayed as a matriarch. Adhering to the model of Ninurta enables the shift from Kingu to Tiāmat. In the first phase Kingu is placed next to Tiāmat and then she is talked into a duel for the second and final phase.

The victory is followed by creation. In Atrahasīs the world was still incomplete, and the conflict ended with the creation of man. Ninurta was born into a civilized and populated world so he created some additional useful things from stones. Ea used the corpse of Apsû to build his abode, but Marduk used the corpse of Tiāmat to create the whole universe.[33] The account of the creation of the world is the final and decisive demonstration of Marduk's superiority. This point is of central importance to the ideology of Enūma eliš, it makes it different from and far more detailed than any other creation story. Such a detailed account invalidates any claim by any other god to be the creator.

The creation of man in the tradition of Atrahasīs is adapted to the course and purpose of Enūma eliš. Mankind was indeed created to work the land and serve the gods, but this happens immediately after the creation of the universe, before the gods had any chance to work themselves and any reason to rebel against that work. The idea to create humanity came from Marduk, a compassionate god, full of initiatives, who conceives plans in advance. Ea, who originally came up with the idea, was left to bring it to realization. There is no role for a mother-goddess and Marduk used a deity's blood and bones, not clay for his work of creation.[34]

The slaughter of Kingu, the leader of a rebellion, who incited Tiāmat to start the war, marks the old tradition re-emerging. Also the statement that Ea imposed the work of the gods on mankind and released the gods echoes the old tradition rather than being an integral element in the plot of Enūma eliš. The text concludes: "By the ingenuity of Marduk Nudimmud created" (VI:38), showing that Marduk surpassed Ea in creativity.

The account of the creation portrays Marduk as a different sort of god than Enlil. He is not a distant irritable ruler. Marduk listens, understands and tries to satisfy his subjects. He created single handedly a perfect world, a world that does not need corrections and additions. Now the universe was ready for the construction of Babylon, a residence for Marduk and the center for the assembled gods.

33 Note that in these examples the opponents of the creators were used as the material for the creation.
34 *eṣemtu* – frame of the body, CAD E, $343^{b}$ with references.

During the rituals around the inauguration of Babylon the gods swore loyalty to Marduk and for the third time reconfirmed his kingship. Anšar proclaimed Marduk's absolute universal supremacy. The recurring proclamations strongly suggest that the promotion of Marduk above Enlil was still seen as something needing to be asserted.

Anšar announced Marduk's name as Asalluhi (VI:101), a reminder that he is not a newcomer but an ancient god, the son of Enki. He then instituted Marduk's hegemony over gods and people with the words "The son, our avenger shall be the highest" (Vl. 105) and "Let him be the shepherd of the black-headed people, his creation" (VI:107). Anšar concluded his announcement with the statement that, no matter by which name he is called, Marduk is the god. In other words, all the local and foreign gods are a manifestation of Marduk. This statement imitates the list of the names of Ninurta at the end of the Anzu story, and comes instead of a similar list. It leads to the last part of the text, the list of 50 names of Marduk[35]. These names evoke the essence of his divine properties, "Let us call his 50 names that his ways shall become apparent, his deeds as well" (VI:121–2). That the sign 50 symbolizes the name of Enlil cryptically indicates the author's agenda.

On the surface these names are a statement about Marduk's divinity, but ideologically, the number of them endows him with absolute powers and indicates that Enlil and Marduk are one and the same god. That it occurred directly after the creation of the universe signifies the antiquity of this identity. On the literary level, thematically and structurally, the giving of these names closes the literary framework of the text, for it began before any name was given and ends with the giving of names, and in the interim the universe was created.

The last names emphasize again the theological objective and the problem confronted by the text. Because Marduk had created heaven and earth Enlil gave him his own name EN.KUK.KUR "lord of all lands", implying that Enlil himself chose Marduk as his successor. Then Ea gives Marduk his own name Ea. This way the divine properties of all the great gods were incorporated in the divinity of Marduk.

The recurring assertions of Marduk's supremacy over the gods suggest that the desire to establish his status was not easily accomplished in the stable religious setting of the region, where traditionally Enlil was the highest authority. Perhaps not even when Nippur was in decline. The use of existing literary traditions from Nippur was probably an attempt to meet that challenge. The well known motifs that depict Marduk as the image of Ninurta give a sense of continuity to his emergence and of natural progress from earlier reality. When the conventional cosmological introduction was stretched to the very beginning of time a window was opened for a different view of the creation of the world, a view that would lead to Marduk's birth, exploits and exaltation. That his father was Ea made this view seem acceptable, for Ea had been linked with creation at least as early as the OB period. As a god of magic his status never really diminished, he was invoked persistently over the course

35 The literary nature of the list of 50 names was treated in A. Seri, The Fifty Names of Marduk in Enūma eliš, Journal of the American Oriental Society 126 (2006), 507–519, with previous bibliography.

of millennia.[36] Slightly modified elements of Enki's mythologies direct the theogony to the birth of Marduk. His premeditated promotion to the head of the pantheon was presented as the natural result of a generation gap that escalated to struggle among the gods. Theomachy and parricide are used as literary devices to skip generations and to raise a minor god to kingship. The occasion of war served to involve the assembly of the gods to officially nominate Marduk and legally promote him to the kingship.

The ideological horizon of Enūma eliš, introducing Babylon in place of the Nippur theology, and the use of battle between gods as a means to reach it, explain the choice of Ninurta as a model. The tales of his triumphs over Asag and Anzu that explain his prominence in Nippur, the respect and authority that he gained from the great gods, are a good precedent for the rise of Marduk. The appropriated and revised Ninurta themes present Marduk as his parallel, his successor and his superior. The idea of Anšar's statement, as encapsulated (VI: 120), implies that Ninurta is actually Marduk.

Ninurta was the model for the creation of Marduk as the only national god, but this position could not be materialized as long as Nippur was the religious centre of Babylonia. In order to establish the national cult in Babylon Marduk had to depose Enlil and take his place completely. The mention of Enlil in the text indicates that the author shunned the option of ignoring him so he minimized his status. Enlil appeared on the scene for the first time only after Marduk had killed Tiāmat so that his appointment as the king was irrefutable.[37] Marduk built the big sanctuary, Ešara for Anu, Enlil and Ea (IV:145–6). Ešara is obviously a deliberate substitute for Enlil's temple Ekur in Nippur, just as Durmah which binds heaven and earth replaces Duranki in the Nippur tradition. By including such details the account prepared the ground to proclaim Babylon as the religious center of Babylonia.

The importance of Babylon in the ideology of Enūma eliš is made clear by the place it occupies in the course of creation. When the world is ready and organized, populated by human beings to work and serve the gods, it was time to build Babylon as the first religious center. It is most important to note that the construction of Babylon derives from the explicit wish of all the gods to erect a shrine, not only for Marduk, but also for themselves. There they would reside when they arrive for the assembly, which would naturally convene in Babylon. The model was obviously taken from Nippur. The Anunaki built Esagil for Marduk, shrines for Anu, Enlil and Ea and then their own shrines in Babylon, the first city, the capital of the world. There is no mention of Nippur, the ancient and well-established religious centre of Mesopotamia. This ideology is the core of an abridged version of the creation story that serves as the cosmological introduction to a Late Babylonian bilingual incantation from Sippar (CT 13, 35:1–10). It portrays the world before it was created by a list of

36 Perhaps because permanence is intrinsic to the practice of magic. Any change, even the slightest modification of the procedure might endanger the efficacy of its power.

37 So also Lambert in: Studies in Marduk, 5.

distinctive features of Babylonia, but negated to signify that they do not exist.[38] The list of non-existent locations ends with Nippur and Ekur, Uruk and Eanna, Apsû and Eridu. At that time, the text says, "All the lands were sea" (l. 10). Then first Eridu and E-sagil were built, "whose foundations Lugal.du$_6$.kù.ga laid within the Apsû". Second, Babylon was built and Esagil completed, and then "he" (i.e. Marduk) created all the Anunaki who pronounced Babylon supreme (ll. 12–16). It was from Babylon that Marduk created the world. Babylon existed even before the Anunaki, and Nippur was not mentioned again.[39] Since Nippur did not exist Babylon could be declared the supreme holy city.

In historical reality, however, Nippur was not neglected by the Babylonian kings, and Babylon did not take the place of Nippur as the privileged religious center of Babylonia. Therefore, in theological terms the ideological objective of Enūma eliš was not achieved. When Babylon, which was already a center of learning, eventually did become the most famous Mesopotamian city and a symbol of high civilization, it was not due to the annual reading of Enūma eliš. Rather it was the result of domestic power struggles inside Babylonia and international ones, especially the fall of the Assyrian empire.

Unlike Nippur, Babylon was a capital city, a political centre. If the ideological objective of Enūma eliš had been achieved the Babylonian king would have had total control of the state, including the administration at Nippur, which usually played an independent role. The ambitious objective of Enūma eliš, the recycled old literary traditions and particularly the repeated assertion of Marduk's legitimacy create the impression that the composition was initiated in a position of political weakness. Presumably, therefore, this text exalting Marduk was meant as a medium to reconstruct the status of the city of Babylon on firmer foundations and to restore its past prominence. It would thereby consolidate the power of the resident ruler who had commissioned the composition.[40]

38 E.g. "Nippur was not built, Ekur was not constructed". The negation was a typical method to describe not yet existing matters.

39 One should infer, therefore, that the city of Nippur was built long after Marduk created the universe.

40 The structure of the plot, the conscious choice of the mythological materials that were used, adapted to the purpose of the text and interwoven with the flow of the narration, suggest that the text of Enūma eliš was carefully thought-out and constructed according to a plan. The deviation of the text from established traditions as of its very beginning, and the political value of its theological objective suggest that it was commissioned. The end of the text supports this possibility, especially VII:157–8 which state that the narrated events were disclosed to the writer personally by an old witness and he wrote it down for the future generations. Thus, the text did not exist before, neither orally nor in writing, and therefore it claims the authority of an ancient original source. A case in point may be the religious reform of the Judean king Josia (640–609 BC), described in 2 kings 23. An allegedly ancient scripture found in the temple of Jerusalem was used by the king as a pretext to consolidate the cult of Yahweh in the capital and to outlaw any cultic activity whatsoever outside Jerusalem. This reform was a success, and ever since Jerusalem is the only holy city for the Jewish people.

*Stefan M. Maul*

# Die Wissenschaft von der Zukunft

## Überlegungen zur Bedeutung der Divination im Alten Orient

Wenn es im Alten Orient[1] wichtige politische Beschlüsse zu fassen galt, so zeigen es die uns zur Verfügung stehenden keilschriftlichen Quellen aus weit mehr als zwei Jahrtausenden, setzten Könige und ihre Beratergremien nicht ausschließlich auf den eigenen Sachverstand. Sie hielten sich nämlich, wie es scheint vom Beginn des zweiten vorchristlichen Jahrtausends an durch alle Jahrhunderte hinweg, konsequent an die Regel, ein von ihnen beschlossenes wichtiges Vorhaben erst dann in die Tat umzusetzen, wenn dessen Tragfähigkeit von einem von Herrscher und Kabinett unabhängigen Sachverständigenrat geprüft und bestätigt worden war.

Die Autorität, die man – aller Machtfülle des Königtums zum Trotz – einer solchen Prüfung beimaß, kann man kaum überschätzen. Sie lässt sich schon allein daran ermessen, dass die Herrschenden sich der Evaluation ihrer Pläne und Absichten ohne jeden Widerspruch unterzogen und in aller Regel bereit waren, dabei auch in Kauf zu nehmen, dass die von ihnen betriebenen Vorhaben gegebenenfalls als aussichtslos oder gar schädlich eingestuft wurden und deshalb nicht zur Durchführung gelangten. Freilich versprach ein Gutachten der unabhängigen Sachverständigen im Gegenzug, verlässliche Prognosen über den Erfolg einer geplanten Unternehmung liefern zu können. Die ausgewiesenen, durch ein langes Studium qualifizierten Fachleute, die ein solches Gutachten erstellten, garantierten nämlich, über einschlägige Kenntnisse und bewährte, zielführende Verfahren zu verfügen, die es ihnen ermöglichten, aus dem Blickwinkel der Zukunft gewissermaßen zurückschauend die Konsequenzen eines beabsichtigten Vorhabens genau zu übersehen. Ihr Fachwissen, davon war man überzeugt, gestattete es ihnen, neben den Planungen und Absichten, die von Erfolg beschieden sein würden, auch jene eindeutig benennen zu können, die – wenn man die Dummheit besaß, sie in die Tat umzusetzen – zu eigentlich ungewollten Fehlentwicklungen und zu Unheil führen würden.

Es liegt auf der Hand, dass die Aussicht auf ein derartiges Wissen um die Zukunft attraktiv und vor allem für politische Entscheidungsträger von unschätzbarem Wert ist. Verspricht ein solches Wissen doch demjenigen, der ein entsprechendes Gutachten über seinen Plan eingeholt und einen positiven Entscheid erhalten hatte, die Gewissheit, mit sei-

1 Eine sehr lesenswerte Einführung in Geschichte und Kultur des Alten Orients bietet A.L. Oppenheim, Ancient Mesopotamia. Portrait of a Dead Civilization, Revised Edition Completed by Erica Reiner, Chicago 1996; siehe auch J.M. Sasson, Civilizations of the Ancient Near East, New York 1995.

nem Vorhaben den richtigen, zukunftsweisenden und alsbald auch von Erfolg gekrönten Weg gewählt zu haben und so all denen gegenüber, die keinen Zugang zu entsprechenden Erfolgsgarantien besaßen, sich schier unüberwindbare Vorteile verschaffen zu können.

So einsichtig und vernünftig es klingt, die Zukunftsfähigkeit geplanter Vorhaben vor deren Umsetzung zu überprüfen, so abwegig, ja geradezu lächerlich und absurd erscheinen aus dem Blickwinkel unseres heutigen Weltbildes die Mittel, mit denen im Alten Orient entsprechende Evaluationen vorgenommen wurden. Die Zukunftsaussichten eines Planes, einer politischen, strategischen oder personellen Entscheidung wurden nämlich an altorientalischen Königshöfen über Jahrhunderte hinweg regelmäßig anhand von Gestalt und Färbung der Eingeweide, und insbesondere der Leber eines eigens zu diesem Zweck geschlachteten Schafes ermittelt.[2]

Bereits im 3. vorchristlichen Jahrtausend wurde in Mesopotamien ein Prüfungsverfahren praktiziert, in dem man das äußere Erscheinungsbild vor allem der Leber, aber auch anderer Innereien eines Schafes in Relation zu Zukünftigem setzte, um Erfolgsaussichten desjenigen herauslesen zu können, der die Schlachtung des Tiers veranlasst und mit einer bestimmten ausformulierten, die Zukunft betreffenden Frage oder Absicht verbunden hatte.[3]

Unter Anwendung eines in sich ganz systematisch und logisch wirkenden Gefüges von Gesetzmäßigkeiten wurden bestimmte Erscheinungen an der Leberoberfläche als günstige oder ungünstige Zeichen gedeutet.[4]

Die eigens hierfür ausgebildeten Fachleute entnahmen dem geschlachteten Schaf die Eingeweide, um sie einer Prüfung zu unterziehen. Ihre besondere Aufmerksamkeit galt der im Tierleib den übrigen Organen zugewandten, sehr stark gegliederten Seite der Leber, die die Veterinäre die *facies visceralis* nennen. Nacheinander, entgegen dem Uhrzeigersinn prüften die Zeichendeuter Vorhandensein und unbeschadeten Zustand von etwa einem Dutzend anatomisch konstitutiver Elemente auf der Leberoberfläche:[5] neben der sehr auffälligen, oft grünlich erscheinenden Gallenblase, inspizierten sie vor allem furchenartige

2 I. Starr, The Rituals of the Diviner (= Bibliotheca Mesopotamica 12), Malibu 1983; J.-W. Meyer, Untersuchungen zu den Tonlebermodellen aus dem Alten Orient, (= Alter Orient und Altes Testament 39), Kevelaer/Neukirchen-Vluyn 1987; R. Leiderer, Anatomie der Schafsleber im babylonischen Leberorakel. Eine makroskopisch-analytische Studie, München/Bern/Wien/San Francisco 1990; I. Starr, Queries to the Sungod. Divination and Politics in Sargonid Assyria (= State Archives of Assyria 4), Helsinki 1990; U. Jeyes, Divination as a Science in Ancient Mesopotamia, Jaarbericht van het Vooraziatisch-Egyptisch Genootschap Ex Oriente Lux 32 (1993), 23–41.

3 A. Falkenstein, Wahrsagung" in der sumerischen Überlieferung, in: La divination en Mésopotamie ancienne et dans les régions voisines, XIVe Rencontre Assyriologique Internationale (Strasbourg, 2–6 juillet 1965), Paris 1966, 45–68.

4 Zusammenfassend: S. M. Maul, Art. „Omina und Orakel. A. In Mesopotamien", in: RlA 10, 1./2. Lieferung Oannes – Pabilsag(a), Berlin/New York 2003, 69–82.

5 U. Koch-Westenholz, Babylonian Liver Omens. The Chapters Manzāzu, Padānu and Pān tākalti of the Babylonian Extispicy Series Mainly from Aššurbanipal's Library, Kopenhagen 2000; dies., Secrets of Extispicy. The Chapter Multābiltu of the Babylonian Extispicy Series and Nisirti bārûti Texts Mainly from Aššurbanipal's Library (= Alter Orient und Altes Testament 326), Münster 2005.

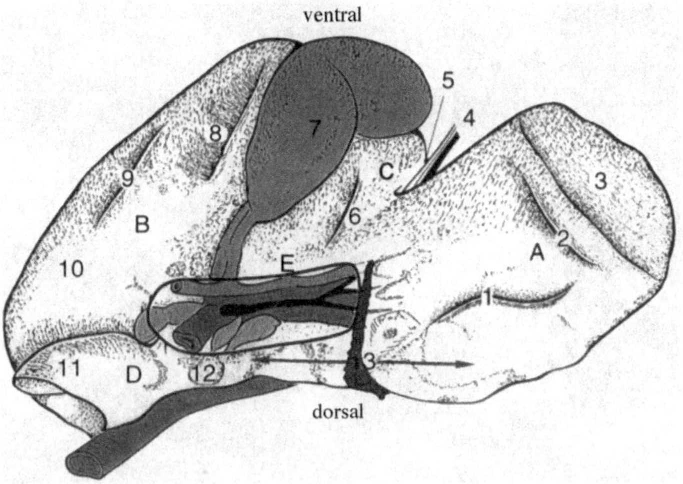


Abb. 1 | Schematische Darstellung einer Schafsleber (nach R. Leiderer, Anatomie der Schafsleber im babylonischen Leberorakel. Eine makroskopisch-analytische Studie, München/Bern/Wien/San Francisco 1990, 158 Abb. 2).

A: Linker Leberlappen (*lobus sinister*). B: Rechter Leberlappen (*lobus dexter*). C: *lobus quadratus*. D: *lobus caudatus*. E: *porta hepatis* (babylonische Bezeichnung: "Fluss der Leber"). Die Zahlen 1–13 bezeichnen – in der Reihenfolge ihrer Prüfung – die für die Leberschau wichtigen Bereiche des Organs, neben der Gallenblase (7) furchenartige Eindrücke auf der Leberoberfläche (1, 2, 6, 8, 9), Einschnitte (5, 13) und markant geformte Oberflächen (3, 10), auffällige Vorsprünge (11, 12) und Reste der Bänder (4), an welchen die Leber im Leib des Tiers befestigt gewesen war.

Zeichnungen und Eindrücke, die von benachbart liegenden Organen herrühren und an der gesunden Schafsleber regelmäßig auftreten, aber auch Einschnitte und markant geformte Oberflächen, auffällige Vorsprünge und Reste der Bänder, an welchen die Leber im Leib des Tiers befestigt gewesen war (Abb. 1).

Der unbeschadete, gesunde, also ganz normale Zustand der einzelnen Leberteile wurde als günstig bewertet. Darüber hinaus spielte nicht zuletzt die Position von bestimmten auffälligen Merkmalen, die allenthalben auf den zwölf Leberregionen auftauchen können, zur Ermittlung des Evaluationsergebnisses eine grundlegende Rolle. Zu solchen Merkmalen zählen etwa stark hervortretende Lymphknoten, aber auch Häutchen, Blasen und Auswüchse sowie nicht zuletzt Löcher im Gewebe der Leber.[6] Diese sind an die Leberoberfläche tretende, offene und bisweilen verkalkte Bohrgänge verursacht von den gar nicht so selten auftretenden Leberegeln, von Bandwurmlarven (Finnen) und anderen Parasiten. Manche dieser Merkmale, wie z.B. ein Loch auf der Leberoberfläche, galten als übles Zeichen: Es ist auch uns nachvollziehbar, dass das Fressloch im Fleisch geradezu zum Inbegriff des Nekrotischen und damit zu einem Unheil signalisierenden Mal wurde. Andere Merkmale, die wir unter einem tiermedizinischen Blickwinkel als durchaus krankhaft ansehen würden, wie z.B. eine weiche, von einer Bandwurmlarve hervorgerufene, mit

6 Hierzu ausführlich mit eindrucksvollen Bildern: R. Leiderer, Anatomie der Schafsleber, München/Bern/Wien/San Fransisco 1990.

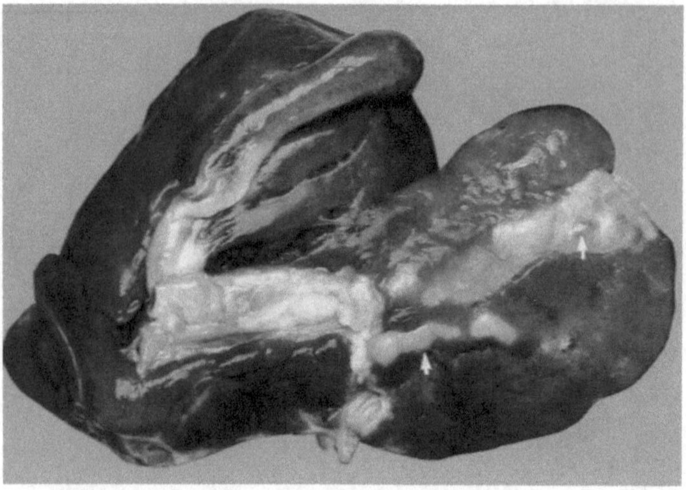

Abb. 2 | Eine frische, von Parasiten befallene Schafsleber (Photo: R. Leiderer, Anatomie der Schafsleber, München u.a. 1990, 161 Abb. 7).

heller Flüssigkeit gefüllte Gewebeblase, wurden jedoch positiv konnotiert. Im heißen, auf Bewässerung angewiesenen Mesopotamien sahen die Zeichendeuter in einer solchen Blase das Bild eines Wasserreservoirs, einer Zisterne, deren Wasser Aussicht auf Ernte und Reichtum versprach – so wie sie, aus Gründen, die sich uns letztendlich entziehen, die Leber des zu untersuchenden geschlachteten Schafes in gewisser Weise als Spiegel der Welt des Fragestellers betrachteten (Abb. 2).

Während schon keine gesunde Schafsleber auch nur annähernd einer anderen gleicht, führen pathologische Erscheinungen von Entzündungen, Parasitenbefall und Nekrosen zu erschütternd unterschiedlichen Befunden, die allesamt ausgedeutet wurden. Die hierfür von den Babyloniern entwickelte anatomische Terminologie erweist sich als bei weitem differenzierter und umfangreicher als es die der zeitgenössischer Veterinäre ist. Für das von den Mesopotamiern praktizierte Eingeweideorakel ist übrigens von nicht unbeträchtlicher Bedeutung, dass man einem lebenden Schaf – so versichern es jedenfalls moderne Tierärzte – keineswegs anzusehen vermag, ob man ihm eine gesunde, oder aber eine von Parasiten und Krankheiten geschädigte Leber entnehmen würde.[7]

Bei der Prüfung einer Leber ging man, um bestimmte Merkmale positiv oder negativ werten zu können, mit überraschender, ja geradezu mathematisch-kühler Präzision vor. Über jedes einzelne der etwa zwölf konstitutiven Bereiche der Leber – wir wählen hier zur Veranschaulichung das Beispiel der Gallenblase (siehe Abb. 3–4) – wurde ein virtuelles

7 Die altorientalischen Gelehrten hingegen glaubten (so zumindest im 1. vorchristlichen Jahrtausend), Parameter gefunden zu haben, die ihnen in einem begrenzten Maße erlaubten, vom äußeren Erscheinungsbild eines Schafes Rückschlüsse auf die Gestalt der Leber zu ziehen (B. Meissner, Omina zur Erkenntnis der Eingeweide des Opfertieres, Archiv für Orientforschung 9 (1933/1934), 118–122, [Nachtrag ebd., 329–330]). Es ist unbekannt, ob sie dieses (wohl nur vermeintliche?) Wissen dazu nutzten, um unbemerkt Einfluss auf das Ergebnis einer Eingeweideschau zu nehmen.

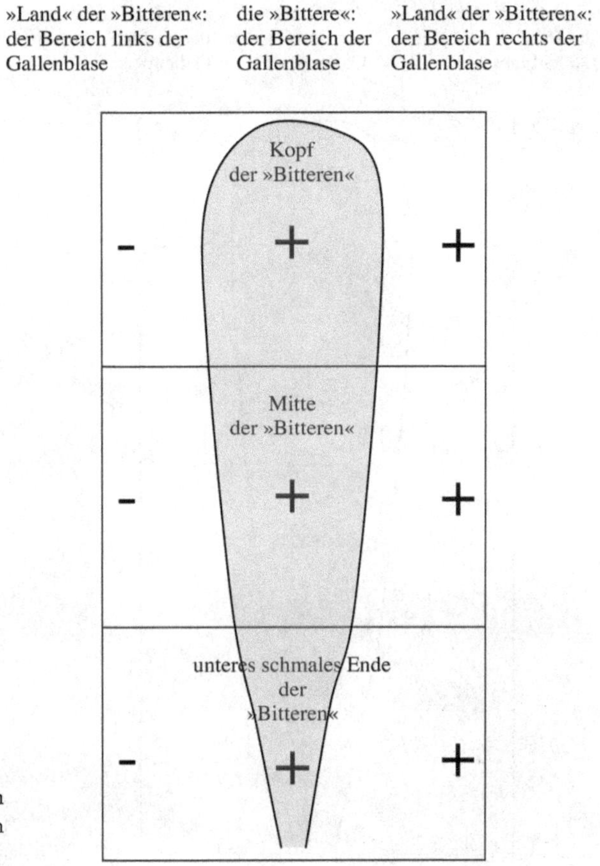


Abb.3 | Die Wertung eines positiv konnotierten Mals auf der Gallenblase (die „Bittere") und in ihrem „Land".

Netz von Feldern gelegt.[8] Keilschriftliche Handbücher der Eingeweideschauer lassen uns erkennen, dass man von einem mittig gelegenen, das Zentrum der jeweiligen Leberregion umfassenden Feld erwartete, Auskunft über die Haltung einer grundsätzlich wohlwollenden Schicksalsmacht zu erlangen. Das rechts davon liegende Feld, so die Lehre der Zeichendeuter, würde die Befürwortung des zu evaluierenden Vorhabens und das links davon befindliche die dem Vorhaben zuwider wirkenden Kräfte anzeigen können.

Wurde nun ein positiv konnotiertes Merkmal wie etwa eine mit heller Flüssigkeit gefüllte Gewebeblase im mittleren Feld, also z.B. in dem schmalen Bereich der Gallenblase beobachtet (vgl. Abb. 3), war dies ein Befund, den man als günstiges Zeichen wertete. Denn die Schicksalsmacht zeigte sich wohlwollend. Wenn das gleiche Zeichen im rechts daran angrenzenden Feld erschien, ergab sich ebenfalls ein positiv gewertetes Zeichen. Auf dem linken Feld hingegen, das den dem Vorhaben zuwider wirkenden Kräften zugeordnet ist,

8 J.-W. Meyer, Untersuchungen zu den Tonlebermodellen aus dem Alten Orient (= Alter Orient und Altes Testament 39), Kevelaer/Neukirchen-Vluyn 1987 *passim*.

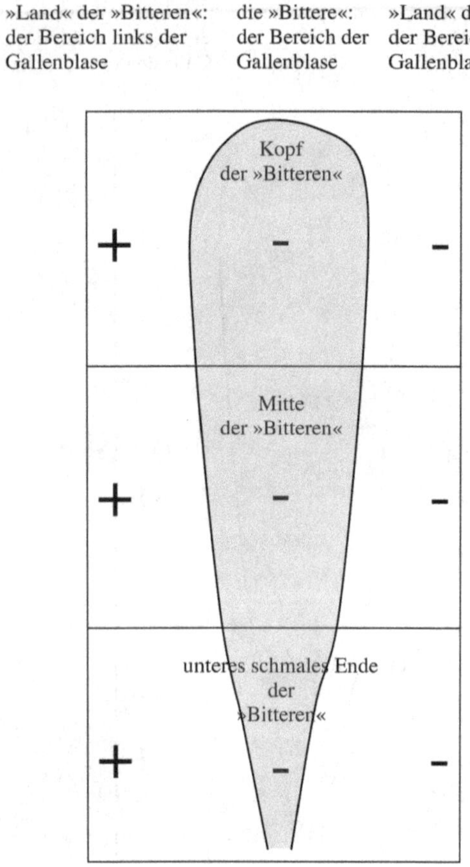


Abb. 4 | Die Wertung des Fraßloches eines Leberparasiten auf der Gallenblase (die „Bittere“) und in ihrem „Land“.

wirkte es sich als Verstärkung dieser Kräfte aus und wurde dementsprechend zu einem als ungünstig eingestuften Zeichen auf der Leber.

Ein negativ konnotiertes Merkmal hingegen (so wie das Fraßloch eines Leberparasiten) wirkte sich dementsprechend bei dem mittleren und dem rechten Feld gemäß der mathematischen Formel $+ \times - = -$ aus, wohingegen im linken Feld eine Schwächung der dem Vorhaben zuwider wirkenden Kräfte Stärkung bedeutet und somit ein negativ konnotiertes Merkmal in diesem negativ geladenen Feld zu einem günstig gewerteten Zeichen geriet (Abb. 4).

In Wahrheit war all dies freilich noch bei weitem komplizierter. Die gelehrten Zeichendeuter unterteilten beispielsweise, wie die Abbildung 3 und 4 zeigen, die drei Felder der Gallenblase ihrerseits jeweils in drei Parzellen, wobei sie diesen Parzellen umso mehr Aussagekraft für den zu ermittelnden Bescheid zubilligten, je näher sie an dem Mittelpunkt der Leber lagen.

Die gelehrten mesopotamischen Eingeweideschauer beschrieben nicht nur die Gallenblase als ein Gefüge von Parzellen mit positiver und negativer Ladung unterschiedlicher

Abb. 5 | Das beschriftete Modell einer Schafsleber (Babylonien, 17. Jh. v. Chr.) (Photo: Trustees of the British Museum).

Intensität, sondern die gesamte Schafsleber.[9] Ein altbabylonisches, zu Studienzwecken angefertigtes Lebermodell aus dem 17. vorchristlichen Jahrhundert dokumentiert dies dem Lernenden, wie auf Abb. 5 deutlich zu erkennen ist, anhand des, wie wir bereits gehört hatten, negativ konnotierten Merkmals „Loch". Die jeweils von der genauen Lage des „Loches" abhängige Bedeutung dieses Zeichens wurde sorgsam in jeder einzelnen Parzelle der Leber in Keilschrift notiert.[10]

Nach den Regeln der Kunst erwies sich ein „Loch" freilich nur auf den Feldern mit positiver Ladung als ein übles Zeichen. Bei jenen Feldern, von denen man annahm, dass sie das Maß der widrigen Kräfte anzuzeigen vermochten, galt ein solches Fraßloch als ein positives, erwünschtes Zeichen. Denn es zeigte die Schwächung des Widrigen und damit eine Stärkung der günstigen Einflüsse an.

Wir können an dieser Stelle nicht auf weitere Vorgehensweisen und Methoden der mesopotamischen Eingeweideschauer eingehen. Aber so viel sei dennoch gesagt: Die gelehrten Zeichendeuter sahen sich, wie uns ihre Handbücher aus dem 1. vorchristlichen Jahrtausend vor Augen führen, sogar in der Lage, die Gültigkeitsdauer ihrer Prognosen an-

9 J. Nougayrol, Le foie „d'orientation" BM 50494, Revue d'assyriologie et d'archéologie orientale 62 (1968), 31–50.

10 J. Nougayrol, Textes hépatoscopiques d'époque ancienne conservés au musée du Louvre, Revue d'assyriologie et d'archéologie orientale 38 (1941), 77–79.

hand einer mathematischen Formel zu errechnen, für die sie die einzusetzenden Zahlenwerte aus Art und Gestalt bestimmter Lebermerkmale ermittelten.[11]

Die endgültige Beurteilung der Erfolgsaussichten eines Planes oder eines Vorhabens ergab sich nach der ausführlichen Prüfung der Leber sowie einer raschen Betrachtung der übrigen Organe und des Knochengerüstes des geschlachteten Tiers durch simples Zusammenzählen der ermittelten positiven und negativen Zeichen. Überwogen die positiven, war das in der eingangs gestellten Orakelanfrage beschriebene Vorhaben als „erwünscht" beurteilt und zur Durchführung freigegeben – und darüber hinaus mit dem Gütesiegel versehen, dass die über dem Menschen stehenden Götter, die der Lehre der Eingeweideschauer zufolge den Orakelentscheid in gewisser Weise als Dank für das ihnen dargebrachte Opfertier in die Organe des Schafes geschrieben hatten, hinter diesem Plan stehen und ihn gemeinsam mit dem gesamten Kosmos fördern würden. Waren hingegen die unguten Zeichen in der Überzahl bedeutete dies, dass die Evaluation negativ ausgefallen war. Das zu beurteilende Vorhaben sollte dann, zumindest in der Weise, wie es in der zumeist schriftlich festgehaltenen Orakelanfrage dargestellt worden war, besser nicht zur Ausführung gelangen, wenn man zukünftigem Missgeschick aus dem Wege gehen wollte.

Auf diese Weise wurde an altorientalischen Königshöfen nicht nur über Personalangelegenheiten und über die Frage, ob man ein umfangreiches Bauvorhaben durchführen solle, entschieden, sondern auch darüber, ob und wann man ohne Schaden gegen einen Feind zu Felde ziehen könne.[12] Ferner gehörte es z. B. im altbabylonischen Mari des 18. Jhs. zur Routine, mit Hilfe der Leberschau monatliche Regelanfragen nach der Sicherheit von König, Stadt und Land zu stellen, mit dem Ziel auf bislang unerkannt gebliebene Gefahren rechtzeitig aufmerksam zu werden und diese bannen zu können, bevor es zu spät war.[13] Die Ursache der potentiellen Gefährdung wurde durch immer erneutes geschicktes Fragen – jedes Mal musste wieder ein Schaf geschlachtet werden – ebenfalls mit Hilfe der Eingeweideschau ermittelt. Im ersten vorchristlichen Jahrtausend standen den gelehrten Zeichendeutern hierfür sogar Sammlungen mit Modellfragen, die sich besonders bewährt hatten, zur Verfügung.[14]

Eine Abrechnung aus dem 18. Jh. v. Chr., die in den Archiven des Königspalastes von Mari gefunden wurde,[15] zeigt, dass dort für allerlei Orakelanfragen, die man im Auftrag des Königs in einem Zeitraum von nur acht Monaten gestellt hatte, insgesamt 4076 Tiere bereitgestellt und geschlachtet worden waren. Der in dem Dokument jeweils festgehaltene

11 U. Koch-Westenholz, Secrets of Extispicy. The Chapter Multābiltu of the Babylonian Extispicy Series and Nisirti bārûti Texts Mainly from Aššurbanipal's Library (= Alter Orient und Altes Testament 326), Münster 2005, 63–66 und 459–479.

12 Mit vielen Beispielen: I. Starr, Queries to the Sungod. Divination and Politics in Sargonid Assyria, State Archives of Assyria 4, Helsinki 1990.

13 J.-M. Durand, Archives épistolaires de Mari I (= Archives royales de Mari 26/I), Paris 1988, 57–58.

14 W. G. Lambert, Babylonian Oracle Questions, Winona Lake/Indiana 2007.

15 J.-M. Durand, a.a.O., 37.

monatliche Verbrauch von für Orakelanfragen benötigten Lämmern liegt zwischen 370 und 707 Tieren. Das Verwaltungszentrum des vergleichsweise kleinen Königreiches entfaltete somit allein für die Eingeweideschauen einen mittleren Monatsbedarf von 510 Lämmern. Dem entspricht ein durchschnittlicher Tagesverbrauch von nicht weniger als 17 Tieren. In den späteren großen Reichen des Alten Orients dürfte der Verbrauch von Tieren, die eigens für das Generieren von Orakelentscheiden geschlachtet wurden, ganz sicher noch um ein Vielfaches höher gewesen sein.

Dem modernen Zeitgenossen ist, so darf man wohl unverhohlen sagen, ein solches Prüfungsverfahren vor allem deshalb ein regelrechtes Skandalon, weil es sich mit Inhalt und Absicht des zu bewertenden Vorhabens in keiner Weise befasst und nicht einmal den Anschein zu erwecken versucht, dieses zu tun. Gleichwohl müssen wir konstatieren, dass über mehr als zwei Jahrtausende hinweg die Mesopotamier selbst, ebenso wie die eifrig um entsprechende Kenntnisse bemühten Völkerschaften im Umfeld des Zweistromlandes, in der Beherrschung von Verfahren der Zeichendeutung einen maßgeblichen Grund für die nachhaltigen kulturellen und machtpolitischen Erfolge von Babyloniern und Assyrern sahen. Nicht nur die Hethiter, die Syrer und die Perser bedienten sich der Technik der Eingeweideschau und ließen ihre Experten eigens zu diesem Zweck das Babylonische und die Keilschrift erlernen. Auch im Heiligen Land erfreute sich, wie Tontafelfunde unzweifelhaft unter Beweis stellen, die babylonische Eingeweideschau großer Beliebtheit.[16] Ja sie stand in dem Ruf, von so großem Nutzen zu sein, dass sie die altorientalischen Kulturen noch weit überdauerte und in der hellenisierten Welt, in Griechenland, Etrurien und Rom ebenfalls als unverzichtbares Mittel beim *political decision making* galt.[17] Lebermodelle aus dem etruskischen Kulturraum im nördlichen Italien zeugen von diesen weitreichenden Einflüssen des Alten Orients.[18] Erst im späten 4. nachchristlichen Jahrhundert, ein halbes Jahrtausend nach dem Untergang der Keilschriftkulturen, kam in Europa die Praxis der Eingeweideschau zum Erliegen. Das von Kaiser Konstantin ausgesprochene und von seinen Nachfolgern Constantius II. (357 n. Chr.) und Theodosius (392 n. Chr.) bekräftigte Verbot des „heidnischen" Tieropfers und der stets damit verbundenen Divination[19] setzte der Jahrtausende alten, ungebrochenen Tradition der Opferschau ein Ende.

16 Zusammenfassend: S. M. Maul, Art. „Omina und Orakel. A. In Mesopotamien", in: RlA 10, 81.

17 C. O. Thulin, Die etruskische Disciplin, Darmstadt 1968; A. J. Pfiffig, Religio etrusca, Graz 1975; D. Collins, Mapping the Entrails: the Practice of Greek Hepatoscopy, American Journal of Philology 129 (2008), 319–345.

18 J. Nougayrol, Les rapports des haruspicines étrusque et assyro-babylonienne et le foie d'argile de Falerii Veteres (Villa Giulia 3786), Comptes Rendus de l'Académie des Inscriptions (1955), 509–519; J. W. Meyer, Zur Herkunft der etruskischen Lebermodelle, Studia Phoenicia 3 (1985), 105–120; L. B. van der Meer, The Bronze Liver of Piacenza. Analysis of a Polytheistic Structure, Amsterdam 1987.

19 Siehe B. Gladigow, Opfer und komplexe Kulturen, in: B. Janowski, M. Welker (Hgg.), Opfer. Theologische und kulturelle Kontexte, Frankfurt am Main 2000, 100ff.

Grundlage der Gewissheit, mit Verfahren wie dem hier beschriebenen Einblick in Zukünftiges gewinnen zu können, war die auch heute noch gültige Vorstellung, dass die erfahrbare Welt Spuren einer sich entfaltenden Zukunft bereithält, die es als solche zu erkennen und zu deuten gilt. Im Alten Orient verstand man jede Form der Bewegung und Veränderung in allen Bereichen des Erfahrbaren (auf der Erde ebenso wie am gestirnten Himmel) jeweils als Teil eines ungeheuer komplexen Vorgangs des sich Entwickelns der Welt durch die Zeit, hin zum Zukünftigen. In der Welt, dem untrennbaren Gefüge von Irdischem und Kosmischem, stehen dieser Vorstellung zufolge alle wahrnehmbaren Erscheinungen, so wenig spektakulär sie im Einzelnen auch sein mögen, miteinander in Verbindung, da sie alle Anteil haben an jener Bewegung des Ganzen hin auf das Zukünftige und gemeinsam diese Bewegung bilden.[20] Einzeln, oder noch viel besser im Verbund beobachtet, lassen sie für den altorientalischen Menschen deshalb in der Gegenwart eine Hochrechnung des Zukünftigen als möglich erscheinen. Die Bewegung des Ganzen hin auf das Kommende ist, wie es etwa Wachsen und Werden in der Natur, der Wechsel von Tag und Nacht, der Jahresablauf und der gestirnte Himmel zeigen, von Gesetzmäßigkeit, von großer Harmonie geprägt und wird als solche wahrgenommen. Jede Abweichung vom Regelmaß in der Natur galt hingegen in dem, zumindest aus unserer Perspektive, fast aberwitzig anthropozentrischen Weltbild, das der altorientalischen Divination zugrunde liegt, als eine vom Menschen hervorgerufene Störung, oder genauer gesagt als eine Reaktion auf menschliches Handeln und wohl auch Wollen. Abweichungen vom Regelmaß wie z. B. Auffälligkeiten bei Pflanzen und Tieren[21], am gestirnten Nachthimmel[22] oder auch auf der Oberfläche einer Schafsleber[23], wurden in diesem Sinne als Botschaft an den Menschen wahrgenommen, die nach Innehalten, nach sich Besinnen und Korrektur verlangen, damit die entstandene Unordnung beseitigt und die Harmonie wiederhergestellt werde.[24] Die gegenwärtige Erfahrung des anthropogenen Klimawandels eröffnet vielleicht auch uns eine Sicht, die erahnen lässt, welche *ratio* hinter derartigen Vorstellungen steht.

20 Vgl. S. M. Maul, Art. „Omina und Orakel. A. In Mesopotamien", in: RlA 10, 45–48.

21 S. M. Moren, The Omen Series „Shumma alu": A Preliminary Investigation, Diss., University of Pennsylvania 1978; S. M. Freedman, If a City is Set on a Height: the Akkadian Omen Series Šumma Alu ina Mēlê Šakin, Vol. I: Tablets 1–21 (= Occasional Publications of the Samuel Noah Kramer Fund 17), Philadelphia 1998; N. P. Heeßel, Divinatorische Texte I. Terrestrische, teratologische, physiognomische und oneiromantische Omina, Keilschrifttexte aus Assur literarischen Inhalts herausgegeben von Stefan M. Maul, Band 1 (= WVDOG 116), Wiesbaden 2007.

22 U. Koch-Westenholz, Mesopotamian Astrology: An Introduction to Babylonian and Assyrian Celestial Divination (= CNI Publications 19), Kopenhagen 1995; F. Rochberg-Halton, The Heavenly Writing: Divination, Horoscopy, and Astronomy in Mesopotamian Culture, Cambridge 2004.

23 U. Koch-Westenholz, Babylonian Liver Omens. The Chapters Manzāzu, Padānu and Pān tākalti of the Babylonian Extispicy Series Mainly from Aššurbanipal's Library, Kopenhagen 2000; dies., Secrets of Extispicy, Münster 2005.

24 S. M. Maul, Zukunftsbewältigung. Eine Untersuchung altorientalischen Denkens anhand der babylonisch-assyrischen Löserituale (Namburbi) (= Baghdader Forschungen 18), Mainz 1994.

Wie dem auch sei: Die grundlegende Überzeugung, dass der gesamte Kosmos auf den Menschen ausgerichtet sei und mit ihm interagiere, beruhigte sich im Alten Orient einerseits in der scheinbar braven Vorstellung von Göttern, die dem Menschen mit Vorzeichen gnädig Leitung geben, obgleich der Mensch sich letztlich göttlichem Willen zu beugen hat. Andererseits aber beflügelte sie einen im Alten Orient nie zur Ruhe gekommenen Forschergeist, der sich zum Ziel gesetzt hatte, die Gesetzmäßigkeiten der Zeichenhaftigkeit der Welt offenzulegen und in ganz unterschiedlichen Systemen wiederzuerkennen.

Im frühen ersten vorchristlichen Jahrtausend hatte sich neben der Lehre von der Eingeweideschau die Lehre von der Bedeutung astraler Zeichen so weit entfaltet, dass babylonische und assyrische Könige sich ihrer zu politischen Zwecken systematisch bedienten.[25] Denn der gestirnte Himmel, der Nacht für Nacht, ganz anders als die Eingeweideschau, *unerbeten* Zeichen hervorbrachte, stellte in Aussicht, in Permanenz Auskunft über Kommendes geben zu können. In neuassyrischer Zeit wurde deshalb ganz Mesopotamien mit einem Netz von Beobachtungsstationen überzogen, nicht nur, weil, wenn hier der Himmel mit Wolken verhangen war, doch noch dort der klare Sternenhimmel beobachtet werden konnte, sondern auch um von ein und demselben astralen Ereignis gleich mehrere Meldungen zu erhalten. Die jeweils unabhängig voneinander geschriebenen Berichte mussten an den Königshof zu Ninive gesandt werden und waren eine Garantie dafür, dass man am Hof nicht böswilligen Falschmeldungen der Astrologen aufsaß. Um Betrug und Irrtümer zu vermeiden, wurden die aus dem ganzen Reich eintreffenden Berichte gesammelt, miteinander verglichen und von den besten Fachleuten unter divinatorischen Gesichtspunkten ausgewertet.[26]

Vom Himmel als dem Abbild der weiten Erde wurden dabei nicht etwa Aussagen über den Einzelnen, sondern über das gesamte Staatswesen, ja sogar Auskünfte globaler Natur erwartet, die auch Prognosen über das Geschick und sogar über geheime Planungen der benachbarten Feindesländer zuließen. Aus diesem Grund war die Astrologie im Alten Orient von höchstem politischem Interesse und eine wohlgehütete Kunst. Denn sie schien imstande zu sein, durch die Auswertung von scheinbaren Unregelmäßigkeiten astraler Bewegungen, Gefahren, aber auch günstige Gelegenheiten so rechtzeitig wahrzunehmen, dass man einerseits ein drohendes Unheil umgehen und andererseits auch von der sich nur kurz auftuenden Chance eines gebotenen *kairós* profitieren konnte.

25 Zusammenfassend S. M. Maul, Art. „Omina und Orakel. A. In Mesopotamien", in: RlA 10, 51–57; H. Hunger, Astrological Reports to Assyrian Kings, (= State Archives of Assyria 8), Helsinki 1992; U. Koch-Westenholz, Mesopotamian Astrology: An Introduction to Babylonian and Assyrian Celestial Divination (= CNI Publications 19), Kopenhagen 1995; D. Brown, Mesopotamian Planetary Astronomy Astrology (= Cuneiform Monographs 18), Groningen 2000; F. Rochberg-Halton, The Heavenly Writing: Divination, Horoscopy, and Astronomy in Mesopotamian Culture, Cambridge 2004.

26 A. L. Oppenheim, Divination and Celestial Observation in the Last Assyrian Empire, Centaurus 14/1 (1969), 97–135; H. Hunger, Astrological Reports to Assyrian Kings (= State Archives of Assyria 8), Helsinki 1992.

Zahlreiche keilschriftliche Traktate, namentlich aus dem 1. vorchristlichen Jahrtausend zeigen, dass mesopotamische Zeichendeuter bestrebt waren, ihre Erkenntnisse über die Aussagekraft von dem äußeren Erscheinungsbild der Leber zu verbinden mit ihrem Wissen über die Bedeutung der Bewegungen am gestirnten Himmel.[27] Ihre noch weitgehend unerforschten Überlegungen führten nicht nur dazu, dass die Leber in gewisser Weise als Emanation des Himmels betrachtet und wie der Tierkreis in zwölf Kreissegmente unterteilt wurde. Die mesopotamischen Gelehrten sahen sich auch in der Lage, Zeichen der Leber gleichbedeutenden astralen und terrestrischen Zeichen zuzuordnen[28] und damit die Gesetzmäßigkeiten von der Dynamik des Weltgeschehens in unterschiedlichen Medien der altorientalischen Divination offenzulegen. Sie sammelten in ihrem Forschergeist dabei nicht nur Zeichen, um auf Zukünftiges zu schließen, sondern betrachteten auch das zur Gegenwart gewordene Zukünftige, um in der Vergangenheit nach den zugehörigen, möglicherweise übersehenen Zeichen Ausschau zu halten. In einem kühnen Vorhaben sollten auch die über Jahrhunderte geführten sog. „*astronomical diaries*"[29] langfristig über das Netz kausaler Zusammenhänge in der Welt genaueren Aufschluss geben. In diesen uns mit Lücken aus der Zeit vom 7. bis zum 1. vorchristlichen Jahrhundert überlieferten Dokumenten wurde in Form von Jahresberichten nicht nur über astrale Zeichen und das Wetter Rechenschaft abgelegt, sondern auch über die Preisentwicklung bestimmter ökonomisch wichtiger Güter, über Wasserstände und bemerkenswerte, als ominös eingestufte terrestrische Vorkommnisse, um Koinzidenzen mit den ebenfalls sorgsam verbuchten einschneidenden zeitgeschichtlichen Ereignissen zu erfassen. Ziel war es, Gesetzmäßigkeiten im Weltgeschehen zu ermitteln, um diese Erkenntnisse für politisches Handeln nutzbar zu machen. Im Dienste dieses Strebens entstand im Babylonien des ersten vorchristlichen Jahrtausends auch die rechnende Astronomie, ein Zweig babylonischer Wissenschaft, der bis heute fortbesteht.[30]

Mit Hilfe der verschiedenen altorientalischen divinatorischen Verfahren sollte erreicht werden, dass Gestalten und Handeln der für das Gemeinwesen Verantwortlichen stets in Harmonie blieb mit dem alles umfassenden Weltgeschehen, welches den ganzen Kosmos beherrscht und dem sich der Mensch deshalb nicht dauerhaft widersetzen kann.

27 U. Koch-Westenholz, Secrets of Extispicy, Münster 2005, 30–31; N. P. Heeßel, Astrological Medicine in Babylonia, in: A. Akasoy, C. Burnett, R. Yoeli-Tlalim (Hgg.), Astro-Medicine: Astrology and Medicine, East and West, Florenz 2008, 1–16.

28 E. von Weiher, Uruk, Spätbabylonische Texte aus dem Planquadrat U 18, Ausgrabungen in Uruk Warka. Endberichte herausgegeben von Rainer Michael Boehmer 12, Mainz 1993, Text Nr. 159; E. Reiner, Astral Magic in Babylonia (= Transactions of the American Philosophical Society 85/4), Philadelphia 1995, 78.

29 H. Hunger – A. Sachs, Astronomical Diaries and Related Texts from Babylonia, I: Diaries from 652 B.C. to 262 B.C. [1988]; II: Diaries from 261 B.C. to 165 B.C. [1989]; III: Diaries from 164 B.C. to 61 B.C. [1996]; V: Lunar and Planetary Texts [2001]; VI: Goal Year Texts [2006], Wien 1988–2006.

30 O. Neugebauer, A History of Ancient Mathematical Astronomy, Berlin 1975; H. Hunger – D. Pingree, Astral Sciences in Mesopotamia (= Handbuch der Orientalistik, Abteilung 1, 44), Leiden 1999.

Die nahezu drei Jahrtausende währende politische und kulturelle Überlegenheit Mesopotamiens ließ einen nachhaltigen Erfolg der uralten Zukunftswissenschaft nicht nur in der Selbstwahrnehmung des Zweistromlandes, sondern auch in den Augen der umliegenden Kulturen unbestreitbar erscheinen. Diesen Eindruck dürfte der nicht unerhebliche Aufwand, der mit den Verfahren der Zukunftsdeutung verbunden ist, ebenso verstärkt haben wie die für alle sichtbare Einbettung der Divination in eine Art wissenschaftliches System, das hohe Fachkompetenz und Professionalität mit dem Umstand zusammenführte, dass die divinatorischen Gutachten ganz offensichtlich unabhängig von Personen und Interessen erstellt wurden.

Aus all diesen Gründen war man überzeugt, mit der Divination ein Mittel in der Hand zu halten, das in hohem Maße Stabilität und Prosperität garantierte, zumindest aber dauerhaft Schutz vor fatalen Fehlentscheidungen lieferte und damit der eigenen Kultur anderen gegenüber einen erheblichen und nachhaltigen Vorteil verschaffte.

Es ist nur folgerichtig, dass die assyrischen und babylonischen Könige versuchten, Wissen und Techniken der Zukunftsschau für sich zu monopolisieren und die besten Fachleute an sich zu binden.[31] Selbst im internationalen Wettstreit wurde die Kenntnis divinatorischer Traktate als so hoch eingestuft, dass im Kriegsgeschehen Tontafeln entsprechenden Inhalts gar auf ausdrücklichen königlichen Befehl hin geraubt und in die eigenen Bibliotheken gebracht wurden.[32]

Das Wissen, das über viele Jahrhunderte in den Familien der Zeichendeuterdynastien erworben, weiterentwickelt und überliefert worden war, wurde im ausgehenden 2. und frühen 1. Jahrtausend v. Chr. an den Königshöfen Babyloniens und Assyriens gesammelt, systematisiert und in sehr umfangreichen, rasch den Charakter des Kanonischen annehmenden Textausgaben zusammengestellt. Der Motor für die Verschriftlichung divinatorischer „Handbücher" und die damit verbundenen höchst bemerkenswerten Redaktionsprozesse dürfte der mit der Komplexität mesopotamischer Herrschaftsstrukturen immer weiter gestiegene königliche Bedarf an divinatorischer Beratung gewesen sein. Die neu entstandenen Editionen bildeten von nun an den königlich autorisierten und verbindlichen Thesaurus geheim gehaltenen divinatorischen Fachwissens, auf den sich die Fachleute im Dienste des Königs zu berufen hatten. Das im 2. vorchristlichen Jahrtausend noch in Familien und Schulen gehütete divinatorische Fachwissen war auf diese Weise fast ganz zentralisiert in königliche Oberhoheit gelangt.

Die Zeichendeuter, die für König und Staat tätig waren, galten als wichtige Geheimnisträger und hatten zu beeiden, dass sie die möglicherweise politisch brisante Kenntnis,

31 B. Pongratz-Leisten, Herrschaftswissen in Mesopotamien: Formen der Kommunikation zwischen Gott und König im 2. und 1. Jahrtausend v. Chr. (= State Archives of Assyria Studies 10), Helsinki 1999.

32 W. G. Lambert, Three Unpublished Fragments of the Tukulti-Ninurta Epic, Archiv für Orientforschung 18 (1957/58), 44; siehe auch S. Parpola, Assyrian Library Records, Journal of Near Eastern Studies 42 (1983), 1–29.

die sie durch ihre Tätigkeit notgedrungen erwarben, unter keinen Umständen an Dritte weitergaben.[33]

Im neuassyrischen Reich des 1. vorchristlichen Jahrtausends war das Divinationswesen so gut organisiert, dass Meldungen über ominöse Vorkommnisse, vor allem astraler, aber auch terrestrischer Natur, regelmäßig aus dem gesamten Herrschaftsgebiet eingingen.[34] Darüber hinaus wurde der Verlauf aller Eingeweideschauen, die von politischer Relevanz waren, sorgsam protokolliert.[35] Die einander ergänzenden divinatorischen Berichte und Gutachten wurden von einer Kommission, die man ein wenig anachronistisch als „Zukunftsministerium" bezeichnen könnte, gesammelt, miteinander abgeglichen, auf Stimmigkeit überprüft und ausgewertet, bevor daraus resultierende politische Maßnahmen beraten, möglicherweise einer neuerlichen divinatorischen Prüfung unterzogen und eingeleitet wurden.

Obgleich aus dem Blickwinkel unseres eigenen Weltbildes die Grundlagen der hier vorgestellten divinatorischen Evaluationsverfahren vollkommen obsolet sind, erweisen sich die mesopotamischen Formen der politischen Entscheidungsfindung ohne Zweifel als erfolgreich – führten sie doch dazu, dass die Kulturen des Zweistromlandes über einen Zeitraum von mehr als drei Jahrtausenden den gesamten Vorderen Orient politisch und kulturell dominierten. So erscheint es uns in dem gleichen Maße beunruhigend wie skandalös, dass ein nach unseren eigenen Maßstäben durch und durch unsinniges Verfahren dauerhaften Erfolg gewährleistet haben soll. Diesem im Grunde nur schwer erträglichen Widerspruch sollen die folgenden Überlegungen gewidmet sein, die ich in fünf Punkten zusammenfasse:

1. Zunächst müssen wir – angesichts der hier ausgebreiteten historischen Evidenzen – ein wenig beunruhigt Folgendes konstatieren: Das Vertrauen in Aussagekraft und Verlässlichkeit prognostischer Verfahren kann ganz offensichtlich auch dann über lange Zeiträume unerschüttert erhalten bleiben, wenn diese Verfahren in Wahrheit keinerlei Einblicke in Zukünftiges zu liefern in der Lage sind, solange nur zwei Bedingungen erfüllt sind:

   1.) müssen ökonomische, politische, militärische und andere Erfolge von einer überzeugten Mehrheit nicht unwesentlich auf Vorteile zurückgeführt werden, die auf der Kenntnis jener prognostischer Verfahren beruhen; oder anders gesagt: die praktizierten prognostischen Verfahren müssen einer Mehrheit der meinungsbildenden Kräfte als plausibel erscheinen und

33 J.-M. Durand, Archives épistolaires de Mari I (= Archives royales de Mari 26/I), Paris 1988, 13–15; siehe auch S. Parpola, Letters from Babylonian and Assyrian Scholars (= State Archives of Assyria 10), Helsinki 1993, Text Nr. 7.

34 A. L. Oppenheim, Divination and Celestial Observation in the Last Assyrian Empire, Centaurus 14/1 (1969), 97–135; U. Koch-Westenholz, Mesopotamien Astrology, Kopenhagen 1995, 180–185.

35 I. Starr, Queries to the Sungod. Divination and Politics in Sargonid Assyria (= State Archives of Assyria 4), Helsinki 1990.

2.) darf die Anwendung dieser prognostischen Verfahren sinnvolle Entscheidungen zumindest nicht dauerhaft oder mehrheitlich behindern.

Vom frühen 2. Jt. v. Chr. bis zum Ende der Keilschriftkulturen um die Zeitenwende scheinen im alten Zweistromland diese beiden Bedingungen stets erfüllt gewesen zu sein.

Angesichts des jüngsten Bankencrashs mit immer noch unüberschaubaren Folgen muss man sich in der Tat ernsthaft die Frage stellen, ob das hier für den Alten Orient Beobachtete nicht ebenso für unsere eigene Zeit gelten muss, in der eine überzeugte Mehrheit nach wie vor unerschütterliches Vertrauen in die prognostischen Fähigkeiten jener Wirtschaftswissenschaftler und Analysten setzt, die mit ihren Methoden die fatale Wirtschaftsentwicklung mehrheitlich nicht voraussahen und doch mit ihren erneuten Prognosen, ohne dass dies ernsthaft kritisiert wird, die Wege zeichnen, die aus der Krise wieder herausführen sollen ...

2. Dem Nachteil, einen möglicherweise sinnvollen Plan nicht durchsetzen zu können, weil die in einem divinatorischen Verfahren eruierten Zeichen dagegen sprachen, stehen die immensen Vorteile gegenüber, die die Billigung eines Vorhabens durch ein divinatorisches Gutachten mit sich bringt. Denn die Wirkung eines solchen Gutachtens darf keinesfalls unterschätzt werden. Unter der Maßgabe, dass das divinatorische Evaluationsverfahren mehrheitlich als plausibel und ein Orakelentscheid als das verlässliche Urteil von über dem Menschen stehenden Mächten wahrgenommen wird, kann ein positiver Orakelentscheid politisches Wollen und Handeln nämlich überzeugend rechtfertigen, indem er dieses als im Einklang mit dem Kosmos und in der Gunst der göttlichen Mächte stehend ausweist. Göttliches Wohlwollen und sich daraus ergebende Erfolge und Vorteile sind dadurch greifbar in Aussicht gestellt. Gelingt es, mittels eines divinatorischen Befundes nicht nur einen umfassenden Konsens zu schaffen, sondern auch das von Vielen getragene Bewusstsein des „Gott mit uns", führt dies zwangsläufig auf allen wichtigen gesellschaftlichen Ebenen zu Optimismus und Selbstsicherheit, zu Entschlusskraft und Handlungsbereitschaft, die ihrerseits eine tragfähige Grundlage bilden für ein beherztes Angehen von sich abzeichnenden Problemen. So wie auch heute noch wissenschaftliche Gutachten politische und ökonomische Entscheidungen Weniger die Gestalt einer letztlich alternativlosen Notwendigkeit annehmen lassen können, sorgte in den altorientalischen Gesellschaften die „Zukunftswissenschaft" dafür, dass folgenreiche Entscheidungen nicht als einsame Beschlüsse einiger Mächtiger wahrgenommen werden konnten, sondern als ein von Göttern und Kosmos legitimiertes „Muss".

3. Im Fall eines durch Zeichen ermittelten deutlich negativen Evaluationsbescheides, galt es einen als „unerwünscht" eingestuften Plan zu überdenken. Hierdurch entstanden Möglichkeit und Notwendigkeit, in den Gremien politischer Entscheidungsfindung trotz eines bereits gefassten Beschlusses Für und Wider erneut zu diskutieren. Es ist nicht unwahrscheinlich, dass dabei Positionen, die zuvor nicht konsensfähig gewesen waren, abermals

erörtert, und dann in der Form eines revidierten Beschlusses Gegenstand einer erneuten Orakelanfrage wurden. Die uns erhaltenen, mit einer Eingeweideschau verbundenen Anfragen sind oft regelrechte Kunstwerke, die ein Vorhaben mit detaillierter Auflistung der einzelnen Schritte benennt. Wurde ein in der Orakelanfrage beschriebenes Vorhaben abschlägig beschieden, hieß das keineswegs zwangsläufig, dass die gesamten Planungen aufgegeben werden mussten. Es bestand die Möglichkeit, einen leicht modifizierten Plan der erneuten Prüfung zu unterziehen. Wurde dieser dann in einem zweiten divinatorischen Prüfungsverfahren positiv evaluiert, konnte man daraus folgern, dass das Detail des Plans, das man in der zweiten Anfrage geändert hatte, Grund der Ablehnung in dem ersten Prüfungsdurchlauf gewesen war. Es ergibt sich hieraus fast notwendigerweise, dass gegebenenfalls diejenigen Segmente eines Vorhabens in einer zweiten Anfrage modifiziert wurden, die bereits anfangs bei der Ausarbeitung des Plans strittig gewesen waren.

Die divinatorische Evaluation, die eine über allen Menschen stehende Autorität bemüht, eröffnet so an ganz ungeahnter Stelle einen Raum für sachbezogene Diskussionen, der weitgehend frei von den Zwängen hierarchischer Strukturen ist. Zunächst überraschend sind es so ausgerechnet divinatorische Verfahren, die einen Interessensausgleich durch Verhandeln ermöglichen. Der in unserer eigenen Kultur gepflegte Mythos vom Ursprung der Demokratie in der griechischen Polis verstellt allzu leicht die Einsicht, dass eine Kultur des Aushandelns nicht zwangsläufig an die Agora und an die uns vertrauten Formen und schon gar nicht an Aufklärung und Säkularisierung gebunden ist.

4. Im theistischen Weltbild Mesopotamiens ist eine durch Divination ermittelte Wertung von Plänen und Vorhaben zusätzlich in einen „Tun-Ergehen-Zusammenhang" gestellt, der Zustimmung und Ablehnung von Vorhaben als ein sich abzeichnendes Lohnen und Strafen durch über dem Menschen stehende Mächte deutet. Es ergibt sich daraus eine zweifache Verantwortlichkeit der Regierenden, die sich nicht nur Menschen gegenüber, sondern auch den über ihnen stehenden Mächten zu rechtfertigen hatten. Aus diesem Grunde hatten sie mit ihrem politischen Handeln nicht zuletzt auch der in religiösen Kontexten nachdrücklich erhobenen Forderung nach sozialer Gerechtigkeit nachzukommen. Wurden mit divinatorischen Mitteln drohende Gefahren und damit auch ein grundsätzlicher Gotteszorn diagnostiziert, musste im engeren Umfeld eines Königs auch dessen rituelles und persönliches Fehlverhalten sowie andere Handlungsweisen zur Sprache gebracht werden, die die Götter verstimmt haben könnten. Zahlreiche Texte zeigen uns, dass gerade der König dabei nicht selten mit der Erkenntnis konfrontiert wurde, Schuld auf sich geladen und die Götter herausgefordert zu haben. Auch wenn uns einschlägige Quellen aus naheliegenden Gründen fehlen, zeigt dies, dass die Diskussion einer ungünstigen Zukunftsprognose sogar einen Freiraum schuf, in dem ein enger Kreis die Rechtmäßigkeit königlichen Handelns hinterfragen konnte.

5. Die Astrologie, die in Permanenz *unerbetene* Zeichen und in der Folge in Permanenz Prognosen generiert, zwingt dazu, die Prognosen permanent mit der Gegenwartssituation abzugleichen und die Gegenwart am Prognostizierten zu messen. In diesem Sinne erweist sich die Astrologie als ein Instrument, das nach permanenter Reflexion politischen Handelns verlangt und so eine Atmosphäre politischer Wachsamkeit hervorbringt. Die auf Himmelsbeobachtung beruhenden Prognosen, die für den neuassyrischen Königshof erstellt wurden, betrafen die innere und äußere Sicherheit des Landes, oft auch Ernteaussichten und die Versorgungssituation. Es liegt in der Natur der Sache, dass Visionen von Sicherheit und Bedrohung eines Landes nicht diskutiert, ja nicht einmal gedacht werden können, ohne dass das Prognostizierte mit dem Gegenwärtigen verbunden würde. Denn aus diesem würde sich ja das Zukünftige entfalten. Prognostiziertes Versagen und Unterliegen zwingt ohne notwendiges Besehen der verantwortlichen Personen und ihres Einflusses zu einer immer wieder neuen Kontrolle der inneren und äußeren Sicherheit, des Zustandes von Militär und Sicherheitskräften, der Vertrauenswürdigkeit von Beratern und Verbündeten, der Versorgungssituation des Landes und vieler anderer Bereiche. Selbst Persönlichkeiten von höchstem Einfluss konnten sich einer divinatorisch motivierten Kontrolle ihres Verantwortungsbereichs nicht entziehen. In diesem Sinne ist die permanente astrologische Analyse des zu Erwartenden in der Tat ein, wie die Texte sagen, „Wachdienst für den König“. Sie erfüllt in gewissem Sinne die Funktion eines politisch-gesellschaftlichen Frühwarnsystems, in dem Aufmerksamkeit auf Fehlentwicklungen schon im frühen Stadium gelenkt werden kann, noch bevor sich schlimme Konsequenzen entwickelt haben.

Die voranstehenden Überlegungen dürften gezeigt haben, dass es unklug und unsachgemäß wäre, die Divination des Alten Orients als Aberglaube und eine kulturgeschichtliche Fehlentwicklung abzutun. Sie erweist sich nämlich, so wie hier gezeigt wurde, in ihrem Kontext als eine ausgesprochen wirkungsvolle Institution, die in der Lage war, der Zukunft eine konkrete Gestalt zu geben, erhebliche Freiräume für das besonnene Aushandeln wichtiger politischer Entscheidungen zu eröffnen und Konsens zu schaffen, um zielstrebiges Handeln zu ermuntern.

*Mark J. Geller*

# Die theoretische Grundlage der babylonischen Heilkunde[1]

Es ist nahezu allgemein anerkannt, dass wissenschaftliche Theorie und Epistemologie die große Entdeckung griechischer Philosophen gewesen sei. Thales, Euklid, Hippokrates, Aristoteles und viele andere Philosophen führten demzufolge eine intellektuelle Revolution herbei, die die Basis für die moderne Naturwissenschaft schuf. Die babylonischen Gelehrten werden demgegenüber gänzlich anders bewertet: Sie hätten kaum theoretisch gedacht, vielmehr meistens praktisch und spezifisch ohne Zuhilfenahme von Verallgemeinerungen oder Regeln spekuliert.[2] Im medizinischen Bereich haben die Babylonier z.B. keine Theorie der Körpersäfte wie die Griechen, sondern führten Krankheiten auf Dämonen und magische Ursachen zurück.

Es ist wahr, dass babylonische Gelehrte kein theoretisches „Oeuvre" und keine eigenen philosophischen Texte hinterlassen haben. Es ist aber kaum vorstellbar, dass ein kompliziertes System wie die babylonische Medizin vollständig ohne Theorie funktioniert haben soll. Im Folgenden soll daher eine Annäherung an jene babylonische Epistemologie und Theorie der Heilkunde versucht werden.

Um eine antike „Theorie" einer „primitiven" (d.h. nicht technologischen) Naturwissenschaft zu rekonstruieren, müssen meiner Auffassung nach drei intellektuelle Vorbedingungen vorhanden sein: **Phantasie**, **Deduktive Logik** und **Beobachtung**.

1) **Phantasie**: Antike Naturwissenschaft suchte nach einer allgemeinen Erklärung der Natur mit dem Ziel, die Wahllosigkeit oder Zufälligkeit (‚Random character') der Umwelt zu vermindern. Es gibt eine große Vielfalt von möglichen Ereignissen im Universum, von der Bewegung der Gestirne, Veränderungen des Wetters und des Klimas, infinite Variationen in der Tier- und Pflanzenwelt, usw. Die Gelehrten suchten dafür eine Erklärung, wobei viele Phänomene relativ wenigen Ursachen zugeschrieben wurden, um – wie bereits angesprochen – die Zufälligkeit des Universums zu reduzieren.

2) **Deduktive Logik**: Die babylonische Logik ist anders geartet als die griechische, die empirisches Denken bevorzugte. Die Griechen untersuchten alle Daten, um eine einfache Regel zu finden. Babylonische Epistemologie hingegen bediente sich der Deduktion und leitete aus vielen Einzelheiten spezielle Schlüsse ab, verwiesen sei hier auf die bereits gut erforschten Protasis- und Apodosis-Sätze der divinatorischen Literatur. Bemerkenswert ist in diesem Zusammenhang, dass die Systematik der Omina die zahllosen Möglichkeiten der Zufälligkeit reduziert. Wenn eine Verbindung zwischen z.B. der Erscheinung eines

1 Vgl. M. J. Geller, Ancient Babylonian Medicine, Theory and Practice, Chichester 2010.
2 So behauptet H.-G. Gadamer, The Enigma of Health, Cambridge/MA 2004, 4.

Sterns und einem Vorfall auf der Erde vermutet wurde, dann stand dieser Zusammenhang musterhaft für weitere erklärungsbedürftige Ereignisse. Die babylonischen Gelehrten unterstellten der Natur eine gewisse Struktur anstelle einer willkürlichen und zufälligen Umwelt. Sie haben ihre Schlussfolgerungen aus den Omina als ihre „Datensammlung" betrachtet. Aus moderner Perspektive ist dieses System ein logischer Fehler oder ein Irrtum (‚post ergo propter hoc').[3] Wir sind heute sicher, dass z. B. Donner im Himmel nichts mit der Zukunft der Regierung zu tun hat. Immerhin ist diese Art babylonischen Denkens jedoch wissenschaftlich und methodisch bedeutungsvoll, weil mit großer Aufmerksamkeit ein Beziehungsgefüge von verschiedenen Gegebenheiten und bestimmten Folgen untereinander festgestellt wird.

3) **Beobachtung**: Babylonische Gelehrte sammelten eine ganze Menge an Daten und Angaben, z. B. in einer relativ primitiven Klassifikation von Gegenständen, Pflanzen, Steinen, Sternen, Götter, Krankheiten, usw., daneben erstellten sie Listen von allerlei Ereignissen in der Welt, von Gestirnsbewegungen und Vogelflug, Klima, Eigenschaften von Tierlebern und vielen anderen Gegebenheiten. Klassifikation und Taxonomie waren somit weitere Methoden, um eine gewisse Ordnung in die chaotische Umwelt zu bringen.[4]

Ein Großteil des naturwissenschaftlichen Denkens bezog sich auf Beobachtungen von Aktion und Reaktion in der Natur, obwohl der antike Mensch keine technologischen Instrumente zur Hand gehabt hat, abgesehen von einfachen Messinstrumenten, um eine aktuelle Kausalverbindung festzustellen.

Diese drei intellektuellen Vorbedingungen (**Phantasie, Deduktive Logik, Beobachtung**) bildeten meiner Meinung nach die intellektuelle Grundlage der antiken Naturwissenschaft insgesamt, und zwar in Babylon selbst, aber auch in Kos oder Pergamon. Diese drei Grundlagen wurden im Rahmen der babylonischen Medizin aktiv benutzt, allerdings auf völlig andere Weise. Die nachfolgend beschriebenen Grundlagen der babylonischen Medizin, die sich den bisher genannten adäquat zuordnen lassen, können als ein Fundament einer babylonischen „Epistemologie der Heilkunde" angesehen werden.

1) In der babylonischen Medizin wurde die Rolle der Phantasie eigentlich von der **Magie** übernommen. Das Problem der Zufälligkeit, das bereits erwähnt wurde, ist ein sehr offensichtlicher Aspekt der babylonischen Heilkunde, wie auch in jedem anderen Heilkundesystem.[5] Magie bietet eine Lösung an. Warum bin ich krank und mein Nachbar nicht? Warum krank im Kopf? Magen? Fuß? Und nirgendwo anders? Warum diese Krankheit und keine andere? Welche Gründe liegen jeder Krankheit zugrunde? Eine zahllose Anzahl von Möglichkeiten kann durch Magie drastisch reduziert werden: Die Götter bestimmen

3 Vgl. F. Rochberg, The Heavenly Writing, Cambridge 2004, 257–259.

4 Vgl. N. Veldhuis, Elementary Education at Nippur, the Lists of Trees and Wooden Objects, Groningen 1997, 7–8.

5 Vgl. H. Sigerist, Primitive and Archaic Medicine, New York 1967, 411ff., und siehe H.-G. Gadamer, The Enigma of Health, Cambridge/MA 2004, 107; laut Gadamer, fragt der deutsche Arzt typischerweise: „Na, wo fehlt's denn?" (ebd., 74).

alle Ursachen des menschlichen Schicksals. Die babylonischen Gelehrten sind auf uralte magische Ideen angewiesen, die erklären, welche Krankheiten durch Dämonen oder wütende Götter verursacht sind, oder die auf speziellere Gründe (z. B. das Essen von verzauberten Speisen) zurückgeführt werden können. Diese allgemeinen Erklärungen bildeten den Hintergrund der babylonischen Heilkunde. Magische Erklärungen boten den Rahmen für eine nützliche allgemeine Theorie von Krankheiten, die der Patient gern annehmen konnte. Es entbehrt nicht einer gewissen Ironie, dass im Vergleich mit den griechischen „Körpersäften" die babylonische Idee von Dämonen als Krankheitsträger viel näher an die modernen Entdeckungen von Viren und Bakterien heranreichen.

Die Beschreibung der Ursächlichkeit von Krankheiten blieb innerhalb der griechischen Medizin sehr umstritten; gegen die Hippokraten und andere philosophische Schulen waren die Methodisten, die jede Untersuchung der Ursache von Krankheiten vermieden, weil sie ein effektives Rezept als viel wertvoller ansahen als philosophische Spekulationen über Krankheiten und Ursachen.[6] Die Methodisten schätzten die Grenzen der Wissenschaft richtig ein, indem sie philosophischen Theorien über Krankheitsgründe wenig praktischen Nutzen für die Heilung eines Patienten zubilligten. Die Methodisten stellten tatsächlich eine Version der babylonischen Medizin im zweiten Jahrhundert vor Christus der graeco-römischen Welt vor.

2) Zweitens ist die Rolle der **Logik** und des analytischen Denkens in der babylonischen Medizin sehr auffällig, allerdings im medizinischen Bereich anders als bei den Omina. In der Medizin gab es eine andere Methode: die Symptome vieler Krankheiten wurden in Form von Omina in Listen eingetragen. Hunderte Symptome sind in dem sogenannten „Handbuch der Diagnose" als prognostische Omina verzeichnet: Wird der Patient weiter leben oder sterben? Wie lange? Nach drei Tagen? Vier Tagen? Einem Monat?[7] In diesem Fall ist die Verbindung zwischen Protasis und Apodosis viel realistischer als bei anderen Omina. Nach der Behandlung von Patienten, hat der Arzt Erfahrung gesammelt, wie lange eine bestimmte Fieberart andauert oder wie gefährlich eine spezielle Infektion ist.[8] Deshalb gilt das allgemeine „Problem des logischen Fehlers" der Omina – post hoc ergo propter hoc – eigentlich nicht für medizinische Omina des „Handbuch der Diagnose": die Verbindung zwischen Annahme und Folgerung reflektiert eine aktuelle Realität, genauer: was nach der Erscheinung eines Zeichens, eines Symptoms, eines Omens passieren wird. Im Gegensatz zu den allgemeinen Omina wurden in der Medizin reale Aktionen und Reaktionen bzw. **Kausalverbindungen** zwischen Krankheit als Ursache und Symptom als Zeichen beobachtet.

6 V. Nutton, Ancient Medicine, London 2004, 187ff, 191; M. Tecusan, Fragments of the Methodists, Methodism Outside Soranus, Leiden 2004, 7ff.

7 N. Heeßel, The Hand of the Gods. Disease Names and Divine Anger, in: I. L. Finkel/M. J. Geller, Disease in Babylonia, Leiden 2007, 121–130.

8 M. Stol, Fevers in Babylonia, in: I. L. Finkel/M. J. Geller, Disease in Babylonia, 1–39, vgl. J. V. Kinnier Wilson/E. H. Reynolds, On Stroke and Facial Palsey in Babylonian Texts, in: ebd., 67–97.

3) Das bringt uns zu unserer dritten Grundlage: **Beobachtung** als Funktion der Epistemologie. Weil die antiken Gelehrten außer einfachen Messinstrumenten keine weiteren Mittel hatten, waren sie völlig auf Beobachtungen angewiesen. Neben der Beobachtung spielte im Rahmen der Medizin ein weiterer Faktor eine Rolle: die **Erfahrung**; besser wäre vielleicht das französische Wort **experiment**, was „Erfahrung" und „Probe" bedeutet.[9] Wie funktionierte diese Kombination in der Medizin?

Anatomie war ein Problem in der antiken Heilkunde, weil weder die Babylonier noch die Hippokraten vor dem dritten Jahrhundert vor Christus Dissektionen oder Autopsien durchgeführt haben.[10] Die babylonischen und griechischen Gelehrten wussten kaum etwas von der inneren Anatomie des Menschen, abgesehen von analogen Informationen, die von untersuchten Tierkörpern auf Menschenkörper übertragen wurden. Ägypten ist eine Ausnahme, weil hier die Mumifikation eine ziemlich genaue Kenntnis der menschlichen Anatomie voraussetzt. Daraus ist aber nicht abzuleiten, dass diese Kenntnisse überall geschätzt waren.[11] Aus dem medizinischen Textkorpus von Ägypten kann nicht festgestellt werden, ob die Trauerarbeiter die anatomische Information an die Ärzte weitergaben. Im Großen und Ganzen hatten antike Gelehrte kein richtiges Verständnis der menschlichen Anatomie. Immerhin erwarben sie viele Kenntnisse aus Erfahrungen und Beobachtungen. Die Heiler hatten bereits eine Idee von Aktion und Reaktion, d.h. welche Veränderungen im Körper in den verschiedenen Stadien der Krankheit auftraten.

Beobachtung und Erfahrung sind noch in einem weiteren Bereich sind wichtig: bei der Therapie, d.h. bei der Nutzung von Pflanzen und Drogen in babylonischen Rezepten. Die Hauptfrage ist hierbei, wie weit die babylonischen Gelehrten damals die Wirkungen von Pflanzen gekannt haben. Es gab zu dieser Zeit kein Labor und auch keine klinischen Proben. Sie studierten aber bereits Pflanzen und Mineralien, zählten sie auf und teilten sie ein. Desweiteren beschrieben die Gelehrten Pflanzen und Mineralien mit einem Klassifikationsprinzip, das akkadisch *šiknu* heißt (*šiknu* – die „Natur" oder der „Charakter" der Pflanze oder des Steins – ist ... usw.). Das Wort ist bedeutungsvoll, weil die Griechen das Wort *phusis* („Natur") in einem vergleichbaren Sinn benutzten. Es bleibt jedoch das Problem bestehen, wie die alten Babylonier die Wirkung von Pflanzen und Steinen entdeckten. Wie konnten sie wissen, dass z.B. eine gewisse botanische oder chemische Eigenschaft eine bestimmte Wirkung im menschlichen Körper hatte?

9 Für „Erfahrung" in der hippokratischen Heilkunde, siehe R. French, Medicine before Science, Cambridge 2003, 37, und H.-G. Gadamer, The Enigma of Health, Cambridge/MA 2004, 2.

10 Siehe V. Nutton, Ancient Medicine, London 2004, 131f., und H. von Staden, The Art of Medicine in Early Alexandria, Cambridge 1998.

11 Ägyptisch *wet* bedeutet „jemand, der verbindet", aber vielleicht kein Arzt, sondern ein Einbalsamierer; die Mumifizierung war mehr eine Kunst der Einbalsamierung als eine Dissektion; vgl. J. Nunn, Ancient Egyptian Medicine, London 1996, 133: „It is not entirely clear whether the word *wet* refers only to an embalmer or whether it can also be used to mean one who bandages the living".

Dass die babylonischen Gelehrten ein Wissen über verschiedene Wirkungen gehabt haben müssen, zeigt der eindeutige Unterschied zwischen einer „simplicia", d.h. einem einfachen, und einem komplexen Rezept im akkadischen therapeutischen Textkorpus. Es ist beides zu finden: einfache Rezepte (eine Droge oder Pflanze bei einer Krankheit) und sehr komplizierte Rezepte mit allerlei Kombinationen von Pflanzen, Mineralien und anderen Zutaten in einer vielseitigen Droge oder Panacea, die bei einer oder mehrerer Krankheiten verabreicht wurde. Wie ist das zu erklären?

Tatsächlich sind die babylonischen Rezepte heutzutage weit gehend unverständlich, weil aus Mangel an Identifikation der Stoffe nur wenige Zutaten nachvollzogen werden können. Außerdem ist die Dosierung in den Rezepten völlig unklar. Im Moment erscheint die Möglichkeit einer Taxonomie babylonischer Drogen hoffnungslos, im Gegensatz zu den Forschungen des Dioskurides oder des Theophrastus. Immerhin ist ersichtlich, dass die Babylonier einen Unterschied zwischen Rezepten bei verschiedenen Krankheiten machten. So benutzten sie beispielsweise für Afterkrankheiten viel mehr Öl, für Nierenkrankheiten viel mehr Mineralien, und für Augenkrankheiten viel mehr Salben. Die Verwendung sowohl von „simplicia" als auch von komplexen Rezepten spiegelt eine bestimmte Einstellung zur Heilkunde wider, obwohl deren Grundlagen heute nicht erklärbar sind. Das babylonische Wissenssystem war jedoch sehr alt und basierte auf einer langen Überlieferung, die bis in das prähistorische Mesopotamien zurückreicht.

Zusammenfassend lässt sich feststellen, dass die verschiedenen Aspekte der babylonischen Medizin innerhalb der babylonischen Wissenschaft – Phantasie, Logik, Beobachtung und Erfahrung – eine bestimmte babylonische Epistemologie im Rahmen der Medizin aufzeigen, die nicht sehr weit entfernt war von der vor-hippokratischen Medizin. Die überlieferten Texte der alt-babylonischen bis zur hellenistischen Zeit stehen nicht am Anfang dieses Prozesses, sondern am Ende einer sehr langen Entwicklung, in der antike Gelehrte ihre Umwelt studiert haben.

*Hans Neumann*

# Bemerkungen zu einigen Aspekten babylonischen Rechtsdenkens im Spannungsfeld von Theorie und Praxis

Die juristische Praxis im Rahmen der altorientalischen Gesellschaftsentwicklung[1] von der ersten Hälfte des 3. Jts. bis zum ausgehenden 1. Jt. v. Chr. wird durch ein umfangreiches Corpus von Rechtsurkunden unterschiedlicher geographischer Provenienz dokumentiert.[2] Die vornehmlich in sumerischer und akkadischer Sprache überlieferten Urkunden lassen darüber hinaus die hinter den jeweiligen rechtlichen Vorgängen stehenden sozialen Strukturen und Zusammenhänge erkennen,[3] wobei hier auch der entsprechende Archivkontext von Bedeutung ist.[4] Die daraus zu gewinnenden Informationen über die verschiedenen Bereiche und Institute des Privatrechts, über Vertragsformulare sowie über die jeweiligen sozioökonomischen Implikationen werden durch die Analyse der Gerichts- und Prozessurkunden ergänzt, die sowohl prozess- als auch materiellrechtlich entsprechende Einblicke vermitteln und hinsichtlich ihrer sozialhistorischen Aussagefähigkeit gleichfalls von nicht geringer Bedeutung sind.[5]

1 Zur Geschichte des Rechts im Alten Orient vgl. jetzt R. Westbrook (Hg.), A History of Ancient Near Eastern Law (= Handbuch der Orientalistik I/72–1,2), Leiden/Boston 2003, den Überblick bei H. Neumann, Recht im antiken Mesopotamien, in: U. Manthe (Hg.), Die Rechtskulturen der Antike. Vom Alten Orient bis zum Römischen Reich, München 2003, 55–122 und 322–327 (jeweils mit Literatur), und den Sammelband M. Liverani – C. Mora (Hgg.), I diritti del mondo cuneiforme (Mesopotamia e regioni adiacenti, ca. 2500–500 a.C.), Pavia 2008. Vgl. darüber hinaus für die präsargonische und altakkadische Zeit C. Wilcke, Early Ancient Near Eastern Law. A History of Its Beginnings. The Early Dynastic and Sargonic Periods, Winona Lake 2007, sowie H. Neumann, Rechtspraktiken und ihr sozialökonomischer Hintergrund in altakkadischer Zeit (im Druck); zu Alalaḫ IV vgl. Ch. Niedorf, Die mittelbabylonischen Rechtsurkunden aus Alalaḫ (Schicht IV) (= Alter Orient und Altes Testament 352), Münster 2008.

2 Vgl. etwa die altorientalischen Textbeispiele in W.W. Hallo – K. Lawson Younger, Jr. (Hgg.), Archival Documents from the Biblical World (= The Context of Scripture III), Leiden/Boston 2003, sowie in B. Janowski – G. Wilhelm (Hgg.), Texte zum Rechts- und Wirtschaftsleben (= Texte aus der Umwelt des Alten Testaments. Neue Folge I), Gütersloh 2004, 1–162.

3 Vgl. dazu auch J. Renger, Who Are all Those People?, in: Or.NS 42 (1973) 259–273.

4 Vgl. dazu etwa K.R. Veenhof (Hg.), Cuneiform Archives and Libraries. Papers Read at the 30ᵉ Rencontre Assyriologique Internationale, Leiden, 4–8 July 1983 (= Publications de l'Institut historique et archéologique néerlandais de Stamboul LVII), Istanbul – Leiden 1986; O. Pedersén, Archives and Libraries in the Ancient Near East, 1500–300 B.C., Bethesda 1998; M. Brosius (Hg.), Ancient Archives and Archival Traditions. Concepts of Record-Keeping in the Ancient World, Oxford 2003.

5 Vgl. in diesem Zusammenhang den Überblick mit Textbeispielen bei F. Joannès (Hg.), Rendre la justice en Mésopotamie. Archives judiciaire du Proche-Orient ancien (IIIᵉ-Iᵉʳ millénaires avant J.-C.), Saint-Denis 2000 (mit Literatur); vgl. auch S. Démare-Lafont – H.A. Hoffner, Jr., Prozess, in: RlA 11 (2006–2008), 72–98.

Was die in sumerischer, akkadischer und hethitischer Sprache überlieferten Rechtssammlungen betrifft,[6] so enthalten diese vielfältige Informationen zu privat- und öffentlich-dienstrechtlichen Gegebenheiten (in Verbindung mit den entsprechenden sozioökonomischen Kontexten) als auch zu den jeweiligen strafrechtlichen Vorstellungen und Praktiken der verschiedenen Perioden altorientalischer Gesellschaftsgeschichte. Ihre Tradition reicht bis in das ausgehende 3. Jt. v. Chr. zurück, wofür der in sumerischer Sprache aus dem 21. Jh. v. Chr. stammende und fragmentarisch in altbabylonischen Abschriften aus Nippur, Ur, Sippar und vielleicht Kiš sowie auf einem (noch unbearbeiteten) Zylinder der Schøyen Collection überlieferte sog. Codex Ur-Namma (CU) steht.[7] Als älteste uns bekannte Rechtssammlung stellt der Codex Ur-Namma selbst das Produkt einer über mehrere hundert Jahre währenden rechts- und sozialhistorischen Entwicklung in Mesopotamien dar,[8] für die sich zwar keine eindeutigen Zeugnisse gesetzten Rechts – Gesetze bzw. Gesetzessammlungen –, jedoch sehr wohl Hinweise auf ‚gesetzliche' Verordnungen der staatlichen Gewalt finden lassen.[9] Aus frühaltbabylonischer Zeit stammen der gleichfalls auf Tontafelabschriften (und zwei Stelenfragmenten)[10] überlieferte sumerischsprachige Codex des Königs Lipit-Eštar (CL) von Isin (1934–1924 v. Chr.)[11] sowie die älteste akkadischsprachige Rechtssammlung, der sog. Codex Ešnunna (CE, 18. Jh. v. Chr.).[12] Einzig für den Codex Ḫammurapi (CḪ) besitzen wir mit der berühmten, sich heute im Louvre befindenden Gesetzesstele das in altbabylonischer Zeit im 18. Jh. v. Chr. entstandene und fast vollständige Original.[13] Hinzu kommen (bislang) acht Fragmente von weiteren Stelen sowie

6 Vgl. M.T. Roth, Law Collections from Mesopotamia and Asia Minor (= SBL Writings from the Ancient World Series 6), Atlanta 1995. Zu den in den vorliegenden Zusammenhang gehörenden Erlässen und Instruktionen vgl. die bei H. Neumann, Zeitschrift für Altorientalische und Biblische Rechtsgeschichte 10 (2004), 74 Anm. 8 notierte Literatur.

7 Zum CU vgl. zuletzt umfassend C. Wilcke, Der Codex Urnamma (CU): Versuch einer Rekonstruktion, in: T. Abusch (Hg.), Riches Hidden in Secret Places. Ancient Near Eastern Studies in Memory of Thorkild Jacobsen, Winona Lake 2002, 291–333.

8 Vgl. dazu bereits die Überlegungen bei H. Neumann, in: P. Vavroušek – V. Souček (Hgg.), Šulmu. Papers on the Ancient Near East Presented at International Conference of Socialist Countries (Prague, Sept. 30–Oct. 3, 1986), Prag 1988, 217f.

9 Wichtigstes Denkmal in dieser Hinsicht sind die sog. Reformtexte des Fürsten Iri-KA-gina; vgl. im vorliegenden Zusammenhang C. Wilcke, Early Ancient Near Eastern Law, 21–25 (mit Literatur).

10 Dazu vgl. M.T. Roth, Law Collections from Mesopotamia, 35 Anm. 1; jetzt auch W. Sallaberger, in: R. Achenbach – M. Arneth (Hgg.), „Gerechtigkeit und Recht üben" (Gen 18,19). Studien zur altorientalischen und biblischen Rechtsgeschichte, zur Religionsgeschichte Israels und zur Religionssoziologie. Festschrift für Eckart Otto zum 65. Geburtstag (= BZAR 13), Wiesbaden 2009, 11f. und 23f.

11 Zu Text und Quellen vgl. F.R. Kraus, Königliche Verfügungen in altbabylonischer Zeit (= Studia et documenta ad iura Orientis antiqui pertinentia XI), Leiden 1984, 20f; M.T. Roth, Law Collections from Mesopotamia, 23–35 und 250 (dazu auch C. Wilcke, in: T. Abusch (Hg.), Riches Hidden in Secret Places, 291 Anm. 2); M. Molina, La ley más antigua. Textos legales sumerios, Barcelona 2000, 77–95; zuletzt W. Sallaberger, Der ‚Prolog' des Codex Lipit-Eštar, in: R. Achenbach – M. Arneth (Hgg.), Festschrift für Eckart Otto, 7–33.

12 Bei den aus Tell Ḥarmal, dem alten Šaduppum, und aus Tell Ḥaddād im Ḥamrīn-Gebiet stammenden Tontafeln mit dem Text (bzw. Textauszügen) handelt es sich um (Schul-)Abschriften. Zu Text und Quellen vgl. M.T. Roth, Law Collections from Mesopotamia, 57–70 und 251.

13 Zum Text vgl. M.T. Roth, Law Collections from Mesopotamia, 71–142.

zahlreiche, vor allem als (fragmentarische) Tontafelabschriften überlieferte Textzeugen.[14] Beispiele für altorientalische Rechtssammlungen sind darüber hinaus die Hethitische Rechtssammlung[15] und die Mittelassyrischen Gesetze[16] aus dem 2. Jt. v. Chr. sowie das sog. Neubabylonische Gesetzesfragment aus dem 1. Jt. v. Chr.[17]

Mit Blick auf die Problemstellung – babylonisches Rechtsdenken im Spannungsfeld von Theorie und Praxis – ist zunächst festzuhalten, dass die Gruppe der Rechts-, Gerichts- und Prozessurkunden eindeutige Zeugnisse der juristischen Praxis darstellen.[18] In Bezug auf die Rechtssammlungen ist dies dagegen wiederholt bestritten worden, indem man den entsprechenden Quellen vornehmlich ihren legislativen Charakter absprach. Ohne dies hier im einzelnen diskutieren zu wollen, sei doch festgestellt, dass trotz immer wieder vorgebrachter Zweifel die Rechtssammlungen in der Mehrzahl der Fälle gewiss auch als Beispiele für Gesetzgebungsakte im Sinne staatlicher Willenskundgebung gewertet werden dürfen, wobei man hinsichtlich des Begriffsinhalts von „Gesetz" keine modernrechtlichen Maßstäbe anlegen kann, da er stets von dem konkreten historischen und politischen Umfeld abhängig war, in dem Rechtsprechung und Gesetzgebung erfolgten.[19] Die Rechts-

14 Zu den Quellen vgl. M.T. Roth, Law Collections from Mesopotamia, 251–253; dazu ergänzend H. Neumann, Zeitschrift für Altorientalische und Biblische Rechtsgeschichte 10 (2004), 74 Anm. 10 (mit Literatur); B. André-Salvini, Le Code de Hammurabi (= Musée du Louvre. Collection Solo 27), Paris ²2008; E. Frahm, Historische und historisch-literarische Texte (= Keilschrifttexte aus Assur literarischen Inhalts 3 / WVDOG 121), Wiesbaden 2009, 123f. Nr. 62 (VAT 1294, mAss).

15 Vgl. H.A. Hoffner, Jr., The Laws of the Hittites. A Critical Edition (= Documenta et monumenta Orientis Antiqui XXIII), Leiden/New York/Köln 1997; ders., Hittite Laws, in: M.T. Roth, Law Collections from Mesopotamia, 213–247. Zur hethitischen Rechtsüberlieferung vgl. im Überblick R. Haase, The Hittite Kingdom, in: R. Westbrook (Hg.), A History of Ancient Near Eastern Law I, 619–656; ders., Recht im Hethiter-Reich, in: U. Manthe (Hg.), Die Rechtskulturen der Antike, 123–150 und 327f. (Literatur).

16 Zu Text und Quellen vgl. M.T. Roth, Law Collections from Mesopotamia, 153–194 und 253 (mit Literatur); vgl. darüber hinaus H. Freydank, Orientalistische Literaturzeitung 80 (1985), 230 [2]; ders., Nachlese zu den mittelassyrischen Gesetzen, in: Altorientalische Forschungen 21 (1994), 203–211. Zur Datierung vgl. ders., Beiträge zur mittelassyrischen Chronologie und Geschichte (= Schriften zur Geschichte und Kultur des Alten Orients 21), Berlin 1991, 68.

17 Zu Text und Quellen vgl. M.T. Roth, Law Collections from Mesopotamia, 143–149 und 253; vgl. darüber hinaus vor allem J. Oelsner, Erwägungen zu Aufbau, Charakter und Datierung des sog. „Neubabylonischen Gesetzesfragments", in: Altorientalische Forschungen 24 (1997), 219–225 (mit Literatur).

18 Die juristischen Aspekte der seit dem 3. Jt. v. Chr. bezeugten zwischenstaatlichen Vereinbarungen aus Mesopotamien, Syrien und Anatolien bleiben im Folgenden unberücksichtigt; zu den entsprechenden Quellen vgl. die Zusammenstellung (mit Literatur) bei H. Neumann, Keilschriftliche Staatsverträge Mesopotamiens und Nordsyriens, in: B. Egger – J. Derlien (Hgg.), Der Neue Pauly. Enzyklopädie der Antike 16: Register, Listen, Tabellen, Stuttgart/Weimar 2003, 321–327 (vgl. zusätzlich jetzt noch C. Günbattı, Two Treaty Texts Found at Kültepe, in: J.G. Dercksen [Hg.], Assyria and Beyond. Studies Presented to Mogens Trolle Larsen [= PIHANS C], Leiden 2004, 249–268 [aAss]; V. Donbaz, An Old Assyrian Treaty from Kültepe, in: Journal of Cuneiform Studies 57 [2005], 63–68 [aAss]; E. Frahm, Historische und historisch-literarische Texte, 127–136 u. Nrn. 65–71 [nAss]) und G.G.W. Müller – Y. Sakuma, Hethitische Staatsverträge, ebenda, 328–337.

19 Vgl. dazu bereits H. Neumann, Zeitschrift für Altorientalische und Biblische Rechtsgeschichte 10 (2004), 72f. mit 6f. Vgl. darüber hinaus auch C. Wilcke, in: T. Abusch (Hg.), Riches Hidden in Secret Places, 297–302; ders., in: C. Wilcke (Hg.), Das geistige Erfassen der Welt im Alten Orient. Sprache, Religion, Kultur und Gesellschaft, Wiesbaden 2007, 211–214.

sammlungen sind damit nicht nur einfach Produkte babylonischer Gelehrsamkeit, die – aus dem Curriculum der Schreiberausbildung erwachsen – in erster Linie aus propagandistischen, aber eben nicht legislativen Gründen verschriftet auf Stelen präsentiert wurden, sondern müssen sehr wohl gleichfalls als Zeugnisse der juristischen Praxis angesehen werden.

Im Kontext der Darstellung einer „Geschichte des wissenschaftlichen Denkens im Altertum" hatte 1982 Helmut Freydank mit Blick auf die altorientalische Rechtsüberlieferung festgestellt, dass „sowohl die Gesetzessammlungen und sonstigen ‚gesetzgeberischen' Akte als auch die differenzierte Erfassung und Regelung oft komplizierter rechtlicher Beziehungen, die sich über nahezu zweieinhalb Jahrtausende in einer Unzahl von Keilschrifturkunden spiegeln, ... die Handhabung von Rechtsnormen (bezeugen), ohne dass sich eine Rechtswissenschaft voll ausgebildet hätte." „Ansätze" hierzu sah er allein „in terminologischen Differenzierungen bestimmter Urkundengattungen und Klauseln nach ihrem sachlichen Hintergrund sowie auch in der Anwendung festliegender Formulare".[20] Ohne den Begriff „Rechtswissenschaft" in seiner komplexen Bedeutung als Geistes- und Sozialwissenschaft mit den die rechtliche Sollensordnung betreffenden Inhalten[21] für die altmesopotamischen Verhältnisse überstrapazieren zu wollen, sei doch darauf hingewiesen, dass gerade die überlieferten Rechtssammlungen, und hier fast paradigmatisch der sog. Codex Ḫammurapi, die Einheit von „Empirie und ... Theorie als konstitutive Momente einer Wissenschaftstotalität"[22] im Bereich der babylonischen Rechtsgeschichte in einer Weise verdeutlichen, die durchaus rechtswissenschaftliches Denken auf einem beachtlichen intellektuellen Niveau vermuten lässt.

Die Komplexität dieses Denkens wird bereits im Zusammenhang mit der Ausarbeitung und Formulierung der Begründung von Rechtsetzung und Rechtsprechung auf entsprechender geistig-weltanschaulicher Grundlage deutlich, was zugleich auf die Legitimierung politischer Herrschaft und ihrer gesellschaftserhaltenden wie auch -gestaltenden Rolle gerichtet war.[23] Zentrale Begriffe im Rahmen des babylonisch-assyrischen Weltbildes, dessen Ausgangspunkte sich bereits in den sumerisch-akkadischen Vorstellungen des 3. Jt. v. Chr. finden lassen, sind *kittu*, sumerisch nì-gi-na, „das, was feststeht, Wahrheit, Recht", als statisches Prinzip, und *mīšaru*, sumerisch nì-si-sá, „Akt oder Instrument des Geradewerdens bzw., um etwas gerade werden zu lassen, Gerechtigkeit" als dynamisches,

20 H. Freydank, in: F. Jürß (Hg.), Geschichte des wissenschaftlichen Denkens im Altertum (= Veröffentlichungen des Zentralinstituts für Alte Geschichte und Archäologie der Akademie der Wissenschaften der DDR 13), Berlin 1982, 79.

21 Vgl. im vorliegenden Zusammenhang etwa M. Rehbinder, Einführung in die Rechtswissenschaft, Berlin – New York [6]1988, 58–145.

22 F. Jürß, in: F. Jürß (Hg.), Geschichte des wissenschaftlichen Denkens im Altertum, Berlin 1982, 11.

23 Vgl. dazu im Einzelnen auch H. Neumann, Schuld und Sühne. Zu den religiös-weltanschaulichen Grundlagen und Implikationen altmesopotamischer Gesetzgebung und Rechtsprechung, in: J. Hengstl – U. Sick (Hgg.), Recht gestern und heute. Festschrift zum 85. Geburtstag von Richard Haase (= Philippika 13), Wiesbaden 2006, 27–43.

aktives Prinzip. Die kosmische Ordnung als göttliche Schöpfung und göttlicher Wille ist zugleich die Voraussetzung für die Existenz und das Funktionieren irdischer Verhältnisse, die im Sinne der oben genannten Ordnungsprinzipien erhalten werden müssen. Nur wenn die politisch-soziale Ordnung im Einklang mit der Ordnung des Kosmos steht, ist das Wohlergehen des Landes gesichert.

Gemäß dem, von Stefan Maul unlängst in der eben beschriebenen Weise thematisierten religiös-politischen Gesellschaftsdenken im alten Mesopotamien[24] kam daher dem König die herausragende Rolle im Bereich der Proklamierung, Wahrung und Durchsetzung ordnungs- und damit auch rechtspolitischer Maßnahmen zu.[25] Durch entsprechende Legitimation, sei es durch göttliche Berufung bzw. Bestätigung, Gotteskindschaft und/oder eigene Göttlichkeit war der Herrscher für die Aufrechterhaltung, den Schutz und die Erneuerung bzw. Ausweitung der (durch die Götter geschaffenen) Ordnung im Lande zuständig. Teil dieses Aufgabenspektrums bildeten die Sorge um Gerechtigkeit (*mīšaru*) und die Wahrung des Rechts (*kittu*) – *mīšaru* und *kittu* im oben genannten Sinne zu verstehen eben auch als zentrale Begriffe altmesopotamischen Rechtsdenkens. Die Vorstellung vom Recht als Mittel der Aufrechterhaltung bzw. Wiederherstellung einer göttlichen Ordnung, in der die durch göttlichen Willen bestimmten Kausalverhältnisse im Rahmen eines geschlossenen Weltbildes einen zentralen Platz einnahmen, begründete die königliche Rechtsetzung und Rechtsprechung,[26] und zwar als Mittel zur Bewältigung von Gegenwartsproblemen in der Gesellschaft wie auch zur Sicherung der gesellschaftlichen Zukunft.[27]

Dies wird nicht zuletzt durch die Formulierung wie auch den Aufbau des juristischen Teils der Rechtssammlungen unterstrichen. Er wird durch sog. Paragraphen gebildet, deren Einteilung modern ist und auf der formalen Stilisierung der Rechtssätze basiert. Die gebräuchlichste Stilform ist dabei die sog. konditionale Form, d.h., die zu regelnden Tatbestände werden als Bedingungssatz mit „wenn" (sumerisch tukumbi, akkadisch *šumma*, hethitisch *takku*) eingeleitet, woraus sich die in einem Hauptsatz formulierten Rechtsfolgen ergeben.[28] Die konditionale Form mit Protasis und Apodosis – das Kausalverhältnis zwischen Vergehen und Sanktion kennzeichnend – entspricht exakt der Formulierung in den akkadischen Omina wie auch in den medizinisch-diagnostischen Texten, was nicht nur

24 Vgl. S.M. Maul, in: K. Watanabe (Hg.), Priests and Officials in the Ancient Near East, Heidelberg 1999, 202f.; vgl. im vorliegenden Zusammenhang auch ders., Im Fadenkreuz von Raum und Zeit. Zum Verhältnis von Weltbild und Herrschaftskonzeption im Alten Orient, in: Heidelberger Jahrbücher 42 (1998), 27–41. Zur sumerischen Begrifflichkeit vgl. G.J. Selz, Wiener Zeitschrift für die Kunde des Morgenlandes 92 (2002), 162–164.

25 Vgl. im vorliegenden Zusammenhang auch E. Otto, Zeitschrift für Altorientalische und Biblische Rechtsgeschichte 11 (2005), 56–58.

26 Vgl. dazu bereits H. Neumann, in: J. Hengstl – U. Sick (Hgg.), Festschrift zum 85. Geburtstag von Richard Haase, 31f. (mit Literatur).

27 Vgl. in diesem Zusammenhang auch die Überlegungen von C. Wilcke, Das Recht: Grundlage des sozialen und politischen Diskurses im Alten Orient, in: C. Wilcke (Hg.), Das geistige Erfassen der Welt im Alten Orient, 209–244.

28 Vgl. dazu im einzelnen H. Petschow, Zu den Stilformen antiker Gesetze und Rechtssammlungen, in: Zeitschrift der Savigny-Stiftung für Rechtsgeschichte 82 (1965), 24–38.

eine formale Nähe zwischen diesen Quellengruppen, sondern wegen der ihnen unter religiös-weltanschaulichen Gesichtspunkten zugrundeliegenden Vorstellung einer durch Kausalität charakterisierten „Gesetzmäßigkeit[29] des (dynamischen) Weltgefüges" durchaus eine Sinngleichheit dokumentiert.[30]

Damit im Zusammenhang steht die Ambivalenz des (auch in juristischen Texten der Rechtspraxis und in den Rechtssammlungen verwendeten) akkadischen Wortes für „Schuld, Vergehen" *arnu* (ähnlich später auch *ḫīṭu*), sumerisch nam-tag, zum Teil sinngleich mit *šērtu*, sumerisch gleichfalls nam-tag bzw. šer$_7$-da. Bestandteil derartiger „Schuld", bei der es sich grundsätzlich um ein „Vergehen" gegen die (göttliche) Ordnung handelt, ist auch jede Form des Vergehens gegen die politisch-soziale Ordnung, was die Verwendung des Begriffs *arnu* sowohl im religiös-magischen als auch im juristischen Kontext erklärt. Da im babylonisch-assyrischen Weltbild ein strenges, durch göttlichen Willen vorherbestimmtes Kausalverhältnis zwischen allen nur denkbaren Erscheinungen, die unprovoziert oder provoziert sein konnten, und den entsprechend vorherseh-, d.h., erkennbaren Folgen bestand,[31] impliziert auch die Kategorie *arnu* die entsprechende Unrechtsfolge, nämlich die „Strafe". D.h., in dem Begriff *arnu* spiegelt sich das (vorherbestimmte) Kausalverhältnis von „Vergehen" (gegen die göttliche Ordnung) und dessen „Ahndung" wider, so dass nam-tag bzw. *arnu* je nach Kontext sowohl das „Vergehen" bzw. die „Schuld" als auch die „Strafe" bezeichnen konnte, was sowohl auf die Beschwörungs-, Ritual- und Gebetsliteratur als auch auf die juristische Überlieferung zutrifft.[32]

Was den im vorliegenden Zusammenhang zu berücksichtigenden Aufbau des juristischen Teils der keilschriftlichen Rechtssammlungen betrifft, so ist paradigmatisch auf die Einteilung der Rechtssätze im altbabylonischen Codex Ḫammurapi zu verweisen.[33] Erinnert sei hier an die Einschätzung von Herbert Petschow, „dass der Rechtstext des CḪ formal und inhaltlich ein zusammenhängendes, im wesentlichen logisch aufgebautes Ganzes darstellt, dessen einzelne Teile ... durch überleitende Normen in meist begründbarer Weise miteinander verknüpft sind."[34] Die Konsequenz wie auch die gedankliche Tiefe, mit der die Redaktoren des Codex Ḫammurapi bei der Abfassung der Rechtssammlung vorgegangen

29 Und damit auch der Zukunftsgültigkeit (H.N.).

30 Vgl. S.M. Maul, in: Der Neue Pauly. Enzyklopädie der Antike 3, Stuttgart/Weimar 1997, 704; ders., RlA 10 (2003–2005), 46. Jüngst erneut thematisiert von J.C. Fincke, Omina, die göttlichen „Gesetze" der Divination, in: Jaarbericht van het Voor-Aziatisch-Egyptisch-Gezelschap Ex Oriente Lux (2006–2007), 131–147.

31 Vgl. dazu zusammenfassend S.M. Maul, RlA 10 (2003–2005), 46–48.

32 Vgl. dazu im Einzelnen und mit Literatur H. Neumann, in: J. Hengstl – U. Sick (Hgg.), Festschrift zum 85. Geburtstag von Richard Haase, 34–36.

33 Vgl. grundlegend H. Petschow, Zur Systematik und Gesetzestechnik im Codex Hammurabi, in: Zeitschrift für Assyriologie 57 (1965), 146–172; vgl. auch zum Codex Ešnunna ders., Zur „Systematik" in den Gesetzen von Eschnunna, in: J.A. Ankum – R. Feenstra – W.F. Leemans (Hgg.), Symbolae iuridicae et historicae Martino David dedicatae, Tomus alter: Iura orientis antiqui, Leiden 1968, 131–143 und B.L. Eichler, Literary Structure in the Laws of Eshnunna, in: F. Rochberg-Halton (Hg.), Language, Literature, and History: Philological and Historical Studies Presented to Erica Reiner (= American Oriental Series 67), New Haven 1987, 71–84.

34 H. Petschow, Zeitschrift für Assyriologie 57 (1965), 169.

sind, begründeten deren große literarische und rechtsgeschichtliche Bedeutung. Es erklärt auch, dass Teile des Codex (juristische und nichtjuristische Bestandteile) über fast 1500 Jahre hinweg mehrfach zu Lehrzwecken oder aus literarhistorischen Gründen abgeschrieben worden sind[35] und damit Vorbildcharakter in Form und Inhalt für die spätere juristische Theorie und Praxis hatten.[36]

Die Gliederung der Rechtssätze lässt eine Systematik erkennen, die auf äußerlichen sachlichen Zusammenhängen beruht. Es handelt sich dabei nicht um eine Systematik nach modernen juristischen Gesichtspunkten, sondern um eine Gliederung „nach einem dem Denken der Zeit gemäßen Ordnungsprinzip", wie es in einer Formulierung von Herbert Petschow heißt,[37] und was auf entsprechende systematische Vorarbeiten seitens der Redaktoren schließen lässt. Die ersten 41 Paragraphen beinhalten Tatbestände und deren Rechtsfolgen, die unmittelbar die Interessen des Palastes und der Tempel sowie die Aufrechterhaltung der öffentlichen Ordnung berühren.[38] Dazu zählen strafrechtliche Bestimmungen, die mit der Durchführung von Prozessen im Zusammenhang stehen und die an den Anfang des juristischen Teils der Rechtssammlung gestellt sind. Mit den §§ 1–5 erfasste der Codex Ḫammurapi „alle für das Finden richtigen Rechts wesentlichen Prozessbeteiligten", namentlich den Anschuldiger, die Zeugen und den Richter. Der Codex Ḫammurapi hob damit auf die Sicherung der Rechtspflege an prominenter Stelle ab,[39] und zwar in § 1 indirekt durch Anwendung einer besonderen Form von Talion mit Bezug auf die Rechtsfolge „Todesstrafe" für Tötung/Mord. Letzteres findet sich explizit als § 1 im Codex Ur-Namma, so dass beide Rechtssammlungen im Rahmen der ihnen jeweils eigenen Systematik ein sog. „Primärdelikt" an den Anfang der Bestimmungen stellen.[40] Wie bekannt, beinhaltet der zweite große Abschnitt des Codex Ḫammurapi (§§ 42–282) dann im wesentlichen Rechtsfälle, die den privaten Lebensbereich des Einzelnen berühren.[41]

Es ist das Verdienst von Herbert Petschow, erkannt zu haben, dass über die Haupteinteilung des Rechtsstoffes hinaus auch im Detail durchaus rechtssystematisches Denken die Redaktoren des Codex Ḫammurapi bei der Gliederung der Rechtssammlung geleitet hat.[42] Dabei sind Empirie und Theorie als Grundlage der Arbeit am Text gleichermaßen nachweisbar. Neben der in die Systematik eingepassten Aufnahme gewohnheitsrecht-

35 Vgl. oben mit Anm. 14.

36 Vgl. im vorliegenden Zusammenhang auch M.T. Roth, Mesopotamian Legal Traditions and the Laws of Hammurabi, in: Chicago-Kent Law Review 71/1 (1995), 13–37.

37 H. Petschow, Zeitschrift der Savigny-Stiftung für Rechtsgeschichte 85 (1968) 3f.; vgl. auch ders., Zeitschrift für Assyriologie 76 (1986), 19–21; unter dem Gesichtspunkt „altorientalische Rationalität" zustimmend D.O. Edzard, in: K. Gloy (Hg.), Rationalitätstypen, Freiburg/München 1999, 261–264.

38 Vgl. H. Petschow, Zeitschrift für Assyriologie 57 (1965), 153.

39 Ebenda, 148.

40 Vgl. dazu H. Petschow, Zeitschrift für Assyriologie 74 (1984), 189f. Anm. 13.

41 H. Petschow, Zeitschrift für Assyriologie 57 (1965), 153 („Individualsphäre des Einzelbürgers").

42 Vgl. vor allem die oben Anm. 33 notierte Literatur.

licher Praktiken ist es zunächst die textkritische Übernahme älterer Bestimmungen unter dem Gesichtspunkt der weiteren Präzisierung von Tatbeständen, die auffällt.[43] Ein äußerst prägnantes Beispiel hierfür ist der § 59, der bestimmt, dass jemand, der *balum bēl kirîm*, „ohne den (= ohne die Erlaubnis des) Garteneigentümer(s)", in dessen Garten einen Baum fällt, eine halbe Mine Silber zu zahlen hat. Vorlage war gewiss der (sumerisch formulierte) § 10 des Codex Lipit-Eštar, der das Baumfällen unter Androhung derselben Sanktion gleichfalls unter Strafe stellte, aber noch ohne den Zusatz „ohne Erlaubnis des Garteneigentümers". Die von den Redaktoren vorgenommene Präzisierung erfolgte gewiss auf Grund der (im Übrigen durch zeitgenössische Quellen bezeugten) Erfahrungen in der Rechts- bzw. Gerichtspraxis, die einen derartigen Zusatz als Erfordernis lehrten.[44]

Von besonderer Bedeutung im vorliegenden Zusammenhang ist die im Codex Ḫammurapi deutlich werdende exakte Trennung von Rechtsfällen im Zusammenhang mit Vertragsverhältnissen von denen mit außervertraglichen Rechtsbeziehungen als Gegenstand.[45] Während z. B. die §§ 253–256 die Unterschlagung von Saat- und Futtergetreide bzw. unerlaubte Vermietung von (Pflug-)Rindern durch die zur Betreuung und Bestellung eines Feldes gemietete Person mit Strafe bedrohen, wird in den §§ 259f. parallel die unerlaubte Wegnahme von Pflug und Egge von der (freien) Flur unter Strafe gestellt. Die (parallelen) Bestimmungen über unerlaubte Handlungen werden also – nach Einschub der Miettarife in den §§ 257f.[46] – den Bestimmungen über die Vertragsverletzungen nachgeordnet und von diesen getrennt platziert.[47] Die hier und andernorts deutlich werdende regelmäßige Scheidung von Normen, die im Zusammenhang mit Verträgen stehen, von denen, die außervertragliche Beziehungen bzw. unerlaubte Handlungen als solche zum Gegenstand haben, setzte mit Herbert Petschow nicht nur juristische Kenntnisse, sondern zum Teil sogar juristisch-dogmatische Überlegungen voraus, was ein bezeichnendes Licht auf das theoretische juristische Denken insgesamt in jener Zeit wirft.[48]

Für jenes Denken wie auch für den damit verbundenen Diskurs im alten Mesopotamien war – bezogen auf die altbabylonische Zeit – das é-dub-ba-a, das „Haus, (in) dem

43 In diesem Sinne bereits P. Koschaker, Rechtsvergleichende Studien zur Gesetzgebung Ḫammurapis, Königs von Babylon, Leipzig 1917 (vgl. dazu auch die Einschätzung von M. Müller, Altorientalische Forschungen 9 [1982], 276). Vgl. im vorliegenden Zusammenhang auch R. Rothenbusch, Die kasuistische Rechtssammlung im ‚Bundesbuch' (Ex 21,2–11.18–22,16) und ihr literarischer Kontext im Licht altorientalischer Parallelen (= Alter Orient und Altes Testament 259), Münster 2000, 148–157.

44 Vgl. dazu H. Petschow, Zeitschrift für Assyriologie 74 (1984), 184f. Anm. 5.

45 Vgl. H. Petschow, Zeitschrift für Assyriologie 57 (1965), 171f.; ders., Zeitschrift für Assyriologie 76 (1986), 20f.

46 Vgl. dazu H. Petschow, ebenda, 36 „... Personenmiete, ... im landwirtschaftlichen Bereich, hier zutreffend zunächst die Vermieter- (§§ 253–256), sodann die Mieterseite (§§ 257, 258) betreffend".

47 In diesem Sinne ausführlich H. Petschow, Zeitschrift für Assyriologie 57 (1965), 167f.; ergänzend ders., Zeitschrift für Assyriologie 76 (1986), 36f.

48 H. Petschow, Zeitschrift für Assyriologie 57 (1965), 172 („juristisch-theoretische Reflexion"); ders., Zeitschrift für Assyriologie 76 (1986), 20 („'juristisch-dogmatische' Gedankenführung").

Tafeln zugeteilt werden", von Bedeutung.[49] Das Edubba'a diente als eine Institution der Lehre vornehmlich der Schreiberausbildung, darf mit Blick auf die literarische Produktion sowie deren Bewahrung und Tradierung im Rahmen besagter Einrichtung mit Jan Assmann jedoch gewiss auch als Träger einer „textgestützten Form des kulturellen Gedächtnisses" im alten Mesopotamien bezeichnet werden.[50]

Im Rahmen des Curriculums der babylonischen Schreiberausbildung wurden neben der (schriftlichen) Tradierung keilschriftlicher Rechtsvorschriften sowie dem Erlernen und dem Üben des Gebrauchs der sumerischen und akkadischen Vertragsterminologie mittels der zweisprachigen (sumerisch-akkadischen) Serie *ana ittišu* auch sog. Musterverträge zum Üben im Formulieren ganzer Verträge in der babylonischen Schule abgeschrieben und tradiert.[51] Im vorliegenden Zusammenhang von besonderer Bedeutung sind nun die gleichfalls aus altbabylonischer Zeit überlieferten sog. literarischen Prozessurkunden in sumerischer Sprache, die eben dieser Schultradition, namentlich von Nippur und vielleicht von Isin, entstammen und die in beeindruckender Weise zeigen, wie anhand fiktiver Fälle und in zum Teil literarischer Formulierung[52] geltendes Recht mit Blick auf zukünftige richterliche Tätigkeit vermittelt und wohl auch diskutiert wurde.[53]

Mit Blick auf die vorliegende Problemstellung scheint ein Beispiel besonders instruktiv zu sein.[54] In mehreren Exemplaren, u.a. als Text auf einer als Schreibübung zu identi-

49 Zu Rolle und Funktion des Edubba'a, der babylonischen ‚Schule', sowie zu den sich damit verbindenden Fragen von Ausbildung und Erziehung im alten Mesopotamien vgl. zuletzt K. Volk, Methoden altmesopotamischer Erziehung nach Quellen der altbabylonischen Zeit, in: Saeculum 47 (1996), 178–216; ders., Edubba'a und Edubba'a-Literatur: Rätsel und Lösungen, in: Zeitschrift für Assyriologie 90 (2000), 1–30; C. Wilcke, Konflikte und ihre Bewältigung in Elternhaus und Schule im Alten Orient, in: R. Lux (Hg.), Schau auf die Kleinen ... Das Kind in Religion, Kirche und Gesellschaft, Leipzig 2002, 10–31; A.R. George, In Search of the é.dub.ba.a. The Ancient Mesopotamian School in Literature and Reality, in: Y. Sefati – P. Artzi – Ch. Cohen – B.L. Eichler – V.A. Hurowitz (Hgg.), „An Experienced Scribe Who Neglects Nothing". Ancient Near Eastern Studies in Honor of Jacob Klein, Bethesda 2005, 127–137.

50 J. Assmann, Das kulturelle Gedächtnis. Schrift, Erinnerung und politische Identität in frühen Hochkulturen, München 1992, 92f.

51 Vgl. den Überblick (mit Literatur) H. Neumann, Zeitschrift für Altorientalische und Biblische Rechtsgeschichte 10 (2004), 75–78; vgl. auch unter dem Gesichtspunkt der Rhetorik und (juristischen) Argumentation W. Sallaberger, in: G. Ueding (Hg.), Rhetorik. Begriff – Geschichte – Internationalität, Tübingen 2005, 256.

52 Vgl. dazu auch H. Neumann, Göttliche Gerechtigkeit und menschliche Verantwortung im alten Mesopotamien im Spannungsfeld von Norm(durch)setzung und narrativer Formulierung, in: H. Barta – R. Rollinger – M. Lang (Hgg.), Recht und Religion. Menschliche und göttliche Gerechtigkeitsvorstellungen in den antiken Welten (= Philippika 24), Wiesbaden 2008, 37–48.

53 Vgl. dazu zuletzt (mit Literatur) H. Neumann, Prozessführung im Edubba'a. Zu einigen Aspekten der Aneignung juristischer Kenntnisse im Rahmen des Curriculums babylonischer Schreiberausbildung, in: Zeitschrift für Altorientalische und Biblische Rechtsgeschichte 10 (2004), 71–92; J. Klein – T.M. Sharlach, A Collection of Model Court Cases from Old Babylonian Nippur (CBS 11324), in: Zeitschrift für Assyriologie 97 (2007), 1–25; vgl. jetzt auch A.R. George, Babylonian Literary Texts in the Schøyen Collection (= Cornell University Studies in Assyriology and Sumerology 10), Bethesda 2009, 123–152 (Nr. 17), betreffend eine akkadische literarische Prozessurkunde der altbabylonischen Zeit.

54 Vgl. dazu bereits ausführlich mit Literatur und Diskussion zu Detailfragen H. Neumann, Zeitschrift für Altorientalische und Biblische Rechtsgeschichte 10 (2004), 79–82.

fizierenden Sammeltafel liegt aus Nippur eine Prozessurkunde vor, die ein Tötungsdelikt zum Gegenstand hat.[55] Einleitend wird zunächst festgestellt, dass drei männliche Personen, darunter wohl ein Sklave, eine vierte getötet hatten. Danach wurde die Ehefrau des Getöteten von den Mördern über die Tat informiert, die darüber allerdings Stillschweigen bewahrte. Die Rechtssache wurde vor den König in Isin gebracht, der sie an die Versammlung (*puḫrum*) von Nippur zur Entscheidung überwies. Neun Mitglieder der Versammlung stellten die Tat als strafbare Handlung fest und forderten die Todesstrafe für die drei Mörder und für die Ehefrau, die die Tat verschwiegen hatte. Daraufhin wurde von zwei weiteren Mitgliedern der Versammlung zwar die (Möglichkeit einer) Schuld der Ehefrau eingeräumt, jedoch die Berechtigung der Todesstrafe wohl mit Verweis auf die eingeschränkten Handlungsmöglichkeiten der/einer Frau in Frage gestellt.[56] Der zur Verteidigung der Ehefrau vorgebrachte Einwand wurde in der Versammlung allerdings verworfen, da in dem Verhalten der Ehefrau eine ehewidrige Motivation erkannt wurde, was ihre Schuld sogar noch größer als die der eigentlichen Täter erscheinen ließ. Nach Abschluss des Verfahrens wurden sowohl die drei Mörder als auch die Ehefrau des Getöteten zum Tode verurteilt.

Grundsätzlich lässt sich zunächst feststellen, dass die Urkunde in all ihren Teilen zeitgenössisches geltendes Recht widerspiegelt, wobei in dem Begriff zeitgenössisch hier das Recht der neusumerischen Periode im ausgehenden 3. Jt. v. Chr. eingeschlossen ist. Als Tötungsdelikt galt die angeklagte Tat als Kapitaldelikt, wurde also mit der Todesstrafe bedroht. Dies erklärt die Tatsache, dass die Rechtssache zunächst vor den König kam.[57] Zeitgenössische Prozessurkunden verdeutlichen jedoch, dass der König ihm vorgebrachte Rechtsangelegenheiten durchaus an lokale Instanzen rückverweisen konnte.

Man kann also konstatieren, dass der Rückverweis der Rechtssache an die Versammlung in dem in Rede stehenden altbabylonischen Text durchaus juristischen Gepflogenheiten jener Zeit entspricht. Dass das Verschweigen der Tat durch die Ehefrau vom Grundsatz her – auch von den beiden Einspruch erhebenden Mitgliedern der Versammlung – als todeswürdiges Verbrechen angesehen wurde, erklärt sich daraus, dass sie de facto die Mörder ihres Mannes begünstigt, ihnen also wissentlich Hilfe geleistet hatte, sie

55 Th. Jacobsen, An Ancient Mesopotamian Trial for Homicide (= Analecta Biblica 12), Rom 1959, 130–150. Zum Text vgl. darüber hinaus die bei H. Neumann, Zeitschrift für Altorientalische und Biblische Rechtsgeschichte 10 (2004), 79 Anm. 43 notierte Literatur sowie jüngst R. Westbrook, A New Look at the Nippur Homicide Trial, in: H. Horowitz – U. Gabbay – F. Vukosavović (Hgg.), A Woman of Valor: Jerusalem Ancient Near Easter Studies in Honor of Joan Goodnick Westenholz (= Biblioteca del Próximo Oriente Antigno 8), Madrid 2010, 195–200. Auf Grund der Erwähnung des Königsnamens Ur-Ninurta von Isin datiert der Text aus der Zeit zwischen 1923–1896 v. Chr. In Duplikaten überliefert wurde die Urkunde bis in die Zeit des Königs Samsuiluna von Babylon (1749–1712 v. Chr.).

56 Hintergrund dieses (vielleicht auf Strafmilderung gerichteten) Einwandes war möglicherweise die Vorstellung, dass für das Schweigen der Ehefrau des Getöteten u. a. auch die Furcht maßgebend war, bei einer entsprechenden Anzeige selbst Opfer eines Tötungsverbrechens seitens der drei Mörder zu werden.

57 Vgl. dazu auch H. Neumann, RlA 11 (2006–2008), 348.

somit de jure als Hehlerin galt und damit dieselbe Strafe wie die Täter zu erwarten hatte. Strafverschärfend kam hinzu, dass nach richterlicher Auffassung die Motive für ihr Verschweigen offensichtlich in einem ehewidrigen Verhalten dem Ehemann gegenüber zu suchen waren, so dass sie möglicherweise sogar als Anstifterin für das Verbrechen gelten konnte.

Während also der rechtliche Gegenstand der Urkunde ohne weiteres mit den entsprechenden juristischen Vorstellungen und Vorschriften sowie den tatsächlichen Gegebenheiten jener Zeit korrespondiert, ist die Diktion des Textes ungewöhnlich. So fehlen jegliche Zeugeneinvernahme und entsprechende Beweiseide, die zumindest im Falle der Motivbeschreibung für die Tat der Ehefrau zu erwarten gewesen wären, und zwar auch dann, wenn man den eigentlichen Prozessgegenstand, nämlich den begangenen Mord, in einer dem vorliegenden Prozess vorausgehenden Verhandlung als erwiesen entschieden sehen möchte. In einer normalen Prozessurkunde der altbabylonischen Zeit hätte das Urteil durchaus mit den hier vorgebrachten Sachargumenten begründet und protokolliert werden können, jedoch kaum in der Art einer Diskussion zwischen Mitgliedern der rechtsprechenden Instanz. Damit erweist sich die vorliegende Urkunde als eine fiktive Fallbeschreibung, die natürlich geltendes und praktiziertes Recht zum Gegenstand hat, dieses aber in einer literarischen Fallkonstruktion darstellt. Das Problem des hinter der Straftat der Ehefrau stehenden Motivs, das in der Konsequenz die Tat explizit als das bereits erwähnte sum. nam-tag (= *arnu*) (Z. 51) mit den entsprechenden religiös-ethischen Implikationen charakterisierte, war hier der gewiss auch als schwierig empfundene Gegenstand, der literarisch-modellhaft im Rahmen des juristischen Curriculums des Edubba'a diskutiert und entschieden wurde.

Der Mordprozess war sozusagen der Ausgangspunkt für eine Diskussion damit zusammenhängender juristischer Probleme, wie man sie sich in der Praxis durchaus hätte vorstellen können. Die Erstellung dieser und anderer, ähnliche Intentionen verfolgender Urkunden war also einer im Edubba'a, in der babylonischen Schule anzunehmenden Kreativität geschuldet. Diese war offensichtlich nicht allein auf die Vermittlung von Kenntnissen in Bezug auf geltende und anzuwendende rechtliche Regelungen gerichtet, sondern bildete gewiss auch die Grundlage für eine juristische Diskussion mit Blick auf Innovationen und möglicherweise Änderungen im Rahmen der juristischen Praxis. Dem dienten im Rahmen des Curriculums des Edubba'a entsprechende Übungen im Bereich der juristischen Spekulation und Improvisation. Ziel war zum einen die möglichst sachgerechte Ausbildung von Schreibern, die als Richter mit den konkreten Fällen betraut waren, und zum anderen gewiss auch die Schaffung der gedanklichen Gestaltungsvoraussetzungen der jeweiligen Rechtssammlungen, wie etwa des Codex Ḫammurapi. Juristische Theorie und Praxis standen hier in einem produktiven Verhältnis zueinander.

Insofern kann man vielleicht auch mit Blick auf die anderen Wissen(schaft)sbereiche Babyloniens in Anlehnung an eine Formulierung in Benno Landsbergers „Eigenbegrifflichkeit" konstatieren, dass insbesondere die „geistigen Funktionen" Beobachtung, Lebenserfahrung und (nunmehr hinzuzufügen) Diskurs „das mächtige Gebäude einer in Form von Bedingungssätzen niedergelegten Wissenschaft errichtet (haben): Omenkunde, Medizin und Rechtswissenschaft".[58]

58 B. Landsberger, Islamica 2 (1926), 370f.

*Nils P. Heeßel*

# „Sieben Tafeln aus sieben Städten"
## Überlegungen zum Prozess der Serialisierung von Texten in Babylonien in der zweiten Hälfte des zweiten Jahrtausends v. Chr.

Die zweite Hälfte des zweiten vorchristlichen Jahrtausends war in Babylonien eine Phase relativer politischer Stabilität. Die Dynastie der eingewanderten Kassiten (ca. 1530–1155 v. Chr.) dominierte nach dem Fall der amoritischen Dynastie von Babylon die Geschicke des Landes fast 400 Jahre lang und spielte im Konzert der damaligen Großmächte eine bedeutende Rolle. In diese Epoche der altorientalischen Geschichte fällt nach der *opinio communis* der Assyriologie jene formative Phase, in der die tradierten Texte zur Literatur, Religion und Wissenschaft einen Prozess der Standardisierung durchlaufen und zu den Serien geformt werden, die wir dann in den Bibliotheken des ersten vorchristlichen Jahrtausends finden.[1]

Leider lässt sich der Prozess, der zur Entstehung der verschiedenen Textserien geführt hat, nur in groben Zügen beschreiben. Dies liegt vor allem daran, dass die Beleglage für die mittelbabylonische Zeit in krassem Widerspruch zu ihrer textgeschichtlichen Bedeutung steht, denn aus dieser Zeit sind bislang nur sehr wenige literarische Texte bekannt geworden, die darüber hinaus nur selten genau zu datieren sind. Es verwundert daher nicht, dass als Beleg für die Bedeutung der mittelbabylonischen Zeit für die Standardisierung und Serialisierung der Texte gerne die Aussagen der babylonischen und assyrischen Gelehrten selbst, vor allem in Form der Kolophone, herangezogen werden. So wird bei der Diskussion um die Frage, wann und wo die Textserien des ersten Jahrtausends v. Chr. entstanden sind, immer wieder eine Hemerologie aus Assur zitiert, die zumeist unter dem Titel „Hemerologie für Nazimaruttaš" firmiert. Den umfangreichsten Textvertreter VAT 9663 hat Erich Ebeling als KAR 177 publiziert. Nach einer Liste von günstigen Tagen für verschiedene Aktivitäten folgt in den Zeilen 25–32 der vierten Kolumne der Anfang eines interessanten Kolophons:

iv 25 U$_4$.ME DÙG.GA.MEŠ KA 7 *ṭ[up-pa-a-n]i*
iv 26 GABA.RI UD.KIB.NUN$^{ki}$ NIBRU$^{ki}$
iv 27 KÁ.DINGIR.RA$^{ki}$ UD.UNUG$^{ki}$
iv 28 ŠEŠ.UNUG$^{ki}$ UNUG$^{ki}$ *u Eri-du*$_{10}^{ki}$
iv 29 *um-ma-a-ni ú-na-as-si-ḫu-ma*

1 Siehe stellvertretend für zahlreiche weitere Äußerungen W. von Soden, Das Problem der zeitlichen Einordnung akkadischer Literaturwerke, Mitteilungen der Deutschen Orientgesellschaft 85 (1953), 22f. und F. Rochberg-Halton, Canonicity in Cuneiform Texts, Journal of Cuneiform Sources 36 (1984), 127f.

iv 30 *ú-na-as-si-qu-ma*
iv 31 *a-na* [I]*Na-zi-múru-*⌈*taš*⌉
iv 32 LUGAL ŠÚ SUM-*nu*

„Günstige Tage nach dem Wortlaut von sieben T[afel]n, Originale aus Sippar, Nippur, Babylon, Larsa, Ur, Uruk und Eridu. Die Gelehrten exzerpierten, wählten aus und überreichten (sie) Nazimaruttaš, dem König der Welt."

Das abgebrochene Ende der ersten Zeile des Kolophons in KAR 177 stellt bis heute eine philologische Krux dar. René Labat, der Erstbearbeiter dieser Hemerologie, ergänzte 7 [*ṭuppāti*], also „sieben [Tafeln]".[2] Diese Lesung hat sich jedoch nicht allgemein durchgesetzt und als Alternativen sind die Lesungen *u[m-ma-a-ni]*[3] und *a[p*?*-kal*?*-le*?*]*[4] vorgeschlagen worden. Nach einer Kollation der Stelle durch Joachim Marzahn[5] wurde deutlich, dass das beschädigte Zeichen nicht AP gelesen werden kann, sondern nur DUB oder UM, so dass die Lesungen *um-ma-a-ni* oder *ṭup-pa-a-ti/ni* im Raum standen. Unter den als „Omina" deklarierten unpublizierten Fragmenten aus Assur konnte ich ein neues Duplikat zu diesem Kolophon identifizieren, durch das sich die Frage nunmehr entscheiden lässt. Das kleine, 26 × 36 mm messende Fragment VAT 11609 (hier im Maßstab 2,25:1 abgebildet) bietet den Text:

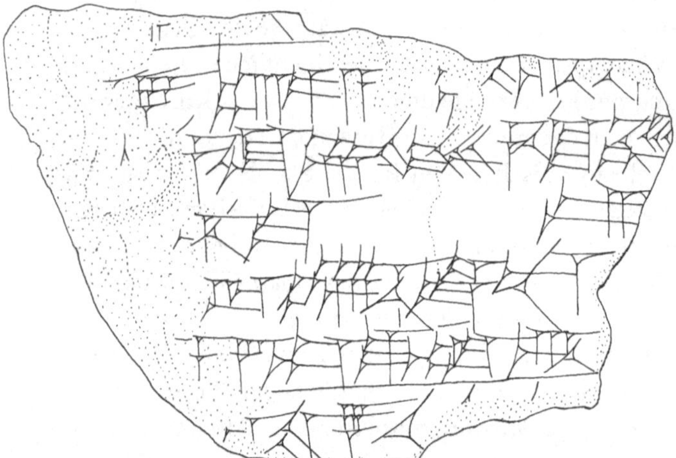

Fig. 1 | VAT 11609: Kopie von N. P. Heeßel.

2 R. Labat, Hémérologies et Ménologies d'Assur, Paris 1939, 162.

3 J. Laessøe, Studies on the Assyrian Ritual and Series *bît rimki*, Kopenhagen 1955, 72, Anm. 155.

4 W.G. Lambert, Ancestors, Authors and Canonicity, Journal of Cuneiform Sources 11 (1957), 8, gefolgt von H. Hunger, Babylonische und assyrische Kolophone, Alter Orient und Altes Testament 2, Neukirchen-Vluyn 1968, 90, Nr. 292. Siehe auch W.G. Lambert, Introduction: The Transmission of the Literary and Scholastic Tradition, in: I. Spar/ W.G. Lambert (Hgg.), Literary and Scholastic Texts of the First Millennium B.C., Cuneiform Texts in the Metropolitan Museum of Art II, New York 2005, XVI.

5 Bei S.J. Lieberman, Canonical and Official Cuneiform Texts: Toward an Understanding of Assurbanipal's Personal Tablet Collection, in: Tzvi Abusch et al. (Hgg.), Lingering over Words: Studies in Ancient Near Eastern Literature. Fs. William L. Moran, Atlanta 1990, 325, Anm. 99.

1' Spuren

---

2' [U$_4$.ME DÙG.GA.MEŠ K]A 7 *ṭup-pa-a-[te/ni]* ⸢GABA.RI⸣ U[D.KIB.NUN$^{ki}$]
3' [NIBRU$^{ki}$ KÁ.DINGIR.RA$^{ki}$ UD.UN]UG$^{ki}$ ŠEŠ.UNUG$^{ki}$ UNU[G$^{ki}$ *u Eri-du*$_{10}$$^{ki}$]
4' [*um-ma-a-ni ú-na-as-si*]*-ḫu-ma* ⸢*ú*⸣*-*[*na-as-si-qu-ma*]
5' [*a-na* $^{I}$*Na-zi-múru*]*-taš* LUGAL ŠÚ SUM*-nu ana* [*ṣu-bu pu-ut-qe-e*]
6' [*za-re-e šèr-re*]*-*⸢*e*⸣ *šá-ba-áš ka-re-*[*e ù mim-ma ṣi-bu-te* DÙG.GA]

---

7' [ ] x U$_4$ 7 *u* x x [ ]

Der Text in Z. 2' zeigt eindeutig, dass Labats ursprüngliche Lesung die richtige ist. Das erste, hier *ṭup* gelesene Zeichen weist nur drei senkrechte Keile auf und sieht damit wie das Zeichen UM aus, doch ist dies bei dem Zeichen DUB häufig zu beobachten.[6] Das nachfolgende Zeichen PA lässt dann keinen Zweifel mehr an der korrekten Lesung.

Dieser Kolophon wird immer wieder als Beleg dafür herangezogen, dass die Kanonbildung bereits zu Zeiten des kassitischen Herrschers Nazimaruttaš, der von 1308–1242 v. Chr. regierte, erfolgte.[7] Doch bereits hier zeigt sich, wie viel Vorsicht bei Schlussfolgerungen aus Kolophonen geboten ist. Zum einen datieren alle Abschriften dieser Hemerologie aus dem ersten vorchristlichen Jahrtausend; der Text ist damit nicht kontemporär, sondern schildert Vorstellungen über die Schaffung des Textes, die auf Quellen aus der Zeit von Nazimaruttaš zurückgehen können, aber nicht notwendigerweise müssen. Zum anderen bietet die Passage zwar Erläuterungen zu diesem einen Text, aber nicht zum Prozess der Serienbildung, da diese Hemerologie in keiner Serie integriert war, sondern ein solitärer Text blieb. Von diesem *caveat* abgesehen bietet der Text jedoch interessante Informationen über die Vorstellungen, die über die Entstehung des Textes im ersten Jahrtausend v. Chr. kursierten. So zeigt der Vermerk über die Übergabe des neu geschaffenen Textes an den babylonischen König Nazimaruttaš, wie die Tätigkeit der Gelehrten im Umfeld des Königshofes gesehen wird, wo „autoritative" Editionen geschaffen werden. Weiterhin wird ihre Arbeitsweise geschildert, indem Vorlagen aus sieben verschiedenen Städten erwähnt werden. Auch wenn die im Alten Orient so bedeutungsträchtige „Siebenzahl"[8] hier eher Zweifel an dem Wahrheitsgehalt des Textes hervorruft, so zeigt sie doch, dass die „Arbeit am Text" unter Berücksichtigung möglichst vieler verschiedener Traditionen den

6 Zu dem wie das Zeichen UM geschriebenen DUB siehe auch die Bemerkungen von R. Borger, Mesopotamisches Zeichenlexikon, Alter Orient und Altes Testament 305, 301.

7 Siehe u.a. W. von Soden, Mitteilungen der Deutschen Orientgesellschaft 85 (1953), 22f., W.G. Lambert, Ancestors, Authors, and Canonicity, Journal of Cuneiform Sources 11 (1957), 8, W.G. Lambert, Introduction: The Transmission of the Literary and Scholastic Tradition, in: I. Spar/W.G. Lambert (Hgg.), Literary and Scholastic Texts of the First Millennium B.C., Cuneiform Texts in the Metropolitan Museum of Art II, New York 2005, XVI.

8 Zur Bedeutung der Zahl 7 im Alten Orient siehe *inter alia* S. Izre'el, Adapa and the South Wind, MC 10, Winona Lake 2001, 132f., J.M. Sasson, About „Mari and the Bible", RA 92 (1998), 110, Anm. 48 und W.L. Moran, UET 6, 402: Persuasion in the Plain Style, JANES 22 (1993), 119f.

Gelehrten durchaus vertraut war. Ob der Reihenfolge der erwähnten sieben Städte eine inhärente Hierarchie zugrunde liegt oder aber die Städte eher grob geographisch von Norden nach Süden aufgelistet werden, bleibt noch unsicher. Leider ist es auch unklar, in welcher Stadt die Gelehrten diesen Text schufen, jedoch dürfte Babylon aufgrund der im Text zum Ausdruck kommenden Nähe zum König eine der wahrscheinlichsten Kandidatinnen sein.

Ein anderer Kolophon auf einer sehr gut erhaltenen mittelbabylonischen Tafel aus dem Kunsthandel hat dazu geführt, dass die Entstehung der terrestrischen Omenserie *šumma ālu ina mēlê šakin* verschiedentlich mit der Zeit des spätkassitischen Königs Melišipak (1186–1172 v. Chr.) in Verbindung gebracht wurde.[9] Der Anfang des Kolophons der Tafel (BM 108874[10]) lautet:

> PAP 25 MU.BI GABA.RI $^{kur}$SU.BIR$_{4}^{ki}$ ITI APIN.DU$_{8}$.A U$_{4}$ 8-KAM MU 3-KAM 2-KAM
> *Me-li-ši-pak* LUGAL

> „Zusammen 25 Omina, Vorlage aus Assyrien, Monat Araḫsamna, achter Tag,
> drittes – zweites[11] Jahr des Königs Melišipak."

Auch dieser Kolophon und seine Interpretationen belegen, wie behutsam man bei der Interpretation von Texten und Serienzugehörigkeiten sein sollte. Allein aufgrund der großen Ähnlichkeit der im Text erscheinenden 25 Vogelomina zu einer neuassyrischen Tafel, die wahrscheinlich zur Omen-Serie *šumma ālu* gehört, wurde diese Tafel als Manuskript dieser Serie aufgefasst, die wiederum allein aufgrund dieses Kolophons bereits in die Zeit des Melišipaks datiert wurde. Jedoch sind die 25 Vogel-Omina nur weitgehend parallel zu dem jüngeren Text und stellen kein echtes Duplikat dar;[12] zudem – und dies ist noch weit schwerwiegender – weist der Text weder eine Stichzeile noch einen Serienvermerk auf, so daß von einer Zugehörigkeit zur terrestrischen Omen-Serie *šumma ālu* nicht ausgegangen werden kann. Dennoch ist dieser Text wiederum ein Beleg für die intensive Arbeit mit wis-

9 So schreibt F. Rochberg-Halton, Aspects of Babylonian Celestial Divination, Archiv für Orientforschung Beih. 22, Horn 1988, 25, Anm. 44: „The Middle Babylonian text BM 108874 (pub. C. J. Gadd, CT 40, 48–49) preserves omens from the series šumma ālu on a tablet dated to the reign of Meli-šipak (1186–1172 B. C.)". Siehe auch J.A. Brinkman, Materials and Studies for Kassite History I, Chicago 1976, 256. Die Bemerkung von S. M. Freedman, If a City Is Situated on a Height. Vol. 1: Tablets 1–21, Occasional Publications of the Samuel Noah Kramer Fund 17, Philadelphia 1998, 13 diese Tafel sei „the earliest Šumma Alu-type text that can be dated", also der erste datierbare Text, der dem Typ der Serientafeln entspricht, wurde von S.M. Maul, Reallexikon der Assyrologie 10, 1./2. Lieferung, 2003, 59a dahingehend interpretiert, daß die Serie „erstmals im 11. (sic) Jh. v. Chr. bezeugt" sei. S. Jakob, Mittelassyrische Verwaltung und Sozialstruktur (= Cuneiform Monographs 29), Leiden/Boston 2003, 527 wertet diese Tafel hingegen als „mB Rezensionen der Serie šumma ālu".

10 Die Tafel wurde von C.J. Gadd, CT 40,48–49 in einer Kompositkopie zusammen mit anderen Manuskripten aus der Kujuncik-Sammlung publiziert. Eine Neuedition dieser Tafel hat Nicla De Zorzi, Bird Divination in Mesopotamia: New Evidence from BM 108874, KASKAL 6 (2009), 85–135 vorgelegt.

11 Zu Doppeldatierungen dieser Art siehe J.A. Brinkman, Materials and Studies for Kassite History I, Chicago 1976, 410f.

12 Siehe hierzu die von C.J. Gadd, CT 40,48 am unteren Rand notierten zahlreichen Abweichungen.

senschaftlichen Texten in der kassitischen Zeit. Weiterhin deutet die Ähnlichkeit des Textes zu Teilen der späteren Serie darauf hin, dass die Textformung, also der Aufbau des Textes und die Anordnung der einzelnen Omina, bereits in kassitischer Zeit einen gewissen Standard erreicht hatte, so dass die Texte mit nur wenigen Änderungen bei der Bildung der Textserien aufgenommen wurden; solche Texte wurden somit wie „Textbausteine" weiter verwendet. Dieser Aspekt wird noch ausführlicher zu untersuchen sein.

Ein weiterer Hinweis auf die kassitische Zeit als Blüte der Gelehrsamkeit findet sich auch in den Familiennamen zahlreicher Gelehrtenfamilien aus dem ersten Jahrtausend. So zeigt sich anhand der in den Kolophonen verzeichneten Genealogien der Schreiber noch aus dem ausgehenden ersten Jahrtausend v. Chr., dass sie sich auf Ahnherren wie Aḫiūtu, Ḫunzu'u, Ekur-zākir oder Sîn-leqe-unnīnī berufen. Diese Ahnherren waren, wie W. G. Lambert gezeigt hat, nachweislich in kassitischer Zeit als Gelehrte tätig.[13]

All das – und noch weitere Evidenz könnte angeführt werden[14] – zeigt, dass zu Zeiten der Kassitenkönige die babylonische Gelehrsamkeit einen hohen Stand erreicht hatte. Trotz all dem gibt es aber nach wie vor keinerlei direkte Evidenz dafür, dass die Textserien, die die Bibliotheken des ersten Jahrtausends v. Chr. füllen, zu ebendieser Zeit geschaffen wurden.

Wenn wir nach den Kolophonen und Zeugnissen über die Gelehrten nun einen Blick auf die religiösen, mantischen und wissenschaftlichen Texte aus der mittelbabylonischen Zeit selbst werfen, dann ergibt sich doch ein recht ernüchterndes Bild, was die Anzahl dieser Texte angeht. Es bleibt zu hoffen, dass die mittelbabylonischen Texte aus Babylon im Vorderasiatischen Museum, die Olof Pedersén so vorbildlich katalogisiert hat, hier mit der Zeit Abhilfe schaffen werden.[15] Welches Potential zum Verständnis von Textgeschichte und Serialisierungsprozess in diesen Texten enthalten ist, soll im Folgenden anhand der Analyse von drei mittelbabylonischen, zu der Bibliothek einiger Opferschauer gehörigen Tafeln aus dem Merkes-Viertel in Babylon (Babylon M4) gezeigt werden.[16] Zwei dieser Tafeln sind zwar von Robert Koldewey in Babylon ausgegraben worden, aber dann in eine amerikanische Privatsammlung gelangt, wo sie von Albert Clay bearbeitet und publiziert wurden.[17]

13 W.G. Lambert, Ancestors, Authors and Canonicity, Journal of Cuneiform Studies 11 (1957), 1–14 und 112 sowie ders., in: I. Spar/W.G. Lambert (Hgg.), Literary and Scholastic Texts of the First Millennium B.C., Cuneiform Texts in the Metropolitan Museum of Art II, New York 2005, XV.

14 Insbesondere wird die Reputation der babylonischen Gelehrsamkeit zur Kassitenzeit auch durch den Wunsch der hethitischen Könige illustriert, babylonische Ärzte und Beschwörer an ihren Hof in Ḫattuša zu holen. Siehe dazu E. Edel, Ägyptische Ärzte und ägyptische Medizin am hethitischen Königshof. Neue Funde von Keilschriftbriefen Ramses' II. aus Bogazköy, Opladen 1976, 112–123 and A. Hagenbuchner, Die Korrespondenz der Hethiter II, Texte der Hethiter 16, Heidelberg 1989, 293.

15 O. Pedersén, Archive und Bibliotheken in Babylon. Die Tontafeln der Grabung Robert Koldeweys 1899–1917, Abhandlungen der Deutschen Orient-Gesellschaft 25, Saarbrücken 2005.

16 Ebd., 78–82.

17 A.T. Clay, Epics, Hymns, Omens and other Texts, Babylonian Records in the Library of J. Pierpont Morgan IV, New Haven 1923, pl. 19–20. Auf die Herkunft dieser beiden Tafeln aus Babylon hat bereits E. Weidner, Keilschrifttexte aus Babylon, Archiv für Orientforschung 16 (1952–53) 71–75 unter „Nr. 4 und 5" hingewiesen.

Obwohl die drei Tafeln einen Kolophon aufweisen, lassen sie sich nicht genau datieren, da die Kolophone jeweils nur die Anzahl der Omina und den Namen des Schreibers verzeichnen. Auch der Fundkontext der Tafeln ermöglicht keine genaue Datierung.[18] Alle drei Tafeln bieten auf den jeweils nur erhaltenen Rückseiten denselben Text, der Omina anhand von Beobachtungen der Darmwindungen eines Opferschafes enthält. Er bezieht sich auf den „Palast der Darmwindungen“ genannten Teil der Gedärme. Auf den ersten Blick ist der Text ein Duplikat zur vierten Tafel der Serie *šumma tīrānu*, die das zweite Kapitel der großen kanonischen *bārûtu*-Serie bildet, welche in Manuskripten aus dem ersten Jahrtausend v. Chr. erhalten ist. Bei genauer Betrachtung zeigt sich jedoch, dass die Texte aus dem zweiten vorchristlichen Jahrtausend und die Manuskripte der vierten Tafel des Kapitels *šumma tīrānu* der *bārûtu*-Serie durchaus einige Abweichungen aufweisen.

## Drei mittelbabylonische Texte zum *ekal tīrāni* und die vierte Tafel des Kapitels *šumma tīrānu* der *bārûtu*-Serie im Vergleich:

Die Tafeln MLC 2614 (BRM IV 15) und MLC 2615 (BRM IV 16) sowie Bab 36400[19] aus Babylon, deren Text sich dupliziert, behandeln Beobachtungen an dem „Palast der Darmwindungen“ (*ekal tīrāni*) genannten Teil der Gedärme. Der Text dieser drei mittelbabylonischen Tafeln ist der vierten Tafel des Kapitels *šumma tīrānu* der *bārûtu*-Serie aus dem ersten vorchristlichen Jahrtausend sehr ähnlich, die sich anhand von sechs Manuskripten aus Ninive und Babylonien fast vollständig rekonstruieren lässt. Da von BRM IV 15–16 jeweils nur die Rückseiten erhalten sind und auch nur die Rückseite von Bab 36400 publiziert ist, wird hier im folgenden der Text dieser drei Tafeln den Omina 34–68 der kanonischen Serientafel in „Partitur-Umschrift“ gegenübergestellt; die ersten 33 Omina der Serientafel bleiben unberücksichtigt.

## Die Textvertreter der vierten Tafel des Kapitels *šumma tīrānu* der *bārûtu*-Serie:

A = BM 32305 (unpub.), fast vollständige, aber sehr abgeriebene babylonische Tafel
B = BM 32607 (unpub.), größeres Fragment aus der Mitte einer babylonischen Tafel
C = K 8272 (teilw. A. Boissier, Choix I 87–88) + Sm 1276 (unpub.), größeres Fragment der Rückseite, assyrische Schrift

18 Die Bibliothek der Opferschauer aus dem Merkes kann nur allgemein in die spätkassitische Zeit datiert werden, siehe O. Pedersén, Abhandlungen der Deutschen Orient-Gesellschaft 25, 78.

19 Wie bereits O. Pedersén, Abhandlungen der Deutschen Orient-Gesellschaft 25, S. 80 erkannt hat, ist die Tafel Bab 36400 (M4, Nr. 20) ein Duplikat zu BRM IV 15 und 16. Ein Foto der Rückseite dieser Tafel ist ebd. S. 79, Fig. 34 veröffentlicht worden.

D = K 6483 + Sm 791 (unpub.), Fragment der oberen linken Ecke einer neuassyrischen Tafel, nur die Rückseite ist erhalten
E = K 3827 (unpub.), Fragment vom linken Rand einer neuassyrischen Tafel
F = BM 65447 (unpub.), untere rechte Ecke einer Tafel, babylonische Schrift

**Die mittelbabylonischen Tafeln:**

X = MLC 2614 (BRM IV/15), Fundnummer Bab 36385
Y = MLC 26155 (BRM IV/16), Fundnummer Bab 36384
Z = Bab 36400 (O. Pedersén, Abhandlungen der Deutschen Orient-Gesellschaft 25, 79, Fig. 34), nur ein Foto der Rs. der Tafel ist publiziert

| | | |
|---|---|---|
| 34 | A35 | [BE SAG] É.GAL ŠÀ.NIGIN ZI-*ma* UGU ŠÀ.NIGIN U$_{5}$ |
| | F13' | [ ŠÀ.NIGI]N U$_{5}$ |
| | XRs1 | [BE SAG É.GAL *t*]*i*-⸢*ra-ni*⸣ [Z]I-*ma* UGU *ti-ra-ni ir-kab* |
| | | |
| | A35 | N[UN ] |
| | F13' | NUN KUR NU UR$_{5}$-*tú* ŠU-*su* KUR-*ád* |
| | X1 | NUN KUR [NU] *šu-a-tú*! Š[U-*su* ] |
| | | |
| 35 | A36 | [BE SAG] É.GAL ŠÀ.NIGIN ZI-*ma* UGU 15 ŠÀ.NIGIN U[$_{5}$ ] |
| | E1' | BE SAG ⸢É.GAL⸣ ŠÀ.[NIGIN ] |
| | F14' | [ 1]5 ŠÀ.NIGIN U$_{5}$ *a-kal* NAM.RA |
| | X2 | [BE SAG É.GAL *t*]*i-ra-ni* Z[I]-*ma* UGU ZAG *ti-ra*-⸢*ni*⸣ *ir-kab* NUN ⸢x⸣ ⸢x⸣ [ ] |
| | | |
| 36 | A37 | [BE SAG É.GAL ŠÀ.NIGIN ZI]-*ma* U[GU 150 ŠÀ.NIGIN U$_{5}$] |
| | E2' | BE SAG É.GAL ŠÀ.N[IGIN ] |
| | F15' | [ UGU] 150 ŠÀ.NIGIN U$_{5}$ |
| | X3 | [BE SA]G ⸢É⸣.G[AL *t*]*i-ra-ni* Z[I-*m*]*a* UGU ⸢GÙB⸣ *ti-ra-ni ir-kab* |
| | Y1 | [BE SAG É.GAL *ti*]-*ra-ni* ZI-*ma* UGU GÙB *ti-ra-ni* U$_{5}$ |
| | | |
| | A37 | [ ] |
| | | Vs. von A bricht ab. |
| | E2' | [ ] |
| | F15' | KAR-*tu*$_{4}$ |
| | X3 | NUN *šá* KUR ⸢lú⸣[KÚR ] |
| | Y1 | KÚR ⸢x⸣ [ ] |

| | | |
|---|---|---|
| 37 | E3' | BE SAG É.GAL ŠÀ.NIGIN ZI-[*iḫ-ma* UGU 15 ] |
| | F16' | [ UG]U 15 ŠÀ.NIGIN $U_5$ |
| | X4 | [BE SA]G É.GAL *ti-ra-ni na-s*[*i*]-*iḫ-ma* UGU ZAG *ti-ra-ni ir-kab* |
| | Y2 | [BE SAG É.GAL *ti*]-*ra-ni na-siḫ-ma* UGU ZAG *ti-ra-ni* $U_5$ |
| | ZRs.1 | [BE SAG *e-kal*] *ti-ra-ni na-siḫ-ma* UGU ZAG *ti-ra-ni* $U_5$ |
| | | |
| | E3' | [ ] |
| | F16' | NUN KUR KÚR-*šú* TI-*qé* |
| | X4 | NUN KUR ⸢$^{\text{lú}}$⸣[KÚR ] |
| | Y2 | NUN KUR KÚ[R ] |
| | ZRs.1 | NUN KUR [ ] |
| | | |
| 38 | ERs1 | BE SAG É.GAL ŠÀ.NIGIN ZI-*iḫ-ma* UGU 150 ŠÀ.NIGIN $U_5$ |
| | F17' | [ UG]U 150 ŠÀ.NIGIN $U_5$ |
| | X5 | [BE SA]G É.GAL *ti-ra-ni na-s*[*i*]-*iḫ-ma* UGU GÙB *ti-ra-ni ir-kab* |
| | Y3 | [BE SAG É.GAL *t*]*i-ra-ni na-siḫ-ma* UGU GÙB *ti-ra-ni* $U_5$ |
| | Z2 | [BE SAG *e-k*]*al ti-ra-ni na-siḫ-ma* UGU GÙ[B *t*]*i-ra-ni* $U_5$ |
| | | |
| | ERs1 | KÚR KUR NUN TI-*qé* |
| | F17' | KÚR KUR NUN TI-*qé* |
| | | F zählt die Omina am Ende der Vs. zusammen: 38-ÀM |
| | X5 | $^{\text{lú}}$KÚR KUR NUN ŠU-⸢*su*⸣! [*i-kaš-šad*] |
| | Y3 | KÚR KUR NUN T[I-*qé*] |
| | Z2 | GABA.RI [0] |
| | | |
| 39 | E2 | BE SAG É.GAL ŠÀ.NIGIN ZI-*iḫ-ma ana* EGIR ŠÀ.NIGIN ŠUB-*ut* |
| | FRs.1 | [BE SAG É.GAL *ti-ra-ni na-siḫ-ma ana* EGIR] ŠÀ.NIGIN ŠUB-*ut* |
| | X6 | [BE SAG] É.GAL *ti-ra-ni na-s*[*i-i*]*ḫ-ma ana* EGIR-*at ti-ra-ni* ŠUB-*ut* |
| | Y4 | [BE SAG É.GAL] *ti-ra-ni na-siḫ-ma ana* EGIR *ti-ra-ni* ŠUB-*ut* |
| | Z3 | [BE SAG] ⸢*e*⸣-*kal ti-ra-ni na-siḫ-ma ana* EGI[R] *ti-ra-ni* ŠUB-*ut* |
| | | |
| | E2 | NUN KUR-*su* BAL-*su* |
| | FRs.1 | NUN KUR-*su* BAL-*su* |
| | X6 | LUGAL *i-da-an-*⸢*ni-in*⸣-*ma* GABA.RI.A NU TUK-[*ši*] |
| | Y4 | LUGAL KALAG-*ma* [ ] |
| | Z3 | LUGAL KALAG-*ma* GABA.RI.A [NU TUK-*ši*] |

| | | |
|---|---|---|
| 40 | E3 | BE *ina* É.GAL ŠÀ.NIGIN DI-*ḫu nu-úr-ru-bu* ŠUB-*di* |
| | F2 | [ ] ŠUB-*di* |
| | X7 | [BE *ina*] É.GAL *ti-ra-ni zi-iḫ-*[*ḫ*]*u* $^{uzu}$ÚŠ SA$_5$ |
| | Y5 | [BE *i+na* É].GAL *ti-ra-ni* DI ÚŠ SA$_5$ |
| | Z4 | [BE *ina*] *e-kal ti-ra-ni* DI GUB-*ma* $^{uz}$[$^{u}$Ú]Š SA$_5$ |
| | | |
| | E3 | *ir-bu ana* É.GAL KU$_4$-*ub ú-lu* SUR-*an* AN-*e* |
| | F2 | *ir-bu ana* É.GAL KU$_4$-*ub ú-lu* SUR-*an* AN-*e* |
| | X7 | É.GAL NUN *is-sà-ap-pa-aḫ*! |
| | Y5 | É.GAL NUN BIR-[*aḫ*] |
| | Z4 | É.GAL NUN BIR-*aḫ* |
| | | |
| 41 | E4 | BE *ina* É.GAL ŠÀ.NIGIN DI-*ḫu* BABBAR ŠUB-*di* ILLU DU-*kam* |
| | F3 | [ ] ILLU DU-*kam* |
| | | |
| 42 | E5 | BE *ina* É.GAL ŠÀ.NIGIN DI-*ḫu* SA$_5$ ŠUB-*di* ŠUB-*ti* ŠÀ.ḪUL ŠUB-*ti* IZI |
| | F4 | [ ] ŠÀ.ḪUL ŠUB-*ti* IZI |
| | | |
| 43 | ARs1' | [BE *ina* É.GAL ŠÀ.NIGIN D]I-*ḫu dan-nu* ŠUB-*di* ŠÈG-*nu* ⸢*ṭaḫ-du* ILLU⸣ [ ] |
| | C1' | BE [*ina* É.GAL ŠÀ.NIGIN DI-*ḫu d*]*an-*⸢*nu* ŠUB⸣-*d*[*i* ] |
| | E6 | [B]E *ina* É.GAL ŠÀ.NIGIN DI-*ḫu dan-nu* ŠUB-*di* ŠÈG-*nu ṭaḫ-du* ILLU *mat-qu* DU-*kam* |
| | F5 | [ ] ŠÈG-*nu ṭaḫ-du* ILLU *mat-qu* DU-*kam* |
| | | |
| 44 | A2' | [BE *ina* É.GAL ŠÀ.NIGIN] DI-*ḫu* SIG$_7$ ŠUB-*di* ÉRIN-*ni* A.MEŠ *i-*[*kal-lu-ú*] |
| | C2' | BE [*ina* É.GAL ŠÀ.NIGIN DI]-*ḫu* SIG$_7$ ŠUB-*di* [ ] |
| | E7 | [BE *ina* É.GAL ŠÀ.NIGIN D]I-⸢*ḫu* SIG$_7$ ŠUB-*di* ÉRIN-*ni* A.MEŠ *i-kal*⸣-*l*[*u-ú*] |
| | | E bricht ab. |
| | F6 | [ ] ÉRIN-*ni* A.MEŠ *i-kal-lu-ú* |
| | | |
| 45 | A3' | [BE *ina* É.GAL ŠÀ.NI]GIN DI-*ḫu* GE$_6$ ŠUB-*di* ŠUB-*ti* [ ] |
| | C3' | BE *ina* É.GAL ŠÀ.NIGIN DI-*ḫu* GE$_6$ ŠUB-*di* [ ] |
| | F7 | [ ] ŠUB-*ti du-ri* |
| | | |
| 46 | A4' | [BE *ina* É.GA]L ⸢ŠÀ⸣.NIGIN *di-ik-šu* GUB-*iz* ZI-[*bu*] |
| | C4' | BE *ina* É.GAL ŠÀ.NIGIN *di-ik-šu* GUB-*iz* [ ] |
| | F8 | [ ] ZI-*bu* |
| | X8 | [B]E *ina* É. ⸢GAL *ti-ra-ni*⸣ [*di*]-*ik-*[*š*]*u iz-zi-iz ti-bu-um* |
| | Y6 | BE *i+na* É.GAL *ti-ra-ni di-ik-šu* GUB-*iz ti-bu-*[*um*] |
| | Z5 | [BE *ina*] *e-kal ti-ra-ni di-ik-šu* G[U]B-*iz ti-bu-um* [o] |

| | | |
|---|---|---|
| 47 | A5’ | [BE *ina* É].GAL ŠÀ.NIGIN *ir-ru* DIRI *it-tab-ši* NUN *ina* É.GAL-*šú* [ ] |
| | C5’ | BE *ina* É.GAL ŠÀ.NIGIN *ir-ru at-ru it-tab-ši* [ ] |
| | F9 | [ ] ⸢NUN⸣ *ina* É.GAL-*šú* Ḫ̮I.GAR.MEŠ-*šú* |
| | X9 | [B]E *ina* É.GAL *ti-ra-ni* $^{uzu}$[ŠÀ] *at-ru it-tab-ši* NUN *ina* É.GAL-*šu i-ba-ar-ru-uš* |
| | Y7 | BE *i+na* É.GAL *ti-ra-ni* ŠÀ *at-rù it-tab-ši* NUN *ina* É.GAL-*š*[*u* Ḫ̮I.GAR]-*šu* |
| | Z6 | [BE] *ina e-kal ti-ra-ni* ŠÀ DIRI *it-*[*ta*]*b-ši* NUN *ina* É.GAL-*šu* Ḫ̮I.GAR-*šu* [o] |
| 48 | A6’ | BE *ina* É.GAL ŠÀ.NIGIN *ir-ru* DIRI ⸢GIB⸣-[*i*]*k* NUN *ina* É.GAL-*šú* [ ] |
| | C6’ | BE *ina* É.GAL ŠÀ.NIGIN *ir-ru at-ru pa-*⸢*ri-ik*⸣ [ ] |
| | F10 | [ NUN] *ina* É.GAL-*šú* GAZ.MEŠ-*šú* |
| | X10 | [B]E *ina* É.GAL *ti-ra-ni* $^{uzu}$⸢ŠÀ *at-ru*⸣ *pa-ri-ik* NUN *ina* É.GAL-*šu* GAZ.MEŠ-*šu* |
| | Y8 | BE *i+na* É.GAL *ti-ra-ni* ŠÀ *at-rù pa-ri-ik* NUN *ina* É.GAL-*š*[*u* GAZ.MEŠ]-*šu* |
| | Z7 | [BE] *ina e-kal ti-ra-ni* ŠÀ DIRI GIB NUN *ina* É.GAL-*šu* GAZ-*šu* [o] |
| 49 | A7’ | BE *ina* É.GAL ŠÀ.NIGIN *ir-ru át*$^?$-*ru um-mat* DÙ |
| | C7’ | BE *ina* É.GAL ŠÀ.NIGIN *ir-ru at-ru um-mat* D[Ù] |
| | F11 | [ ] |
| | X11 | [B]E *ina* ⸢É.GAL⸣ *ti-*[*ra-ni*] $^{uzu}$ŠÀ *a*[*t*]-*ru u*[*m*]-*ma-*[*ta*] *ib-ni* |
| | Y9 | BE *i+na* É.GAL *ti-ra-ni* ŠÀ *at-rù um-ma-ta*$_5$ *ib-ni* |
| | Z8 | [BE] *ina e-kal ti-ra-ni* ŠÀ DIRI *um-*[*m*]*a-ta*$_5$ *ib-ni* |
| | A7’ | *šá* NU.MU.SU AŠ.TE [ ] |
| | C7’ | [ ] |
| | F11 | [*šá* NU].MU.SU AŠ.TE DAB-*bat* |
| | X11 | DUMU *al-ma-at-ti* $^{giš}$GU.ZA ⸢DAB⸣-*bat* |
| | Y9 | DUMU NU.MU.SU $^{gi}$[$^{š}$G]U.ZA DAB-*bat* |
| | Z8 | DUMU NU.MU.SU $^{giš}$GU.ZA DAB-*bat* [o] |
| 50 | A8’ | BE *ina* É.GAL ŠÀ.NIGIN BABBAR *ina* 15 ŠUB-*di* É.GAL NUN KÚR [TI-*qé*] |
| | C8’ | BE *ina* É.GAL ŠÀ.NIGIN BABBAR *ina* 15 ŠUB-[*di* ] |
| | F12 | [ N]UN KÚR TI-*qé* |
| | X12 | BE *ina* É.GAL *ti-ra-ni* BABBAR *ina* ZAG GAR-*in* [É.GAL] NUN $^{lú}$KÚR *i-leq-qé* |
| | Y10 | BE *i+na* É.GAL *ti-ra-ni* BABBAR *ina* ZAG GAR-*in* É.GAL N[UN K]ÚR TI-*qé* |
| | Z9 | [BE] *ina e-kal ti-ra-ni* BABBAR *ina* ZAG GAR-*in* É.GAL NUN KÚR TI-*qé* [o] |

51 A9’ BE *ina* É.GAL ŠÀ.NIGIN BABBAR *ina* 150 ŠUB-*di* É.GAL KÚR NUN [TI-*qé*]
C9’ BE *ina* É.GAL ŠÀ.NIGIN BABBAR *ina* 150 ŠUB-[*di* ]
F13 [ KÚ]R NUN TI-*qé*
X13 BE *ina* É.GAL *ti-ra-ni* BABBAR *ina* GÙB ⸢GAR⸣-*in* [É.GAL] $^{\text{lú}}$KÚR NUN *i-leq-qé*
Y11 BE *i+na* É.GAL *ti-ra-ni* BABBAR *ina* GÙB GAR-*in* É.GAL KÚR N[UN] TI-*qé*
Z10 [BE] *ina e-kal ti-ra-ni* BABBAR *ina* GÙB GAR-*in* É.GAL KÚR NUN TI-*qé* [o]

52 A10’ BE *ina* É.GAL ŠÀ.NIGIN ZI-*bu* KI.TA-*nu* ŠÀ.NIGIN GAR-*in*
C10’ BE *ina* É.GAL ŠÀ.NIGIN ZI-*bu* KI.TA-*nu* [ ]
F14 [ ]
X14 BE *ina* É.GAL *ti-ra-ni ti-bu* KI.⸢TA⸣-*nu* [*ti-ra-ni*] GAR-*in*
Y12 BE *i+na* É.GAL *ti-ra-ni ti-bu* K[I].TA-*nu ti-ra-ni* GAR
Z11 [BE] *ina e-kal ti-ra-ni ti-bu* KI.⸢TA-*nu*⸣ *ti-ra-ni* GAR

A10’ LUGAL *a-a-bi-šú* ŠU-*s*[*u* KUR-*ád*]
C10’ [ ]
F14 [LUGAL *a*]-*a-bi-šú* ŠU-*su* KUR-*ád*
X14 NUN *ja-a-bi-šu* ŠU-*su i-kaš-šad*
Y12 NUN *a-a-b*[*i*] ⸢ŠU⸣-*su* KUR-*ad*
Z11 NUN *a-a-bi-šu* ŠU-*su* KUR-*ad* [o]

53 A11’ BE *ina* É.GAL ŠÀ.NIGIN $^{\text{giš}}$TUKUL *a-ri-du* E$_{II}$-*ma* UGU ŠÀ.NIGIN U$_5$
C11’ BE *ina* É.GAL ŠÀ.NIGIN $^{\text{giš}}$TUKUL *a-ri-du* E$_{II}$-*ma* [ ]
F15 [ ŠÀ.NIGI]N U$_5$
X15 BE *ina* É.GAL *ti-ra-ni* $^{\text{giš}}$TUKUL *a-ri-du* ⸢È⸣-[*ma* UG]U *ti-ra-ni* U$_5$-*ma*
Y13 BE *i+na* É.GAL *ti-ra-ni* $^{\text{giš}}$TUKUL *a-ri-du* È-*ma* UGU *ti-ra*-[*ni*] ⸢U$_5$⸣
Z12 [BE] *ina e-kal ti-ra-ni* $^{\text{giš}}$TUKUL *a-ri-du* ⸢È⸣-*ma* UGU ŠÀ.NIGIN U$_5$

A11’f. ZI-*bu šu*[*r-ru-ur* ] | ÌR LUGAL SIG-*ú* [ ]
C11’ [ ]
F15f. ZI-*bu šur*-⸢*ru*⸣-*ur u* ŠUB.ŠUB-*ut* | [ *ra-b*]*i-ú* ZÁḪ-*bit*
X15 *šur-ru-ur ù* ŠUB.MEŠ-*ut* ÌR LUGAL ⸢SIG⸣ [ ]
Y13 *šur-rù-ur u* ŠUB.MEŠ ÌR LUGAL SIG *in-na*-⸢*bit*⸣
Z12 *šur-ru-ur u* ŠUB.ŠUB [ ]

54 A13’ BE *ina* É.GAL ŠÀ.NIGIN $^{\text{giš}}$TUKUL *a-ri-du* E$_{II}$-*ma* UGU 15 ŠÀ.NIGIN U$_5$
C12’ BE *ina* É.GAL ŠÀ.NIGIN $^{\text{giš}}$TUKUL *a-ri-du* E$_{II}$-*ma* UGU [ ]
F17 [ ŠÀ.NIG]IN U$_5$
X16 BE *ina* É.GAL *ti-ra-ni* $^{\text{giš}}$TUKUL *a-ri-du* ⸢È⸣-*m*[*a* U]GU ZAG *ti-ra-ni ir-kab*
Y14 BE *i+na* É.GAL *ti-ra-ni* $^{\text{giš}}$TUKUL *a-ri-du* È-*ma* UGU ZAG *ti-ra*-[*ni*] U$_5$
Z13 [BE] *ina e-kal ti-ra-ni* $^{\text{giš}}$TUKUL *a-ri-du* È-*ma* UGU ZAG ŠÀ.NIGIN U$_5$

A13' *a-kal* [ ]
C12' [ ]
F17 *a-kal* NAM.RA
X16 *a-kal* NAM.RA
Y14 GU$_7$ NAM.RA
Z13 GU$_7$ NAM.RA [o]

55 A14' [BE *ina*] É.GAL ŠÀ.NIGIN $^{giš}$TUKUL *a-ri-du* E$_{11}$-*ma* UGU 150 ŠÀ.NIGIN U$_5$
C13' BE *ina* É.GAL ŠÀ.NIGIN $^{giš}$TUKUL *a-ri-du* È-*ma* UGU 1[50 ]
F18 [ ŠÀ.NIGI]N U$_5$
X17 BE *ina* É.GAL *ti-ra-ni* $^{giš}$TUKUL *a-ri-du* È-*ma* [UGU] GÙB *ti-ra-ni ir-kab*
Y15 BE *i+na* É.GAL *ti-ra-ni* $^{giš}$TUKUL *a-ri-du* È-*ma* UGU GÙB *ti-ra-*[*n*]*i* U$_5$
Z14 [BE] *ina e-kal ti-ra-ni* $^{giš}$TUKUL *a-ri-du* È-*ma* UGU GÙB ŠÀ.NIGIN U$_5$

A14' KAR-[*tu*$_4$]
C13' [ ]
F18 KAR-*tu*$_4$
X17 MUNUS.KAR-*tum*
Y15 KAR-*tum*
Z14 KAR-*tum* [o]

56 A15' [BE É.GAL] ⸢ŠÀ.NIGIN 2⸣ DUMU LUGAL KÚR AŠ.TE AD-[*šu* DAB-*bat*]
C14' BE É.GAL ŠÀ.NIGIN 2 ]
D1' [ ] ⸢x x⸣ [ ]
F19 [ ] AD-[*š*]*u* DAB-*bat*
F bricht ab.
X18 BE É.GAL *ti-ra-ni* 2-*na* DUMU L[UGAL $^{g]iš}$GU.ZA *a-bi-šu i-ṣa-bat*
Y16 BE É.GAL *ti-ra-ni* 2 DUMU.MEŠ L[UGAL ] $^{giš}$GU.ZA *a-bi-šu-nu iš-ta-an-na-an-nu*
Z15 [BE] *e-kal ti-ra-ni* 2 DUMU.MEŠ LUGAL *ana* $^{giš}$GU.ZA *a-bi-šu-nu* SÁ.S[Á-*nu*]

57 A16' [BE É.GAL ŠÀ.NIGIN 2-*m*]*a* [SAG.UŠ-*tu*$_4$ UGU BAR-*ti* U$_5$]
C15' [BE É.GAL ŠÀ.NIGIN 2-*ma*] ⸢SAG⸣.UŠ-*tu*$_4$ UGU BAR-*ti* U[$_5$]
D2' [BE É.GAL ŠÀ.NIGIN 2-*ma* ]
X19 BE É.GAL *ti-ra-ni* 2-*ma* SAG.UŠ-*tum* UGU *a-*[*ḫu*]*-u ir-kab*
Y17 BE É.GAL *ti-ra-ni* 2-*ma* SAG.U[Š-*tum* U]GU *a-ḫi-i* U$_5$
Z16 [BE] *e-kal ti-ra-ni* 2-*ma* SAG.UŠ-*tum* UGU *a-ḫi-i* U$_5$

A16’ [ ]
C15’ [ ]
C bricht ab.
D2’ NUN LUGAL-*ut* [ ]
X19 NUN LUGAL-*ut kiš-šu-ti i-ṣa-bat*
Y17 ⸢NUN⸣ LU[GAL]-*ut kiš-šu-ti* DAB-*bat*
Z16 NUN LUGAL-*ut kiš-šu-*⸢*ti*⸣ [DAB-*bat*]

58 A17’ [BE É.GAL ŠÀ.NIGIN] ⸢*2-ma*⸣ BAR-*t*[$u_4$ U]GU SAG.UŠ $U_5$
D3’ [BE É.GAL ŠÀ.NIGIN *2-ma* ] ⸢SAG.UŠ $U_5$⸣
X20 BE É.GAL *ti-ra-ni 2-ma a-ḫi-tum* UGU ⸢SAG.UŠ⸣-*ti* $U_5$
Y18 BE É.GAL *ti-ra-ni 2-ma a-ḫu-*⸢*ú*⸣ [UG]U SAG.UŠ $U_5$
Z17 [BE] *e-kal ti-ra-ni 2-ma a-ḫu-ú* UGU SAG.UŠ $U_5$

A17’ ⸢GAR⸣-*nu* KUR-⸢*su*⸣ [ ]
D3’ GAR-*nu* KUR-*s*[*u* ]
X20 NUN *ana* ⸢$^{lú}$⸣[KÚR-*šú*] KA LAL GAR-*an*
Y18 NUN *ana* [KÚ]R-*šú* KA LAL GAR-*an*
Z17 NUN *ana* KÚR-*šu* KA LAL GAR-[*an*]

59 A18’ [BE É.GAL ŠÀ.NIG]IN *2-ma im-nu-ú* LUGAL.MEŠ UŠ.MEŠ-*ma* KUR *i-*[*be-lu*]
BRs.1’ [BE É.GAL ŠÀ.NIGIN *2-ma im-nu*]-⸢*ú* LUGAL⸣.MEŠ ⸢UŠ.MEŠ⸣-[*ma* ]
D4’ [BE É.GAL ŠÀ.NIGIN *2-ma im-nu*]-⸢*ú*⸣ LUGAL.MEŠ *in-nim-me-du-m*[*a* ]
X24 BE É.GAL *ti-ra-ni 2-ma im-nu-ú* LUGAL.E.NE *in-ni-im-mi-du-ma* KUR EN.MEŠ-*lu*
Y22 BE É.GAL *ti-ra-ni 2-ma im-nu-*⸢*ú*⸣ 2 LUGAL.MEŠ ⸢*in*⸣*-ni-im-mi-du-ma* KUR EN-*lu*
Z21 [BE] *e-kal ti-ra-ni 2-ma im-nu-ú* 2 LUGAL.MEŠ *in-nim-mi-du-ma* KUR EN-[*lu*]

60 A19’ [BE É.GAL ŠÀ].NIGIN *2-ma šu-me-lu-ú* LUGAL.MEŠ UŠ.MEŠ-*ma*
B2’ [BE É.GAL ŠÀ.NIGIN *2-ma šu-me*]-*lu-ú* LUGAL.MEŠ UŠ.MEŠ-*m*[*a*]
D5’ [BE É.GAL ŠÀ.NIGIN *2-ma šu-me-lu*]-*ú* LUGAL.MEŠ *in-nim-me-du-ma*
X25 BE É.GAL *ti-ra-ni 2-ma šu-me-lu-ú* LUGAL.E.NE *in-ni-im-mi-du-ma*
Y23 BE É.GAL *ti-ra-ni 2-ma šú-me-lu-ú* 2 LUGAL.MEŠ *in-ni-im-mi-du-ma*
Z22 [BE] ⸢*e*⸣*-kal ti-ra-ni 2-ma šú-me-lu-ú* 2 LUGAL.MEŠ *in-nim-mi-du-ma*

| | | |
|---|---|---|
| | A19’ | KUR NU *i-[be-lu]* |
| | B2’ | [ ] |
| | D5’ | [ ] |
| | X25 | KUR NU EN.MEŠ-*lu* |
| | Y23 | KUR NU EN-[*lu*] |
| | Z22 | KUR NU EN-[*lu*] |
| | | |
| 61 | A20’ | [BE É.GAL ŠÀ].NIGIN 2-*ma ina* 15 [G]AR.MEŠ AB.BA.MEŠ UŠ.MEŠ-*ma* |
| | B3’ | [BE É.GAL ŠÀ.NIGIN 2-*ma ina* 1]5 GAR.MEŠ AB.BA.MEŠ UŠ.ME[Š-*ma*] |
| | D6’ | [BE É.GAL ŠÀ.NIGIN 2-*ma ina* 15 GAR].MEŠ A.AB.BA.MEŠ *in-nim-me-du-ma* |
| | X26 | BE É.GAL *ti-ra-ni* 2-*ma ina* ZAG GAR.MEŠ-*nu ši-bu-tu in-ni-im-mi-du-ma* |
| | Y24 | BE É.GAL *ti-ra-ni* 2-*ma ina* ZAG GAR.MEŠ-*nu ši-bu-*$tu_4$ *in-ni-im-mi-du-ma* |
| | Z23 | [BE *e*]-⸢*kal ti-ra-ni* 2⸣-*ma ina* ZAG GAR.MEŠ-*nu ši-bu-*$tu_4$ *in-nim-mi-du-ma* |
| | | |
| | A20’ | KUR *i-[be-lu]* |
| | B3’ | [ ] |
| | D6’ | [ ] |
| | X26 | KUR *i-bé-lu* |
| | Y24 | KUR EN-[*lu*] |
| | Z23 | KUR EN-[*lu*] |
| | | |
| 62 | A21’ | [BE É.GAL ŠÀ].NIGIN 2-*ma ina* 150 GAR.MEŠ AB.BA.MEŠ UŠ.MEŠ-*ma* |
| | B4’ | [BE É.GAL ŠÀ.NIGIN 2-*ma ina*] 150 GAR.MEŠ AB.BA.MEŠ UŠ.MEŠ-[*ma*] |
| | D7’ | [BE É.GAL ŠÀ.NIGIN 2-*ma ina* 150 G]AR.MEŠ A.AB.BA.MEŠ *in-nim-me-du-ma* |
| | X27 | BE É.GAL *ti-ra-ni* 2-*ma ina* GÙB GAR.MEŠ-*nu* MIN-*ma* |
| | Y25 | BE É.GAL *ti-ra-ni* 2-*ma ina* GÙB GAR.MEŠ-*nu ši-bu-*$tu_4$ *in-ni-im-mi-du-ma* |
| | Z24 | [BE *e-kal ti-ra-ni* 2-*m*]*a ina* GÙB ⸢GAR.MEŠ-*nu*⸣ *ši-bu-*$tu_4$ *in-nim-mi-du-ma* |
| | | |
| | A21’ | KUR NU ⸢*i*⸣-[*be-el-lu*] |
| | B4’ | [ ] |
| | D7’ | [ ] |
| | X27 | KUR NU *i-be-el-lu* |
| | Y25 | KUR NU EN-[*lu*] |
| | Z24 | KUR NU EN-*l*[*u*] |
| | | |
| 63 | A22’ | [BE É.GAL ŠÀ].NIGIN 3 DUMU.MEŠ LUGAL *ana* AŠ.TE AD-*šú-nu* |
| | B5’ | [BE É.GAL ŠÀ.NIGIN 3] DUMU.MEŠ LUGAL *ana* AŠ.TE AD-[*šú-nu*] |
| | D8’ | [BE É.GAL ŠÀ.NIGIN 3] DUMU.MEŠ LUGAL *ana* AŠ.TE [AD-*šú-nu*] |
| | X21 | BE É.GAL *ti-ra-ni* 3 DUMU.MEŠ LUGAL *ana* ⸢$^{\text{giš}}$⸣⸢GU.ZA *a*⸣*-bi-šu-nu* |
| | Y19 | BE É.GAL *ti-ra-ni* 3 DUMU.MEŠ LUGAL [*ana* $^{\text{g}}$]$^{\text{iš}}$GU.ZA *a-bi-šu-nu* |
| | Z18 | [BE] *e-kal ti-ra-ni* 3 DUMU.MEŠ LUGAL *ana* $^{\text{giš}}$GU.ZA *a-bi-šu-nu* |

A22’ SÁ.[SÁ-*nu*]
B5’ [ ]
D8’ [ ]
X21 *ana* ⸢ḪUL⸣-*ti i-te-eb-bu-ú*
Y19 SÁ.SÁ-*nu*
Z18 SÁ.S[Á-*nu*]

64 A23’ [BE É.GAL Š]À.NIGIN 3-*ma* 2 *ina* 15 1 *ina* 150 GAR.MEŠ
B6’ [BE É.GAL Š]À.⸢NIGIN⸣ 3-*ma* 2 *ina* 15 1 *ina* 150 GAR.MEŠ
D9’ [BE É.GAL ŠÀ.NIGIN 3]-⸢*ma*⸣ 2 *ina* 15 1 *ina* 150 GAR-*in*
X22 BE É.GAL *ti-ra-ni* 3-*ma* 1-*et ina* ZAG 1-*e*[*t ina*] GÙB [GAR-*at*]
Y20 BE É.GAL *ti-ra-ni* 3-*ma* 1 *ina* ZAG 1 *ina* [GÙB] GAR-*at*
Z19 [BE] *e-kal ti-ra-ni* 3-*ma* 1 *ina* ZAG ⸢1⸣ *ina* GÙB GAR-*at*

A23’ NIN-*tu*$_4$ AŠ.TE [DAB-*bat*]
B6’ NIN-*t*[*u*$_4$ AŠ.TE DAB-*bat*]
D9’ NIN-*tu*$_4$ AŠ.T[E DAB-*bat*]
X22 *be-el-tum* $^{giš}$GU.ZA *i-ṣa-bat*
Y20 NI[N] $^{giš}$G[U.ZA] DAB-*bat*
Z19 NIN $^{giš}$G U.ZA DAB-⸢*bat*⸣

65 A24’ [BE É.GAL Š]À.NIGIN 3-*ma* 2 *ina* 150 1 *ina* 15 GAR.MEŠ AŠ.TE [KÚR-*ir*]
B7’ [BE É.GAL Š]À.NIGIN 3-*ma* 2 *ina* 150 1 *ina* 15 GAR.MEŠ AŠ.T[E KÚR-*ir*]
D10’ [BE É.GAL ŠÀ.NIGIN] ⸢3-*ma* 2 *ina* 150 1 *ina* 15 GAR-*in* AŠ.TE KÚ[R-*ir*]
X23 BE É.GAL *ti-ra-ni* 3-*ma* 1-*et* ⸢*i*⸣-[*ri-ik*] 1-*et* ⸢*ik*⸣-*ru* $^{giš}$GU.ZA *i-na-ak-ki-ir*
Y21 BE É.GAL *ti-ra-ni* 3-*ma* 1 *i-ri-ik* [1] *ik-ru* $^{giš}$⸢GU⸣.[Z]A *ina-kir*
Z20 [BE] *e-kal ti-ra-ni* 3-*ma* 1 GÍD.DA 1 *ik-ru* $^{giš}$GU.ZA *ina-kir*

66 A25’ [BE É.GAL Š]À.NIGIN ZI-*iḫ-ma ina* 15 ŠÀ.NIGIN *um-mat* DÙ
B8’ [BE É.GAL ŠÀ.N]IGIN ZI-*iḫ-ma ina* 15 ŠÀ.NIGIN *um-mat* DÙ
D11’ [BE É.GAL ŠÀ.NIGIN Z]I-*iḫ-ma ina* 15 ŠÀ.NIGIN *um-mat* DÙ
X28 BE É.GAL *ti-ra-ni na-si-iḫ-ma ina* ZAG *ti-ra-ni um-ma-ta ib-ni*
Y26 BE É.GAL *ti-ra-ni na-siḫ-ma ina* ZAG *ti*-⸢*ra-ni*⸣ *um-ma-ta*$_5$ *ib-ni*
Z25 [BE *e-kal ti-ra-ni na-siḫ-ma ina* ZAG *ti-ra-ni*] *um-ma-ta*$_5$ *ib-ni*

A25’ NUN *ina* URU-*šú iṭ-ṭar-rad-m*[*a* ]
B8‘ NUN *ina* URU-*šú* [*iṭ-ṭar-rad-ma* ]
D11‘ NUN *ina* URU-*šú iṭ-ṭar-rad-ma* [ ]
X28f. NUN *ina* URU-*šu iṭ-ṭa-rad-ma* | *ina* KUR MAN-*ma it-ta-nam-giš*
Y26 NUN *ina* URU-*šú iṭ-ṭa-rad-ma ina* KUR MAN-*ma it*-<*ta*>-*nam-giš*
Z25 NUN *ina* URU-*šu iṭ-ṭar-rad-ma ina* KUR [MAN-*ma it-ta-nam-giš*]

67 A26‘ [BE É.GAL ŠÀ].NIGIN ZI-*iḫ-ma* *ina* 150 ŠÀ.NIGIN *um-mat* DÙ
B9‘ [BE É.GAL ŠÀ.NIGIN Z]I-*iḫ-ma* *ina* 150 ŠÀ.NIGIN *um-mat* DÙ
D12‘ [BE É.GAL ŠÀ.NI]GIN ZI-*iḫ-ma* *ina* 150 ŠÀ.NIGIN *um-mat* DÙ
X30 [BE] ⸢É⸣.GAL *ti-ra-ni na-si-iḫ-ma* *ina* GÙB *ti-ra-ni* *um-ma-ta ib-ni*
Y27 BE É.GAL *ti-ra-ni na-siḫ-ma* *ina* GÙB *ti-*[*ra-ni*] *um-ma-ta*$_5$ *ib-ni*
Z26 [BE *e-kal* *ti-ra-ni na-siḫ-ma* *ina* GÙB *ti-ra-ni* *u*]*m-ma-ta*$_5$ *ib-ni*

A26‘ NUN *šá* KUR KÚR ⸢*ina* URU-*šú*⸣ *i*[*ṭ-ṭar-rad-ma* ]
B9‘ NUN *šá* KUR ⸢KÚR⸣ *in*[*a* URU-*šú* ]
D12‘ NUN *šá* KUR KÚR *ina* URU-*šú* *iṭ-*[*ṭar-rad-ma* ]
X30 NUN *šá* KUR $^{lú}$KÚR GABA.RI
Y27 GABA.RI
Z26 GABA.RI

68 A27‘ [BE É.GAL ŠÀ.NI]GIN *ana* 2 *i-bé-eš* UŠ$_4$ KUR MAN-*ni* :: [ (x x) ](Strich)
B10‘ [BE É.GAL ŠÀ.NIGIN *ana*] ⸢2⸣ *i-bé-eš* UŠ$_4$ [KUR MAN-*ni*] (Strich)
D13‘ [BE É.GA]L ŠÀ.NIGIN *ana* 2 *i-bé-eš* UŠ$_4$ KUR [MAN-*ni*] (Strich)
X31 [BE É].GAL *ti-*<*ra-ni*> *ana* 2-*na* *i-*⸢*zu*⸣*-uz* *ṭe*$_4$*-em* KUR *i-*MAN-*ni* (Strich)
Y28 BE É.GAL *ti-ra-ni a-na* 2 ⸢*i*⸣*-zu-*[*uz* UŠ]$_4$ KUR MAN-*ni* (Strich)
Z27 [BE *e-kal* *ti-ra-ni a-na* 2 *i-zu-uz* UŠ$_4$ K]UR MAN-*ni* (Strich)

---

Stichzeile und Kolophon von A:

28‘ [IM 4-KAM.MA B]E ŠÀ.NIGIN ÉŠ.GÀR NAM.AZU NU A[L.TIL]

---

29’ [BE *um-mat* ŠÀ.NI]GIN 2-*ta* ZI-*ib* MUNUS.ḪUL [o]
30’ [(GABA.RI ON$^{ki}$) GIM SUMUN-*šú* SA]R-*ma* IGI.KÁR *u up-pú-uš* IM $^{Id}$+EN-A-ŠEŠ DUMU *šá* $^{I}$MU-ŠEŠ A x [ ]
31’ [*pa-liḫ* GN *in*]*a qé-reb* NU GIŠ-*šú* *ina me-reš-tu*$_4$ NU ŠÀ-*bi la ú-še-l*[*i* ]
32’ [(ON$^{ki}$) $^{iti}$MN] U$_4$ 2-KAM MU 32-KAM $^{I}$*Ar-tak-šat-su* [LUGAL KUR.KUR]
A endet.

Stichzeile und Kolophon von B:

11’ [IM 4-KAM.MA BE ŠÀ.NIG]IN ÉŠ.GÀR NAM.[AZU NU AL.TIL]

---

12’ [BE *um-mat* ŠÀ.NIGIN 2-*ta*] ZI-*ib* [MUNUS.ḪUL]
13’ [(GABA.RI ON$^{ki}$) GIM SUMUN-*šú* SAR-*ma* IGI.KÁ]R *u up-pu-uš* I[M ]
14’ [(ON$^{ki}$) $^{iti}$MN U$_4$ x-KA]M MU 40$^{?}$-KA[M $^{I}$*Ar-tak-šat-su*$^{?}$ LUGAL KUR.KUR]
B bricht ab.

Stichzeile und Kolophon von D:

14' BE *um-mat* ŠÀ.NIGIN 2-*ta* ZI-*ib* MUNUS.ḪU[L]
15' DUB-*pi* 4-KÁM.MA ÉŠ.GÀR BE ŠÀ.NIGIN GIM SUMUN-*šá* SAR-*m*[*a bari*]
16' KUR $^{I}$AN.ŠÁR-DÙ-[IBILA MAN ŠÚ MAN KUR AN.ŠÁR$^{ki}$]
D endet.

Kolophon von X:

32 ŠU.NÍGIN 1 ŠU.ŠI BE É.GAL *ti-ra-ni* ŠU $^{I}$*Ì-lí-ma*-⌈ŠEŠ $^{lú}$ḪAL⌉
X endet.

Kolophon von Y:

29 ŠU.NÍGIN 1 ŠU É.GAL *ti-ra-*[*ni*]
30 ŠU $^{I}$ŠE.GA-GIŠ.M[I-$^{d}$AMAR.UTU]
31 U$_4$ [ ]
Y endet.

Kolophon von Z:

28 [ ] *En-bi-lu-lu*
Z endet.

**Übersetzung:**

34 [Wenn die Spitze] des Palastes der Darmwindungen aufgerichtet ist und auf den Darmwindungen reitet: Der Fürst wird ein Land, das ihm nicht gehört, erobern.
35 Wenn die Spitze des Palastes der Darmwindungen aufgerichtet ist und über der rechten Seite der Darmwindungen reitet: Genießen von Beute (X: Der Fürst ... ...).
36 Wenn die Spitze des Palastes der Darmwindungen aufgerichtet ist und über der linken Seite der Darmwindungen reitet: Verwüstung (X: Der Fürst des [Feindes]landes [... ...]; Y: Der Feind ... [... ...]).
37 Wenn die Spitze des Palastes der Darmwindungen verschoben ist und über der rechten Seite der Darmwindungen reitet: Der Fürst wird das Land seines Feindes erobern.
38 Wenn die Spitze des Palastes der Darmwindungen verschoben ist und über der linken Seite der Darmwindungen reitet: Der Feind wird das Land des Fürsten erobern.
39 Wenn die Spitze des Palastes der Darmwindungen verschoben ist und hinter den Darmwindungen abfällt: Das Land des Fürsten wird gegen ihn revoltieren (X, Y und Z: Der König wird stark werden und keinen Gegner haben).

40 (E und F:) Wenn im Palast der Darmwindungen eine weiche Finnenblase liegt: Einnahmen werden in den Palast kommen oder es wird regnen.
(X, Y und Z): Wenn im Palast der Darmwindungen eine Finnenblase da ist (Z: sitzt) und sie voller Blutgefäße[?] ist: Der Palast des Fürsten wird zerstreut werden.

41 Wenn im Palast der Darmwindungen eine weiße Finnenblase liegt: Die Flut wird kommen.

42 Wenn im Palast der Darmwindungen eine rote Finnenblase liegt: Trauer, Blitzschlag.

43 Wenn im Palast der Darmwindungen eine harte Finnenblase liegt: Überreichlicher Regen, eine angenehme Flut wird kommen.

44 Wenn im Palast der Darmwindungen eine gelb-grüne Finnenblase liegt: Mein Heer wird von Wasser zurückgehalten.

45 Wenn im Palast der Darmwindungen eine schwarze Finnenblase liegt: Fall der Mauer.

46 Wenn im Palast der Darmwindungen eine Schwellung sitzt: Aufstand.

47 Wenn im Palast der Darmwindungen ein überzähliges Gedärm(teil) erscheint: Man wird gegen den Fürsten in seinem Palast revoltieren.

48 Wenn im Palast der Darmwindungen ein überzähliges Gedärm(teil) quer liegt: Man wird den Fürsten in seinem Palast töten.

49 Wenn im Palast der Darmwindungen ein überzähliges Gedärm(teil) eine Masse formt: Der Sohn einer Witwe wird den Thron besteigen.

50 Wenn im Palast der Darmwindungen ein weißer Fleck auf der rechten Seite liegt: Den Palast des Fürsten wird der Feind erobern.

51 Wenn im Palast der Darmwindungen ein weißer Fleck auf der linken Seite liegt: Den Palast des Feindes wird der Fürst erobern.

52 Wenn im Palast der Darmwindungen eine Erhebung unterhalb der Darmwindungen liegt: Der König wird seine Feinde gefangen nehmen.

53 Wenn im Palast der Darmwindungen eine absteigende Keule hervorsteht und auf den Darmwindungen reitet, (A und C: eine Erhebung aufweist,) sehr wacklig ist und abfällt: Ein hochgestellter Diener des Königs wird fliehen.

54 Wenn im Palast der Darmwindungen eine absteigende Keule hervorsteht und auf der rechten Seite der Darmwindungen reitet: Genießen von Beute.

55 Wenn im Palast der Darmwindungen eine absteigende Keule hervorsteht und auf der linken Seite der Darmwindungen reitet: Verwüstung.

56 Wenn der Palast der Darmwindungen zweifach vorhanden ist: Der Sohn des (A: eines feindlichen) Königs wird den Thron seines Vaters besteigen (Y und Z: Die Söhne des Königs werden um den Thron ihres Vaters miteinander kämpfen).

57 Wenn der Palast der Darmwindungen zweifach vorhanden ist und der reguläre auf dem abnormen reitet: Der Fürst wird das Königtum der Welt ergreifen.

58 Wenn der Palast der Darmwindungen zweifach vorhanden ist und der abnorme über dem regulären reitet: ... ... sein Land [... ...] (X, Y und Z: Der Fürst wird seinem Feind demütig daherkommen).

59 Wenn der Palast der Darmwindungen zweifach vorhanden ist und sie rechts sind: (Y und Z: Zwei) Könige werden sich zusammentun und das Land beherrschen.

60 Wenn der Palast der Darmwindungen zweifach vorhanden ist und sie links sind: (Y und Z: Zwei) Könige werden sich zusammentun, aber das Land nicht beherrschen.

61 Wenn der Palast der Darmwindungen zweifach vorhanden ist und sie auf der rechten Seite liegen: Die Alten werden sich zusammentun und das Land beherrschen.

62 Wenn der Palast der Darmwindungen zweifach vorhanden ist und sie auf der linken Seite liegen: Die Alten werden sich zusammentun, aber das Land nicht beherrschen.

63 Wenn der Palast der Darmwindungen dreifach vorhanden ist: Die Söhne des Königs werden um den Thron ihres Vaters miteinander rivalisieren (X: gegen den Thron ihres Vaters in böser Absicht revoltieren).

64 Wenn der Palast der Darmwindungen dreifach vorhanden ist und zwei auf der rechten und einer auf der linken Seite liegen (X, Y und Z: und einer auf der rechten und einer auf der linken Seite liegt): Eine Herrin wird den Thron besteigen.

65 Wenn der Palast der Darmwindungen dreifach vorhanden ist und zwei auf der linken und einer auf der rechten Seite liegen (X, Y und Z: und einer lang und einer kurz ist): Der Thron wird sich ändern.

66 Wenn der Palast der Darmwindungen verschoben ist und eine Masse zur Rechten der Darmwindungen formt: Der Fürst wird aus seiner Stadt vertrieben werden und in einem anderen Land umherziehen.

67 Wenn der Palast der Darmwindungen verschoben ist und eine Masse zur Linken der Darmwindungen formt: Der Fürst des Feindeslandes wird aus seiner Stadt vertrieben werden und in einem anderen Land umherziehen.

68 Wenn der Palast der Darmwindungen sich in zwei Teile spaltet (X, Y und Z: teilt): Die Stimmung des Landes wird sich ändern.

---

Stichzeile und Kolophon von A:

28' [Vierte Tafel (des Kapitels) „We]nn die Darmwindungen" der Serie *bārûtu*, nicht be[endet].

---

29' [Wenn die Masse der Darmwin]dungen zweifach da ist: Aufkommen von Unheil.

30' [(Vorlage aus ...), wie sein Original geschr]ieben, kollationiert und hergestellt. Tafel des Bēl-apla-uṣur, des Sohnes des Nādin-aḫi, der Sohn des ... [... ...].

31' [Wer (die Gottheit) ... fürchtet, der soll] sie aus dem Inneren nicht wegtragen?, mit Absicht soll er sie nicht daraus entfernen [... ...].

32' [(Ort ...) Monat ...], zweiter Tag, 32. Jahr des Artaxerxes, [des Königs der Länder].

Stichzeile und Kolophon von B:

11' [Vierte Tafel (des Kapitels) „Wenn die Darmwindung]en" der Serie *bār*[*ûtu*, nicht beendet].

---

12' [Wenn die Masse der Darmwindungen zweifach da ist]: Aufkommen [von Unheil].
13' [(Vorlage aus ...), wie sein Original geschrieben, kollationie]rt und hergestellt. Ta[fel des ... ...].
14' [(Ort ...) Monat ...], ...-ter Tag, 40$^{?}$. Jahr [des Artaxerxes$^{?}$, des Königs der Länder].

Stichzeile und Kolophon von D:

14' Wenn die Masse der Darmwindungen zweifach da ist: Aufkommen von Unheil.
15' Vierte Tafel der Serie „Wenn die Darmwindungen". Wie sein Original geschrieben un[d kollationiert].
16' Palast des Assurbani[pal, des Königs der Welt, des Königs des Landes Assur].

Kolophon von X:

32 Summe: 60 (Omina von) Wenn der Palast der Darmwindungen. Hand des Ilīma-aḫu, des Opferschauers.

Kolophon von Y:

29 Summe: 60 (Omina von) Wenn der Palast der Darmwindun[gen].
30 Hand des Tab-Ṣill[i-Marduk (... ...)].
31 Tag$^{?}$ [... ...].

**Philologische Anmerkungen:**

38 In BRM IV 16 (Y) ist am Ende der Apodose noch ein waagerechter Keil erhalten, so daß hier wahrscheinlich mit den Manuskripten E und F zu T[I-*qé*] und nicht mit Manuskript X zu ŠU-⌈*su*⌉! [...] zu ergänzen ist.
43 Vgl. U. Koch-Westenholz, Babylonian Liver Omens, Kopenhagen 2000, Nr. 27/5. Zur Lesung *mīlu matqu* oder *mīlu šatqu* siehe CAD M/I 413bf.
53 Zu *šurrur* siehe CAD Š/II 58b und I. Starr und F.N.H. al-Rawi, Tablets from the Sippar Library VIII. Omens from the Gall-Bladder, Iraq 61 (1999), 184, die auf die Bedeutung „to wobble" hinweisen.

Kolophon von A:

Auffällig, aber für Kolophone nicht ungewöhnlich, ist die Schreibung *up-pú-uš* mit dem seltenen Lautwert *pú*, siehe dazu von W. von Soden und W. Röllig, Syllabar$^{4}$, 56,

Nr. 83. Vergleiche zu Zeile 31' die sehr ähnlichen Kolophone bei H. Hunger, Babylonische und assyrische Kolophone, Neukirchen-Vluyn 1968, Nr. 124–126 und Nr. 146, Z. 4. Zu dem unklaren Logogramm GIŠ in diesem Kontext siehe ebd., 13.

Durch diesen Vergleich einer Tafel der kanonischen *bārûtu*-Serie mit dem Text dreier leider nur fragmentarisch erhaltener mittelbabylonischer Tafeln zeigen sich sowohl die grundlegenden Übereinstimmungen, als auch die zahlreichen Abweichungen im Detail.[20] Von den 30 erhaltenen Omina der mittelbabylonischen Tafeln Bab 36400, BRM IV 15 und 16 wird in der vierten Tafel des Kapitels *šumma tīrānu* der *bārûtu*-Serie zwar keines ausgelassen, doch fünf Omina werden an einer Stelle hinzugefügt;[21] eine Protase und fünf Apodosen werden ersetzt,[22] je zweimal werden Wörter in der Protase und Apodose ersetzt[23] und zahlreiche kleinere Veränderungen und Hinzufügungen[24] sind zu beobachten. Zudem werden drei Omina der mittelbabylonischen Texte in der kanonischen Tafel an einer anderen, späteren Stelle eingeschoben; in der vierten Tafel des Kapitels *šumma tīrānu* wird also die Reihenfolge der Omina gegenüber dem mittelbabylonischen Text leicht verändert.[25] Zudem gab es auch Unterschiede im Umfang des Textes. So geben sowohl BRM IV 15 als auch BRM IV 16 die Anzahl der Omina im Text mit 60 an, während die vierte Tafel des Kapitels *šumma tīrānu* 68[26] Omina aufweist. Das heißt, dass an einer oder mehreren Stellen die kanonische Tafel drei weitere Omina aufweist, die in den mittelbabylonischen Tafeln nicht erscheinen.[27]

Diese Veränderungen zeigen zusammen mit den Kolophonen, dass die mittelbabylonischen Tafeln Bab 36400, BRM IV 15 und 16 keineswegs als Serientafeln der *bārûtu*-Serie angesehen werden dürfen. Der Text dieser drei Tafeln weist zwar eine sehr große Ähnlichkeit im Aufbau und in der Anordnung der Omina zu der kanonischen Tafel der Serie *bārûtu* auf, doch die zahlreichen Omenumstellungen, -hinzufügungen, -veränderungen

20 Die geringen Abweichungen (Protase von Omen 40, Apodosen der Omina 36, 56 und 63) innerhalb des Textes der drei mittelbabylonischen Tafeln können hier vernachlässigt werden.

21 Die Omina 41–45 erscheinen nur in der vierten Tafel des Kapitels *šumma tīrānu*.

22 Die Protase von Omen 65 und die Apodosen der Omina 35–36, 39–40 und 58.

23 Ersetzungen von einzelnen Wörtern finden sich in den Protasen der Omina 50–51 und den Apodosen von 38 und 49.

24 Kleinere Veränderungen, Auslassungen und Hinzufügungen finden sich in den Protasen der Omina 53, 64 und 68 sowie in den Apodosen der Omina 52, 59 und 60.

25 So erscheinen die Omina 63–65 in den mittelbabylonischen Tafeln bereits nach dem 58. Omen. Hier ist die Anordnung der kanonischen Tafel der *bārûtu*-Serie durchdachter, da in den mittelbabylonischen Tafeln die logische Zahlenfolge 2, 2-*ma*, 3, 3-*ma* durchbrochen wird und sich damit die Reihenfolge 2, 2-*ma*, 3, 3-*ma*, 2-*ma* ergibt.

26 Die Zahl ergibt sich aus der Rekonstruktion aller Textvertreter, kann sich aber aufgrund eines kleinen Bruches in BM 32305 (unpub.), dem vollständigsten Textvertreter, noch um ein Omen erhöhen oder verringern.

27 Da eventuell auch in den mittelbabylonischen Tafeln mit Omina gerechnet werden muß, die in der kanonischen Tafel der *bārûtu*-Serie ausgelassen wurden, können auch noch mehr Omina an anderer Stelle in die kanonische Tafel eingefügt worden sein.

und die unterschiedliche Gesamtzahl der Omina bezeugen die ebenso vorhandenen Abweichungen. Außerdem wird in den Kolophonen von BRM IV 15 und 16 die Anzahl der Omina direkt auf den „Palast der Darmwindungen" bezogen; ein Vermerk über eine eventuelle Serienzugehörigkeit fehlt völlig. Damit gleichen diese Kolophone den Kolophonen der altbabylonischen Opferschau-Kompendien, obwohl sie in der Textgestalt schon den kanonischen Tafeln der *bārûtu*-Serie des ersten vorchristlichen Jahrtausends nahestehen.

So erweist sich die Annahme, daß es sich um direkte Duplikate handelt, als Trugschluss. Auch fehlen in diesen mittelbabylonischen Texten die für Serientafeln typische Stichzeile zu einer folgenden Tafel sowie der Vermerk der Serienzugehörigkeit. Ganz ähnliche Beobachtungen lassen sich auch bei den anderen mittelbabylonischen Texten zur Opferschau machen. Diese Texte haben große Ähnlichkeiten zu den Serien des ersten Jahrtausends, zeigen im Detail aber auch einige Abweichungen und weisen niemals reguläre Serienvermerke in den Kolophonen auf, die sich mit den Serien des ersten vorchristlichen Jahrtausends in Einklang bringen lassen.[28] Die große Ähnlichkeit dieser mittelbabylonischen Texte zu den späteren kanonischen Tafeln zeigt jedoch, dass sie zu dieser Zeit bereits weitgehend „durchgearbeitet", d.h. in eine Form gebracht waren, die für mehr als ein Jahrtausend verbindlich bleiben sollte. Diese Texte wurden dann bei der Schaffung der Textserien wie „Textbausteine" weiterverwendet, wobei oft ganze Texte mit einigen wenigen Anpassungen als kanonische Tafeln in die Serie eingegliedert wurden.[29] Es zeigt sich somit, dass die spätkassitischen Opferschau-Texte den serialisierten Texten des ersten Jahrtausends v. Chr. weitaus ähnlicher sind, als den altbabylonischen Opferschau-Texten, die sich in Aufbau, Länge und Anordnung der Omina noch sehr voneinander unterscheiden.[30] Inwieweit sich diese anhand von Opferschau-Texten gewonnen Erkenntnisse auch an anderen Textgruppen bestätigen lassen, werden zukünftige Untersuchungen klären müssen; insbesondere die zahlreichen lexikalischen Texte dürften sich hierzu anbieten.

Diese Ergebnisse zeigen, dass die mittelbabylonischen – und auch die sehr ähnlichen mittelassyrischen – Texte zur Divination, Religion und Wissenschaft nicht ohne weiteres als kanonische Texte angesehen und zur Rekonstruktion von Textserien verwendet werden

28 Siehe z. B. die ganz ähnlichen Beobachtungen von U. Jeyes, A Compendium of Gall-Bladder Omens Extant in Middle Babylonian, Niniveh and Seleucid Versions in: A.R. George und I.L. Finkel (Hgg.), Wisdom, Gods and Literature, Festschrift W.G. Lambert, Winona Lake/Indiana 2000, 345–373. Zahlreiche weitere Beispiele finden sich unter den mittelbabylonischen Opferschau-Texten aus Assur, die demnächst vom Verfasser in der Serie *Keilschrifttexte aus Assur literarischen Inhalts* publiziert werden.

29 Dies lässt sich auch an anderen Serien beobachten, vgl. z. B. die Einfügung von schon existierenden Texten als „kanonische" Tafeln in die medizinisch-diagnostische Serie *sakikkû*, siehe dazu N.P. Heeßel, Babylonisch-assyrische Diagnostik (= Alter Orient und Altes Testament 43), Münster 2000, 107.

30 U. Jeyes, Old Babylonian Extispicy, Publications de l'Institut historique-archéologique néerlandais de Stamboul 64, Leiden 1989, 9 und Th. Richter, Untersuchungen zum Opferschauwesen, I. Überlegungen zur Rekonstruktion der altbabylonischen *bārûtum*-Serie, OrNS 62 (1993) 122.

dürfen.[31] Sie stellen vielmehr inhaltlich weitgehend „durchgearbeitete" und standardisierte Texte dar, die erst später und mit einigen Veränderungen in die Textserien integriert wurden, die in den Bibliotheken des ersten vorchristlichen Jahrtausends zu finden sind.[32]

Auch der auf die Kassitenzeit folgenden Periode der Isin II-Dynastie, die von 1157–1026 v. Chr. Babylonien regierte, wurde im ersten vorchristlichen Jahrtausend eine hohe literarische Aktivität zugeschrieben. Das belegt zum Beispiel die sog. „Liste der Weisen und Gelehrten",[33] die von dem Gelehrten Anu-bēlšunu aus Uruk in seleukidischer Zeit geschrieben wurde. Sie bietet eine Zusammenstellung bedeutender Weiser (*apkallu*) und Gelehrter (*ummânu*) von vor der Flut, angefangen bei Uanna/Oannes, bis hin zu Aḫiqar, dem *ummânu* des neuassyrischen Königs Asarhaddon. In der Liste werden zwei Gelehrte aufgeführt, die zu Zeiten der Isin-II-Dynastie aktiv waren, jedoch keiner, der während der kassitischen Zeit lebte. Bei diesen Gelehrten handelt es sich um Esagil-kīn-apli und Esagil-kīn-ubba:

16 [*ina tarṣi* $^{\text{Id}}$IŠKUR-IBILA-MU.SUM] LUGAL $^{\text{I}}$*É-sag-gil-ki-i-ni*-IBILA *um-man-nu*
17 [*ina tarṣi*] $^{\text{Id}}$IŠKUR-IBILA-MU.SUM LUGAL $^{\text{I}}$*É-sag-gil-ki-i-ni-ub*$^{!}$*-ba um-man-nu*
18 [*ina tarṣ*]*i* $^{\text{Id}}$AG-NÍG.DU-ŠEŠ LUGAL $^{\text{Id}}$*É-sag-gil-ki-i-ni-ub-ba* LU$^{\text{sic}}$ *um-man-nu*

16 [Während der Regierungszeit des] Königs [Adad-apla-iddina] war Esagil-kīn-apli der Gelehrte.
17 [Während der Regierungszeit] des Königs Adad-apla-iddina war (auch) Esagil-kīn-ubba der Gelehrte.
18 [(Auch) während der Regierungs]zeit des Königs Nebukadnezar (I.) war Esagil-kīn-ubba (bereits) der Gelehrte.[34]

31 So wurde z. B. VAT 9966 (KAR 403) von E. Leichty, The Omen Series *šumma izbu*, Texts from Cuneiform Sources 4, Locust Valley 1970, 91 als Manuskript der siebten Tafel der teratologischen Omenserie *šumma izbu* bearbeitet, obwohl dieser mittelassyrische Text keinen Kolophon aufweist und auch keinen der anderen Textvertreter dupliziert, siehe hierzu ausführlich N.P. Heeßel, KAL 1, 9 Anm. 80.

32 Zu dem sehr problematischen Begriff „Vorläufer", mit denen diese Texte auch gerne bezeichnet werden, siehe bereits W. Farber, „Forerunners" and „Standard Versions": A Few Thoughts about Terminology, in: M. Cohen et al. (Hgg.), The Tablet and the Scroll, Festschrift. W.W. Hallo, Bethesda 1993, 95–97.

33 Die Liste wurde von J.J. van Dijk, Die Inschriftenfunde, in: UVB 18, Berlin 1962, 44–52 und I.S. Klotchkoff, The Late Babylonian List of Scholars, in: H. Klengel (Hg.), Gesellschaft und Kultur im Alten Vorderasien, Berlin 1982, 149–154 ediert. Siehe auch A. Lenzi, The Uruk List of Kings and Sages and Late Mesopotamian Scholarship, Journal of Ancient Near Eastern Religions 8.2 (2008) 137–169. Zu den assyrischen Gelehrten in dieser Liste siehe auch P.-A. Beaulieu, The Cult of AN.ŠÁR/Aššur in Babylonia after the Fall of the Assyrian Empire, State Archives of Assyria. Bulletin 11 (1997) 55–73, besonders 67.

34 Zur Abfolge dieser Namen und zum zweimaligen Erscheinen von Esagil-kīn-ubba in dieser Liste siehe ausführlich N.P. Heeßel, Neues zu Esagil-kīn-apli. Die ältere Version der physiognomischen Omenserie *alamdimmû*, in: S.M. Maul/N.P. Heeßel (Hgg.), Assur-Forschungen, Wiesbaden 2010, 139–187, besonders 160 f.

Sowohl Esagil-kīn-apli als auch Esagil-kīn-ubba sind uns gut bekannt. Esagil-kīn-ubba ist der Autor der babylonischen Theodizee, der sich durch ein Akrostichon selbst als Autor zu erkennen gibt: *a-na-ku Sa-ag-gi-il-ki-[i-na-am-u]b-bi-ib ma-áš-ma-šu ka-ri-bu ša i-li ú šar-ri* „Ich bin Saggil-kīnam-ubbib, der *mašmaššu*, ein Beter des Gottes und des Königs." Esagil-kīn-apli[35] ist uns noch weitaus besser bekannt, da er sich in einem Text ausführlich als Editor der medizinisch-diagnostischen Serie *sakikkû*(SA.GIG) und der physiognomischen Omenserie *alamdimmû* beschreibt:[36]

„Betreffs dessen, was seit alters her keine Serialisierung erfahren hatte und aufgrund von widersprüchlichen Traditionen, für die keine Textduplikate verfügbar waren – in der Zeit der Herrschaft von Adad-apla-iddina, dem König von Babylon, um es neu zu schaffen, hat Esagil-kīn-apli, der Nachkomme des Asalluḫi-mansum, des Weisen von König Ḫammurapi, der ‚*ummatu*' von Sîn, Lisi und Nanaja, ein vornehmer Bürger der Stadt Borsippa, ... der Hauptgelehrte von Sumer und Akkad, durch die geniale Intelligenz, die die Götter Ea und Asalluḫi ihm verliehen haben, (alles) gründlich erwogen und die Neufassung von *sakikkû* geschaffen, „Von der Schädeldecke zu den Füßen", und (sie) für die Lehre etabliert. ... Möge der Beschwörer, der die Entscheidungen fällt, der über das Leben der Menschen wacht, der die Serie *sakikkû* gänzlich kennt, den Patienten untersuchen und die Texte überprüfen, möge er alles gründlich erwägen und erst dann seine Diagnose für den König treffen!".

Dieser Text, der auf Abschriften aus dem siebten vorchristlichen Jahrhundert erhalten ist, gibt den ersten direkten Hinweis auf die Schaffung einer der Serien, die auch in das erste Jahrtausend v. Chr. weitertradiert werden. Weiterhin lässt sich der Text durch die Erwähnung der Regierungszeit des Adad-apla-iddina recht genau auf 1068–1047 v. Chr. datieren. Der Kompilator Esagil-kīn-apli hielt hohe Ämter in Borsippa, der eng mit Babylon verflochtenen Stadt des Gottes der Schreibkunst Nabû, und war der *ummân šarri*, der wichtigste Gelehrte des babylonischen Königs. Ein weiteres wichtiges Dokument desselben Gelehrten ist der Katalog der Beschwörungskunst, der in mehreren Manuskripten aus Assur und Babylonien überliefert ist.[37] Dieser Katalog bietet tatsächlich zwei Kataloge, von denen der zweite laut Überschrift von Esagil-kīn-apli stammen soll.

Schließlich sei noch auf ein weiteres Dokument verwiesen, in denen Gelehrte und ihre Werke erwähnt werden. Dies ist der „Katalog der Texte und Autoren", der durch Ab-

35 Siehe zu diesem Gelehrten ausführlich N.P. Heeßel, ebd.

36 Zu diesem Text siehe I. L. Finkel, Adad-apla-iddina, Esagil-kīn-apli, and the Series SA.GIG, in: E. Leichty et al. (Hgg.), A Scientific Humanist, Gedenkschrift A. Sachs, Philadelphia 1988, 143–159 und N.P. Heeßel, Diagnostik, 13–17 und 104–110.

37 Den wichtigsten Textvertreter VAT 8275 hat E. Ebeling in Keilschriftkopie als KAR 44 publiziert und H. Zimmern, Zu den Keilschrifttexten aus Assur religiösen Inhalts, Zeitschrift für Assyriologie 30 (1915/16), 184–229, besonders 204–229 sowie J. Bottéro, Mythes et Rites de Babylone, Genf/Paris 1985, 65–112 haben den Text umschrieben, bearbeitet und kommentiert. M.J. Geller, Incipits and Rubrics, in: A.R. George/ I.L. Finkel (Hgg.), Wisdom, Gods and Literature, Festschrift W.G. Lambert, Winona Lake 2000, 225–258, bes. 242–258 hat eine vollständige Neubearbeitung dieses Textvertreters zusammen mit weiteren zwischenzeitlich identifizierten Textvertretern vorgelegt.

schriften aus der Bibliothek Assurbanipals in Ninive bekannt ist.[38] In diesem Text wird vielen Werken der Keilschriftliteratur ihr Autor zugeordnet sowie sein Beruf und seine Herkunft angegeben. In einigen Fällen wird auch der König genannt, unter dessen Herrschaft der Autor lebte. So findet sich in diesem Dokument auch die babylonische Theodizee erwähnt und wir erfahren, dass Saggil-kīnam-ubbib, dessen Name leider nicht erhalten ist, aus Babylon stammte.[39] Überhaupt ist es beeindruckend, wie viele der genannten Gelehrten in diesem Text aus Babylon stammten. Auch wenn wir den sehr fragmentarischen Zustand des Textes berücksichtigen, so ist doch auffällig, dass von zehn Gelehrten, deren Herkunftsort genannt wird, sieben aus Babylon kommen.[40] Dies dürfte die Bedeutung der Stadt als Zentrum der Gelehrsamkeit in Babylonien, zumindest für den Autor dieses Textes, spiegeln.

Nach der Evidenz dieser Texte, aber auch nach den Kolophonen und Texten von und über die babylonischen und assyrischen Gelehrten zeigt sich, dass die Standardisierung der Texte, also ihre inhaltliche Durcharbeitung zur Zeit der Kassitenherrscher nicht nur in vollem Gange, sondern bereits sehr weit fortgeschritten war. Die wissenschaftlichen Texte dieser Zeit sind den Serien des ersten Jahrtausends v. Chr. weitaus ähnlicher, als den vergleichbaren Texten der altbabylonischen Periode. Die endgültige Serienbildung hingegen dürfte erst – jedenfalls für die überwältigende Mehrheit der Texte – *nach* der Kassitenzeit eingesetzt haben. Hierfür finden sich in der späten Isin-II-Zeit, also im 11. Jh. v. Chr. mit Berichten über die Serialisierung von divinatorischen und medizinischen Texten erstmals direkte Belege.

38 W.G. Lambert, A Catalogue of Texts and Authors, Journal of Cuneiform Studies 16 (1962) 59–77.
39 Ebd. 66 V 2.
40 Ebd. 64–66 II 6; III 9; V 2 und 5; VI 8 und 14; VII 8.

*Wilfred van Soldt*

# The Role of Babylon in Western Peripheral Education

## 1. Introduction

In this article I will discuss the role that Babylon played in the transmission of knowledge to western peripheral areas. We are primarily concerned with school material, most of which derives from the Middle Babylonian period and has been found at sites in Anatolia, Syria, the Levant and Egypt, in particular at such important cities as Hattuša[1], Emar[2], Ugarit[3] and Amarna[4]. Babylonian sites have produced relatively few texts[5]; only at Ashur was a sizable corpus uncovered[6], the oldest texts of which can be dated to the middle of the 12th century, that is, at the end of the period that we are concerned with here.

## 2. Babylon and Babylonia in the peripheral archives

One of the ways to see whether or not Babylon and its inhabitants were directly involved in the education process in the periphery is a survey of place names and personal names of these areas. In this overview I will limit myself to data from texts found in Syria, the Levant and Egypt. As for Hattuša, in 1993 Gary Beckman gave an overview of Mesopotamians at this city.[7] He showed that direct contact between Hatti and Mesopotamia must have existed before the fall of Mittani in the late 14$^{th}$ century and that at least eight Babylonians and ten

1 See E. Laroche, Catalogue des textes hittites, Paris 1971, 299–316. The lexical texts from Hattuša are the subject of a dissertation to be published by Tobias Scheucher (see below, 3).

2 D. Arnaud, Recherches au pays d'Aštata. Emar VI/4: Textes de la bibliothèque: transcriptions et traductions, Paris 1987. A dissertation on the Emar lexical texts will be published by Merijn Gantzert (see below, 3).

3 See W. H. van Soldt, Babylonian Lexical, Religious and Literary Texts, in: M. Dietrich/O. Loretz (eds.), Ugarit. Ein ostmediterranes Kulturzentrum in Alten Orient (= Abhandlungen zur Literatur Alt-Syrien-Palästinas 7), Münster 1995, 171–212. New material has been published by B. André-Salvini; 1. Textes lexicographiques, in: M. Yon/ D. Arnaud, Études ougaritiques. I. Travaux 1985–1995 (= Ras Shamra-Ougarit 14), Paris 2001, 237f. and Textes lexicographiques de Ras Shamra-Ugarit (campagnes 1986–1992), Studi Micenei ed Egeo-Anatolici 46 (2004), 147–154.

4 See most recently S. Izre'el, The Amarna Scholarly Tablets (= Cuneiform Monographs 9), Groningen 1997.

5 For an overview, see N. Veldhuis, Elementary Education at Nippur. The List of Trees and Wooden Objects, Groningen 1997, 67f.

6 For more information on the lexical texts from Ashur, see most recently B. Groneberg, F. Weiershäuser, Th. Linnemann and D. Ullrich, http://www.mpg.de/bilderBerichteDokumente/dokumentation/jahrbuch/2005/gwdg/forschungsSchwerpunkt1/pdf.pdf.

7 G. Beckman, Mesopotamians and Mesopotamian Learning at Hattusha, Journal of Cuneiform Studies 35 (1983), 97–114.

Assyrians are attested at Hattuša, most of them during the Empire period. Only three among these are explicitly called scribes. One is the well-known Anu-šar-ilāni, who is mentioned as the father of the scribe Hanikuili during the Middle Hittite period, and the other two are Assyrians in Hittite service during the Empire period. Alp discussed a number of scribes with Akkadian names in the Mashat letters[8] and Mascheroni gave a prosopography of Hurrian scribes at Hattuša.[9]

In the city of Emar, Babylon and Babylonia are rarely attested. Three times the land of Karduniaš is mentioned in connection with a person originating there, first with Ibašši-ili[10], then with someone whose name is not fully preserved[11], then with a messenger from Karduniaš whose name is not fully given[12]. In a will we find references to a man from Babylon and a man from Ḫatti. The gentilic of the first is written syllabically as (lú-*lum*) kur *ba-bu-li-ia*.[13] The term *kaššû* is not attested. As shown by Cohen, at least one Babylonian scribe was active in Emar. His name was Kidin-Gula and he worked as a teacher in the house of the diviner's family (temple $M_1$).[14] As suggested by Cohen[15], he probably should be identified with the Kidin-Gula who occurs in an archive from House 5, which is noteworthy for its Babylonian script and language and for the Babylonian names of a number of persons attested. He seems to have belonged to this group of people with a foreign background. He probably also wrote a few of the documents found in House 5.[16]

Finally, in Amarna Karduniaš is attested in eleven letters, both in the international correspondence and in letters sent by vassals.[17] The term *kaššû* is attested three times, all in the expression *šar māt kaššî*, in three letters sent to the Pharaoh by Rib-Addi. The former accuses Aziru of Amurru of striving to be an equal to the kings of Mittani and Hatti, and with the king of the Kassites.[18]

8 S. Alp, Akkadian Names of Some Scribes in the Mashat-Letters, XXXIVème Rencontre Assyriologique Internationale, 6–10/VII/1987, Istanbul 1998, 47–61.

9 L. M. Mascheroni, Scribi hurriti a Bogazköy: una verifica prosopografica, Studi Micenei ed Egeo-Anatolici 24 (1984), 151–173.

10 [p]*i-ba-aš-ši*–dingir-*lì* lú kur *kar-du-ni-ia-aš*, in a list of personal names from the house of the diviner (Emar VI.3, 336:84; temple M[I]).

11 [...] *-mu-un-du* lú *kar-du-ni-ia* NA x, Arnaud reads *kar-du-ni-ia-«na»-á*[*š*][!] (Emar VI.3, 251:4; temple M[I]).

12 [lú]dumu.kin-*ri ša* kur *kar-du-ni-ia-«an»-aš* in a letter from the [lú]ugula.kalam.ma to a certain Agiya (Emar VI.3, 262:15; temple M[I]).

13 D. Arnaud, Mariage et remariage des femmes chez les syriens du moyen-Euphrate, à l'âge du Bronze Récent d'après deux nouveaux documents, Semitica 46 (1996), 13:12.

14 Y. Cohen, Kidin-Gula – The Foreign Teacher at the Emar Scribal School, RA 98 (2004), 81–100; for the family of Zū-Ba'la the diviner, see also M. Yamada, The Family of Zū-Ba'la the Diviner and the Hittites, Israel Oriental Studies 18 (1998), 323–334.

15 Y. Cohen/I. Singer, A Late Synchronism between Ugarit and Emar, in: Y. Amit/E. Ben Zvi/I. Finkelstein/O. Lipschits (eds.), Essays on Ancient Israel in its Near Eastern Context. A Tribute to Nadav Na'aman, Winona Lake 2006, 129f.

16 Ibid.

17 For the various spellings, see RGTC XII/2, 153. Letters to and from the Babylonian kings in which Kara(n)duniaš occur are EA 1–3, 5, 7–11. Letters from vassals are EA 200 (origin unknown) and 255 (Mut-Ba'li of Piḫilu).

18 EA 76:15, 104:19f., 116:71, see RGTC XII/2, 159.

Unfortunately, none of the attestations discussed so far give us any clue to the relations of the city of Babylon with the west. One can only say that Babylonians were able to travel to these parts of the ancient world and to do business there or to find employment as specialized craftsmen. Mesopotamian scribes are attested in the archives of the large cities and some of them also trained students. Direct evidence for this and for their relations with Babylon, however, is lacking. Nevertheless, Anu-šar-ilāni in Hattuša and Kidin-Gula in Emar are possible candidates.

## 3. The lexical and literary material

Since the data examined so far is of minimal value it is appropriate to look more closely at the texts themselves, that is, at the school texts that were found in Babylon and on the peripheral sites. Only then may it be possible to say more about the possible ways the transfer of knowledge took place.

Two years ago a project started in Leiden which aims at reconstructing the curriculum followed in the peripheral cities on the basis of the many school texts that were written there. Since its inception the lexical and the literary material from Hattuša has been studied by Tobias Scheucher and Jeanette Fincke and their results will be published in due course. At the same time Merijn Gantzert has been working on the Emar material and this combined effort will hopefully lead to an integrated approach to the evaluation of how and when the school material was transferred and what the impact of literacy was on the western periphery.

## 4. The Old Babylonian Period

However, I would first like to make a brief comparison between the most important features of education in Babylonia, in particular of the Old and Middle Babylonian periods, and the peripheral schools. I will focus on the types of tablets that were in use and on the order in which the exercises were studied.

We will start with the Old Babylonian period. The nature of the school texts and the curriculum during this period have been discussed by Miguel Civil, Niek Veldhuis, Steve Tinney and Eleanor Robson.[19]

19 M. Civil, Ancient Mesopotamian Lexicography, in: J. M. Sasson (ed.), Civilizations of the Ancient Near East IV, New York 1995, 2308f; N. Veldhuis, Elementary Education at Nippur. The List of Trees and Wooden Objects, Groningen 1997; S. Tinney, On the Curricular Setting of Sumerian Literature, Iraq 61 (1999), 159–172; E. Robson, The Tablet House: a Scribal School in Old Babylonian Nippur, RA 95 (2001), 39–67.

## 4.1. The dating of the texts

Veldhuis divides the Old Babylonian period into three different parts[20]:

1) Early Old Babylonian, until the fall of Larsa in 1763, when we have texts from Larsa, Uruk, Sippar and Kisurra,

2) Middle Old Babylonian, from the fall of Larsa till the abandonment of the south, from ca. 1763 to 1720, roughly the later years of Hammurabi and the reign of Samsuiluna. We have texts from Nippur, especially from the time close to the abandonment of the city, and from Isin and Ur. Texts from the last two cities are quite different from those found at Nippur.

3) Late Old Babylonian, from the abandonment of the south till the end of the period around 1600. Texts come from Babylon, Sippar, Sippar-Amnānum, Dilbat, Kiš, Tell Harmal, Meturan and some other places. The traditions in the north are rather different from those in the south, a subject to which I shall return later.

## 4.2. The typology of the lexical texts

As for the typology it is agreed that there are at least four different types of lexical texts[21]:

**Type I** refers to large multi-column tablets, with 2–6 columns. The tablets contain an entire composition, or a part thereof. The texts are written in relatively small script.[22]

**Type II** are the large teacher-student copies, which have on the obv. 2–3 col., with 10–30 lines. The left contains the teacher's example, the right the student's copy. The reverse has a long extract (3–6 col.) of the same or another composition, written in a smaller, cursive script.[23] The first column is usually preserved, whereas the student's columns are usually erased.

**Type III** refers to small single column tablets (Sum. im.gíd.da), with one column on each side containing an extract of 10 to 30 lines. Sometimes the line of the next section or composition of the series is given.[24]

**Type IV** are round or lenticular tablets, with 2 or 4 lines of a composition written by the teacher and copied by the student.[25]

20 N. Veldhuis, Elementary Education at Nippur. The List of Trees and Wooden Objects, Groningen 1997, 18f.

21 E. Reiner/M. Civil, Materials for the Sumerian Lexicon XII, The Series lú = ša and Related Texts, Rome 1969, 27f.; M. Civil, Ancient Mesopotamian Lexicography, in: J. M. Sasson (ed.), Civilizations of the Ancient Near East IV, New York 1995, 2308f.; N. Veldhuis, Elementary Education at Nippur. The List of Trees and Wooden Objects, Groningen 1997, 28f.; E. Robson, The Tablet House: a Scribal School in Old Babylonian Nippur, RA 95 (2001), 46f.

22 N. Veldhuis, Elementary Education at Nippur. The List of Trees and Wooden Objects, Groningen 1997, 29f.

23 Ibid., 34f.; E. Robson, The Tablet House: a Scribal School in Old Babylonian Nippur, RA 95 (2001), 46.

24 N. Veldhuis, Elementary Education at Nippur. The List of Trees and Wooden Objects, Groningen 1997, 38.

25 Ibid., 39. In the terminology of Tinney these are Type S tablets (1999, 160).

Robson distinguishes a special **Type P** for the prisms.[26] Since their contents are similar to the tablets of Type I, Civil and Veldhuis regard the prisms as part of the latter. The prisms have 4 or 6 sides, with 1 or 2 columns per side, and they have a hollow central axis through which a stick can be inserted.

## 4.3. The curriculum[27]

The Old Babylonian curriculum has two different phases. Phase one consists of lexical lists, model texts and proverbs, the second phase consists of literary texts.[28]

As for the first phase, Veldhuis has shown that the order of the compositions can be inferred from the Type II tablets. On the reverse of these tablets the student wrote a text which is either part of the same composition as the one on the obverse or part of another composition. When he wrote the same composition it is clear that the text on the reverse was studied at an earlier stage. On the basis of this we can distinguish the following four steps in the first phase:

Step 1: *writing techniques*: sign forms, Syll. Alph. B (sign forms), *tu-ta-ti* and personal names.

Step 2: *thematic noun lists* (Hh): trees/wood, other manufactured materials, animals, stones, etc.

Step 3: *advanced lists*: metrology, Ea, Lu (acrographic list), Izi, Kagal (acrographic), Nigga, Diri, multiplication & reciprocal tables.

Step 4: *introductory Sumerian*: model contracts and proverbs.

In the second phase the student studied Sumerian literary material and he was less intensively supervised by his teacher. On the basis of Nippur catalogs and the material in House F in Nippur, Tinney and Robson give the following reconstruction of the curriculum in the second phase[29]:

Step 1: The tetrad, which consists of four elementary hymns (Lipit-Eštar B, Iddin-Dagan B, Enlil-bani A and Nisaba A[30]).

26 E. Robson, The Tablet House: a Scribal School in Old Babylonian Nippur, RA 95 (2001), 46; cf. also S. Tinney, On the Curricular Setting of Sumerian Literature, Iraq 61 (1999), 160 (literary texts).

27 N. Veldhuis, Elementary Education at Nippur. The List of Trees and Wooden Objects, Groningen 1997, 40f.; E. Robson, The Tablet House: a Scribal School in Old Babylonian Nippur, RA 95 (2001), 46f.

28 N. Veldhuis, Elementary Education at Nippur. The List of Trees and Wooden Objects, Groningen 1997, 40f.

29 S. Tinney, On the Curricular Setting of Sumerian Literature, Iraq 61 (1999); E. Robson, The Tablet House: a Scribal School in Old Babylonian Nippur, RA 95 (2001), 39–67.

30 N. Veldhuis, Elementary Education at Nippur. The List of Trees and Wooden Objects, Groningen 1997, 65; S. Tinney, On the Curricular Setting of Sumerian Literature, Iraq 61 (1999), 159, 162f.; E. Robson, The Tablet House: a Scribal School in Old Babylonian Nippur, RA 95 (2001), 53.

Step 2: The decad: 10 important literary compositions, like the Hymn to the Hoe, some hymns to kings, the Kesh temple hymn, Gilgamesh & Huwawa, etc.[31]

Step 3: In Nippur House F, fourteen compositions follow the decad, including Eduba texts, Gilgamesh, Enkidu and the Netherworld, the Curse of Agade, the Lament over the destruction of Ur, Dumuzi's Dream.[32] There are many more compositions in House F.

### 4.4. Differences between north and south[33]

The traditions of the north and the south are different in two respects:

1) The texts in the northern tradition are shorter than those from the south, like, for example Ḫarra-ḫubullu.

2) Different texts were in use; in the south Ea was an important elementary school exercise, in the north Syllabary A (Sa) held that position.

According to Veldhuis, the destruction of southern Old Babylonian cities brought about a mixture of northern and southern traditions, because refugees came to the north, for example, from Uruk to Kiš.

## 5. The Middle Babylonian period

### 5.1. Find spots and dates

After the Old Babylonian period there is a break in the school material from Babylonia that lasts for several centuries. Lexical material was found in Nippur, Babylon, Kiš, Ur, Qal'at el-Bahrain and possibly Sippar.[34] The oldest tablet seems to be a fragment of Diri that was found on Bahrain[35]. Another tablet that was found together with this tablet is dated to a king Agum[36] and from this, one could infer that this text possibly dates to the 15th century.[37] In Nippur a number of school tablets were excavated and they were partly published by Veldhuis[38];

31 S. Tinney, On the Curricular Setting of Sumerian Literature, Iraq 61 (1999), 168f.; E. Robson, The Tablet House: a Scribal School in Old Babylonian Nippur, RA 95 (2001), 53.

32 E. Robson, The Tablet House: a Scribal School in Old Babylonian Nippur, RA 95 (2001), 54.

33 N. Veldhuis, Elementary Education at Nippur. The List of Trees and Wooden Objects, Groningen 1997, 22f.

34 N. Veldhuis, Kassite Exercises: Literary and Lexical Extracts, Journal of Cuneiform Studies 52 (2000), 67.

35 Ibid., 70.

36 B. André-Salvini, 1. Textes lexicographiques, in: M. Yon/ D. Arnaud, Études ougaritiques. I. Travaux 1985–1995 (= Ras Shamra-Ougarit 14), Paris 2001, 126, 163.

37 For the various Agums in early Kassite history, see J.A. Brinkman, Materials and Studies for Kassite History, Vol. I. A Catalogue of Cuneiform Sources Pertaining to Specific Monarchs of the Kassite Dynasty, Chicago 1976, 9f. and 95f.

38 N. Veldhuis, Kassite Exercises: Literary and Lexical Extracts, Journal of Cuneiform Studies 52 (2000), 67–94.

one of these texts is dated by him to the second half of the 13th century.[39] The whole group has to be dated between 1400 and 1225.[40] In Babylon the excavators retrieved 154 tablets and about 1000 fragments from a house in the Merkez and thanks to the work of Olof Pedersén we now at least have an idea of their contents and their whereabouts.[41] According to Pedersén a date in the Late Kassite period (ca. 13th century) seems likely. However, since the texts are not dated this cannot be verified.

## 5.2. Typology

The number of school texts that can be dated to the Middle Babylonian period is considerably less than that for the Old and Neo-Babylonian periods. We are therefore not certain that all types that were current in the Middle Babylonian period have actually come down to us. Veldhuis has studied the school texts of this period that were found at Nippur and he concluded that there were two clearly identifiable types. First, there are the so-called pillow-shaped tablets, which often have a literary text on the obverse and an excerpt of a lexical text on the reverse. His catalogue contains 54 tablets of this type from Nippur and 15 from other places. Second, there are the so-called round tablets which were discussed by Sassmannshausen and Veldhuis.[42] Twenty of the twenty-two texts published by Sassmannshausen are administrative in nature, the round tablets listed by Veldhuis contain texts similar to the pillow-shaped tablets.

Recently I have had the opportunity to study a number of the school texts from Merkez 25n1 in Babylon, archive M6 in Pedersén's recent book on the archives at Babylon. The contents of these texts are relevant enough for my overview to be mentioned.[43] Seven more texts from Babylon have been published in Vorderasiatische Schriftdenkmäler 24 and they have been discussed by Veldhuis.[44] These texts provide more evidence on the typology of the school material of this period. If we compare them with the Old Babylonian texts we can identify the following types:

39 See Veldhuis (ibid., 68), who dates one of the tablets from Nippur to the second half of the 13th century.

40 The earliest text from the Nippur archive is dated to Kadašman-Enlil I, year 15, that is around 1360, see J.A. Brinkman, Materials and Studies for Kassite History, Vol. I. A Catalogue of Cuneiform Sources Pertaining to Specific Monarchs of the Kassite Dynasty, Chicago 1976, 36, 139 (J.2.45), and 387 (text no. 15). Since the later king Kadašman-Enlil II reigned only eight years a date in the reign of that king can be ruled out.

41 O. Pedérsen, Archive und Bibliotheken in Babylon: Die Tontafeln der Grabung Robert Koldeweys 1899–1917 (= Abhandlungen der Deutschen Orient-Gesellschaft 25), Berlin/Saarbrücken 2005, 85f., archive M6.

42 N. Veldhuis, Kassite Exercises: Literary and Lexical Extracts, Journal of Cuneiform Studies 52 (2000); L. Sassmannshausen, Mittelbabylonische runde Tafeln aus Nippur, Bagdader Mitteilungen 28 (1997).

43 Of the 44 texts kept in the Vorderasiatische Museum in Berlin I was able to study 28. I would like to thank Dr. J. Marzahn for his permission and his kind assistance. One of the texts turned out to be a contract (VAT 21945). The MB lexical texts from Babylon in the Museum will be published by Mrs. Alexa Bartelmus (München).

44 N. Veldhuis, Kassite Exercises: Literary and Lexical Extracts, Journal of Cuneiform Studies 52 (2000), 67–94; see the list on p. 86.

**Elementary exercises.** Three small tablets contain scribal exercises that must have been written at the very beginning of the curriculum. Two of these are roundish, the third one has an irregular form. On one (VAT 13235[45]) there are verticals and horizontal rulings, on the second (VAT 13454[46]) verticals, horizontals and winkelhaken, and on the third (VAT 13399[47]) there are three lines in archaic script and two occurrences of the sign that has been interpreted by Sassmannshausen as the number 10. They can be compared to the Neo-Babylonian exercises of the three basic wedges, called by Petra Gesche '1.BAD exercises', in which all three elements of the script, verticals, horizontals and winkelhaken, are practised.[48]

**Type I**, the large multi-column tablets, with 2–6 columns of the OB period are also attested in this period. From archive M6 there are at least five such tablets, which contain one division of a series or a just a part of it (VAT 13372[49]). The script is that of an advanced student. Attested series are Hh, tablets 2, 4 and 5/7, Diri 5 and Erimḫuš 2. Prisms are so far not attested.

**Type II** tablets of the OB shape are so far not attested in the Middle Babylonian period. However, it would seem that the pillow-shaped tablets have taken over their function as exercise tablets during this phase of education. They are referred to as **Type V** by Civil.[50] As Veldhuis has pointed out, this type of exercise is already attested in the OB period[51], but then the tablets did not have a lexical text on the reverse. The Kassite pillow-shaped tablets from Babylon that I was able to study cover a wide variety of subjects and distinguishing the obverse and the reverse is not always straightforward. Half of the 16 tablets of this type are inscribed on only one side.[52] Six of them have a Sumerian or an Akkadian text or a bilingual, normally of a religious or literary nature, one has a multiplication table and one a few entries from Hh 3 and 5. Another text has omens on one side and what looks like a drawing of a bird on the other.[53] The remaining texts have a literary or religious excerpt on one side and a lexical excerpt on the other.[54] The first is normally written parallel to the long side of the

45 O. Pedérsen, Archive und Bibliotheken in Babylon: Die Tontafeln der Grabung Robert Koldeweys 1899–1917 (= Abhandlungen der Deutschen Orient-Gesellschaft 25), Berlin/Saarbrücken 2005, 88, no. 23.

46 Ibid., 89, no. 41.

47 Ibid., 89, no. 52.

48 P. D. Gesche, Schulunterricht in Babylonien im ersten Jahrtausend v. Chr., Münster 2000 (= Alter Orient und Altes Testament 275), 58f.

49 O. Pedérsen, Archive und Bibliotheken in Babylon: Die Tontafeln der Grabung Robert Koldeweys 1899–1917 (= Abhandlungen der Deutschen Orient-Gesellschaft 25), Berlin/Saarbrücken 2005, 89, no. 36.

50 M. Civil, Ancient Mesopotamian Lexicography, in: J. M. Sasson (ed.), Civilizations of the Ancient Near East IV, New York 1995, 2308.

51 N. Veldhuis, Kassite Exercises: Literary and Lexical Extracts, Journal of Cuneiform Studies 52 (2000), 67.

52 Note that among the pillow-shaped tablets published in Vorderasiatische Schriftdenkmäler 24 6 out of 7 have a text only on one side, see N. Veldhuis, Kassite Exercises: Literary and Lexical Extracts, Journal of Cuneiform Studies 52 (2000), 86.

53 VAT 13141, O. Pedérsen, Archive und Bibliotheken in Babylon: Die Tontafeln der Grabung Robert Koldeweys 1899–1917 (= Abhandlungen der Deutschen Orient-Gesellschaft 25), Berlin/Saarbrücken 2005, 89, no. 45.

54 Compare also Vorderasiatische Schriftdenkmäler 24, 15.

tablet, the second parallel to the short side[55], but one text[56] has both texts parallel to the short side and turns like a book. On another, one side has been completely erased (VAT 13182[57]). The religious and literary excerpts are either bilingual or in Akkadian (for example, incantations). The excerpts on the other side are lists of plants or numbers and there is one with an excerpt from Hh 9 (VAT 13357[58]). One tablet repeats the same part of Hh 2. First it was written by the teacher and then copied by a student.[59]

**Type III**, the small one-column tablets of the OB period, are rarely attested in the MB period. There is one tablet from Babylon which seems to fit the description[60], a text containing part of Diri IV. Although the left edge is broken it most probably has only a single column which covers the whole of the obverse, continues on the bottom edge and covers part of the reverse. The text has 27 lines, which comes close to the OB maximum of 30 and the average number of entries found on peripheral excerpt texts.[61]

**Type IV** are the OB round or lenticular tablets. In the MB period there are a large number of such texts from Nippur. Most of these have recently been published by Veldhuis and Sassmannshausen. The ones published by Veldhuis contain the same sort of excerpts as the pillow-shaped texts, including the different angles of orientation on obverse and reverse.[62] The ones edited by Sassmannshausen[63] contain administrative texts which were meant as exercises. Only four more or less lentil-shaped exercise tablets have been found at Babylon so far.[64]

55 See already N. Veldhuis, Kassite Exercises: Literary and Lexical Extracts, Journal of Cuneiform Studies 52 (2000), 67.
56 VAT 17176, O. Pedérsen, Archive und Bibliotheken in Babylon: Die Tontafeln der Grabung Robert Koldeweys 1899–1917 (= Abhandlungen der Deutschen Orient-Gesellschaft 25), Berlin/Saarbrücken 2005, 91, no. 115. The obverse of this tablet contains an Akkadian religious text, the reverse a list of gods in archaic script.
57 O. Pedersén, Archive und Bibliotheken in Babylon: Die Tontafeln der Grabung Robert Koldeweys 1899–1917, Berlin/Saarbrücken 2005, 89, no. 37.
58 Ibid., 88, no. 17.
59 VAT 13124 (ibid., 89, no. 38), the text has lines 277–282, but they are not identical with the canonical version. The Akkadian text on the other side is almost completely broken.
60 VAT 17100 (ibid., 91, no. 137), published as $S^{21}$ in MSL 15, 160f. (the number was unknown to the editors).
61 There is also a broken list of personal names among the lexical tablets (VAT 19255, O. Pedersén, Archive und Bibliotheken in Babylon: Die Tontafeln der Grabung Robert Koldeweys 1899–1917, Berlin/Saarbrücken 2005, 88, no. 7). However, the names are preceded by numbers an the text is surely administrative, possibly from the Old Babylonian period (note the spelling *a-wi-la-tum* in obv. 1' and the omission of the *personenkeil*).
62 N. Veldhuis, Kassite Exercises: Literary and Lexical Extracts, Journal of Cuneiform Studies 52 (2000), 67f., 85.
63 L. Sassmannshausen, Mittelbabylonische runde Tafeln aus Nippur, Bagdader Mitteilungen 28 (1997), 185–208.
64 O. Pedersén, Archive und Bibliotheken in Babylon: Die Tontafeln der Grabung Robert Koldeweys 1899–1917, Berlin/Saarbrücken 2005, 87.

### 5.3. The curriculum

Not much can be said about the curriculum at this point. We assume that the order of the tablets probably was more or less like the one in the Old Babylonian period and what we find in the peripheral areas, but this cannot be deduced from the Kassite texts themselves. One can safely assume, however, that the elementary exercises should be put at the beginning of the scribal education and the Type I tablets at the end. From contemporary parallels at the peripheral sites it is clear that the Type I texts belong to the last stage of the scribal training and that they were the last step towards the ultimate goal of the training, the memorization of the entire corpus. Whether this was also the case in Kassite Babylonia is difficult to ascertain. Perhaps the scribes were already relying on standardized written copies of lexical series more than on memorization, but additional material will have to be found before this can be decided.

Judging by the proficiency with which they have been written, the texts that should be placed between the elementary exercises and the Type I texts are the pillow-shaped tablets (Type II) and the lentils (Type IV). Their exact position in the curriculum, however, is unclear. The same goes for the few examples of Type III tablets, which one would expect to come after Type II.

## 6. The peripheral archives of the Late Bronze Age

### 6.1. Findspots and dates

Our richest sources for school material in this period are found in the western periphery. Here at least five major sites have produced enough school texts to allow a typology and a reconstruction of the curriculum. These sites are Hattuša in Anatolia, Alalah, Emar and Ugarit in Syria, and Amarna in Egypt. Other sites have produced only a few texts and cannot be used for a reconstruction of the curriculum.[65] As for the city of Ashur, which by some scholars is also regarded as a town peripheral to Babylonian learning, the archives there have to be dated at least half a century after the destruction of the peripheral sites. I hope to come back to them on another occasion.[66]

65 For tablets from Palestine, see W. Horowitz/T. Oshima/S. Sanders, Cuneiform in Canaan. Cuneiform Sources from the Land of Israel in Ancient Times, Jerusalem 2006 and A Bibliographical List of Cuneiform Inscriptions from Canaan, Palestine/Philistia, and the Land of Israel, Journal of the American Oriental Society 122 (2002), 753–766. See also K. van der Toorn, Cuneiform Documents from Syria-Palestine: Texts, Scribes and Schools, Zeitschrift des Deutschen Palästina-Vereins 117 (2000), 97–113.

66 The texts have been edited in the *Digitale Keilschriftbibliothek Lexikalischer Listen aus Assur* of Göttingen University, see http://keil.uni-goettingen.de/.

The site with the most lexical material is Ugarit, closely followed by Emar and Hattuša. Amarna has produced a few dozen texts and Alalah only a handful. The dates of the texts vary. Those from Hattuša date to the Middle Hittite and to the Empire periods[67], those from Alalah to the Hittite period (14th and 13th centuries)[68], those from Amarna to the mid-14th century and those from Emar and Ugarit to the latter part of the 13th and the early 12th centuries.[69]

## 6.2. Typology

The Syrian sites have produced tablets that more or less follow the typology and the curriculum known from Mesopotamia, but with some significant differences. For the sake of comparison I will retain the terminology used for the OB and MB texts.

**Elementary exercises.** Tablets with just a few wedges, such as those found in Babylonia, have not turned up at peripheral sites. One can only speculate that probably such tablets must have existed but that all of them have been reused for other purposes.

**Type I.** The large multi-column tablets with 2–6 columns that are known from the OB and MB periods are the best attested type in the periphery. Almost all these texts show a complete composition or a section of a composition. In Ugarit the unilingual texts are the majority and in view of the proficient way in which they were written they belong to the last stage of the curriculum.[70] The bilingual texts, which also form an important part of the total, were written less carefully and usually show a hand that was able to write cuneiform but still had to advance far in quality. Also the fact that the scribe sometimes needed more than one translation for a Sumerian word points to a somewhat lower level than that of the unilingual texts. All the major series are attested, although not every single one of them at all peripheral sites. Prisms seem to be attested at Hattuša only.[71]

**Type II** tablets in the OB format are not attested in the western periphery. However, we do have combinations of unilingual elementary exercises like *tu-ta-ti*, Syllable Alphabet A,

67 See T. Scheucher, forthcoming, The First Phase of the School Curriculum in Syria and Anatolia.
68 The tablets were found in levels I–III, which are to be dated in the Hittite period; see E. von Dassow, Archives of Alalaḫ IV in Archaeological Context, Bulletin of the American Society of Oriental Research 338 (2005), 19, 30 and 66; E. von Dassow, State Society in the Late Bronze Age. Alalah under the Mittani Empire (= Studies on the Civilization and Culture of Nuzi and the Hurrians 17), Bethesda 2008, 63f.
69 See also footnotes 1–6.
70 W. H. van Soldt, Babylonian Lexical, Religious and Literary Texts, in: M. Dietrich/O. Loretz (eds.), Ugarit. Ein ostmediterranes Kulturzentrum in Alten Orient (= Abhandlungen zur Literatur Alt-Syrien-Palästinas 7), Münster 1995, 174f.
71 Note the use of a cylinder at Hammam at-Turkman during the Late OB period, W. H. van Soldt, Three Tablets from Tell Hammām et-Turkman, in: Th. P. J. van den Hout/J. de Roos (eds.), Studio Historiae Ardens. Ancient Near Eastern Studies Presented to Philo H.J. Houwing ten Cate on the Occasion of his 65th Birthday, Leiden 1995, 279f.

Syllabary A, and paradigms/names on a single tablet. As far as I am aware there are no tablets where we find an example written by a teacher and a copy by a student.

**Type III**, the small one-column tablets of the OB period. As we have seen these also occur in the MB period, but they are rarely attested. The peripheral schools in Syria used a type of tablet that was more or less pillow-shaped but a little bigger than the Middle Babylonian ones (**Type V**[72]). The texts consist of about 30 lines from a serial composition, written in landscape format. Some OB examples are also attested with 30 lines. In the periphery, the left half was reserved for the Sumerian column and the right half for the Akkadian translation. Between the Sumerian and the Akkadian columns we often find a pronunciation column which begins in the middle of the tablet.[73] In about a quarter of the cases the right half is broken off and one wonders whether that was done deliberately for the purpose of memorizing the text. Since these tablets show no signs of repeated erasures and reuse as in the OB period, it is unlikely that the disappearance of the right half can be explained in the same way.[74]

**Type IV.** The number of lentils found in the periphery is small. The only one from Ugarit contains a list of names.[75]

## 6.3. Curriculum

The curriculum as we know it for the Syrian texts and the texts from Amarna is mainly based on the school texts from Ugarit. Here we find enough catchlines and colophons to ensure a reconstruction.[76] As it was in the OB period, the education in the periphery was divided into two different phases. In the first phase the student wrote exercises and word lists, in the second phase he wrote religious and literary texts. Since the structure of the second phase is hard to determine I will focus on the first phase and use the terminology for the OB period:

Step 1: *writing techniques*: *tu-ta-ti*, Syllable Alphabet A (sign forms), Syllabary A (with its paleographic version[77]), followed by personal names or other exercises (like paradigms).[78] The formats of Syllable Alphabet A and Syllabary A can be both unilingual and multilingual

72 See above, paragraph 5.2.

73 W. H. van Soldt, Babylonian Lexical, Religious and Literary Texts, in: M. Dietrich/O. Loretz (eds.), Ugarit. Ein ostmediterranes Kulturzentrum in Alten Orient (= Abhandlungen zur Literatur Alt-Syrien-Palästinas 7), Münster 1995, 175.

74 W. H. van Soldt, Studies in the Akkadian of Ugarit: Dating and Grammar (= Alter Orient und Altes Testament 40), Neukirchen-Vluyn 1991, 751 f.

75 RS 15.54, see PRU 3, 213.

76 See W. H. van Soldt, Three Tablets from Tell Hammām et-Turkman, in: Th. P. J. van den Hout/J. de Roos (eds.), Studio Historiae Ardens. Ancient Near Eastern Studies Presented to Philo H.J. Houwing ten Cate on the Occasion of his 65th Birthday, Leiden 1995, 172 f. For a reconstruction of the curriculum at Emar, see M. Gantzert, The Emar Lexical Texts: 1. Text Edition, 2. Composite Edition, 3. Structural Analysis, Vol. 3, Maastricht 2008, 134 f.; see also Y. Cohen, The Scribes and Scholars of the City of Emar in the Late Bronze Age, Cambridge/Mass, 2009, 46 f.

77 See M. Gantzert, Syrian Lexical Texts, Ugarit Forschungen 38 (2006), 269 f.

78 For the Emar versions, see ibid., 283 f.

(Syllable Vocabulary, Sa vocabulary). The unilingual versions were often combined on larger tablets.[79] A bilingual version of Syll. A is followed by the Weidner God List.[80]

Step 2: *various lists*: the Weidner God List, Metrological tables and Grammatical texts. The position of the last two in the curriculum is uncertain.[81]

Step 3: *advanced lists*: Hh, Lu, Izi, Diri, probably also Nigga.[82] The series Izi and Diri are only attested in bilingual copies. There are also two texts that belong to Erimḫuš and, probably, to Ea.[83] Where they should be placed is uncertain.[84]

## 6.4. The origin of the peripheral texts

The question of where the peripheral school texts came from probably has many different answers. For example, Beckman has shown that the texts in Hattuša came partly through Hurrian mediation, and partly directly from Mesopotamia. Also in Ugarit texts differ from school to school and some show strong Hurrian while others show Assyrian or Babylonian influence.[85] Due to the lack of sufficient material for comparison from contemporary Babylonia it has usually been assumed that the peripheral texts were borrowed from early Middle Babylonian or even Old Babylonian examples. This is probably true for a part of these texts, but we will also have to reckon with direct borrowing from contemporary Mesopotamia. In this connection I would like to end with text VAT 13372 from the M6 archive at Babylon[86], probably to be dated to the 13th century, that is, in about the same period as the texts from Hattuša, Emar and Ugarit. It is a Type I tablet probably with a complete version

79 W. H. van Soldt, Babylonian Lexical, Religious and Literary Texts, in: M. Dietrich/O. Loretz (eds.), Ugarit. Ein ostmediterranes Kulturzentrum in Alten Orient (= Abhandlungen zur Literatur Alt-Syrien-Palästinas 7), Münster 1995, 750.

80 RS 20.123+, Ugaritica 5, no. 137. For a reconstruction of the peripheral versions and Assur, see M. Gantzert, Syrian Lexical Texts, Ugarit Forschungen 38 (2006), 299f.

81 W. H. van Soldt, Three Tablets from Tell Hammām et-Turkman, in: Th. P. J. van den Hout/J. de Roos (eds.), Studio Historiae Ardens. Ancient Near Eastern Studies Presented to Philo H.J. Houwing ten Cate on the Occasion of his 65th Birthday, Leiden 1995, 205f.

82 Ibid., 173f. The only text that may belong to the series Nigga is RS 20.221, see W. H. van Soldt, Babylonian Lexical, Religious and Literary Texts, in: M. Dietrich/O. Loretz (eds.), Ugarit. Ein ostmediterranes Kulturzentrum in Alten Orient (= Abhandlungen zur Literatur Alt-Syrien-Palästinas 7), Münster 1995, 205. For Nigga at Emar, see M. Gantzert, The Emar Lexical Texts: 1. Text Edition, 2. Composite Edition, 3. Structural Analysis, Vol. 1, Maastricht 2008, 266f.

83 W. H. van Soldt, Three Tablets from Tell Hammām et-Turkman, in: Th. P. J. van den Hout/J. de Roos (eds.), Studio Historiae Ardens. Ancient Near Eastern Studies Presented to Philo H.J. Houwing ten Cate on the Occasion of his 65th Birthday, Leiden 1995, 205f.

84 Erimḫuš is attested at Ḫattuša, see CTH 301 and T. Scheucher, forthcoming.

85 W. H. van Soldt, The Syllabic Akkadian Texts, in: W. G. E. Watson/N. Wyatt (eds.), Handbook of Ugaritic Studies I/39, Leiden 1999, 42f.

86 O. Pedersén, Archive und Bibliotheken in Babylon: Die Tontafeln der Grabung Robert Koldeweys 1899–1917 (= Abhandlungen der Deutschen Orient-Gesellschaft 25), Berlin/Saarbrücken 2005, 89, no. 36. I want to thank the Vorderasiatisches Museum for their kind permission to quote part of this text.

of Ḫarra-ḫubullu 5. Since the text will be published in the near future I will limit myself to the third column of the tablet:

| | VAT 13372 | Hh can. | Emar | Ugarit | OB Ni. |
|---|---|---|---|---|---|
| R.i'.1' | [giš].p[isan$^?$] | | 4306 | 1' | – |
| | [giš.gan.d]u$_7^?$ | VIIb:298 | 4307 | 2' | – |
| | [giš.gan.du$_7$.ká.n]a$^?$ | 299 | 4308 | 3' | – |
| | [gi]š.[gà]r.ba | V:296 | 4309 | 4' | – |
| 5' | [g]iš.gag.gàr.ba | 297 | 4310 | 5' | – |
| | giš.si.gar | 294 | – | 6' | – |
| | giš.gag.si.[gar] | 295 | – | 7' | – |
| | giš.ugu.x.x.x | | 4311(gú.gàr.ba) | 8'(gú.ne.zag) | – |
| | giš.tukul | VIIa:6 | 4312 | 9' | 496 |
| 10' | giš.tukul.šu | 14 | [...] | 10' | 497 |
| | giš.tukul.úr.[ra] | 15 | [...] | 11' | 498 |
| | giš.tukul.dingir.[ra] | 9 | 4313 | 12' | 500 |
| | giš.tukul.dingir [7$^?$] | 17 | 4314 | | – |
| | | | 4315(lugal) | 13'(lugal) | |
| | [gi]š.tukul.da.[mu] | 16 ($^d$) | 4316 | 14' | – |
| 15' | [gi]š.tukul.giš$^?$.[ma.nu] | – | 4317 | 15' | – |
| | [gi]š.tukul.[ | | | | |
| | [gi]š.[ | | | | |
| | ... | | | | |

The table shows that the differences between this tablet and the versions from the Old Babylonian period and the first millennium are considerable. The versions that we have from Emar and Ugarit, however, are practically the same text.

Direct contacts between Babylonia (and Assyria) and the western periphery are well attested in the texts from Emar and Ugarit. Several letters found at Ugarit testify the Babylonian presence in the region[87] and the small archive found at House 5 in Emar mentioned earlier are even stronger evidence[88]. In the Lamaštu-archive at Ugarit there was a teacher who taught his students to write in Babylonian script and orthography[89] and in Emar there was a teacher called Kidin-Gula.[90]

87 Cf. P. Bordreuil (ed.), Une bibliothèque au sud de la ville, Ras Shamra-Ougarit VII, Paris 1991, nos. 39 and 40, and see Y. Cohen/I. Singer, A Late Synchronism between Ugarit and Emar, in: Y. Amit/E. Ben Zvi/I. Finkelstein/O. Lipschits (eds.), Essays on Ancient Israel in its Near Eastern Context. A Tribute to Nadav Na'aman, Winona Lake 2006, 123–139.

88 Ibid., 128f.

89 See above.

90 Y. Cohen, Kidin-Gula – The Foreign Teacher at the Emar Scribal School, RA 98 (2004), 81–100.

It would seem then, that Babylon could have played an important role in the transmission of school material and it confirms the observation made by Veldhuis in his book on Elementary Education at Nippur: „Babylon is a plausible centre for the export of school compositions to the west, particularly because of the well-attested relations between Babylon and Ḫattuša. Extant lexical texts from Babylon, however, are so few that there is no way of proceeding beyond conjecture."[91]

In my opinion, the Ḫarra-ḫubullu text from Babylon can be used as evidence that the periphery borrowed at least part of its material from contemporary Babylonia and possibly even from Babylon itself. This material could have been brought by the teachers who are attested in the peripheral sources, most of it in memorized form.

91 N. Veldhuis, Elementary Education at Nippur. The List of Trees and Wooden Objects, Groningen 1997, 71.

*Mathieu Ossendrijver*

## Science in Action: Networks in Babylonian Astronomy

Like no other cultural achievement, astronomy stands out as a legacy of Babylon. The Chaldaean astronomer, a stock character of the biblical and classical literature, is the personification of that legacy.[1] In this contribution I investigate the astronomers of Babylonia as participants of a network of interacting scholars. With this approach I hope to expand our knowledge of the practice and context of Babylonian astronomy, which is still rather limited. Babylonian astronomers created a large and diverse body of literature consisting of observational astronomy, mathematical astronomy, zodiacal astrology, and omen astrology.[2] Especially the observational texts, by far the most numerous group, imply that astronomy was the collaborative effort of a community of scholars. Since most astronomical diaries contain observations for 6 months, they must have been compiled from individual reports of different scholars. The same holds for other compilations extracted from the diaries, some of which contain data for many decades. Also the fact that observational texts, unlike other scholarly texts, are anonymous (with very few exceptions; cf. below), can be viewed as a consequence of the collaborative effort by which they were produced. Astronomical diaries and related texts are therefore testimonies of a systematic, collaborative program of observation. This program is believed to have existed continuously from the Neo-Babylonian era, perhaps as early as the middle of the 8th c. BC, until the very end of cuneiform writing in the 1st c. AD.

Next to nothing is known about the astronomers of the Neo-Babylonian era, since we have only anonymous observational texts from that period.[3] The earliest reference to astronomers is contained in an Achaemenid-era administrative document of the Esagila temple concerning food rations for 14 astronomers.[4] This suggests that astronomers were by now – in surprisingly large numbers – employed by the temples. In spite of political upheavals, the observational program continued throughout the Achaemenid, Seleucid and Parthian eras (450 BC–50 AD), as evidenced by thousands of diaries and related texts,

1 Astronomer is to be understood in a broad sense as a specialist of the astral sciences, i.e. astronomer/astrologer.

2 The four groups are composed as follows: 1) Observational astronomy: astronomical diaries, planetary reports, eclipse reports, Lunar-Six reports, Goal-Year texts, almanacs, normal-star almanacs. 2) Mathematical astronomy: procedure texts, auxiliary tables, synodic tables, template tables and daily motion tables. 3) Zodiacal astrology: horoscopes, calendar texts, and various other texts 4) Omen astrology: the series 'When Anu and Enlil' and its commentaries.

3 We do not know, for instance, whether they were primarily associated with the court, like the astrologers of the Neo-Assyrian era, or with the temples, as is evident in the later periods.

4 P.-A. Beaulieu, The Astronomers of the Esagil Temple in the Fourth Century BC, in: A. K. Guinan (ed.), If a Man Builds a Joyful House, Leiden 2006, 5–22.

nearly all from Babylon. However, the Achaemenid era was also a time of fundamental innovations which led to mathematical astronomy and zodiacal astrology. It is therefore legitimate to ask whether these different forms of astronomy and astrology, which require rather different skills, were practiced by a single network of scholars. The answer to this question is provided by several texts from Babylon. A few astronomical diaries from the Seleucid era do mention the name of a scribe.[5] One of them is Bēl-apla-iddin/Mušallim-Bēl from the Mušēzib-clan, who is known to have written two tablets with mathematical astronomy.[6] Secondly, a Parthian-era protocol of the Esagila counsel lists mathematical astronomy (*tērsītu* = 'computed table') and observation, using the technical term for diaries (*naṣāru ša ginê* = 'regular watch'), among the duties of an astronomer.[7] Incidentally, this document also states that the astronomers were expected to perform their duties in collaboration.

The evidence from Babylon therefore suggests that the various astral disciplines were pursued by a single community of collaborating scholars. In order to study this community in more detail we may to turn our attention to the library of the Rēš temple in Uruk.[8] Unlike the scholarly tablets from Babylon, those from the Rēš usually document a collaboration between two individuals, in the sense that each colophon mentions an 'owner', identified by the phrase 'tablet of $PN_1$' (*ṭuppi* $PN_1$), and, in second position, a scribe, identified by the phrase 'hand of $PN_2$' (*qāt* $PN_2$). By collecting this information, and combining it with data from non-scholarly documents, we can reconstruct the biographies of the astronomers, and the network of their scholarly and economic interactions.

Elsewhere I have presented first results of this approach.[9] In particular, I have argued that the activities of scribe and 'owner' of scholarly tablets correspond to strictly consecu-

5 ADRT I, –372A; –361; –321; ADRT V, 60.

6 He wrote the procedure texts ACT 816 (O. Neugebauer, Astronomical Cuneiform Texts: Babylonian Ephemerides of the Seleucid Period for the Motion of the Sun, the Moon, and the Planets, I–III, London 1955) and BM 33552 (J. P. Britton/C. B. F. Walker, A 4th century Babylonian model for Venus: B.M. 33552, Centaurus 34 [1991], 97–118), the diary ADRT I –321 (BAK 180), and perhaps, as proposed by Oelsner (Von Iqīšâ und einigen anderen spätgeborenen Babyloniern, in: S. Graziani (ed.), Studi sul Vicino Oriente Antico dedicati alla memoria di Luigi Cagni, Naples 2000 [= Istituto universitario orientale dipartimento di Studi Asiatici Series Minor 61], 797–814), also ADRT V 60, a compilation with Jupiter observations.

7 CT 49, 144: G. J. P. McEwan, Priest and Temple in Hellenistic Babylonia, Wiesbaden 1981 (= Freiburger Altorientalische Studien 4), 18–21; cf. also R. van der Spek, Biliotheca Orientalis 42 (1985), 551–552; F. Rochberg, in: J. Marzahn/H. Neumann (eds.), Assyriologica et Semitica. Festschrift für Joachim Oelsner anläßlich seines 65. Geburtstages am 18. Februar 1997 (= Alter Orient und Altes Testament 252), Münster 2000, 359–375.

8 About 150 tablets, many of a scholarly nature, were found during the German excavations in room 79b near the SE gate. It is believed they are the remainder of the temple library (Uruk 4 in the list of O. Pedersén, Archives and Libraries in the Ancient Near East 1500–300 BC, Bethesda/ML 1998), and that many other scholarly tablets from Uruk, e.g. those published in TCL 6 (Textes cunéiformes. Musée du Louvre. Département des Antiquités Orientales), originate from this library.

9 M. Ossendrijver, Babylonian astronomers in context: a network approach, in: Proceedings of the 53rd RAI: Babel und Bibel (in press) and Id., Exzellente Netzwerke: die Astronomen von Uruk, in: G. Selz (ed.), The Empirical Sciences in Mesopotamia, Vienna, 2011, 631–644.

tive, non-overlapping phases in the biography of a scholar. Secondly, the scribe phase involves advanced, specialised education. Thirdly, the transition to the 'owner' phase may correspond to the assumption of a temple position. Fourthly, the scholarly tablets were not physically possessed by the 'owners', since most of them were kept in the library of the Rēš. It rather appears that 'ownership' of a scholarly tablet reflects a responsibility for the correctness of its content, i.e. an aspect of supervision. Fifthly, the scholarly library of the Rēš is a product of advanced scribal education, since nearly all of its tablets were written by students (the scribes) under the guidance of a supervisor (the 'owners'). In other words, very few tablets from the Rēš were written by scholars who are in their 'owner' phase. Sixthly, legal acts, which clearly belong to a different professional realm than the scholarly tablets, were written by scholars in both phases. Finally, by following the collaborations between 'owners' and scribes it is possible to reconstruct the network of the astronomers of the Rēš, the result of which is shown in Fig. 1.

In what follows I will exemplify these results by considering the biography of one scholar, the exorcist Nidinti-Anu/Anu-bēlšunu of the Ekur-zākir-clan, who has not yet received much attention. I will then expand and refine the interpretation of the network of astronomers. Table 1 lists the tablets that can be attributed to Nidinti-Anu with a certain likelihood.[10] Nidinti-Anu is attested as scribe of two scholarly tablets, both of which are omen texts. Text 1, which is unedited, may contain birth omens. Text 2 is concerned with extispicy, which was to become his favorite subject (Texts 2, 4, 6–9, 11). The 'owner' of Text 2, probably also of Text 1, was Anu-balāssu-iqbi/Anu-aḫa-ittannu of the Aḫûtu-clan. Like other members of that clan, Anu-balāssu-iqbi occupied high positions in the city, namely 'chief mayor of Uruk' (*rabû rēš āli ša Uruk*)[11] and 'delegate of the temple' (*paqdu ša bīt ilāni*).[12] Hence Nidinti-Anu studied with an elite person from another clan. Whether he was also instructed by his own father, as is true for many of his colleagues, we do not know, since his father is not attested.

10 This table is based on lists compiled by Wallenfels (Seleucid Archival Texts in the Harvard Semitic Museum [= Cuneiform Monographs 12], Groningen 1998), p. 26, Boiy (Hellenistic Legal Documents from Uruk in the "Royal Museums of Art and History" [Brussels], Akkadica 124 [2003], 19–64), p. 27, and Clancier (Les bibliothèques en Babylonie dans la deuxième motié du 1er millénaire av. J.-C. [= Alter Orient und Altes Testament 363], Münster 2009), p. 62. Boiy (2003) also mentions RIAA2 299 = O 196 (duplicate: Bibliotheca Mesopotamica 24, 24), a deed of sale dated SE 101 and written by a Nidinti-Anu/Anu-bēlšunu//[NN], who might be the same individual.

11 In the colophon of Text 2.

12 As a party in the legal act Forschungen und Berichte 16, 1 (VAT 9175), a land lease dated SE 91.

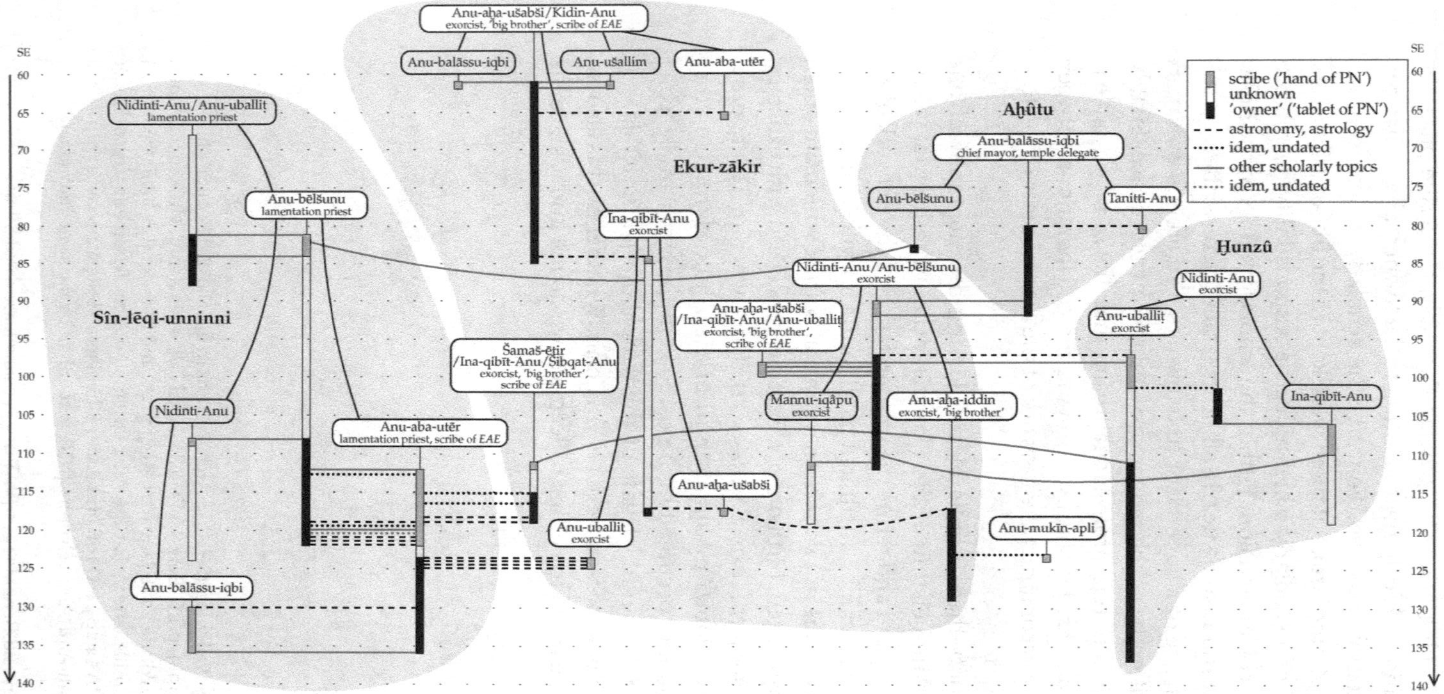


Fig. 1 | Network of the astronomers and associated other scholars of the Rēš temple in Uruk during the Seleucid period. Time increases downward from SE 60 to SE 140. Each scholar is represented by a vertical bar, the colour of which indicates his role as scribe (grey) or 'owner' (black) of tablets, or whether this could not be established (white). Vertical curved lines between two individuals indicate paternal links. Each horizontal line (for clarity sometimes an arc) represents an instance of collaboration between scribe and 'owner' as evidenced in the colophon of a scholarly tablet. The dotted ones are undated, i.e. their vertical position is uncertain.

Table 1 | Tablets mentioning Nidinti-Anu/Anu-bēlšunu//Ekur-zākir[13]

| | tablet[14] | find spot | content | fi-liat.[15] | role | title | collaborator[16] | date (SE)[17] |
|---|---|---|---|---|---|---|---|---|
| 1 | *TU* 10 | ? | birth omens? | 12a | scribe | exorcist of A&A | [Anu-balāssu-iqbi] | 90 |
| 2 | *TU* 1 | ? | *Multābiltu* T.14 | 12a | scribe | exorcist of A&A | Anu-balāssu-iqbi | 4/II/91 |
| 3 | *TU* 16 | ? | *EAE* T.56 | 12a | 'owner' | exorcist of A&A | Anu-uballiṭ | 30/VII/97 |
| 4 | *BRM* 4 12 | ? | *Barûtu* T.55 | 12a | 'owner' | exorcist of A&A | Anu-uballiṭ | 25/II/98 |
| 5 | *TU* 35 | ? | *Erimḫuš* T.5 | 12a | 'owner' | exorcist of A&A | Anu-aḫa-ušab ši | 5/II/99 |
| 6 | *BRM* 4 13 | ? | *Barûtu* T.7 | 12a | 'owner' | exorcist of A&A | Anu-aḫa-ušab ši | 7/II/99 |
| 7 | *TU* 4 | ? | *Barûtu* T.48 | 12a | 'owner' | exorcist of A&A | Anu-aḫa-ušab ši | 27/II/99 |
| 8 | *TU* 2 | ? | *Šumma martu* T.2 | 12a | 'owner' | exorcist of A&A | Anu-aḫa-ušab ši | 1/X/99 |
| 9 | *TU* 3 | ? | *Pān tākalti* T.6 | 12 | 'owner' | absent | Anu-aḫa-ušab ši | 13/X/99 |
| 10 | *TU* 7 | ? | sacrificial omens | [12a] | 'owner'? | exorcist of A&A | Anu-aḫa-ušab ši | 15/[m]/99 |
| 11 | *TU* 5 | ? | *Rikis girri* T.20 | 12a | 'owner' | exorcist of A&A | Ina-qibīt-Anu | absent |
| 12 | NCBT 2306 | ? | deed of sale | ? | scribe | – | – | 14/V/104 |
| 13 | MLC 2188 | ? | deed of sale | ? | scribe | – | – | 14/IX/106 |
| 14 | *RIAA2* 294 | ? | deed of sale | 12a | scribe | – | – | 10/X/107 |
| 15 | MLC 2165 | ? | deed of sale | ? | scribe | – | – | 18/VII/108 |
| 16 | *VS* 15 18 | ? | deed of sale | 12a | scribe | – | – | 8/[m]/108 |
| 17 | NCBT 1971 | ? | deed of sale | ? | scribe | – | – | 3/IX/108 |
| 18 | *CM* 12 4; *BRM* 2 30 | ? | deed of sale | 12a | scribe | – | – | 16/VIII/109 |
| 19 | *BiMes* 24 51 | ? | hymn to Adad? | 12a | 'owner' | [exorcist] of A&A | Mannu-iqâpu (1) | [d/m]/111 |
| 20 | *OECT* 9 47 | ? | deed of sale | 12a | scribe | - | - | 11/V/112 |
| 21 | *SpTU* 2 33 | Ue XVIII[18] | *Šumma ālu* excerpt | 12a | 'owner' | exorcist of A&A | Mannu-iqâpu (2?) | [d]/VII/[yr] |
| 22 | *BaM Beih* 2 65 | SE gate[19] | medical omens | [1]2a | 'owner'? | broken | [NN]/Nidinti-Anu | [d/m/yr] |

During Nidinti-Anu's 'owner' phase at least 4, perhaps 5 individuals are attested as scribes of his tablets. Texts 3 and 4 were written by the exorcist Anu-uballiṭ/Nidinti-Anu/Ina-qibīt-Anu from the Ḫunzû clan.[20] Text 3 is Tablet 56 of the astrological omen series *Enūma Anu Enlil*, wich contains omens about planets. It is the only astrological text on Nidinti-Anu's list, and the reason why I include him in the network of astronomers. According to the colophon, this tablet was copied from an original from Babylon – a rare explicit hint at scholarly interactions between Babylon and Uruk. Text 4 is the 55th tablet of the extispicy series *Barûtu*, also labeled as the 6th tablet of its subseries *Ubānu* ('Finger'). Text 11 is also about extispicy. It was written by Ina-qibīt-Anu/Nidinti-Anu//Ḫunzû, a brother of the aforementioned. He is known to have written only one other scholarly tablet, the mathematical text *TU* 31, which was 'owned' by his father. Neither of these tablets preserves a date, but on 18 IX SE 118 Ina-qibīt-Anu appears as a witness in the legal act BRM 2, 31, which was written by Nidinti-Anu's son Mannu-iqâpu (cf. below).[21]

Texts 5–10 were written by the exorcist Anu-aḫa-ušabši/Ina-qibīt-Anu/Anu-uballiṭ// Ekur-zākir, all within a period of less than a year. His kinship relation with other scholars

13 The conventional notation for filiations involves a dash (/) for parental links and a double-dash (//) preceding the ancestral (clan) name.

14 The sigla in this column refer to published copies as far as available. Museum numbers, references to the colophons in BAK, and editions: **1** AO 6466: BAK 96B. **2** AO 6452: BAK 96A; Koch (Secrets of Extispicy, Münster 2005), 190–210. **3** AO 6470: BAK 90B; Largement (Contribution à l'Étude des Astres errants dans l'Astrologie chaldéenne (1), Zeitschrift für Assyriologie 52 [1957]), 235–264). **4** MLC 1865: BAK 95A. **5** O 171: BAK 90C; MSL 17, 65–72, Ms. A. **6** MLC 1874: BAK 95B. **7** AO 6468: BAK 90A; Jeyes (in: A. R. George/ I. Finkel (eds.), Wisdom, Gods and Literature. Studies in Assyriology in Honour of W.G. Lambert, Winona Lake 2000, 345–373), Ms. C. **8** AO 6453: BAK 104B. **9** AO 6457: BAK 104C; Koch (op.cit., 2005), 343–353, Ms. A. **10** AO 6468; BAK 103G. **11** O 172: BAK 104D; Koch (op. cit., 2005), 297–312. **12,13,15,17** to appear in Yale Oriental Studies 20 (2010); dates taken from Doty (Cuneiform Archives from Hellenistic Uruk, Yale 1977). **14** O 198: Corò (Prebende Templari in Età Seleucide, Padova 2005 [= History of the Ancient Near East. Monographs 8]), 166–168; Boiy (Hellenistic Legal Documents from Uruk in the "Royal Museums of Art and History"[Brussels], Akkadica 124 [2003], 19–64). **16** VAT 7758: Corò (op. cit., 2005), 194–195; Funck (Uruk zur Seleukidenzeit, Berlin 1984 [= Schriften zur Geschichte und Kultur des Alten Orients 16]), 200–202. **18** Cuneiform Monographs 12,4 = HSM 913.1.2 (7310); duplicate: BRM 2 30 = MLC 2130. **19** A 3673. **20** Ash 1923.727. **21** W 22729/7; edition in Spätbabylonische Texte aus Uruk 2. **22** W 20030/18.

15 The code indicates which generations are present in the filiation, such that 1 denotes the indivual, 2 the father, 3 the grandfather, and a the ancestral name.

16 In the case of a scholarly tablet, the collaborator is the other individual mentioned in the colophon.

17 The year SE 1 corresponds to 3 April 311 BC–21 April 310 BC of the Julian calendar.

18 Excavated in the living quarters of area Ue XVIII, in the same house as the library of the exorcist Iqīšâ// Ekur-zākir (UVB 29–30, pp. 95,102). Since Iqīšâ's tablets are about 100 years older, this tablet must belong to a younger layer, as suggested by Oelsner (Aus dem Leben Babylonischer "Priester" in der 2. Hälfte des 1. Jahrtausends v. Chr. [am Beispiel der Funde aus Uruk], in: J. Zablocka/S. Zawadzki [eds.], Šulmu 4. Everyday Life in the Ancient Near East, Papers presented at the International Conference, Poznań, 19–22 September 1989, Poznan 1993).

19 Found in a niche in room 79b (G. Lindström, Ausgrabungen in Uruk-Warka. Endberichte 20, p. 211).

20 For his biography cf. M. Ossendrijver, Exzellente Netzwerke: die Astronomen von Uruk, in: G. Selz (ed.), The Empirical Sciences in Mesopotamia, Vienna 2011, 638–640.

21 Ina-qibīt-Anu/Nidinti-Anu/Ina-qibīt-Anu//Ḫunzû (Rev. 24). This document is partly translated by Funck (Uruk zur Seleukidenzeit [Schriften zur Geschichte und Kultur des Alten Orients 16], Berlin 1984, p. 289.

of the Ekur-zākir clan remains unclear. He was not only a productive scribe, but a high-ranking person, judging from his title 'Big Brother of the Rēš' (*aḫu rabû ša Rēš*). This high priestly function is attested for three other scholars of the Rēš, all members of the Ekur-zākir clan, among which Nidinti-Anu's son Anu-aḫa-iddin (cf. below).[22] His third title, 'scribe of *Enūma Anu Enlil*' (*ṭupšar Enūma Anu Enlil*), implies that Anu-aḫa-ušabši was an astronomer, but thus far the tablets do not confirm this. Except Text 5, which is tablet 5 of the bilingual lexical series *Erimḫuš*, all tablets written by Anu-aḫa-ušabši belong to the realm of extispicy. Hence we do not know what competence, if any, corresponds to his use of that title.

The scribe of Text 19 is Mannu-iqâpu, 'exorcist of Anu and Antu', and very likely a son of Nidinti-Anu.[23] That Nidinti-Anu had a son called Mannu-iqâpu is confirmed by the mentioned legal act, BRM 2 31, which was written by 'Mannu-iqâpu, son of Nidinti-Anu, descendant of Ekur-zākir'. Text 19 contains a hymn, perhaps to Adad, which remains unedited. Text 21 is an omen text about encounters with a fox. According to the colophon it is the 38th excerpt of the series *Šumma ālu ina melê šakin*, 'When a city is positioned on a height'. Its scribe is also a Mannu-iqâpu, 'exorcist of Anu and Antu', but he appears to be a grandson of Nidinti-Anu.[24] This tablet is known to have been stored outside the Rēš, in the same private house where the library of the exorcist Iqīšâ was located.[25] Text 22 is a small fragment of an unidentified omen text. The badly damaged colophon mentions [NN]/Anu-bēlšunu//Ekur-zākir as the 'owner'. As suggested by R. Wallenfels[26], he may be our Nidinti-Anu, since no other Urukean scholar known to us matches this filiation. The scribe of the tablet is '[... son of] Nidintu-Anu, chief lamentation priest (*galamaḫḫu*)'.[27] I am not aware of any scholar who carried this title around SE 100, so he remains unidentified.

22 Anu-aḫa-ušabši/Kidin-Anu//Ekur-zākir (attested SE 0–85) and Šamaš-ēṭir/Ina-qibīt-Anu/Šibqat-Anu//Ekur-zākir (attested SE 110–120).

23 Less likely a grandson, if one assumes that one generation is missing from the 'owner's' filiation. Although the left part of the colophon is damaged, the filiation of the 'owners' appears to complete, i.e. that of Nidinti-Anu.

24 His filiation is given as $^{m}$*Man-nu-i-qa-pu* A *ša*$_2$ $^{m}$DUMU.A.NI, 'Mannu-iqâpu, son of his son'. The unexpected personal wedge in front of DUMU.A.NI (overlooked by Hunger, Spätbabylonische Texte aus Uruk 2, p. 146) may be explained as an error, since A *ša2* is usually followed by a personal name. Of course, if one assumes that A *ša2* is an erronenous insertion, than this Mannu-iqâpu is the son of Nidinti-Anu as in Text 19. Yet another Mannu-iqâpu, the son of Anu-aḫḫē-iddin/Anu-bēlšunu//Ekur-zākir (a brother of Nidinti-Anu) appears as a witness in the legal act TCL 13, 243 dated 29 VII SE 116 (Rutten, op. cit. 1935, p. 226–232; Pl. 2; Funck, op. cit.1984, 148–153).

25 For this library, 'Uruk 10' in Pedersén (op. cit. 1998), cf. Spätbabylonische Texte aus Uruk 1–5; Oelsner (Von Iqīšâ und einigen anderen spätgeborenen Babyloniern, in: S. Graziani (ed.), Studi sul Vicino Oriente Antico dedicati alla memoria di Luigi Cagni, Naples 2000 [= Istituto universitario orientale dipartimento di Studi Asiatici Series Minor 61], 797–814).

26 R. Wallenfels, Seleucid Archival Texts in the Harvard Semitic Museum, Groningen 1998 [= Cuneiform Monographs 12]

27 The clan name appears to be omitted, which is very unusual.

There is a gap of more than 6 years between Nidinti-Anu's last attestation as a scribe of scholarly tablets (Text 2), and his first attestation as an 'owner' of such tablets (Text 3). His transition from scribe to 'owner' must have occurred in between these dates. We cannot currently assess whether this might be correlated with an event in the biography of his father Anu-bēlšunu, since he is not attested. We may also look for a correlation between Nidinti-Anu's last attestation as an 'owner' and the careers of his sons. While Mannu-iqâpu is attested only as a scribe, Nidinti-Anu's had another son, Anu-aḫa-iddin, who was an 'exorcist of Anu and Antu' and a 'Big Brother of the Rēš', and 'owned' scholarly tablets. His earliest attestation as 'owner', in 26 XI SE 117,[28] is several years after his father's last attestation in that capacity. This phenomenon, which can be observed for other scholars as well, is compatible with the mentioned hypothesis that the transition to the 'owner' phase involves the assumption of a temple position, since these positions were often inherited from father to son.

Apart from his scholarly work, Nidinti-Anu was active as a notary, as evidenced by the numerous legal acts that were written by him. A discussion of the network of interactions that is implied by these acts is beyond the scope of the present investigation. He is thus far not attested as an active party in legal acts, unlike some of his colleagues. Hence we know nothing about his own economic transactions.

Although Nidinti-Anu is connected with the network of astronomers (Fig. 1), the content of the scholarly tablets (Table 1) suggests that he does not belong to the 'hard-core' astronomers – those who were active in the field of mathematical astronomy, and presumably observational astronomy. Indeed Nidinti-Anu does not use the traditional title of an astronomer, 'scribe of *Enūma Anu Enlil*' (*ṭupšar Enūma Anu Enlil*), but neither do many other astronomers.[29] In fact, among the 17 astronomers that make up the network only 4 are known to use that title, and of these only 2 are known to have been active in the field of mathematical astronomy.[30] Conversely, 7 astronomers who were active in the field of mathematical astronomy or Goal-Year astronomy do not seem to have carried the title 'scribe of *Enūma Anu Enlil*'.[31]

In my final remarks I want to reflect on the fact that the astronomical texts on which the reconstruction of the network in Fig. 1 is based are probably not representative for the astronomical corpus as a whole. As mentioned in the beginning, the evidence from Babylon suggests that observational texts constitute the bulk of the astronomical corpus. Although very few of these texts were found in Uruk,[32] this is probably a coincidence. It can

28 TCL 6, 15 + Archiv für Orientforschung 14, Tf 2 (BAK 93), a catalogue of the astrological omen series *Enūma Anu Enlil*.

29 For his biography cf. Ossendrijver (op. cit., in press).

30 Anu-bēlšunu//Sîn-lēqi-unninni and Šamaš-ēṭir/Ina-qibīt-Anu/Šibqat-Anu//Ekurzākir.

31 Anu-balāssu-iqbi/Nidinti-Anu/Anu-bēlšunu//Sîn-lēqi-unninni, Ina-qibīt-Anu/Anu-aḫa-ušabši//Ekur-zākir and his two sons Anu-aḫa-ušabši and Anu-uballiṭ, and Nidinti-Anu//Ḫunzû and his son Anu-uballiṭ.

32 Diaries: ADRT I –463 (= W 20030,142; excavated in room 79b of the Rēš); planetary observations:

be assumed that the astronomers of the Rēš, like their colleagues in Babylon, carried out regular observations, and wrote astronomical diaries and related texts. Hence the actual number of collaborations (linkages) between the astronomers is certainly far greater than in Fig. 1. However, we can be confident that many of the astronomers in Uruk who carried out observations were also active in the field of mathematical astronomy, because this was expected from their colleagues in Babylon. Since the tablets of mathematical astronomy usually mention an 'owner' and a scribe, many of the astronomers who wrote observational texts may already be represented in the network of Fig. 1. We may therefore have some confidence that, in spite of the lack of observational texts, the actual size of the network of the astronomers in Uruk may not be fundamentally different from this reconstruction.

T.G. Pinches/J.N. Strassmaier /A. Sachs, Late Babylonian Astronomical and Related Texts, Providence 1955: 1377 (= A 3456; Hunger, Fs Sachs, 201–223); ADRT V 42, V 82; Normal-Star Almanacs (to appear in a future volume of ADRT): Pinches/Strassmaier/Sachs (op. cit.): 1004*, 1025*, 1030*a, 1031*, 1032*; Almanacs (to appear in a future volume of ADRT): Pinches/Strassmaier/Sachs (op. cit.): 1124*.

# Babylon – Wirkungsgeschichten

*Michael Erler*

# Chaldäer im Platonismus

## 1. Platon und die Weisheit der Alten

„Griechenland ist groß, Kebes, und es gibt dort tüchtige Männer. Groß sind auch die Nationen der Barbaren, die ihr ... alle durchforschen müsst, und dabei dürft ihr weder Geld noch Mühen scheuen, gibt es doch nichts, wofür ihr euer Geld besser ausgeben könntet" (Üb. Ebert).[1]

Sokrates' Aufforderung ist in der Antike als Auftrag verstanden worden, für fremde Traditionen – gerade auch aus dem Orient – offen zu sein, fremde Völker aufzusuchen und sich belehren zu lassen.[2] Platon selbst führt diese Offenheit in seinen Dialogen vor. An nicht wenigen Stellen beruft sich Sokrates auf Lehren der ‚Alten', auf Dichter oder Weise der Vorzeit – auch aus östlichen Ländern – und führt sie als Beleg für eigene – d.h. platonische – Lehren an, freilich nicht ohne sie auf eine Weise zu modifizieren und zu transformieren, so dass sie ins eigene Gedankengebäude passen.[3] Denn die Dialoge zeigen auch: Die Bedeutung alter Weisheit wird zwar akzeptiert, dies jedoch nicht kritiklos und ungeprüft. Ein Automatismus „alter Logos gleich Wahrheit" besteht für Platon nicht. Dieses kritische Rezeptionsverhältnis gilt für Platon, aber auch in anderen Bereichen des kulturellen Lebens seiner Zeit: Ein Satz aus der pseudo-platonischen *Epinomis* bringt es auf den Punkt. Demnach waren sich die Griechen bewusst, vieles der Weisheit der Barbaren, der *barbaros philosophia* wie es heißt, zu verdanken, aber auch dieses Übernommene zu schönerer Vollendung geführt zu haben.[4] In dieser Haltung gegenüber der *barbaros philosophia* spiegelt sich jenes Selbstbewusstsein, das seit Marathon, Salamis und Plataiai im 5. Jahrhundert v. Chr. zu beobachten ist und das Christian Meier glücklich als „Könnensbewusst-

1 Vgl. Platon, Phaidon 78a (T. Ebert, Phaidon. Übersetzung und Kommentar von T. Ebert, [= Platons Werke Bd. I 4], Göttingen 2004, 39); zur Stelle vgl. M. Erler, Kontexte der Philosophie Platons, in: Platon. Handbuch, hg. von C. Horn u.a., Stuttgart 2009, 61–99, bes. 61ff.; noch in der Spätantike wurde dieser Forschungsauftrag des Sokrates aufgegriffen, vgl. L.G. Westerink (Hg.), Olympiodorus Alexandrinus. Olympidori in Platonis Gorgiam commentaria, Leipzig 1970, 41, 6.

2 Vgl. M. Baltes, Der Platonismus und die Weisheit der Barbaren, jetzt in: M. Baltes, EPINOHMATA. Kleine Schriften zur antiken Philosophie und homerischen Dichtung, hg. v. M.-L. Lakmann, München/Leipzig 2005, 1–26, bes. 11; H. Dörrie, Platons Reisen zu fernen Völkern. Zur Geschichte eines Motivs der Platon-Legende und zu seiner Verwendung durch Lactanz, in: W. den Boer u.a. (Hgg.), Romanitas et Christianitas, Studia I.H. Waszink, Amsterdam/London 1973, 99–118.

3 Vgl. M. Erler, Legitimation und Projektion. Die ‚Weisheit der Alten' im Platonismus der Spätantike, in: D. Kuhn/H. Stahl (Hgg.), Die Gegenwart des Altertums, Heidelberg 2001, 313–326.

4 Vgl. Pseudo-Platonische Epinomis 986e.

sein" charakterisiert hat.[5] Es passt z. B. ins Bild, wenn Platon auf der Suche nach Vorbildern für die ideale Verfassung Athens im Dialog *Timaios* zunächst auf die alte ägyptische Stadt Sais verweist. Was freilich wie ein Beitrag *ad maiorem gloriam Aegyptiorum* aussieht, erweist sich schnell als Beleg für den Ruhm Athens. Denn jenes Sais in Ägypten galt Platon als eine Kopie des 1000 Jahre zuvor gegründeten Ur-Athen.[6] Hier kommt ein Selbstbewusstsein zum Ausdruck, das Offenheit gegenüber fremdem Wissen – vornehmlich aus dem Osten – mit souveräner Fähigkeit zu dessen Anverwandlung und Transformation verbindet. Schon Hesiod überträgt in den *Werken und Tagen* auf die Göttin Dike Vorstellungen östlichen, ‚magischen' Königtums, wie sie u. a. in Ägypten, aber auch in Mesopotamien zu beobachten sind, indem er die Verehrung Dikes für das Wohlergehen von Mensch und Natur in der Polisgemeinschaft verantwortlich macht.[7] Indem er jedoch diese Verehrung nicht mehr einer Person – dem König – anvertraut, wie dies in der ägyptischen oder in östlichen Gesellschaften der Fall war, sondern einem jeden Bürger, formuliert er mit Hilfe alter östlicher Vorstellungen etwas Neues und für das Verständnis griechischer Polisgemeinschaft Grundsätzliches: Jedes Mitglied trägt Verantwortung für ‚Wohl und Weh' aller anderen Mitglieder und der Natur: In Sophokles' *Oidipous Tyrannos* wird deshalb nicht automatisch der König für jene Pest in Theben verantwortlich gemacht, mit der das Stück beginnt, sondern es wird der Verursacher unter allen Bürgern gesucht – man darf darüber spekulieren, ob im mesopotamischen oder ägyptischen Kontext ein „Oidipous Tyrannos" wie der des Sophokles möglich gewesen wäre.[8]

## 2. Babylon als Quelle alter Weisheit

Als wohl wichtigste Quelle alter Weisheit galt Ägypten; doch auch Babylonien spielt eine wichtige Rolle, vor allem jenes Babylon, das uns schon früh in der griechischen Literatur begegnet. Der Bruder des Dichters Alkaios, Antimenidas, dient dort als Söldner. Babylon galt als bedeutende Metropole, die großen Mauern und die ‚hängenden Gärten' als Weltwunder. Herodot berichtet von ihrer Größe und über seltsame Bräuche.[9] Auch wenn seine Angaben wenig zuverlässig scheinen, – später wird manches von Ktesias zurechtgerückt und auch

5 Vgl. C. Meier, Die Entstehung des Politischen bei den Griechen, Frankfurt 1983, 469 ff.

6 Vgl. Platons Timaios 23d; dazu M. Baltes, Der Platonismus und die Weisheit der Barbaren, jetzt in: M. Baltes, EPINOHMATA. Kleine Schriften zur antiken Philosophie und homerischen Dichtung, hg. v. M.-L. Lakmann, München/Leipzig 2005, 12.

7 Vgl. dazu M. Erler, Das Recht (Dike) als Segensbringerin für die Polis. Die Wandlung eines Motivs von Hesiod zu Kallimachos, in: Studi Italiani di Filologia Classica 5 (1987), 5–36.

8 Vgl. M. Erler, Die drei Gebote des Aischylos im ‚Oidipous Tyrannos' des Sophokles in: Würzburger Jahrbücher für die Altertumswissenschaft 28a (2004), 9–19.

9 Vgl. Alkaios, Fragment 48. 50. 350 (Voigt, in: Sappho et Alcaeus. Fragmenta, hg. v. E.-M. Voigt, Amsterdam 1963); dazu W. Burkert, Königs-Ellen bei Alkaios. Griechen am Rand der östlichen Monarchien, Journal of Medieval History 53/2 (1996), 69–72; vgl. Herodot 1, 178 ff.

Aristophanes zeigt sich weniger beeindruckt, vielmehr macht er Babylon zum Gegenstand des Spottes – so zeigt dies alles doch: Babylon war Teil des literarischen Diskurses und Gegenstand des Interesses.[10] Es wurde zudem nicht nur wegen seines Reichtums und seiner Macht, sondern auch als Ort der Weisheit bewundert. Mit Babylon verband man Errungenschaften der Astronomie und Astrologie ebenso wie mit solchen der Psychologie, der Theologie und der Mathematik – denn der grundlegende Satz des Pythagoras war offenbar schon seit Jahrhunderten bei den Babyloniern im Gebrauch.[11] Nach Pausanias stammt sogar die Lehre von der Unsterblichkeit der Seele von den Chaldäern und den indischen Magiern.[12] Noch bei Macrobius findet sich eine Planetenordnung (Saturn, Jupiter, Mars, Sonne, Venus, Merkur, Mond), die nicht der platonischen (zu finden in Platons *Politeia*), sondern der chaldäischen Reihung folgt, die so auch bei Cicero und den Platonikern zu finden ist.[13] Rechentechniken hatten eine große Blüte erreicht und gingen wie später bei Pythagoras mit Zahlenmystik Hand in Hand – aber auch die Philosophie wird mit Babylon in Verbindung gebracht.[14] Lukian lässt die personifizierte Philosophie dem Gott Zeus von ihrem Weg nach Griechenland berichten.[15] Zuerst sei sie zu den Barbaren gegangen, zu den Brahmanen, den Indern und Äthiopiern, zu den Ägyptern und dann schließlich zu den Chaldäern und Magiern in Babylon. Noch Diogenes Laertios weiß von solchen, nach deren Meinung die Entwicklung der Philosophie ihren Anfang bei den ‚Barbaren' genommen hat.[16] Demnach hatten die Perser ihre Magier, die Babylonier und Assyrer ihre Chaldäer, und die Gallier ihre so genannten Druiden und Semnotheten. Es wundert vor diesem Hintergrund nicht, dass Babylon als ein Quell der *barbaros philosophia* zur Station jener Bildungsreise wurde, die man griechischen Weisen teilweise andichtete, um die Herkunft ihrer Weisheit zu erklären und die Wahrheit ihres Wissens zu legitimieren. Wir hören z.B., dass sich Pythagoras zunächst in Ägypten, danach aber zwölf Jahre in Babylon aufgehalten habe und dort von den Chaldäern, also den babylonischen Priestern, Unterricht in Mathematik, Astronomie, Astrologie, Musik und Traumdeutung erfahren sowie sich ihre Theologie und ihre Rituale angeeignet habe.[17] Ähnliches hören wir von weiteren griechischen Weisen, darunter von

10 Ob in Aristophanes' *Aves* 549 ff. und 1124–38 wirklich eine Parodie der Beschreibung der Stadtmauer von Babylon wie bei Herodot vorliegt (vgl. Herodot 1, 179), ist umstritten. Zu Ktesias' „Persika" und seiner – durchaus problematischen – Herodotkritik vgl. F. Jacoby, s.v. Ktesias, RE 11 (1922), 2032–2073; vgl. auch F.W. König, Die Persika des Ktesias, Graz 1972.

11 Vgl. Diogenes Laertius 8, 12; dazu C. Riedweg, Pythagoras, München 2002, 119.

12 Vgl. Pausanias 4, 32, 4 (Dörrie-Baltes 66.1, in: H. Dörrie/M. Baltes, Der Platonismus in der Antike Bd. 2, Stuttgart/Bad Cannstatt 1990, 176 mit 454).

13 Vgl. Platon, De re publica, 616d ff.; vgl. Cicero, De re publica, 6, 17 und Macrobius, In Somnium Scipionis 2, 3, 13; J. Flamant, Macrobe et le néo-platonisme latin à fin du IV siècle, Leiden 1977, 421 ff.; H. Dörrie/M. Baltes, Der Platonismus in der Antike Bd. 6, 2, Stuttgart/Bad Cannstatt 2002, 191.

14 Vgl. C.H. Kahn, Pythagoras and the Pythagoreans. A Brief History, Indianapolis 2001, 342f.

15 Vgl. Lukian, fug. 6.

16 Vgl. Diogenes Laertios 1, 1.

17 Vgl. z.B. Porphyrius, Vita Pythagorae 6 (= Dörrie-Baltes 67, 6, in: H. Dörrie/M. Baltes, Der Platonismus in der Antike, Bd. 2, Stuttgart/Bad Cannstatt 1990, 182 mit 465).

keinem geringeren als Platon. Freilich, Platon war so bedeutend, dass chaldäische Priester auch zu ihm pilgerten, wie Philipp von Opûs uns wissen lässt, dies freilich erst kurz vor seinem Tod, so dass er nicht mehr viel von ihnen lernen konnte.[18] Seneca zufolge sollen Magier in Athen dem Verstorbenen Platon geopfert haben.[19] Gewiss, diese Geschichten sind zumeist Fiktion. Zudem protestiert der antike Philosophiehistoriker Diogenes Laertios gegen die Orientverbindung und lässt die Philosophie in Griechenland beginnen – eine Position, die besonders im 19. Jahrhundert der bedeutende Philosophiehistoriker Eduard Zeller wieder vehement vertritt, wenn er alles sich aus den „Voraussetzungen des griechischen Volkslebens entwickeln lässt".[20] Gleichwohl sind die Versuche, bedeutende Lehren mit der Weisheit des Ostens in Verbindung zu bringen und eigenen Errungenschaften auf diese Weise Autorität zu verleihen, ein Beleg für den Respekt, den die Griechen, insbesondere auch Philosophen der klassischen Zeit, dem Wissen der Alten entgegen gebracht haben und wie sehr sie es gewürdigt haben.[21] Babylon und Babylonien erweisen sich in diesem Kontext als wichtige Station und als Bezugspunkt – auch im Bereich der Philosophie.

## 3. Homer und der Beginn der Philosophie

Ein Beispiel für eine solche Beziehung lässt sich in der Tat sogar für den Anfang der griechischen Philosophie ausmachen und dies an prominenter Stelle des ersten Werkes europäischer Literatur, in Homers *Ilias*. Es handelt sich um eine Partie im Kontext jener berühmten Geschichte über den Betrug Heras an Zeus (der *Dios Apate*) im 14. Gesang. Hier hat die moderne Forschung Reminiszenzen mesopotamischer theologischer Vorstellungen erkannt.[22]

18 Vgl. Philodem, Academicorum ind. col. 3, 36ff. (= K. Gaiser, Philodems Academica, Stuttgart/Bad Cannstatt 1988, 176ff. mit 421ff.; Dorandi, 3, 36ff., in: Filodemo, Storia dei Filosofi. Platone e l'Academia (PHerc. 1021 e 164), hg., übers. und komm. von T. Dorandi, Neapel 1991); Philipp von Opûs, Fragment 5 Tarán (in: L. Tarán, Academica. Plato, Philip of Opus and the Pseudo-Platonic Epinomis, Philadelphia 1975); Fragment 14a Lasserre (in: F. Lasserre, De Léodamas de Thasos à Philippe d'Oponte, Neapel 1987); dazu H. Dörrie, Die Wertung der Barbaren im Urteil der Griechen. Knechtsnaturen? Oder Bewahrer und Künder heilbringender Weisheit?, in: R. Stiehl/A. Lehmann (Hgg.), Antike und Universalgeschichte, Festschrift für H.E. Stier, Münster 1972, 146–175; zu Platon vgl. M. Erler, Platon, in: Grundriss der Geschichte der Philosophie. Die Philosophie der Antike Bd. 2,2, hg. v. H. Flashar, Basel 2007, 48.

19 Vgl. Seneca, Epistulae morales 58, 31.

20 Vgl. E. Zeller, Die Philosophie der Griechen in ihrer geschichtlichen Entwicklung, Bd. 1, 7. Auflage, Leipzig 1923, 21–52; interessant die „nota addizionale" von R. Mondolfo in: E. Zeller/R. Mondolfo, La filosofia dei greci nel suo sviluppo storico, Bd. 1,1, 3. Auflage, Florenz 1959, 63–99.

21 Vgl. Seneca, Epistulae morales 58, 31; vgl. U. Hölscher, Anaximander und der Anfang der Philosophie, Hermes 81 (1953), 257–277. 385–418; erweiterte Fassung in: ders., Anfängliches Fragen, Göttingen 1968, 9–89, dort 43; insbesondere M.L. West, The east face of Helicon. West Asiatic elements in Greek poetry und myth, Oxford 1997; ders., Early Greek philosophy and the Orient, Oxford 1971; W. Burkert, Die Griechen und der Orient, München/Zürich 2003.

22 Vgl. W. Burkert, Die Griechen und der Orient, München/Zürich 2003, 36ff.

In der berühmten Passage versucht Hera durch eine Täuschung Zeus vom Kampfgeschehen vor Troja abzulenken, weil dies sich so gar nicht nach ihren eigenen Vorstellungen und zum Nachteil der Griechen zu entwickeln scheint.[23] Hera plant deshalb ein Schäferstündchen mit ihrem Gemahl. Zuvor freilich holt sie sich bei der zuständigen Göttin und Expertin in Liebesdingen, Aphrodite, taktischen Rat und Hilfe. Natürlich wäre es nicht zuträglich, wenn sie der anderen Göttin zu erkennen gäbe, wer das Opfer ihres Verführungsvorhabens sein soll. Hera ‚improvisiert' deshalb eine Erklärung für ihren Wunsch, der edel klingt. Sie gibt vor, ihren Eltern, Okeanos und Tethys, helfen zu wollen. Sie sei deshalb auf dem Weg zu ihnen, die an den Grenzen der bewohnten Erde lebten:

> „Denn ich mache mich auf zu den Grenzen der fruchtbaren Erde,
> Will nach Okeanos sehn, dem Ahnen der Götter, und Tethys,
> Die mich in ihrem Palast erzogen und hüteten, beide; [...]
> Diese will ich besuchen, die dauernden Händel zu schlichten,
> Denn schon lange Zeit vermeiden sie einer des andern
> Ehelager und Liebe, getrennt durch bittere Feindschaft" (Üb. Rupé).[24]

Der Trug gelingt, und Hera kann Zeus mit Hilfe von Aphrodites Liebesmittel zu einem Schäferstündchen überreden.

> „Also sprach der Kronide und schloß in die Arme die Gattin.
> Unten erblühte die heilige Erde von sprießenden Gräsern,
> Tauigem Lotosklee, Hyazinthen und saftigem Krokos,
> Dicht und üppig und weich, die über den Boden sie hoben.
> Beide lagerten dort und deckten sich zu mit den schönen
> Goldenen Wolken; es fielen herab die Tropfen des Taues" (Üb. Rupé).[25]

Diese Stelle ist wegen stilistischer Besonderheiten, die nicht so recht zum mündlichen Stil passen wollen,[26] besonders aber aus inhaltlichen Gründen von Interesse. Okeanos und Tethys werden vorgestellt als *Ursprung der Götter* und *Mutter* und Okeanos als *Ursprung für alles.* Wenn es heißt, dass sich beide des Beilagers im Streit enthielten, wird signalisiert, dass das Werden – auch der Götter – zum Stillstand kommt.[27] Zwar findet sich die Bemerkung im Kontext einer Trugrede. Doch bleibt der Eindruck, dass die Vereinigung von Zeus

23 Vgl. Homer, Ilias 14, 153ff.

24 Vgl. Homer, Ilias 14, 200ff. Übersetzung: H. Rupé, Homer. Ilias, übertragen von dems., München 1989, 477.

25 Vgl. Homer, Ilias 14, 346ff. Übersetzung: H. Rupé, Homer. Ilias, übertragen von dems., München 1989, 485.

26 Vgl. A. Dihle, Homer-Probleme, Opladen 1970, 83–93.

27 Vgl. R. Cornford, A Ritual Basis for Hesiod's Theogony, in: The Unwritten Philosophy and other Essays, Cambridge 1950, 95–116; Vgl. W. Burkert, Die orientalisierende Epoche in der griechischen Religion und Literatur, Heidelberg 1984, 7ff. und ders., Die Griechen und der Orient, München/Zürich 2003, 57ff; Bezüge stellt bereits her O. Gruppe, Die griechischen Culte und Mythen in ihren Beziehungen zu den orientalischen Religionen, Leipzig 1887, 614–622.

und Hera, die ein Aufblühen der Natur bewirkt, eine kosmische Dimension hat und geradezu Teil einer Kosmogonie wird. Diese Vorstellung ist Homers Anthropomorphismus ansonsten fremd und nur hier zu beobachten. Schon Gladstone hat dies 1890 erkannt, und nach ihm haben andere darauf hingewiesen, dass unsere Partie in der Tat enge Parallelen zu Wasserkosmogonien orientalischer Herkunft, insbesondere aber zum babylonischen Weltschöpfungsmythos „Enuma Elisch" aufweist. Man denkt an Apsu – wie Okeanos ein Süßwassermeer, das die Welt umgibt und Untergrundquelle aller Quellen und Flüsse ist, der, wie es heißt, da war, als der *Himmel oben* und die Erde noch nicht existierten; man denkt an Tiamat, das Salzmeer als *Gebärerin von allem.* Wenn beide ihre Wasser vermischen, bringen sie die Götter hervor, den Himmel, und die Erde.[28] Beide aber vermischten ihre Wasser bis Apsu eingeschläfert und Tiamat erschlagen wird. Weitere Entsprechungen im Kontext der Homerischen *Dios Apate* bestärken zudem den Eindruck, dass hier, am Beginn der griechischen Literatur, ein Echo dieses babylonischen Mythos zu vernehmen ist – ein Echo, das nicht nur die literarische Darstellung Homers beeinflusst hat. Denn schon die Homerleser Platon[29] und Aristoteles bringen Heras Hinweis auf Okeanos und Tethys mit den Anfängen der griechischen Philosophie, mit Thales, in Verbindung. In der *Metaphysik* des Aristoteles lesen wir:

„Manche meinen auch, dass auch die Alten, welche lange vor unserer Zeit und zuerst über die göttlichen Dinge geforscht haben, derselben Ansicht seien; denn den Okeanos und die Tethys machten sie zu Erzeugern der Entstehung ..." und dann: „Thales jedoch soll sich auf diese Weise über die Grundursache ausgesprochen haben" (Üb. Bonitz).[30]

Homer also als Kosmologe und Vorgänger des Vorsokratikers Thales. Dass Aristoteles bei Homer Theologisch-Philosophisches zu finden glaubt, kann nicht erstaunen, ist er doch der Meinung, dass Philosophie und Theologie schon in der Urzeit existiert haben, durch Sintfluten aber immer wieder in Vergessenheit geraten und nach der letzten Sintflut in rudimentärer mythischer Form tradiert worden seien, wie z. B. Kosmologien bei Autoren wie Orpheus, Homer, Hesiod oder Pherekydes aus Syros bezeugten.[31] An eben jener Stelle also erkennt Aristoteles erste Ansätze der Philosophie, an der moderne Forschung ein Echo altorientalisch-babylonischer Vorstellungen ausgemacht hat. Auch bei Thales selbst hat man beobachtet, dass seine Vorstellungen – soweit wir das erkennen können – mit solchen aus dem Orient, u. a. aus Babylonien konvergieren, aber auch, dass bei ihm entsprechende kosmologische Vorstellungen in mythologisch-anthropologischer Form

28 Vgl. W. Burkert, Die orientalisierende Epoche in der griechischen Religion und Literatur, Heidelberg 1984, 88; vgl. Enuma Elisch 1, 1ff.; A. Heidel, The Babylonian Genesis, 2. Auflage, Chicago 1951; J.B. Pritchard, Ancient Near Eastern Texts Relating to the Old Testament, 3. Auflage, Princeton 1969, 61.

29 Vgl. Platon, Kratylos 402b (orphisches Gedicht), vgl. auch Platon, Timaios 40e (mit Unterschieden), dazu R. Janko, The Iliad. A Commentary, Books 13–16, Bd. 4, Cambridge 1992, 180ff.

30 Vgl. Aristoteles, Metaphysik 983b27; vgl. Aristoteles, Metaphysik. Übersetzt und herausgegeben von H. Bonitz u. a., Hamburg 1966, 16.

31 Vgl. Aristoteles, Metaphysik 1074a38ff.

gleichsam ‚physikalisch' werden.[32] Offenheit für Altes, verbunden mit Transformation also auch hier – Babylonisches am Beginn der griechischen Philosophie: Gewiss, Aristoteles wie auch Platon mag wohl kaum bewusst gewesen sein, dass Heras improvisierte Trugrede Bezüge zu babylonischen Vorstellungen erkennen lässt.

## 4. Eudemos und Enuma Elisch

Ein Schüler des Aristoteles jedoch, Eudemos aus Rhodos, der in der zweiten Hälfte des 4. Jahrhundert v. Chr. wirkte, stellt einen solchen Bezug her. In seinem Werk *Geschichte der Theologie* – einer Pionierarbeit der Religionsgeschichte – hat dieser Aristoteles-Schüler nämlich die Grundlagen der griechischen und der orientalischen Theologie dargestellt. Das Werk ist uns durch den verkürzten Bericht des Neuplatonikers Damaskios aus dem 5. Jahrhundert n. Chr. bekannt.[33] Demnach ist Eudemos in seinem Werk bei der Darstellung von Prinzipienlehren vorphilosophischer, antiker Dichter offenbar chronologisch vorgegangen. Er hat Orpheus (bei dem die Nacht als Prinzip gilt), dann Homer mit Hinweis auf Okeanos und Tethys – d.h. unsere Stelle im 14. Gesang –, darauf Hesiod (mit dem Chaos), Akusilaos (auch mit dem Chaos), Epimenides (mit Aer und Nyx) und Pherekydes (mit Zeus und Chronos) behandelt. Bei Damaskios folgt sodann separat eine Besprechung der kosmologischen Prinzipien bei den Babyloniern, bei den persischen Magiern und den Ägyptern.

Moderne Interpreten haben nun darauf hingewiesen, dass sich Damaskios' Darstellung der babylonischen theologischen Prinzipienlehre konkret auf den Anfang des *Enuma Elisch* bezieht, wenn er nämlich von zwei Prinzipien spricht:[34] Von Tauthe, der Mutter der Götter, und Apason; Tauthe entspricht dabei Tiamat, dem Salzwasser, Apson entspricht Apsu, dem Süßwasser. Die Stelle bei Damaskios wird deshalb gern als ein – freilich spätes – Zeugnis für Kenntnisse babylonischer Mythologie im Griechischen herangezogen. Doch darf darauf hingewiesen werden, dass wir uns mit diesem Zeugnis keineswegs nur in der Philosophie der Spätantike bewegen. Ein Hinweis des Damaskios auf Eudemos macht nämlich deutlich, dass auch das Referat über babylonische kosmologische Vorstellungen schon von diesem stammt. Wir befinden uns also auch mit seinem Hinweis auf die Kos-

32 Vgl. z. B. die Sonnenfinsternis 586/5 v. Chr. nach Herodot 1, 74, 9; zu Problemen dieser Stelle und zu babylonischen Vorbildern vgl. zu Thales als Astronom, M. L. Gemelli Marciano, Die Vorsokratiker Bd. 1, Auswahl der Fragmente und Zeugnisse. Übersetzung und Erläuterungen von ders., Düsseldorf 2007, 22 ff. und zu Anaximander, U. Hölscher, Anaximander und der Anfang der Philosophie, Hermes 81 (1953), 257–277, 385–418.

33 Vgl. Eudemos, Fragment 150 Wehrli (in: F. Wehrli, Die Schule des Aristoteles. Text und Kommentar, Bd. 6–7, Basel 1953) = Damascius 124 (in: Damascius Diadochus, dubitationes et solutiones in Platonis Parmenidem, Bd. 1, hg. von C. A. Ruelle, Amsterdam 1966); siehe dazu auch Damascius 123–125 (ebenfalls bei Ruelle); zu Eudemos aus Rhodos und Babylon vgl. F. Wehrli u. a., Grundriss der Geschichte der Philosophie. Die Philosophie der Antike, Bd. 3. Peripatos, hg. von H. Flashar, 2. Auflage, Basel 2004, 558–563.

34 Vgl. Enuma Elisch 1, 30–72; dazu W. Burkert, Die Griechen und der Orient, München/Zürich 2003, 56.

mogonie des *Enuma Elisch* bereits in der Zeit des vierten bis dritten Jahrhundert v. Chr., jener Zeit also, in der wir ein generelles Interesse für fremde Kulturen im Frühhellenismus, nicht zuletzt infolge der Alexanderzüge und der mit ihnen verbundenen ‚Globalisierung' vornehmlich bei Historikern, antreffen.

Gerade auch die Schule des Aristoteles folgte Sokrates' Aufforderung zur Offenheit gegenüber Fremdem und sammelte Material der *barbaros philosophia*. Theophrast und Straton z. B. haben in ihren Werken *Über Erfindungen* auch diese Barbaren berücksichtigt.[35] Dikaiarch behandelt neben den Ägyptern auch die Leistungen der Chaldäer; Hekataios von Abdera schreibt ein Buch mit dem Titel *Über die Philosophie der Ägypter*. Megasthenes verfasst *Berichte über Indien*, Ktesias setzt sich kritisch mit Herodots Bericht über Babylon auseinander und Berossos, ein Priester des Bel, verfasst ein Werk mit dem Titel *Chaldaika* in drei Bänden und widmet es Antiochos Soter – ein Werk, das wiederum als Quelle für Alexander Polyistor, Apollodoros und Juba diente.[36] Auch in der Literatur wird Babylon zum Thema: In einem Roman des Chariton aus Aphrodisias, mit dem Titel *Chaireas und Kallirhoe* wird Babylon zu einem wichtigen Schauplatz des Geschehens.

## 5. Diogenes aus Seleukeia am Tigris

In der griechischen Philosophie begegnen wir zu dieser Zeit sogar bedeutenden Figuren, die aus Babylon oder dem Umland nach Griechenland kamen. Stellvertretend sei hier auf die Stoa und jenen Diogenes aus Seleukia hingewiesen, der auch der Babylonier genannt wurde, weil die Einwohner dieser Stadt der Landschaft den Namen gaben, wie wir bei Diogenes Laertios lesen.[37] Jener Diogenes war Schüler des ersten Schulgründers Zenon, wurde dann Chrysipps Nachfolger in der Schulleitung der Stoa und nahm in hohem Alter gemeinsam mit Karneades und Kritolaos (156/5 v. Chr.) an jener Philosophengesandtschaft teil, die in Rom in schwieriger politischer Lage für Athen gut Wetter machen sollte

35 Vgl. zu Aristoteles selbst, Aristoteles, De philosophia Fragment 9 Rose (in: V. Rose, Aristoteles Pseudepigraphus, Hildesheim/New York 1971) = Diogenes Laertios 1,8 mit H. Flashar, Aristoteles. Fragmente zu Philosophie, Rhetorik, Poetik, Dichtung, übersetzt und erläutert von. H. Flashar/U. Dubielzig/B. Breitenberger, Berlin 2006, 133, wo Aristoteles anders als in der „Metaphysik" die Anfänge der Philosophie u.a. bei den Magiern suchte; vgl. Aristoteles, De bon. Fragment 29 Rose (in dems.) = Diogenes Laertios 1,1 und dazu Aristoteles, De bono Fragment 30 (in dems.) = Diogenes Laertios 1, 6–7 = Sotion, Fragment 36 Wehrli (in: F. Wehrli. Die Schule des Aristoteles. Texte und Kommentar, Supplement 2, Basel 1978).

36 Vgl. Dikaiarch, Fragmente 55–58 Wehrli (in: F. Wehrli, Die Schule des Aristoteles. Texte und Kommentar, Bd. 1, Basel 1944); Hekataios von Abdera, FGrH 264 F 25 (in: F. Jacoby, Die Fragmente der griechischen Historiker, Bd. 3.1, Leiden 1940); Berossos, FGrH 680 (in: F. Jacoby, Die Fragmente der griechischen Historiker, Bd. 3.C1, Leiden 1958); zu Berossos vgl. P. Schnabel, Berossos und die babylonisch-hellenistische Literatur, Leipzig u.a. 1923; S. Mayer Burstein, The Babyloniaca of Berossus, 2. Auflage, Malibu 1980; G. P. Verbrugghe/J. M. Wickersham: Berossos and Manetho. Introduced and Translated. Native traditions in Ancient Mesopotamia and Egypt, Ann Arbor 1996.

37 Vgl. Diogenes Laertios 6, 81.

und in der Tat einen nachhaltigen Eindruck hinterließ. Mit Diogenes begegnen wir einem Stoiker, der diese Lehre vorsichtig weiterentwickelte, sich in Sprachphilosophie, vor allem aber in Ästhetik und insbesondere in der Musiklehre einen Namen machte. Denn er trat vehement für eine musisch-ethische und besonders für eine ‚psychagogische' Funktion der Musik ein. Er unterschied bei ihr zwei Wahrnehmungsvermögen: Die Wahrnehmung der Sinneseindrücke und die Wahrnehmung der Lust oder Unlust, die diese Sinneswahrnehmung belgeitet. Mit letzterer erklärt er, warum es unterschiedliche, individuelle Empfindungen bei der Musik geben könne. Neben der Ästhetik setzte Diogenes auch in der Theologie eigene Akzente, insofern er monotheistische Aspekte stärkte. Wir sind gerade über seine Musiktheorie und ihre Bedeutung gut unterrichtet, weil sich der Epikureer Philodem intensiv und kritisch mit seiner Auffassung auseinandersetzte. Diese Ausführungen – und damit die Positionen des Diogenes – sind uns jetzt durch eine neue Edition mit verbesserten Lesungen in ihrer Bedeutung besser kenntlich.[38] Ob man bei Diogenes Einflüsse jenes orientalisch-babylonischen Hintergrundes ausmachen kann, von dem er stammt, scheint zwar fraglich. Doch belegt die bedeutende Rolle, die ein Babylonier wie er im intellektuellen Leben Griechenlands spielen konnte, einmal mehr jene Offenheit, die Platons Sokrates einfordert, und die Rolle, die Babylonien dabei spielte. Jedenfalls wird immer wieder deutlich, dass Babylon auch *in philosophicis* keinen schlechten Namen hatte.

## 6. Das positive Babylon-Bild im Platonismus der Kaiserzeit

Babylon also als eine Quelle alter Weisheit: Nachdem im Hellenismus babylonisches Wissen durchaus gewürdigt worden war, begegnen wir in der Kaiserzeit einem sogar noch wachsenden Interesse an orientalischem Gedankengut – an ägyptischer Offenbarungsweisheit ebenso wie am babylonisch-chaldäischem Bereich.[39] Freilich weicht nun gerade in der Philosophie jene von Platon propagierte kritische Rezeptionshaltung einer Bereitschaft, ‚altes Wissen' mit einer geradezu religiösen Verehrung zu verbinden und anzunehmen, eine Form der Rezeption, der es weniger um transformierende Anverwandlung als vielmehr um Legitimation des Eigenen mit Hilfe eines Rekurses auf die Autorität des Alten geht.[40] Gerade im Platonismus der Kaiserzeit lässt sich dieser Umgang mit der *barabaros philosophia* beobachten. Programmatisch ist, was im 2. Jahrhundert n. Chr. der Platoniker

38 Vgl. Philodème de Gadara, Sur la musique Livre IV, hg., übersetzt und kommentiert von D. Delattre, Paris 2007, besonders Bd. 2, 91ff.

39 Vgl. M. Baltes, Der Platonismus und die Weisheit der Barbaren, jetzt in: M. Baltes, EPINOHMATA. Kleine Schriften zur antiken Philosophie und homerischen Dichtung, hg. v. M.-L. Lakmann, München/Leipzig 2005, 8ff.

40 Vgl. M. Erler, Legitimation und Projektion. Die ‚Weisheit der Alten' im Platonismus der Spätantike, in: D. Kuhn/H. Stahl (Hgg.), Die Gegenwart des Altertums, Heidelberg 2001, 313–326.

Kelsos aus Alexandrien in seiner Schrift *Wahrer Logos*, in der er Philosophie und Bildung der Hellenen gegen das Christentum verteidigte, ausführt. Kelsos spricht ausdrücklich von einem Urwissen bei den Alten, auch und vor allem bei barbarischen Völkern. Schon vor Platon verfügten demnach Völker wie die Ägypter, die Assyrer, die Perser oder die Inder u. a., über eine *barbaros philosophia* von höchster Autorität.[41] Denn jene weisen Männer und Völker verdankten ihre Weisheit göttlicher Inspiration. Sie waren einer derartig intuitiven Erkenntnis zugänglich, weil sie weniger verdorben waren als spätere Menschen, wie es heißt. Das Verdienst der Griechen habe dann darin bestanden, diese traditionelle Weisheit in systematischer Weise zu behandeln. Und wieder wird Mesopotamien, gerade auch Babylon, als Ort des Wissens gewürdigt.[42]

Merkmal der neuen, intensivierten Offenheit für alte Weisheit ist also die Suche nach Wahrheit und Erkenntnis, aber auch das Verlangen nach Autoritäten. Gerade im Platonismus dieser Zeit ist zu beobachten, dass jenes ‚Könnensbewusstsein' immer mehr abnimmt, welches Platon und noch später hellenistische Philosophen auszeichnete und das zu souveräner Offenheit, aber auch zu einer durchaus kritischen Anverwandlung fremden Wissens befähigte.

In der Kaiserzeit hingegen finden sich immer mehr Philosophen, die den früheren Optimismus, wonach der Mensch aus eigener Kraft Glück erlangen kann, nicht mehr teilen, sondern die Schwäche des Menschen konstatieren und von menschlicher Nichtigkeit sprechen. Das Bewusstsein des Unvermögens, sich aus eigener Kraft Wahrheit und Erkenntnis wenigstens annähern zu können, ist platonischen Philosophen der Kaiserzeit Anlass und Grund, sich religiösem Kult und sogar magisch-theurgischen Praktiken und Gebeten zu öffnen und hier Hilfe zu suchen.[43] Da die Platoniker zudem kulturelle Gegebenheiten jedweder Epoche als Manifestation einer den Menschen von alter Zeit her zugeteilten und zugänglichen Wahrheit verstanden, erhielten im 1. und 2. Jahrhundert n. Chr. alte religiöse Vorstellungen im Kontext der *barbaros philosophia* wachsenden Einfluss, nicht zuletzt wohl auch infolge der Konkurrenz mit dem wachsenden Einfluss des Christentums.

Eben in diesem Kontext treten Babylonier, genauer, jene Chaldäer in den Vordergrund, die als Fachleute der Astronomie, des Kalenderwesens und anderer Wissenschaften galten, deren Weltsicht von Determinismus und göttlicher Providenz geprägt war, die eng mit Orakelkunst in Verbindung gebracht wurden und die nicht zuletzt für die zeitweilige,

41 Vgl. Kelsos 1, 14. 16 Bader (in: R. Bader, Der Alethes Logos des Kelsos, Stuttgart/Berlin 1940, 44–45); vgl. dazu M. Frede, Celsus philosophus Platonicus, in: Aufstieg und Niedergang der römischen Welt, Bd. 2, 36, 2, Berlin/New York 1994, 5183–5213, besonders 5194.

42 Vgl. Kelsos 7, 45 Bader und auch Kelsos 1, 2 Bader (in: R. Bader, Der Alethes Logos des Kelsos, Stuttgart/Berlin 1940).

43 Vgl. M. Erler, Hilfe der Götter und Erkenntnis des Selbst. Sokrates als Göttergeschenk bei Platon und den Platonikern, in: Th. Kobusch/M.Erler (Hgg.), Metaphysik und Religion. Zur Signatur spätantiken Denkens, München/Leipzig 2002, 387–414.

intellektuelle Hegemonie Babylons verantwortlich waren. Wortorakel und Physiognomik erlaubten ihnen Aussagen über die Zukunft der Menschen und wurden mit ethischen Empfehlungen verbunden.[44] Diese Chaldäer wurden infolge ihrer umfassenden Gelehrsamkeit und Reputation als Philosophen bezeichnet und als Lehrer späterer Philosophen angesehen – auf ihre Beziehung zu Platon hatte ich bereits hingewiesen.

Es ist kein Zufall, dass besonders die philosophisch-theologische Seite ihrer Lehre bei Platonikern starke Beachtung gefunden hat, und dass man glaubte, diese Aspekte in den sogenannten „chaldäischen Orakeln" zu finden.[45]

Im Bemühen der Platoniker, Symphonie zwischen theologischen und religiösen Anschauungen bei Griechen und Barbaren herzustellen, spielen die Chaldäer eine wichtige Rolle.[46] Denn im kaiserzeitlichen Platonismus zählten die Chaldäer, die mit Assyrern gleichgesetzt wurden, als Quelle auch von religiöser Weisheit. Deshalb wurden insbesondere die chaldäischen Orakel gerne herangezogen und interpretiert. Porphyrios z. B. widmete der Auslegung dieser Orakel eine Schrift mit dem Titel *Über die aus den Orakeln zu gewinnende Philosophie* (*De philosophia ex oraculis haurienda*).[47]

Diese chaldäischen Orakel sind vermutlich im 2. oder 3. Jahrhundert n. Chr. von einem gewissen Julian, der den Beinamen ‚der Theurge' trug, und seinem Vater, der ein Chaldäer war, verfasst worden. Diese Orakel sind also Produkte spätantiker Religiosität, stellen aber einen Bezug zur alten babylonischen Weisheit her. Deshalb trugen sie für die Platoniker der Zeit den Nimbus ‚alten Wissens', wurden von ihnen auf ihrer Suche nach ‚alter Weisheit' als hilfreiche Elemente ‚alter Religiosität' empfunden und galten wie orphische Dichtung, homerische Hymnen oder Hesiods Gedichte als Zeugnisse von den Göttern eingegebener Weisheit. Neuplatonische Philosophen glaubten, in ihnen ein dem eigenen verwandtes philosophisches System zu erkennen, das auf diese Weise gleichsam göttliche Legitimation erfuhr. Man kann mit einigem Recht sagen, dass die Sammlung der chaldäischen Orakel gleichsam die Bibel des ausgehenden, platonisierenden Heidentums darstellen – an Bedeutung gleich dem platonischen Dialog *Timaios*. Auch hier ist in der Tat die Konkurrenzsituation zum Christentum greifbar. Man wollte ein ‚Heiliges Buch' des Heidentums der Bibel entgegensetzen.[48]

Die ursprüngliche Sammlung der in Hexametern abgefassten Orakel ist freilich nicht erhalten. Einzelne Orakel sind uns in unterschiedlichen Kontexten überliefert, weil neuplatonische Philosophen sie eingehender Interpretation unterzogen. In der Tat stellen sie

44 Lukrez 5, 726 f.

45 Vgl E. des Places, Oracles chaldaïques, Paris 1971; W. Kroll, De oraculis Chaldaicis, Breslauer philologische Abhandlungen, Bd. 7, 1, Breslau 1894, ND Hildesheim 1962; H. Lewy, Chaldaean Oracles and Theurgy, in: Recherches d' Archeologie, de Philologie et d' Histoire, Bd. 13, Le Caire 1956.

46 Vgl. Athenaios 12, 529 f. 530b.; Cicero, De divinatione 1, 2, dazu W.J.W. Koster, Art. „Chaldäer", in: Reallexikon für Antike und Christentum, Bd. 2 (1954), 1006–1021, besonders 1010–1013.

47 Porphyrius, De philosophia ex oraculis haurienda, hg. von G.G. Wolff, Hildesheim 1962 (unveränderter reprographischer Nachdruck der Auflage Berlin 1856).

48 Vgl. Marinos, Vita Procli, hg. von J.F. Boissonade, Amsterdam 1966, 170, 12.

eine Art Heilslehre dar, in der orientalische Elemente mit platonischen Vorstellungen auf eine Weise verbunden sind, dass platonische Philosophen Grundlagen ihrer Philosophie wieder erkennen konnten. Denn in diesen Orakeln, die als Gottesgeschenke galten, fanden die Platoniker jene triadische Struktur der Realität (Vater oder erstem Intellekt, zweitem Intellekt und Seele) wieder, die im späteren Platonismus zum durchgehenden Strukturelement des Seins wird. Hier wie dort geht es um die Rettung der menschlichen Seele aus der Finsternis der sinnlichen Welt und ihre Rückführung zum Licht des geistigen Bereiches. Die Orakel handeln von Gott, der von der sinnlichen Welt getrennt ist, von dämonischen Wesen, die Göttliches und Sinnliches verbinden, von der Seele und ihrer Rückkehr in ihre geistige Heimat und vom Kosmos. Wenn sich die Seele an ihre Herkunft erinnert, kann sie sich vom körperlichen Bereich lösen und zum göttlichen Ursprung zurückkehren, wo sie befreit vom Zwang der Bestimmung Erlösung und Ruhe findet. Doch kann sie dies nach Überzeugung kaiserzeitlicher Platoniker eben nicht mehr allein aus eigener Kraft tun. Und eben hier machen die chaldäischen Orakel ihr Hilfsangebot: So hören wir[49], dass der griechische Theologe Gregor aus Nazianz einerseits die Seele (noch) zum Göttlichen steigen lässt mit Hilfe der Vernunft. Die Chaldäer hingegen behaupteten, dass wir nur mit Hilfe materieller Riten zu Gott gelangen können. Eben diese aber bieten jene chaldäischen Orakel an, indem sie das Reden über Gott (Theologie) durch Einwirken auf Gottes Handeln (Theurgie) ergänzen und magische Praktiken für religiöse Aspekte funktionalisieren, um die Seele vom Körper zu befreien.

Auch wenn die chaldäischen Orakel also keine ‚alten' Orakel sind, so zeigt doch ihre Bedeutung für kaiserzeitliche Platoniker die nachhaltige Wirkung, welche ‚Babylon' oder ‚Babylonisches' generell als Symbol für einen Hort ‚alten Wissens' behalten hat. Freilich, dieses ‚Babylon' ist jetzt nicht mehr nur Quelle anregender Informationen, die der Prüfung bedürfen, sondern wird verstanden als Autoritäts- und Legitimationsinstanz, ja geradezu als Projektionsfläche, mit deren Hilfe man eigene Vorstellungen und deren Wahrheitsanspruch gleichsam sanktionieren zu können glaubte. Einer historisch analysierenden Sichtweise mag dieses Vorgehen, das für die Platoniker der Kaiserzeit bei ihrem Umgang mit alten Wissensinstanzen gängige Praxis war, problematisch scheinen. Doch ist zu bedenken, dass derartige Projektionen in dieser Zeit nicht zuletzt der Etablierung eigener Identität in einer schwierig gewordenen Welt, aber auch der Distanzierung gegenüber dem aufstrebenden Christentum dienten. Babylon als Hort ‚alter Weisheit' steht in dezidiertem Kontrast zum Babylon-Bild der jüdisch-christlichen Tradition, in der Babylon durchweg negativ konnotiert ist. Das Alte Testament macht Babylon in der Erzählung vom Turmbau zum Symbol menschlicher Hybris und zum Begriff von Gottesfeindschaft.[50] Im

49 Vgl. Psellos, De Anima p. 132 O'Meara (D.J. O'Meara, Michaelis Pselli philosophica minora, Bd. 2, Leipzig 1989, 1–164); dazu O. Geudtner, Die Seelenlehre der chaldäischen Orakel, Meisenheim am Glan 1971. Vgl. auch Proclus, Ek tes chaldaikes philosophias, hg. von A. Hahn, Halle 1891.

50 Vgl. Genesis 11,1,9; Jesaja 13, 19–22.

christlichen Bereich wird die Stadt zum Decknamen für Rom und die negativ bewertete diesseitige Welt. Babylon als Hure, als Hort des Luxus und der Verkommenheit: Die Platoniker hatten eine klare Begründung für diese verfehlte Sichtweise. Diese hängt nach Ansicht der paganen Platoniker damit zusammen, dass Juden und Christen mit der Tradition aller alten Völker gebrochen hätten, seien doch die Juden von den Ägyptern, die Christen von den Juden abgefallen.[51] Nicht Eintracht und Akzeptanz ‚alten Wissens', sondern im Gegenteil Aufruhr (*stasis*) und Neuerungssucht (*kainotomia*) seien für ihr Verhalten kennzeichnend[52] – also eine ganz andere Haltung als diejenige, die Platon und die späteren Platoniker propagierten und die ihre kritische Offenheit für Orientalisches begründete. Dieses Streben nach Neuerung machte dann offenbar den Weg frei für eine negative Sicht Babylons, als Quelle geistiger Fehlentwicklungen und Fehlverhaltens – eine Sichtweise, die sich auch da durchsetzte, wo ansonsten Platons Einfluss in anderer Hinsicht durchaus kenntlich blieb, wie Augustinus zeigt.[53] Doch sollte darüber das positive Bild von Babylon nicht vergessen werden, das im pagan platonischen Kontext wirksam war.

51 Vgl. dazu Kelsos 8, 2 Bader (in: R. Bader, Der Alethes Logos des Kelsos, Stuttgart/Berlin 1940), anders als z. B. Numenios sieht Kelsos den „Abfall" und die Innovationssuche als Rezeptionshaltung der Juden und dann der Christen; M. Baltes, Der Platonismus und die Weisheit der Barbaren, jetzt in: M. Baltes EPINOHMATA. Kleine Schriften zur antiken Philosophie und homerischen Dichtung, hg. von M.-L. Lakmann, München/Leipzig 2005, 6.

52 Vgl. Kelsos 3, 5–14. 7, 53. 8, 2 Bader (in: R. Bader, Der Alethes Logos des Kelsos, Stuttgart/Berlin 1940), dazu M. Frede, Celsus philosophus Platonicus, in: Aufstieg und Niedergang der römischen Welt, Bd. 2, 36, 2, Berlin/New York 1994, 5195.

53 Vgl. Offenbarung 14, 8. 16, 19. 17, 5. 18, 2. 10. 21; B. Altaner, Art. „Babylon. D. Christlich", in: Reallexikon für Antike und Christentum, Bd. 1 (1950), 1130–1134; E. Lamirande, Art. „Babylon(ia)", in: Augustinuslexikon, Bd. 1 (1986–1994), 566–569.

*Daniel T. Potts*

# The *politai* and the *bīt tāmartu*: The Seleucid and Parthian Theatres of the Greek Citizens of Babylon

## Introduction

Towards the end of his biography of Marcus Licinius Crassus (c. 115–53 BC), Plutarch described an extraordinary scene that has been commented upon by generations of ancient historians. Plutarch had just given a lengthy and very detailed account of Crassus' campaign of 53 BC against the Parthians and his defeat at Carrhae (ancient Harran, near Urfa in southeastern Turkey) while the Parthian king Orodes II was in Armenia. We are then told (Plutarch, *Crassus* 33. 1–5):

> 'During these transactions, Orodes was reconciled to Artavasdes the Armenian, and had agreed to a marriage between that prince's sister and his son Pacorus. On this occasion they freely went to each other's entertainments, in which many of the Greek tragedies were presented. For Orodes was not unversed in the Grecian literature; and Artavasdes had written tragedies himself, as well as orations and histories, some of which are still extant. In one of these entertainments, while they were yet at table, the head of Crassus was brought to the door. Jason, a tragedian of the city of Tralles [modern Aydın Güzel-Hissar, near Izmir], was rehearsing the *Bacchae* of Euripides, and the tragical adventures of Pentheus and Agave. All the company were expressing their admiration of the pieces, when Sillaces entering the apartment prostrated himself before the king, and laid the head of Crassus at his feet. The Parthians welcomed it with acclamations of joy ...'

This episode has fascinated readers for centuries. The very notion that the Parthian and Armenian kings were familiar with 'many of the Greek tragedies', that Artavasdes himself wrote tragedies, and that a performance of Euripides' *Bacchae* was being rehearsed when the messenger arrived with news of the Romans' defeat, is astonishing, even if this engagement with Greek drama reflects the predilections of Artavasdes the Armenian king far more than Orodes the Parthian, notwithstanding the propagandistic epithet *philhellenos* on Parthian coinage.[1] A love of and engagement with Greek drama in the East, of course, naturally leads one to consider the evidence of Greek theatres themselves and this brings us to one of the most interesting, late monuments at Babylon, namely its theatre.

1 Huyse doubts that Greek drama was significant for the Parthians, underscoring the fact that Orodes was merely a guest at the court of Artavasdes, who was the real *philhellene* here (P. Huyse, Die Rolle des Griechischen im 'hellenistischen' Iran, in: B. Funck (ed.), Hellenismus, Tübingen 1996, 68ff.).

Mignan's plan of Babylon[2] identified an 'eliptical mound', marked C, which is none other than the theatre excavated by Felix Langenegger, and planned by Arnold Nöldeke under the supervision of Robert Koldewey, in the spring of 1904. The theatre lay under one of three mounds in the northern part of the site called 'Homera' (Arab. *himr*, 'red', a term given by the local inhabitants because of the colour of the soil there). Although Koldewey himself only commented briefly on the theatre,[3] Erich Schmidt discussed it at length in the Sitzung of the Archäologische Gesellschaft zu Berlin on 27 May 1941[4] and in 1957 Alfred Mallwitz[5] published a detailed report on the 1904 excavations.

## Architectural history

The theatre at Babylon was built from recycled bricks taken from earlier construction(s) of the Neo-Babylonian king Nebuchadnezzar II, most probably the Etemenanki itself.[6] Koldewey noted that along with bricks bearing the stamp of Nebuchadnezzar, a fragment of a Nebuchadnezzar cylinder relating to the construction of Etemenanki was found amongst the material used to construct the theatre.[7]

Excavations revealed two major building periods, the first of which was divided into three sub-phases. In Phase $I_1$ (Fig. 1) the theatre consisted of a simple, horizontal stage building (*skene*) with three doorways opening onto the *orchestra*, the principle performance space. The stage building had small rooms at either end, as well as two *paraskenia*, side

2 R. Mignan, Travels in Chaldaea, Including a Journey from Bussorah to Bagdad, Hillah and Babylon, Performed on Foot in 1827, London 1829, 138.

3 R. Koldewey, Das wieder erstehende Babylon. Leipzig 1913.

4 E. Schmidt, Die Griechen in Babylon und das Weiterleben ihrer Kultur, Archäologischer Anzeiger 56 (1941), 786–843.

5 A. Mallwitz, Das Theater von Babylon, in: F. Wetzel/E. Schmidt/A. Mallwitz, Das Babylon der Spätzeit (= WVDOG 62), Berlin 1957, 3–22.

6 H.J. Lenzen, Die Ausgrabungen des Deutschen Archäologischen Instituts Abteilung Baghdad in Babylon und Uruk-Warka, Atlantis 33 (1961), 387; cf. S.R. Hauser, Babylon in arsakidischer Zeit, in: J. Renger (ed.), Babylon: Focus mesopotamischer Geschichte, Wiege früher Gelehrsamkeit, Mythos in der Moderne (= Colloquien der Deutschen Orient-Gesellschaft 2), Saarbrücken 1999, 216.

7 R. Koldewey, Das wieder erstehende Babylon, Leipzig 1913, 300: „Die Ziegelbrocken sind vielfach noch mit dem alten Asphalt oder mit Kalkmörtel behaftet. Es kommen auch ungebrannte vor, und namentlich die feineren Schichten enthalten viel Lehm. Gefunden wurden Nebukadnezar-Stempel, keine Scherben, einige griechische Terrakotten und ein Stück eines Nebukadnezar-Zylinders, dessen Inschrift sich auf den Bau von Etemenanki, dem Turm von Babylon, bezieht ... Demnach stammt der Schutt von einem babylonischen Gebäude, ist in griechischer Zeit hier aufgeschichtet und enthält ein Dokument von Etemenanki. Bei der Ruine von Etemenanki ist uns bereits das Fehlen des Schuttes aufgefallen ... Wir haben also in der 'Sachn' die unscheinbaren Reste eines kolossalen Gebäudes ohne Schutt, und in 'Homera' eine kolossale Schuttmasse ohne Gebäude, und können also mit größtmöglicher Sicherheit sagen, daß in Homera der Schutt von Etemenanki ruht. Das vereinigt sich vortrefflich mit der Nachricht griechischer Autoren (Strabo XVI I, 5), wonach Alexander der Große den damals verfallenen Turm wieder herzustellen beabsichtigte und auf die Wegschaffung des Schuttes 600 000 Tagelöhne verwendete".

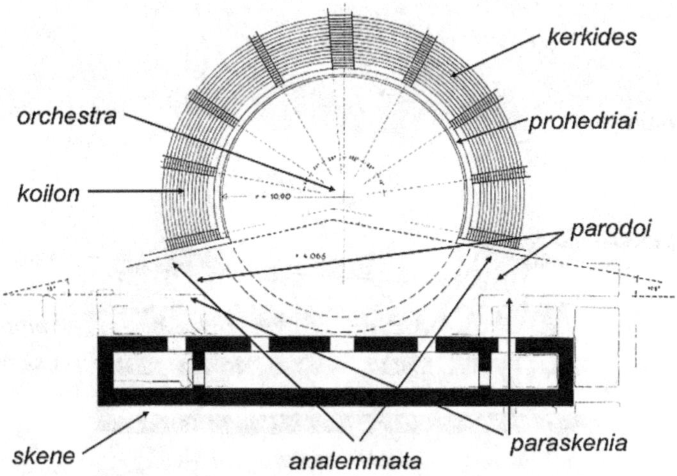


Fig. 1 | The theatre of Babylon, Period I[1] (after Mallwitz 1957).

wings projecting at each end of the stage building. The southern side of the *orchestra* actually touched the front of the stage building. The outer retaining walls or *analemmata* on the eastern and western sides of the *koilon*, adjacent to the *paradoi* or passageways through which the chorus and audience entered and exited, are not parallel to the stage-building but angled. Seven staircases divided the *koilon* or seating area into eight *kerkides* or wedges, the central one of which was wider than the others, while two additional stairways bounded the southwestern and southeastern ends of the *koilon*, as in the theatres at Sagalassos and Patara.[8]

In Phase $I_2$ (Fig. 2) the stage building was enlarged with the addition of a *proskenion* or front structure built between the original *paraskenia* which were now enclosed[9], providing an elevated area for action above the level of the *orchestra*. A pedestal was erected in the orchestra, and the *parodoi* were fitted with doors. Twelve Doric half-columns adorned the length of the *proskenion* which overlapped the orchestra by more than 1 m., a feature consistent with Vitruvius' description of Greek theatres.[10] The use of the Doric order here followed standard practice in Greek theatres, and the eleven intercolumniations are matched in the theatres at Oiniadai and Priene.[11]

In Phase $I_3$ (Fig. 3) the orchestra was raised and the diameter of the orchestra was further expanded. Three more new pedestals were erected but the scene building remained largely unchanged.

8 W.B. Dinsmoor, The Architecture of Ancient Greece: An Account of its Historic Development, New York 1975, 317.
9 'where it may have been felt that the use of mud bricks for walls and baked brick for columns required heavy abutments, there are solid parascenia aligning exactly with the proscenium' (ibid., 304).
10 Ibid., 313.
11 Ibid., 303.

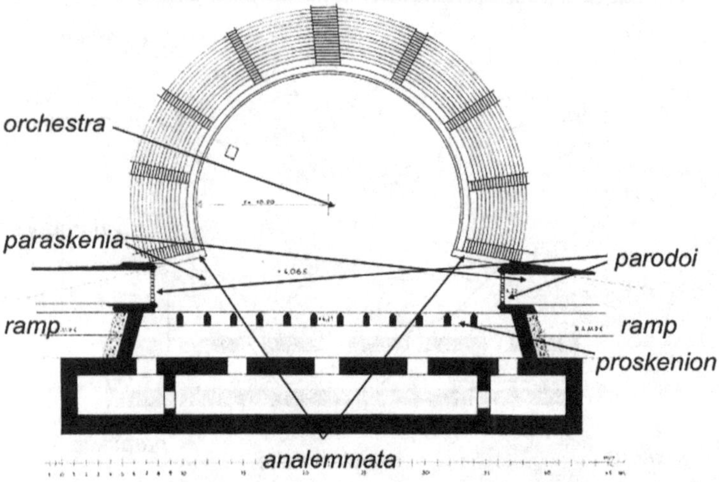


Fig. 2 | The theatre of Babylon, Period I[2] (after Mallwitz 1957).

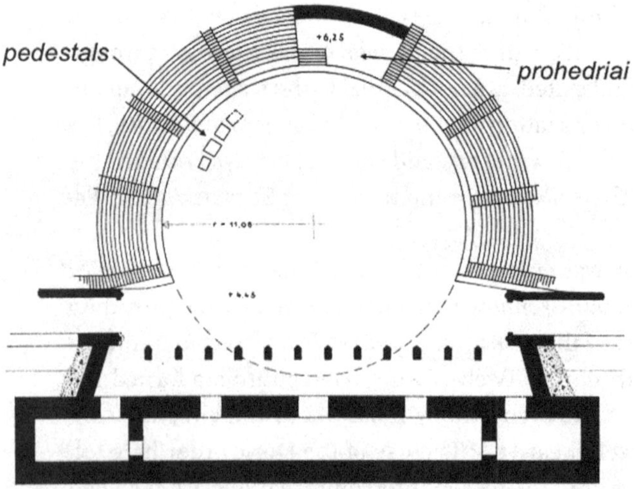


Fig. 3 | The theatre of Babylon, Period I[3] (after Mallwitz 1957).

This theatre was thoroughly destroyed and eventually rebuilt on a much grander scale in a style more reminiscent, in Mallwitz's opinion, of Roman than Greek theatres, in Phase II (Fig. 4).

The orchestra was enlarged to a diameter of 24.8 m and the front colonnade of the *proskenion* was replaced by a solid wall with three doors,[12] decorated above with a carved stucco frieze of ivy leaves and running dogs. Above this were half-columns, neither the bases nor

12 Ibid., 308.

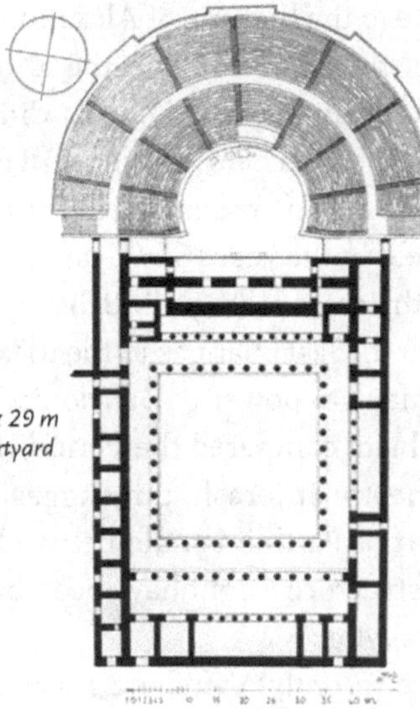


Fig. 4 | The theatre of Babylon, Period II (after Mallwitz 1957).

the capitals of which have survived. A carved, three register, stucco architrave with a meander pattern at the bottom, grape vine and clusters in the middle and Lesbian *kyma* at the top, bore traces of red, yellow-brown and possibly blue paint. How elaborate the upper part of the front façade of the scene building may have been is unclear. Four plinths, three of which were preserved, stood directly in front of the scene building, but whether these supported statues, or served as altars, is unclear. The *parodos* doors were removed and in their place walls of ashlar masonry were erected.

The greatest difference with respect to the Period I theatres, however, was the construction of a large, rectangular building with a variable configuration of rooms on the eastern, western and southern sides, and a columned portico all the way around it opening onto an open courtyard in the middle. The remains of one and possibly two staircases were recovered. This has been generally identified as a palaestra, although not the one alluded to in an inscription of 110/109 BC (SEG VII.39) which is a 'list of ephebes and *neoi* (young men), winners of various athletic competitions'[13], all of whom have Greek names.[14] Although roof tiles were not recovered, Schmidt reconstructed the building with a tiled roof, unlike Mallwitz who assumed a standard flat, Mesopotamian roof.

The secondary literature contains many different conflicting views on the date of the various sub-phases of the theatre. Koldewey expressed the view that the first construction

13 S. Sherwin-White/A. Kuhrt, From Samarkhand to Sardis: A New Approach to the Seleucid Empire, London 1993, 157.

14 B. Haussoulier, Inscriptions grecques de Babylone, Klio 9 (1909), 352–363, K. Szelényi-Graziotto, Der Kult in Babylon in seleukidischer Zeit – Tradition oder Wandel?, in: B. Funck (ed.), Hellenismus, Tübingen 1996, 187.

could date to the lifetime of Alexander[15] while Mallwitz, too, suggested that the original construction may have dated to the 4th century, noting that the wars of the Diadochoi made a later construction unlikely, as did Seleucus I's foundation of Seleucia-on-the-Tigris.[16] Mallwitz dated Period $I_2$ to the reign of Antiochus IV, i.e. to the 2nd quarter of the 2nd century BC (175–164 BC), based on the unproven belief that OGIS 253, which commemorates games held in the year 166/5, came from Babylon; and Period $I_3$ to the reign of the Parthian ruler Mithridates II (123–88 BC)[17], not for any good archaeological reason but rather in the belief that the last changes to the theatre prior to its destruction might reflect the coming of the Parthians to power in Babylonia.

Mallwitz compared the Period II theatre to the theatres of Marcellus in Rome and the South Theatre at Jerash, but suggested a date in the mid-2nd century AD, coinciding with the floruit of Parthian settlements elsewhere in Babylonia, and raised the possibility that the new structure might have been built as a result of either Trajan's or Septimius Severus' visit to Babylon.[18]

More recently, Van der Spek has discussed the date of the theatre. He has suggested that the original structure may have been built 'in the days of Alexander or shortly afterwards, when Babylon was again the capital of Asia, ca. 300 BC'[19]. A recently published chronicle fragment might also provide a context for the construction. The text states that 'Antiochus, the crown prince, settled the Macedonians, as many as there were in Babylon', and while the text is undated, it probably pre-dates 294 BC, as that was the year in which Antiochus I became co-ruler with his father Seleucus I[20]. The settlement of Macedonians in Babylon could have been the trigger for the construction of a theatre there.

Like Mallwitz, van der Spek dated the next phase to the introduction, by Antiochus IV, of a new group of Greek citizens to the city.[21] While the significance of the reference to Anti-

15 R. Koldewey, Das wieder erstandene Babylon, Leipzig 1913, 301. Cf. D.G. Lyon who wrote, 'The original construction may well belong to the time of the occupation of the city by Alexander' (D.G. Lyon, Recent Excavations at Babylon, The Harvard Theological Review 11 (1918), 31) and later Anonymous, The Greek Theatre at Babylon, Sumer 12 (1956), 92 discussing the 10-day German excavation campaign of November, 1956: 'In this short period of investigation Prof. Lenzen became sure that this theatre was built by Alexander the Great from the ruins of the staged tower of Babylon'. Cf. E. Weidner, Ausgrabungen in Nippur und Babylon: Babylon, Archiv für Orientforschung 18 (1967), 174.

16 A. Mallwitz, Das Theater von Babylon, in: Wetzel/Schmidt/Mallwitz, Das Babylon der Spätzeit, Berlin 1957, 20.

17 Cf. F. Sear, Roman Theatres: An Architectural Study (= Oxford Monographs in Classical Archaeology), Oxford 2006, 5.

18 A. Mallwitz, Das Theater von Babylon, Berlin 1957, 21.

19 Cf. S.R. Hauser, Babylon in arsakidischer Zeit, in: J. Renger (ed.), Babylon: Focus mesopotamischer Geschichte, Wiege früher Gelehrsamkeit, Mythos in der Moderne (= Colloquien der Deutschen Orient-Gesellschaft 2), Saarbrücken 1999, 216: 'um 300 v. Chr.'.

20 R.J. van der Spek, The Size and Significance of the Babylonian Temples under the Successors, in: P. Briant and F. Joannès (eds.), La transition entre l'empire achéménide et les royaumes hellénistiques (vers 350–300 av. J.-C.) (= Persika 9), Paris 2006, 293, Reverse 6'–7'.

21 R.J. van der Spek, The Theatre of Babylon in Cuneiform, in: W.H. van Soldt (ed.), Veenhof Anniversary Volume, Leiden 2001, 446.

ochus IV in OGIS 253 as 'founder of the city' has diminished since the provenance of the text is uncertain[22], a recently discovered chronicle (BCHP 14), dating to 149 SE or 163/2 BC, i.e. just a year before Antiochus IV's death, explicitly mentions the "so-called Greeks, the *politai*, who in the past at the command of king Antiochus had entered Babylon and who anoint with oil just like the *politai* who are in Seleucia, the royal city".[23] Clearly there was a community of Greek citizens (*politai*) at Babylon and the implication here is that this particular group had been introduced by Antiochus himself.

Erich Schmidt, on the other hand, worked from the Period II building backwards in his dating of the structures. He pointed to parallels between the stucco decoration found at both Babylon and Assur, and believed that the letter forms in the only Greek building inscription found in the theatre corroborated these parallels. Inscribed on an alabaster plaque, the text reads (trans. R. J. van der Spek, The Theatre of Babylon in Cuneiform, 446):

| | |
|---|---|
| Διοσκουρ[ίδης- – -] | Dioskouri[des……] |
| ὁ φιλόδοξ[ος- – -] | showing a zeal for glory [in his disposition towards the people/city] |
| τὸ θέατρο[ν- – -] | ….repaired (?) the theatre […..] |
| καὶ σκην[ήν- – -] | and the stage building [….] |

Schmidt adduced parallels of 1st and 2nd century A.D. date for individual letter forms[24], but this is now considered entirely unsound and highly spurious (Ch. Habicht, pers. comm.). Finally, Schmidt was unable to adduce any clear evidence for the date of the Period I theatre beyond noting that a) it was obviously older than Period II; b) it was unlikely to have been built during the lifetime of Alexander or in the confusion that followed his death, as marked by the wars of the Diadochoi; and c) it was most likely to have been built during the reign of Antiochus I, who rebuilt the Esagila.[25]

22 S. Sherwin-White/A. Kuhrt, From Samarkhand to Sardis: A New Approach to the Seleucid Empire, London 1993, 157–158.
23 R.J. van der Spek, Ethnic Segregation in Hellenistic Babylon, in: W.H. van Soldt (ed.), Ethnicity in Ancient Mesopotamia: Papers Read at the 48th Rencontre Assyriologique Internationale, Leiden, 1–4 July 2002, Leiden 2005, 396, 403–404.
24 Specifically the *phi* – which he said was unattested in that form before the middle of the 1st century BC and was attested in ephebe lists of 40/41 and 53/54 AD – and *kappa*, for which he adduced a parallel in an inscription of Vibius Hipparchus from 118/119 AD. Further parallels for the two letters together in this form were known from Attic inscriptions dating to 138 and 200 AD, respectively, leading him to conclude that a date between 130 and 200 seemed likely (E. Schmidt, Die Griechen in Babylon und das Weiterleben ihrer Kultur, Archäologischer Anzeiger 56 (1941), 843; F. Wetzel/E. Schmidt/A. Mallwitz, Das Babylon der Spätzeit, Berlin 1957, 50).
25 E. Schmidt, Die Griechen in Babylon und das Weiterleben ihrer Kultur, 843.

The presence of the honorific title *philodoxos* in the Babylon inscription is chronologically significant, inasmuch as this is a typical *philo*-composite of the imperial period.[26] Unfortunately, the use of this term cannot be any more closely dated. A 1st/ 2nd century A.D. date for the last phase, based on the Dioskurides inscription, has been suggested by some scholars subsequent to Schmidt[27] but a 1st century date is preferred by others (e.g. John Ma, pers. comm.). At best one can simply say that the Roman architectural parallels and the use of the honorific *philodoxos* support a 1st and/or 2nd century A.D. date for the theatre's last phase.

None of the ancient authors who described Babylon (Quintus Curtius, Arrian, Diodorus Siculus, Strabo, Pausanias and Pliny) mentioned a theatre there but recently, building on a suggestion by John Ma, Bert Van der Spek has shown that it is attested in the astronomical diaries as *bīt tāmartu* (É IGI.DU$_8$.MEŠ)[28], a term derived from the verb *āmaru*, 'to watch', 'to see'. Thus the *bīt tāmartu* is the 'house of observation' or 'house of watching', an almost literal translation of the Greek term *theatron*, meaning a 'place for seeing'. What might have been performed in the theatre of Babylon? As discussed earlier, the famous passage of Plutarch's *Crassus* attests to a familarity in Armenia with Euripides' *Bacchae*, and according to a further statement of Plutarch's (On the fortune of Alexander, 328), 'When Alexander civilised Asia Homer became common reading, and the sons of Persians, Susians and Gedrosians learned to intone the tragedies of Sophocles and Euripides'. That plays were staged at Babylon seems highly likely, particularly given the Dioskourides inscription attesting to the restoration of the stage building, but we have no precise records of which may have been staged.

Yet theatres had multiple functions and in Greece they were intimately linked to the institution of the *polis* and indeed to democracy itself, at least in Athens. As Simon Goldhill has observed, 'The institution of theatre is analogous to the two other great Athenian democratic institutions for the staging of speeches, the law-court and the assembly. Each congregates a body of citizens, constituted in a privileged way as the collective of the *polis* ... More important than the precise demography of the audience, however, is the way that the theatre seating constructed a political map of the city'[29]. Particular seats were reserved for members of the *boulê* or Council; for ambassadors; priests; tribes (*kerkis*); and *ephebes*, or younger men about to become full citizens. It is in this political realm that we have some evidence for the use of the theatre at Babylon, specifically as the meeting place of the *politai*.

26 C. Veligianni, Philos und philos-Komposita in den griechischen Inschriften der Kaiserzeit, in: M. Peachin, Aspects of friendship in the Graeco-Roman world: Proceedings of a conference held at the Seminar für Alte Geschichte, Heidelberg, on 10–11 June, 2000 (Journal of Roman Archaeology Suppl. 43), Portmouth 2001, 63–80.
27 E.g. F. Canali di Rossi, Iscrizioni dello Estremo Oriente Greco (= Inschriften griechischer Städte aus Kleinasien 65), Bonn 2004, 72, no. 112; queried by A. Invernizzi, Hellenism in Mesopotamia: A View from Seleucia on the Tigris, Al-Rafidan 15 (1994), 6.
28 R.J. van der Spek, The Theatre of Babylon in Cuneiform, in: W.H. van Soldt (ed.), Veenhof Anniversary Volume, Leiden 2001, 447.
29 S. Goldhill, Greek Drama and Political Theory, in: C. Rowe/M. Schofield (eds.), The Cambridge History of Greek and Roman Political Thought, Cambridge 2000, 62.

The theatre is mentioned in eight diaries, beginning in April/May 162 BC, i. e. late in the reign of Antiochus V.[30] The context is unfortunately unclear as the text is badly damaged, but the text mentions the 'theatre, which is in the district ...' and then unfortunately breaks off.

Twenty years later, however, we have a valuable reference from 2 July 141 BC, i.e. just a few months after the Parthian king Mithradates I. had conquered Babylonia.[31] There we read, 'That month, 22nd day, a parchment letter concerning the *pahatu* (governor) of Babylon and the *politai*, who are in Babylon, which to the *pahatu* and the *politai* aforementioned was written ... was read in the theatre'[32]. As noted already, the *politai* are the Greek citizenry of Babylon, while Van der Spek suggests that *pahatu*, or governor, was the Akkadian translation of the Greek term *epistates*.

A further diary from January/February 133 BC refers to the reading of a parchment letter to the *pahatu* and *politai* in the theatre[33], while one from April/May 125 BC talks of 'parchment letters of the king to the *politai*, who are in Babylon', being 'read in the theatre', specifically 'parchment letters of the king in the aforementioned place were read concerning the appointment of one from the *politai*'.[34]

At about this time part of Babylonia came under the control of one Hyspaosines (Akk. Aspasine), a former Seleucid governor in Messene who took the opportunity to seize power and establish the so-called kingdom of Characene, based at his capital Spasinou Charax. In an important and detailed diary from January/February 124 BC, we learn that 'the letter of the king, which was written to the governor of Babylon and the *politai* who are in Babylon was read in the theatre', and it goes on to quote verbatim from the letter concerning a battle with 'Pittit, the Elamite enemy'.[35] According to a diary from October/November 119 BC, letters of Mithradates II to 'the governor of Babylon and the *politai*, who are in Babylon, were ... read in the theatre' detailing a campaign against Gutium.[36] A further reference to the reading of letters in the theatre is recorded in a diary of February/March 88 BC[37], and five years later, in December 83/January 82 BC, we have a final fragmentary entry that simply says 'in the theatre'[38].

30 A.J. Sachs/H. Hunger, Astronomical Diaries and Related Texts from Babylonia, Vol. III. Diaries from 164 B.C. to 61 B.C. (= Denkschriften der Österreichischen Akademie der Wissenschaften, phil.-hist. Kl. 247), Vienna 1996, 30–31, No. –161 A[1], A[2] 'Obv.' 25'.

31 R.J. van der Spek, The Theatre of Babylon in Cuneiform, Leiden 2001, 450.

32 A.J. Sachs/H. Hunger, Astronomical Diaries and Related Texts from Babylonia, Vienna 1996, 134–135, No. –140 A: 'Rev. 6'.

33 Ibid., 230–231, No. –132 D[2] 'Rev.' 15'.

34 Ibid., 264–264, No. –124 A: 'Obv.' 6'–7'.

35 Ibid., 278–279, No. –124 B: 'Rev.' 17'.

36 Ibid., 326–327, No. –118 'Rev. A19'.

37 Ibid., 454–455, C: 'Rev.' 30'

38 Ibid., B: 'Obv.' 21'.

We thus have a fairly continuous set of attestations, extending from 162 to 82 BC, in which the use of the theatre is attested epigraphically. It is tempting to suggest that it was at this point that the Period I theatre was destroyed or went out of use until the new theatre, which Mallwitz likened to Roman theatres, was constructed, in the 1$^{st}$ or 2$^{nd}$ century AD.

In Greece, of course, theatres served a crucial role within the context of 'democratic polis ideology'. As Goldhill says, 'The theatre was a space in which all the citizens were actors – as the city itself and its leading citizens were put on display'.[39] Moreover, 'Theatre is a fundamental factor in the politicization of the Athenian citizen, in putting political reflection in the public domain'[40]. For Babylon's Greek *epistates* and *politai*, the theatre may have functioned in precisely the same manner. Unfortunately, we have little evidence, beyond its physical structure, that can shed light on this side of its existence. What the astronomical diaries, however, show us, is another, equally political side to the theatre, one perhaps less romantic than visions of Babylonian Greeks listening and watching transfixed to *Philoctetes* or *Antigone*, but one no less important. McDonald's 1943 study of *The political meeting places of the Greeks* contains numerous references to the use of theatres for large, political gatherings, a point borne out in a more recent study by Rune Frederiksen of the role of the Greek theatre in the *polis*.[41]

References to a decree announced in an assembly in the theatre at Athens (e.g. from 163/2 BC; SEG 38 [1988], No. 108.4–5) show that the theatre there was used for public assemblies. In cities like Ephesus, Megalopolis, Samos, Erythrae and Methymna, citizens were divided into divisions of 1000. In the theatre at Samos, sign posts indicated where each division was meant to sit, while in the theatre at Megalopolis the names of individual tribes were inscribed on the backs of the *prohedriai* around the *orchestra*.[42]

Writing of Athenion or Aristion's speech to the Athenians in 87/86 BC, a speech in which the orator tried to persuade his audience to abandon Rome and support Mithridates of Pontus, a deprecating reference is made to 'our theatre deserted by the assembly'.[43] When the apostle Paul was in Ephesus (Acts 19.21–41), Demetrius the silversmith denounced the new teachings and there ensued a riot leading to the mob rushing into the theatre there, dragging with them two Macedonian companions of Paul's. Much later, according to Livy and Polybius (Livy 34.22.6, 34.48.3), the Romans convened the Panhellenic assemblies of 195 and 194 BC in the theatre of Corinth.

39 S. Goldhill, Greek Drama and Political Theory, in: C. Rowe/M. Schofield (eds.), The Cambridge History of Greek and Roman Political Thought, Cambridge 2000, 64.

40 Ibid., 65.

41 R. Frederiksen, The Greek Theatre: A Typical Building in the Urban Centre of the Polis?, in: T.H. Nielsen (ed.), Even More Studies in the Ancient Greek Polis: Papers from the Copenhagen Polis Centre 6 (= Historia Einzelschriften 162), Stuttgart 2002, 65–124.

42 W.A. McDonald, The Political Meeting Places of the Greeks, Baltimore 1943, 62.

43 Athenaeus 5.213D, R.J. van der Spek, The Theatre of Babylon in Cuneiform, in: W.H. van Soldt (ed.), Veenhof Anniversary Volume, Leiden 2001, 447.

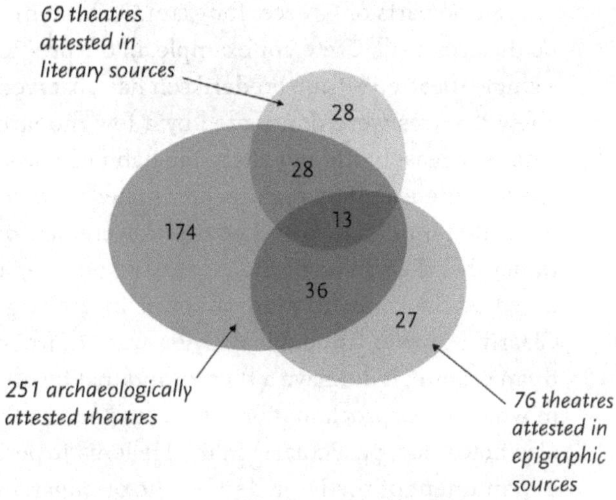


Fig. 5 | Greek theatres from Cyrenaica to Bactria (data drawn from Frederiksen 2002).

Another riotous gathering is documented at Antioch-on-the-Orontes in 361/2 AD, when Julian the Apostate says he 'arrived among you the populace in the theatre', as he wrote in his satirical *Misopogon* (368C). As some scholars have pointed out[44], question marks hover over some instances cited in the literature for the use of theatres as places of assembly, but Cicero (106–43 BC) (*Pro Flacco,* 16) was explicit in asserting that it was the norm for the Greeks to convene assemblies in theatres.[45] Indeed Frank Kolb went so far as to declare, 'Die griechischen Theatergebäude waren wohl sogar von vorneherein in erster Linie für die Volksversammlung gedacht'[46]. Whether or not this was the primary *raison d'être,* Mogens Hermann Hansen and Tobias Fischer-Hansen admit, 'we believe that the Greeks built theatres for theatrical purposes but that these could conveniently and frequently be used as meeting places for the city's popular assembly'.[47] The use of the theatre at Babylon for the purpose of public, political communication is therefore interesting in light of broader patterns of Greek theatre usage. But this topic is far from straightforward.

To begin with, the notion that a theatre was an essential part of any Greek *polis* is patently incorrect (Fig. 5).

44 M.H. Hansen/T. Fischer-Hansen, Monumental Political Architecture in Archaic and Classical Greek poleis. Evidence and Historical Significance, in: D. Whitehead (ed.), From Political Architecture to Stephanus Byzantius: Sources for the Ancient Greek Polis, Stuttgart 1994 (= Historia Einzelschriften 87), 48ff.
45 F. Kolb, Agora und Theater, Volks- und Festversammlung (= Archäologische Forschungen 9), Berlin 1981, 88–89; M.H. Hansen/T. Fischer-Hansen, Monumental Political Architecture in Archaic and Classical Greek poleis. Evidence and Historical Significance, Stuttgart 1994, 50–51.
46 F. Kolb, Agora und Theater, Volks- und Festversammlung (= Archäologische Forschungen 9), Berlin 1981, 90.
47 M.H. Hansen/T. Fischer-Hansen, Monumental Political Architecture in Archaic and Classical Greek poleis. Evidence and Historical Significance, Stuttgart 1994, 53.

Some parts of Greece, for example, have many *poleis* and few theatres, and some have no theatres at all. Crete, for example, in the pre-Roman period, had around 60 *poleis* and not a single theatre.[48] Rune Frederiksen has observed that 'Some of the regions with no or only a few theatres were dominated by a few monarchically governed *poleis*; perhaps they, for whatever reason, did not share the habit of mass meetings of people, religious or political, to the same extent as regions dominated by democratic and oligarchic *poleis*? ... My point is not that monarchies and tyrannies were not fond of drama and theatres .... The claim is rather that the spread of the theatre in some regions governed by kings or tyrants was hindered by the concentration of power in the larger urban centres, perhaps deliberately'.[49] Clearly, however, although Babylon was an urban centre in the Seleucid and later the Parthian empire, it did have a theatre and that theatre was explicitly used for social assemblies in which news proclamations were read out, as the astronomical diaries attest. Frederiksen also notes that, particularly in the Hellenistic period, 'the monumental theatre seems to be a monument of the large democratic or oligarchic *polis* in regions in which dramatic performances were an integrated part of the political and religious culture. Such *poleis* are at the same time the places where we would expect most other types of assembly activity – such as *ekklesia* – to have taken place, and for which the theatre was a convenient architectural frame'. This seems to be an apt appraisal of the situation we have been considering at Babylon.

The decline of Babylon is usually seen as the complement to the rise of Seleucia-on-the-Tigris, particularly in light of Strabo (*Geographie* 16.1.5), who says that Babylon 'was neglected and thrown into ruins, partly by the Persians and partly by time and by the indifference of the Macedonians to things of this kind, and in particular after Seleucus Nicator had fortified Seleuceia on the Tigris near Babylon, at a distance of about 300 stadia'. Interestingly, the head of the Italian mission there, Antonio Invernizzi, has suggested that the quite circular mound known as Tell 'Umar, in the north of the site, might be the location of Seleucia's own theatre. Strabo's famous assertion that 'the greater part of Babylon is so deserted that one would not hesitate to say ... "the great city is a great desert"', seems, however, to be a great exaggeration.

The *Diadochoi Chronicle* makes it clear that around 310/9 BC, in the course of the conflict between Seleucus I and Antigonus-the-one-eyed, Babylon was attacked and plundered, while in 274 BC Antiochus I moved some of the population of the city to Seleucia as reinforcements during a time of war. But as the evidence of the theatre at Babylon suggests, and many other sources confirm[50], life there went on and the theatre played an integral part in it, at least for the city's Greek inhabitants. Throughout the second and early first cen-

48 R. Frederiksen, The Greek Theatre: A Typical Building in the Urban Centre of the Polis?, Stuttgart 2002, 91.
49 Ibid.
50 T. Boiy, Late Achaemenid and Hellenistic Babylon, Leuven 2004, (= OLA 136); S.R. Hauser, Babylon in arsakidischer Zeit, in: J. Renger (ed.), Babylon: Focus mesopotamischer Geschichte, Wiege früher Gelehrsamkeit, Mythos in der Moderne, Saarbrücken 1999 (= Colloquien der Deutschen Orient-Gesellschaft 2), 207–239.

turies BC the theatre served as an important locus where political communiqués were read out, informing the *politai* of recent events involving their rulers. And when Babylon became a Parthian city, the rebuilding of the theatre – whether or not it involved the arrival of a Roman emperor – signaled its ongoing importance. Plutarch may have overstated the philhellenism of Orodes II, but the Greek citizens of Babylon, of whom only Dioskurides has come down to us as an historical identity, continued a tradition which may have extended back as far as the late 4$^{th}$ century BC and which, through the use of recycled bricks from Etemenanki, brought something tangible if recycled from a more ancient Babylon into the new millennium.


## Acknowledgements

I would like to thank a number of scholars who helped me with advice, references or a critical reading: Mark Geller (University College London); Christian Habicht (Institute for Advanced Study); John Ma (Corpus Christi College, Oxford); Pantelis Nigdelis (Aristotle University of Thessaloniki); Bert van der Spek (Free University, Amsterdam); and Peter Wilson (Univ. of Sydney). Naturally none are responsible for the views expressed here. It will be obvious to anyone who has ever dealt with this subject, however, just how indebted the present modest contribution is to van der Spek's article of 2001.

*Reinhard G. Kratz*

# Nabonid in Qumran

*Hermann Spieckermann zum Sechzigsten*

## 1. Einführung

Wie schon in der Antike, so haben sich auch in der Neuzeit immer wieder einmal Hirtenknaben in die Höhlen nahe der Siedlung Khirbet Qumran am Toten Meer verlaufen und sind dort auf hebräische oder aramäische Handschriften gestoßen: Bibelhandschriften, aber auch Handschriften nichtbiblischer Texte aus dem dritten vor- bis zum zweiten nachchristlichen Jahrhundert. Im achten Jahrhundert soll es ein entlaufener Hund, im zwanzigsten Jahrhundert eine Ziege gewesen sein, die ihre Herren zu den Höhlen und den dort versteckten Texten geführt haben.[1]

Wie dem auch sei, niemand hätte an diesem Ort und in der Zeit, aus der die Handschriften datieren, mit irgendwelchen sensationellen Nachrichten über Babylon gerechnet. Doch im Jahre 1956 veröffentlichte Józef Tadeus Milik, einer der Erstherausgeber der Texte vom Toten Meer, Fragmente einer Handschrift, die sich ein Jahr zuvor in Höhle 4 von Khirbet Qumran gefunden hatten und den Namen des letzten neubabylonischen Königs Nabonid (Nabuna'id) enthielten.[2] Milik gab dem aramäischen Text nach seinem *incipit* den Titel „Prière de Nabonid". Auch wenn die Schreibung des Namens נבני etwas ungewöhnlich ist und wir nicht genau wissen, wie er ausgesprochen wurde (vermutlich: Nabunaj), ist die Identität doch eindeutig: Nabunaj wird als מלך בבל „König von Babylon" bezeichnet, und zweimal ist תימן, die Wüstenoase Tema auf der arabischen Halbinsel, als sein Aufenthaltsort genannt.[3] Auch die Geschichte von der Bekehrung des Königs zum höchsten Gott erinnert an den letzten neubabylonischen König und die Überlieferung, die sich um seine Person gebildet hat und in den Erzählungen des Danielbuches (und weit darüber hinaus) ihren Nachhall gefunden hat.

1 Zur Fundgeschichte von 1947 vgl. J.C. VanderKam, Einführung in die Qumranforschung: Geschichte und Bedeutung der Schriften vom Toten Meer (übers. von M. Müller), Göttingen 1998, 20–32. Über antike Funde, von denen eine Mitteilung des Patriarchen Timotheus I. von Seleukia (726–819 n. Chr.) an den Metropoliten von Elam, Sergius, spricht, berichtet P.E. Kahle, Die Kairoer Genisa: Untersuchungen zur Geschichte des hebräischen Bibeltextes und seiner Übersetzungen, hg. von R. Meyer, Berlin 1962, 16–17.

2 J.T. Milik, „Prière de Nabonide" et autres écrits d'un cycle de Daniel. Fragments Araméens de Qumrân 4 (Pl. I), Revue biblique 63 (1956), 407–411.415 (im Folgenden J.T. Milik, «Prière de Nabonide» [1956]).

3 Zu den Schreibweisen der Namen vgl. R. Meyer, Das Gebet des Nabonid: Eine in den Qumran-Handschriften wiederentdeckte Weisheitserzählung, Sitzungsberichte der Sächsischen Akademie der Wissenschaften, Philologisch-Historische Klasse 107/3, Berlin 1962, 17–18.20–21 (im Folgenden R. Meyer, Das Gebet des Nabonid, Berlin 1962); neuerdings F. García Martínez, The Prayer of Nabonidus. A New Synthesis, in: Ders., Qumran and Apocalyptic: Studies on the Aramaic Texts from Qumran (= Studies on the Texts of the Desert of Judah 9), Leiden 1992, 116–136, hier 120–122 (im Folgenden F. García Martínez, The Prayer of Nabonidus [1992]).

Da dieser Band des begleitenden Symposiums zur Berliner Ausstellung „Babylon: Mythos und Wahrheit" im Besonderen an der Rezeption Babylons interessiert ist, will ich der Frage nachgehen, wie Nabonid in die Texte vom Toten Meer gelangt ist und welche Bedeutung ihm dort zukommt. Bekanntlich ist die Gemeinschaft von Qumran, die sich selbst היחד „die Gemeinschaft" nennt, eine jüdische Gruppierung, die sich ausschließlich der Tora und den übrigen biblischen Schriften verpflichtet weiß. In den biblischen Schriften aber kommt Nabonid, jedenfalls dem Namen nach, nicht vor. Dies unterscheidet das *Gebet des Nabonid* von anderen Qumrantexten, die in aller Regel einen mehr oder weniger direkten Bezug zur biblischen Überlieferung aufweisen. Zwar erinnert das *Gebet des Nabonid* in manchem an die Kapitel 4 und 5 des Danielbuches, doch sind hier Nebukadnezar und Belsazar und nicht Nabonid die Protagonisten. Insofern dürften die Beziehungen komplexer sein. Sie sollen hier exemplarisch für die Rezeption Babylons in den Schriften vom Toten Meer untersucht werden.

## 2. Der Text

Nach der Erstveröffentlichung von Milik wurde der Text verschiedentlich bearbeitet und liegt inzwischen auch in der offiziellen Edition der *Discoveries of the Judean Desert* vor.[4] Er-

4 J.J. Collins, 4QPrayer of Nabonid ar (4Q242), 4Qpseudo-Daniel[a-c] (4Q243–245), in: Qumran Cave 4.XVII, DJD XXII, Oxford 1996, 83–93. 95–164 + Pl. VI, VII–X (im Folgenden J.J. Collins, DJD XXII, Oxford 1996). An weiteren, dort nicht aufgeführten oder später erschienenen Beiträgen vgl. noch A. Dupont-Sommer, Les écrits esséniens découverts prés de la mer morte, Paris 1959, zitiert nach der deutschen Ausgabe: Die essenischen Schriften vom Toten Meer (übers. von W.W. Müller), Tübingen 1960 (im Folgenden A. Dupont-Sommer, Die essenischen Schriften, Tübingen 1960); R. Meyer, Das Qumranfragment Gebet des Nabonids, Theologische Literaturzeitung 85 (1960), 831–834 (im Folgenden R. Meyer, Qumranfragment [1960]); E. Haag, Die Errettung Daniels aus der Löwengrube. Untersuchungen zum Ursprung der biblischen Danieltradition (= Stuttgarter Bibelstudien 110), Stuttgart 1983, 62–73; R.G. Kratz, Translatio imperii: Untersuchungen zu den aramäischen Danielerzählungen und ihrem theologiegeschichtlichen Umfeld (= Wissenschaftliche Monographien zum Alten und Neuen Testament 63), Neukirchen-Vluyn 1991, 99–111 (im Folgenden R.G. Kratz, Translatio imperii, Neukirchen-Vluyn 1991); K. Koch, Gottes Herrschaft über das Reich des Menschen: Daniel 4 im Licht neuer Funde, in: A.S. van der Woude (Hg.), The Book of Daniel in the Light of New Findings (= Bibliotheca Ephemeridum Theologicarum Lovaniensium 106), Louvain 1993, 77–119; ders., Daniel 1–4, Biblischer Kommentar AT XXII/1–6, Neukirchen-Vluyn 2005 (im Folgenden K.Koch, Daniel 1–4, Neukirchen-Vluyn 2005); K. Beyer, Die aramäischen Texte vom Toten Meer. Ergänzungsband, Göttingen 1994, 104; A. Lange/M. Sieker, Gattung und Quellenwert des Gebets des Nabonid, in: H.-J. Fabry et al. (Hgg.), Qumranstudien: Vorträge und Beiträge der Teilnehmer des Qumran-Seminars auf dem internationalen Treffen der Society of Biblical Literature, Münster, 25.–26. Juli 1993 (= Schriften des Institutum Judaicum Delitzschianum 4), Göttingen 1996, 3–34 (im Folgenden A. Lange/M. Sieker, Gattung und Quellenwert, Göttingen 1996); S. Ackerman, The Prayer of Nabonidus, Elijah on Mount Carmel and the Development of Monotheism in Israel, in: W.G. Dever/J.E, Wright (Hgg.), The Echoes of Many Texts: Reflections on Jewish and Christian Traditions, Essays in Honor of Lou H. Silberman (= Brown Judaic Studies 313), Atlanta 1997, 51–65 (im Folgenden S. Ackerman, The Prayer of Nabonidus, Atlanta 1997); P.W. Flint, The Daniel Tradition at Qumran, in: C.A. Evans/P.W. Flint (Hgg.), Eschatology, Messianism and the Dead Sea Scrolls, Michigan/Cambridge 1997, 41–60, hier 55–59 (im Folgenden P.W. Flint, The Daniel Tradition at Qumran, Michigan/Cambridge 1997) und ders., The Daniel Tradition at Qumran, in: J.J. Collins/P.W. Flint (Hgg.), The Book of Da-

halten sind drei – ihrerseits aus mehreren Teilen zusammengesetzte – Fragmente (Fr. 1, 2a, 2b und 3), die sich zum rechten Teil einer Kolumne zusammenfügen lassen und die Anfänge von 9 Zeilen bieten. Da es sich in Z. 1 um eine Art Überschrift handelt, ist damit offensichtlich der Anfang des Werkes erhalten. Ein physischer Kontakt besteht zwischen den diversen Teilen in Fr. 1 sowie zwischen Fr. 2a, 2b und 3. Zwischen Fr. 1 und 2–3 besteht kein physischer Kontakt, doch lässt sich der Abstand anhand von Z. 4 bestimmen. Hier folgen die in Fr. 1 und 2–3 lesbaren Wörter aller Wahrscheinlichkeit nach unmittelbar aufeinander (vgl. *Est 3,4*), so dass man die beiden Hälften sehr nahe aneinander rücken muss und sich von hier aus der Platz für Ergänzungen in den übrigen Zeilen ermessen lässt.[5]

4Q242 Fr. 1–3

1 מלי צ[ל]תא די צלי נבני[a] מלך [ב]בׄל[a] מלכׄ[א רבא כדי כתיש הוא]
2 בשחנא באישא בפתגם אׄ[לה]א בתימן[ אנה נבני בשחנא באישא]
3 כתיש הוית שנין שבע ומן[a] [די] שוי[b] א[להא עלי אנפוהי ואסא לי][b]
4 וחטאי שבק לה גזר[a] והוׄא יהודי מׄ[ן[b] בני גלותא[c] על לי ואמר][c]
5 החוי[a] וכתב למעבד יקר ורׄ[בו] לשם[b] א[להא עליא[b] [c]וכן כתבת אנה][c]
6 כתיש הוית בשחנא ב[אישא] בתימן [בפתגם אלהא עליא]
7 שנין שבע מצלא[a] הוי[ת קדם] אׄלׄהׄיׄ כספא ודהבא [נחשא פרזלא]
8 אעא אבנא חספא מן די [הוית סב]ר די אלהין ה[מון
9a ] [a]או◦[ ד שׄלׄ◦[ ]◦[a]
9 [a]תׄ[ ]מיהׄון[a]

niel: Composition and Reception (= Vetus Testamentum Supplementum 83/II), Leiden 2001, 329–367, hier 332–338 (im Folgenden P.W. Flint, The Daniel Tradition at Qumran, Leiden 2001); E. Eshel, Possible Sources of the Book of Daniel, in: ebd., 387–394, hier 387–388; A. Steinmann, The Chicken and the Egg: A New Proposal for the Relationship Between the *Prayer of Nabonidus* and the *Book of Daniel*, Revue de Qumrân 20 (2002), 557–570 (im Folgenden A. Steinmann, The Chicken and the Egg [2002]); L.T. Stuckenbruck, The Formation and Re-Formation of Daniel in the Dead Sea Scrolls, in: J.H. Charlesworth (Hg.), The Bible and the Dead Sea Scrolls, The Second Princeton Symposium on Judaism and Christian Origins, Waco 2006, 101–130, hier 104–106, sowie die Handausgaben J. Maier, Die Qumran-Essener. Die Texte vom Toten Meer II, Basel 1995, 185–186; F. García Martínez/E.J.C. Tigchelaar, The Dead Sea Scrolls Study Edition I, Leiden 1997, 486–489 (im Folgenden F. García Martínez/ E.J.C. Tigchelaar, Study Edition I, Leiden 1997); A. Steudel, Die Texte aus Qumran II, Darmstadt 2001, 159–165.

5 Anders noch R.G. Kratz, Translatio imperii, Neukirchen-Vluyn 1991, 99 Anm. 91. Doch schon die Transkription von R. Meyer (Das Gebet des Nabonid, Berlin 1962, 16), der das von J.T. Milik („Prière de Nabonide" [1956], 415) in einem Addendum erwähnte Fr. 2b (hier „Nachtragsfragment" genannt) berücksichtigte, sowie sein Kommentar zu Z. 6 (Meyer a.a.O., 26) führen auf die richtige Spur. Die Konsequenz haben – wohl unabhängig voneinander – K. Beyer (Die aramäischen Texte vom Toten Meer, Göttingen 1984, 223–224) und F.M. Cross (Fragments of the Prayer of Nabonidus, Israel Exploration Journal 34 [1984], 260–264, im Folgenden F.M. Cross, Fragments [1984]) gezogen. Letzterer nahm Z. 6 zum Ausgangspunkt (F.M. Cross, a.a.O., 260–261; in der Transkription auf S. 263 ist die Ergänzung in ב[אישא] nicht notiert), doch ist seine Rekonstruktion nicht unwidersprochen geblieben (F. García Martínez, The Prayer of Nabonidus, Leiden 1992, 118–119; unentschieden J.J. Collins, DJD XXII, Oxford 1996, 84). Neue Argumente für den Zusammenschluß bringt E. Puech, La Prière de Nabonide (4Q242), in: K.J. Cathcart/M. Maher (Hgg.), Targumic and Cognate Studies, Essays in Honour of Martin McNamara, Journal for the Study of the OT 230 (1996), 208–227 (im Folgenden E. Puech, La Prière de Nabonide [1996]) bei.

1[a–a] Puech (La Prière de Nabonide [1996]) מלך ד[י ב]בל, supralinear korrigiert zu מלכ[א ד]י ב[בל 3[a] B. Jongeling/C.J. Labuschagne/A.S. van der Woude, Aramaic Texts from Qumran (= Semitic studies series 4), Leiden 1976: וכן, doch ist כ paläographisch ausgeschlossen[6] 3[b–b] Andere: א]נה כדי הודית עויתי] (Milik, „Prière de Nabonide“ [1956]), א]נה לחיוא וצלית קדם עליא] (Cross, Fragments [1984]), א]גרי] (K. Beyer, Die aramäischen Texte vom Toten Meer, Göttingen 1984) 4[a] Lange/ Sieker (Gattung und Quellenwert, Göttingen 1996): גיר „Schutzbürger“, doch ist י paläographisch ausgeschlossen[7] 4[b] Beyer (a.a.O.): ב]אדין] „dann“, doch ist ב paläographisch ausgeschlossen[8] 4[c–c] Milik (a.a.O.): בבבל והוא] 5[a] Beyer (a.a.O.): חו[י]ני „er tat mir kund“, doch ist נ paläographisch ausgeschlossen[9] 5[b–b] Oder: א]להא שמיא 5[c–c] Milik (a.a.O.): וכן כתב כדי] 7[a] Lange/Sieker (a.a.O.): מוצלא (Part. Huf'al) scheint auf einer groben Fehllesung zu beruhen[10] 9a[a–a] Puech (a.a.O.): אוב[לו תורין ל]דבח שלם קד[מיהון] „Fu[rent] ame[nés *des taureaux en*] sacrifice de salut dev[ant *eux*“, Lange/Sieker (a.a.O.): און [ב◦◦ ◦ ]◦[ ]לם 9[a–a] Puech (a.a.O.): שמעין בע[ית רח]מיהון [מן קדמיהון „*entend(r)aient.]* J'*[ai imploré]* leur [miséri]corde[ *devant eux*]“

1 Worte des Ge[be]ts, das gebetet hat NBNY, der König von [Babe]l, der [*Groß*]könig, [*als er geschlagen war*] 2 mit einem bösartigen Geschwür auf Befehl G[ott]es in Teman: [*Ich, NBNY, mit einem bösartigen Geschwür*] 3 war ich geschlagen sieben Jahre lang. Und nach[dem] erhoben hatte G[*ott sein Angesicht über mich, heilte er mich*] 4 und meine Sünde, er hat sie vergeben. Ein Wahrsager – er war ein Judäer v[*on den Exilierten – kam zu mir und sprach:*] 5 Tu es kund und schreib es auf [a], um Ehre und Gr[öße] zu geben dem Namen G[*ottes, des Höchsten*[b]. [c]*Und so habe ich geschrieben/so schreibe ich hiermit: Ich,*[c]] 6 geschlagen war ich[a] mit einem b[ösartigen] Geschwür in Teman [*auf Befehl des höchsten Gottes*] 7 sieben Jahre lang, ich betete[a] [zu] den Göttern aus Silber und Gold [*Bronze, Eisen*] 8 Holz, Stein, Ton, denn/solange [ich dacht]e[a], dass Götter s[ie seien …] 9 […] ihr […] 9a supralinear […]

5[a] Oder: [und er] teilte mit und schrieb 5[b] Oder: des Himmels 5[c–c] Oder: Und so hat er geschrieben:] 6[a] Oder: warst du 7[a] Oder: du betetest 8[a] Oder: du dachtest

Die Lesung des erhaltenen Textes ist bis auf wenige Ausnahmen sicher. Unsicher ist an einigen Stellen hingegen die Textauffassung. Die Unsicherheiten betreffen zum einen Z. 3–4, zum anderen Z. 5–8.

Die Wiedergabe von Z. 3b mit „Und nach[dem] erhoben hatte G[*ott sein Angesicht über mich, heilte er mich*]“ ist rein hypothetisch und beruht auf der oben begründeten engen Anordnung der Fragmente, der Auffassung von מן די als temporaler Konjugation und der

6 E. Puech, La Prière de Nabonide (1996), 215 Anm. 23, hier auch zu anderen Alternativlesungen.

7 A. Steudel, Die Texte aus Qumran II, Darmstadt 2001, 265.

8 E. Puech, La Prière de Nabonide (1996), 218 Anm. 42.

9 Ebd., 219 Anm. 47.

10 A. Lange/M. Sieker, Gattung und Quellenwert des Gebets des Nabonid, Göttingen 1996, 11–12. Wurde das Wort von links nach rechts gelesen (*'alef – nun*), das צ als נ und der Abstrich des ל fälschlicherweise als „schräger Strich“ (= *waw* oder *yod*) identifiziert?

Ableitung der Verbform von der Wurzel שוי II „setzen, stellen, legen", interpretiert als Pa'el Aktiv mit Gott als Subjekt.[11] Doch das ist nicht die einzige Möglichkeit.[12] Andere setzen eine größere Textlücke vor dem Verbum voraus und fassen die Form als Partizip Passiv (qātil) auf: „Von [den Menschen] wurde i[ch] entfernt [bis ich meine Sünden bekannte (o.ä.)]".[13] Wieder andere denken – ob mit großer oder kleiner Textlücke – an die Wurzel שוי I „gleich sein": „Und seit[dem] glich i[ch einem Tier, da betete ich zum höchsten Gott (o.ä.)]".[14] Eigene Wege geht schließlich K. Beyer (Die aramäischen Texte vom Toten Meer, Göttingen 1984), der מן די – anders als in Z. 8 – nicht temporal oder kausal (מִן), sondern relativisch (מַן) auffasst: „Aber derjenige, [welcher] bestimmt hat [meinen Lohn] und meine Strafe". Sämtliche Vorschläge sind aufgrund der Lückenhaftigkeit des Textes mit großen Unsicherheiten behaftet. Die meisten Vorschläge ergänzen nach Dan 4–5 (bes. 4,30; 5,21) und tragen so Motive in den Text ein, die dort nicht stehen.

Hinzu kommt das syntaktische Problem des Übergangs von Z. 3 zu Z. 4, das ebenfalls mit Fragen der Semantik zusammenhängt. Die Wiedergabe von Z. 4a mit „und meine Sünde, er hat sie vergeben. Ein Wahrsager –" nimmt eine *casus pendens*-Konstruktion mit rückbezüglichem Pronomen (לה) an, versteht das Verbum שבק „lassen, ablassen" im Sinne der Sündenvergebung mit Gott als Subjekt und lässt mit גזר „Wahrsager", einer Nominalbildung der gleichlautenden Wurzel „schneiden, bestimmen", einen neuen Satz beginnen.[15] Zu sämtlichen Punkten sind jedoch auch andere Deutungen möglich, die in der Literatur vertreten und auf verschiedene Weise miteinander kombiniert werden.[16] So zieht Milik חטאי „meine Sünde(n)" zur vorausgehenden Zeile und erkennt darin das Objekt zu einem dort ergänzten Verbum des Schuldeingeständnisses; שבק interpretiert er im Sinne von „überlassen, gewähren" mit Gott als Subjekt und dem Wahrsager als Objekt, wobei er לה in לי konjiziert: „[Mais, quand j'eus *confessé mes péchés*] [4]et mes fautes, (Dieu)

11 Vgl. J.J. Collins, DJD XXII, Oxford 1996, 89–90, der seinerseits P. Grelot, La prière de Nabonide (4Q Or Nab), Revue de Qumran 9 (1978), 483–495 folgt; P.W. Flint, The Daniel Tradition at Qumran, Leiden 2001.

12 Einen Überblick über die Vorschläge gibt F. García Martínez, The Prayer of Nabonidus, Leiden 1992, 123–124.

13 J.T. Milik, „Prière de Nabonide" (1956); R. Meyer, Das Gebet des Nabonid, Berlin 1962; F. García Martínez, The Prayer of Nabonidus, Leiden 1992; F. García Martínez/E.J.C. Tigchelaar, Study Edition I, Leiden 1997.

14 F.M. Cross, Fragments (1984); E. Puech, La Prière de Nabonide (1996); vgl. schon E. Vogt, Precatio regis Nabonid in pia narratione Iudaica (4Q), Biblica 37 (1956), 532–534 (im Folgenden E. Vogt, Precatio [1956]); H.M.I. Gevaryahu, The Qumran Fragment of the Prayer of Nabonidus, in J. Liver (Hg.), Studies in the Dead Sea Scrolls: Fs Sukenik, Jerusalem 1957, 12–33 (im Folgenden Qumran Fragment, Jerusalem 1957); J.D. Amusin, Kumranskij fragment „molitvy" vavilonskogo carja Nabonida, Vestnik drevnej istorii 66 (1958), 104–117; A. Dupont-Sommer, Die essenischen Schriften, Tübingen 1960; B. Jongeling/C.J. Labuschagne/A.S. van der Woude, Aramaic Texts from Qumran, Leiden 1976; A.S. van der Woude, Bemerkungen zum Gebet des Nabonid, in: M. Delcor (Hg.), Qumrân: Sa piété, sa théologie et son milieu (= Bibliotheca Ephemeridum Theologicarum Lovaniensium 46), Louvain 1978, 121–129.

15 P. Grelot, La prière de Nabonide (1978); F.M. Cross, Fragments (1984); E. Puech, La Prière de Nabonide (1996); J.J. Collins, DJD XXII, Oxford 1996; P.W. Flint, The Daniel Tradition at Qumran, Leiden 2001; mit Ausnahme des *casus pendens* so auch schon R. Meyer, Das Gebet des Nabonid, Berlin 1962, gefolgt von W. Dommershausen, Nabonid im Buche Daniel, Mainz 1964, 69–70 und A. Mertens, Das Buch Daniel im Lichte der Texte vom Toten Meer (= Stuttgarter Biblische Monographien 4), Stuttgart 1971, 35–36.

16 Einen kleinen Überblick gibt wieder F. García Martínez, The Prayer of Nabonidus, Leiden 1992, 125–126.

*m'*accorda un devin".[17] Den entgegengesetzten Interpretationstypus hat Dupont-Sommer begründet, der חטאי als Objekt des folgenden Verbums שבק auffasst und in dem Wahrsager das Subjekt erkennt, mit לה als sogenanntem *dativus ethicus*: „[Aber ich betete zum höchsten Gott,] 4und meine Sünden erließ ein Beschwörer".[18]

Hält man sich strikt an den erhaltenen Text, ist in beiden Fällen eine sichere Entscheidung nicht möglich. In Z. 3 wird man sich damit begnügen müssen, dass mit ומן [די] שוי א] irgendein Umstand ausgedrückt wird, der die Voraussetzung zur Besserung in Z. 4 darstellt. Welcher Art dieser Umstand war, lässt sich nicht sagen, obwohl die Parallele mit Dan 5,21 – nicht zuletzt aufgrund der sonstigen Berührungen mit Dan 4–5 (s.u.) – schon sehr verlockend ist. In Z. 4 scheint mir mit שבק חטאי eine geprägte Wendung vorzuliegen.[19] Sie spricht dafür, dass man es mit einem Satzanfang zu tun hat; die betonte Voranstellung des Objekts zieht fast notwendig das rückbezügliche Pronomen לה nach sich, so dass sich die oben gebotene Übersetzung von selbst ergibt. Wer die Sünde vergibt, Gott oder der anschließend eingeführte judäische Wahrsager, muss hingegen offen bleiben; die Wortstellung deutet allerdings eher in Richtung der ersten Möglichkeit, die auch die übliche ist.

Weitere Schwierigkeiten des Textverständnisses bereiten die Verbformen in Z. 5–8. Die beiden Verben am Beginn von Z. 5 können als Imperativ oder als Perfekt der 3. Pers. Sg. aufgefasst werden, die Verben in Z. 6–8 als Perfekt der 1. oder als 2. Pers. Sg. Das bedeutet: Entweder folgt der König der Aufforderung des judäischen Wahrsagers und spricht bzw. schreibt wie in Z. 2–4 weiter im Ich-Stil; oder der Wahrsager selbst teilt etwas schriftlich mit und redet den König im Du-Stil an.[20]

Auf die Ergänzungen des Textes ist in dieser wie in den anderen Fragen nichts zu geben. Einigermaßen sicher sind nur die Ergänzungen zwischen Fr.1 und 2–3, also in der rechten Texthälfte, wo die Zwischenräume feststehen und sich meist auch ein Anhalt im erhaltenen Text findet. Ohne Anhalt sind die Ergänzungen der Zeilenenden, die in der Textwiedergabe nur als Umriss und in der Übersetzung kursiv dargestellt sind. Geboten werden die in DJD aufgenommenen und einige alternative Vorschläge, die sich leicht ver-

17 J.T. Milik, „Prière de Nabonide" (1956), 408; E. Vogt, Precatio (1956); ähnlich K. Beyer, Die aramäischen Texte vom Toten Meer, Göttingen 1984, 224: „Aber derjenige, [welcher] bestimmt hat [meinen Lohn] 4und meine Strafe, sparte sich einen Wahrsager auf".

18 A. Dupont-Sommer, Die essenischen Schriften, Tübingen 1960; ders., Exorcisms et guérisons dans les écrits de Qoumrân, in: J.A. Emerton (Hg.), Congress Volume Oxford 1959 (= Vetus Testamentum Supplementum 7), Leiden 1960, 246–261 (im Folgenden A. Dupont-Sommer, Exorcisms, Leiden 1960); vgl. auch J.A. Fitzmyer/D.J. Harrington, A Manual of Palestinian Aramaic Texts (= Biblica et Orientalia 34), Rom 1978 (לה „for Him" = Gott); F. García Martínez, The Prayer of Nabonidus, Leiden 1992; F. García Martínez/E.J.C. Tigchelaar, Study Edition I, Leiden 1997. Eine Art Kompromiss zwischen Dupont-Sommer und Milik hat A.S. van der Woude (Bemerkungen zum Gebet des Nabonid, Louvain 1978, 124–125) gefunden: „[Ich betete zum allerhöchsten Gott] 4und Er verzieh meine Sünden ... Er (sc. Gott) hatte einen Weissager ..."; vgl. schon B. Jongeling/C.J. Labuschagne/A.S. van der Woude, Aramaic Texts from Qumran, Leiden 1976, 128.

19 Vgl. in *11Q10 XXXVIII,2–3*; vielleicht auch *4Q547 6,3*.

20 Vgl. J.T. Milik, „Prière de Nabonide" (1956) und R. Meyer, Das Gebet des Nabonid, Berlin 1962 einerseits, A. Dupont-Sommer, Die essenischen Schriften, Tübingen 1960 und die meisten nach ihm andererseits.

mehren ließen und allesamt ins Belieben gestellt sind. Wie lang die Zeilen waren, wissen wir nicht. Eine gewisse Richtlinie gibt der Übergang von Z. 7 zu Z. 8 mit der Reihung der Materialien, aus denen die Götterbilder hergestellt werden.[21]

Die Frage des Sprechers in den Z. 5–8 muss folglich allein aufgrund des erhaltenen Textbestands entschieden werden. Da der Text als *Gebet des Nabonid* überschrieben ist, tendiere ich heute dazu, auch in Z. 6–8 die 1. Pers. Sg. zu lesen.[22] Ob der Seher dem König geschrieben hat oder ihn dazu auffordert, in einem Schriftstück dem Namen Gottes die Ehre zu geben, ist demgegenüber zweitrangig. In jedem Fall dürfte das Bekenntnis des Königs Nabonid, das er selbst in den Z. 6–7 – ob mündlich oder schriftlich – ablegt, identisch sein mit den „Worten des Gebets, die NBNY gebetet hat". Auch dass Gott nirgends direkt angesprochen, sondern immer nur über ihn die Rede ist, spricht nicht dagegen, das Ganze als *Gebet des Nabonid* aufzufassen. Hierfür gibt es einschlägige Parallelen.[23]

Somit ergibt sich eine klare Zweiteilung von Überschrift in Z. 1 und Ich-Bericht (Gebet) in Z. 3–9, mit Übergang in Z. 1–2 oder, sicher greifbar, Z. 2–3. Innerhalb des Ich-Berichts handeln die Z. (2)3–4a von der Krankheit, dem Aufenthalt in Tema und der Vergebung der Sünde des Königs, die Z. 4b–5a von dem Auftreten eines – anonymen[24] – judäischen Wahrsagers, der den König zum Lobpreis Gottes bewegt, und die Z. 6–9 von der Krankheit in Verbindung mit dem Götzendienst des Königs, vermutlich gefolgt von einem Bekenntnis zum wahren Gott.

*4Q242 Fr. 4*

[ ]ל̊ב̇ר[a] המון אחלמת 1
[ ][a]מ̊נה אח̊[ל]ך̊ ש̊ל̊ם̇ של[ותי יתוב עלי][a] 2
[ ]◦[ ]נו[a] רחמי לא יכלת ] 3
[ ]כ̊מה[a] דמא אנתה[b] ל ◦][b] 4
[ ]◦ ◦◦[ ] 5

1[a] Puech (La Prière de Nabonide [1996]): מעבד[, Beyer (Die aramäischen Texte vom Toten Meer, Göttingen 1984): גבר[ 2[a–a] Beyer (a.a.O.): אנתה ארז ו[ל]עלם של[ם „Du bist (wieder stark wie) eine Zeder und [für] immer gesund", Puech (a.a.O.):
[גזר חד יהודי מן בני גלותא ובכ]מתנה אה[ל]ך שלם של[ותי „[*un devin, un juif d'entre les membres de la déportation, et en/comme*] cadeau il apporta la paix [de ma] tranqui[llité]" 3[a] Puech (a.a.O.):

21 Die Ergänzung folgt der Reihe in *Dan 5,23* (etwas anders auch 5,4 „Gold und Silber …"!), wo allerdings der „Ton" fehlt, der in den üblichen Reihungen der Götzenpolemik (*Ps 115,4*; *Dtn 4,28* u.ö.; *Dtn 29,16* und *11QT[a] LIX, 3–4*) auch sonst nicht vorkommt. Eine Parallele zum „Ton" in diesem Zusammenhang ist *Sapientia Salomonis 15,7–10*; ansonsten vgl. *Dan 2,31–33*.

22 Anders noch R.G. Kratz, Translatio imperii, Neukirchen-Vluyn 1991, 100 Anm. 93.

23 *Ps 151* (*LXX* und *11QPs[a] XXVIII 3–12*); *Sir 51,12a–o*. Zur Gattung des Texts als „Gebet" vgl. auch A.S. van der Woude, Bemerkungen zum Gebet des Nabonid, Louvain 1978.

24 Der Name hätte bei dieser Gelegenheit genannt werden müssen. Es ist daher weder „wahrscheinlich" (A. Dupont-Sommer, Die essenischen Schriften, Tübingen 1960, 348–349 Anm. 4) noch „sicher" (A. Steudel, Die Texte aus Qumran II, Darmstadt 2001, 265), dass es sich um Daniel handelt.

יערצו/ו[ 4[a] Puech (a.a.O.): כמא[ 4[b–b] Milik („Prière de Nabonide“ [1956], 410) vermutet hier den Namen Daniel: „Comme tu es semblable à *D*[aniel]!“; dagegen Puech (a.a.O.): ](?)לה

1 [...] fern von ihnen (oder: ihnen zu dienen), ich träumte/gesundete/suchte Heilung[25]
2 [...] von ihm ging ich vorüber/ließ er vorübergehen (oder: ein Geschenk, er brachte), das Wohlbefinden [meiner] Ru[he ...
3 [...] (und plötzlich kamen) meine Freunde (oder: mein Erbarmen), ich konnte nicht [
4 [...] wie ähnlich bist du dem [...
5 [...]

Fr. 4 ist äußerst schlecht erhalten und daher schwer zu verstehen. Zwar handelt es sich um dieselbe Hand wie in Fr. 1–3, aber nicht um dasselbe Leder. Das bedeutet, dass das Fragment entweder von einer anderen Kolumne oder sogar aus einem anderen Werk stammt. Es ist daher nur mit großem Vorbehalt, wenn überhaupt, für die Interpretation heranzuziehen. Dies sei ausdrücklich betont, da sich viele Deutungen auf dieses Fragment beziehen und mit ihm auch strittige Fragen in den Fr. 1–3 entscheiden.

## 3. Babylon in Qumran

Das Vorkommen des Namens Nabonid in den Texten von Qumran ist eine kleine Sensation. Nicht nur, dass er biblisch nicht belegt ist. Auch Babylon selbst spielt in den Texten von Qumran keineswegs eine derart dominierende Rolle, dass man die Bekanntschaft mit dem letzten neubabylonischen König Nabonid erwarten könnte. Die beiden einzigen babylonischen Könige, die namentlich erwähnt sind, heißen erwartungsgemäß Nebukadnezar und Belsazar. Erstgenannter begegnet in den erhaltenen Texten zwei Mal: einmal in der sogenannten Damaskusschrift (CD I,6 par. *4Q266* und *4Q268*), einer wichtigen Programmschrift der Gemeinschaft von Qumran, das andere Mal in einer aramäischen Apokalypse, die an das Danielbuch erinnert und darum in der Forschung Pseudo-Daniel genannt wird (*4Q243 13/4Q244 12*). Auch Belsazar ist in einem der Fragmente von Pseudo-Daniel (*4Q243 2*) erwähnt und gleich unterhalb des Namens Daniel gut zu erkennen.[26]

Die Damaskusschrift und Pseudo-Daniel sind nichtbiblische Werke, doch sind die Erwähnungen der babylonischen Könige in beiden Werken von der biblischen Literatur abhängig. Beide beziehen sich in einer biblischen Formulierung auf die Auslieferung Israels

25 Je nachdem, ob man אחלמת ableitet von חלם I „stark, gesund sein“ oder חלם II „träumen“. Vgl. J.T. Milik, „Prière de Nabonide“ (1956) einerseits, A. Dupont-Sommer, Die essenischen Schriften, Tübingen 1960, 349 Anm. 4 andererseits.

26 Angaben nach M. Broshi, The Damascus Document Reconsidered, Jerusalem 1992; J.M. Baumgarten, The Damascus Document (4Q266–273), Qumran Cave 4. XIII, DJD XII, Oxford 1996 für CD; J.J. Collins, DJD XXII, Oxford 1996 für Pseudo-Daniel.

in die Hand Nebukadnezars, des Königs von Babylon, also auf das babylonische Exil und Babylon als erstes von mehreren aufeinanderfolgenden Reichen, die an die vier Reiche von *Dan 2* und *7* erinnern.[27] Bei der Erwähnung Belsazars ist offenbar die Situation von *Dan 5* vorausgesetzt, eine Szene bei Hofe, wie sie auch in *4Q243 1* und *4Q244 1–3,4* durchscheint.

Die Belege sind daher nicht so sehr als Beispiele für die Rezeption Babylons von Interesse, sondern als Beispiele der biblischen Rezeptionsgeschichte. So fügt sich der Beleg in CD bestens in die Tendenz dieser Schrift, die Regeln der Gemeinschaft von Qumran, wie sie in den älteren Fassungen der „Gemeinschaftsregel" *Serekh ha-Yachad* (*1QS* par. *4QS*) überliefert sind, „biblisch" zu untermauern. Die Regeln selbst werden durch Rückgriff auf die Tora sowie andere biblische Schriften begründet und ergänzt, die Gründung der Gemeinschaft wird in die biblische Heilsgeschichte eingeordnet. Letzterem dient insbesondere der Rekurs auf das babylonische Exil als Ausgangspunkt der Geschichtsschau, die dem bekannten deuteronomistischen Schema folgt.[28]

In Pseudo-Daniel wiederum scheint es sich um eine Auslegung und Weiterführung des Danielbuches zu handeln.[29] Die aus *Dan* (1–)5 genommene Hofszene bildet den erzählerischen Rahmen für eine weit ausgreifende Geschichtsschau, die von der Flut über den Turmbau und den Exodus und das babylonische Exil bis zur hellenistischen Zeit reicht und in ihren Grundzügen ebenfalls dem biblischen, genauer deuteronomistischen Geschichtsbild verpflichtet ist.[30] Letzteres verbindet Pseudo-Daniel mit der Geschichtsauffassung der makkabäischen Teile des Danielbuches (*Dan 8–12*), besonders dem Gebet in *Dan 9*.[31]

Ansonsten ist in den Texten vom Toten Meer nicht mehr sehr häufig von Babel die Rede. Eine Gruppe von Texten, in denen Babel Erwähnung findet, sind die sogenannten Pesharim, das sind Kommentarwerke zu den biblischen Propheten. Im Pesher zum Buch Micha (*4Q168 1 und 3, Z. 3*) ist *Mi 4,10* zitiert, wo Babel erwähnt wird, doch ist die Auslegung dazu nicht erhalten. Im Pesher zum Buch Jesaja verhält es sich, wenn die Anordnung der Fragmente

27 In *4Q243 16* sind vielleicht die Jeremianischen „70 Jahre" erwähnt, die auch in Dan 1–6 den chronologischen Rahmen der vier Reiche von *Dan 2* bilden und in *Dan 9* eine Neuinterpretation erfahren. Doch die Lesung ist nicht sicher. Noch unsicherer ist die vermutete Zählung der Reiche in demselben Fragment Z. 4; vgl. J.J. Collins, DJD XXII, Oxford 1996, 108–109.

28 Vgl. O.H. Steck, Israel und das gewaltsame Geschick des Propheten: Untersuchungen zur Überlieferung des deuteronomistischen Geschichtsbilds im Alten Testament (= Wissenschaftliche Untersuchungen zum Alten und Neuen Tewsatment 23), Neukirchen-Vluyn 1967, 165–170 (im Folgenden O.H. Steck, Israel, Neukirchen-Vluyn 1967).

29 Anders J.J. Collins, DJD XXII, Oxford 1996, 136, der zwar richtig bemerkt, dass „the mention of the relative obscure Belshazzar is a significant parallel to the biblical book", aber auch eine „common tradition" für denkbar hält und im übrigen den Zusammenhang von Pseudo-Daniel (und anderen apokalyptischen Geschichtsrückblicken) mit Dan 8–12 über das deuteronomistische Geschichtsbild nicht erkennt.

30 Vgl. O.H. Steck, Israel, Neukirchen-Vluyn 1967, 155–156 Anm. 10; 191 Anm. 2; R.G. Kratz, Translatio imperii, Neulirchen-Vluyn 1991, 75–76.220.

31 Vgl. R.G. Kratz, ebd., 37–42.265–267; ders., Die Visionen des Daniel, in: ders. u. a. (Hgg.), Schriftauslegung in der Schrift (Fs. O.H. Steck) (= Beihefte zur Zeitschrift für die alttestamentliche Wissenschaft 300), Berlin 2000, 219–236; engl. Fassung: The Visions of Daniel, in: J.J. Collins/P.W. Flint (Hgg.), The Book of Daniel: Composition and Reception (= Vetus Testamentum Supplementum 83/I), Leiden 2001, 91–113.

stimmt, so, dass Assur als feindliche Fremdmacht, die Jerusalem bedrängt, im Kommentar auf Babel und die Zerstörung Jerusalems unter Nebukadnezar bezogen wird.[32] Das ist ein Verständnis, das sich schon im Jesajabuch selbst findet (*Jes 13–14*) und sich in den Pesharim fortzusetzen scheint. Immerhin lässt sich dieser Deutung in den Pesharim entnehmen, dass Babylon im Denken der Gemeinschaft vom Qumran durchaus präsent war. Hier steht der Name Babel (und Nebukadnezar) gewissermaßen als Chiffre für Fremdherrschaft und Exil schlechthin und repräsentiert das göttliche Gericht und die (anhaltende) Zeit des Zorns.

Dieselbe Auffassung findet sich schließlich noch in einer anderen Gruppe von Texten, die ebenfalls von der biblischen Tradition abhängig sind, aber – anders als die Damaskusschrift und die Pesharim – vermutlich nicht in der Gemeinschaft von Qumran selbst entstanden sind, sondern aus einem anderen Milieu importiert und in Qumran nur tradiert wurden. Es handelt sich um sogenannte (hebräische und aramäische) Apokryphen: das aramäische Genesis-Apokryphon (*1QapGen ar*), ein Jeremia-Apokryphon (*4Q385 a 18 i a–b,4.6* und in der Erg. 10), ein Pseudo-Ezechiel (*4Q385 c A,2; 4Q386 1 iii,1*), ein weiterer von Daniel inspirierter Text über die vier Reiche (*4Q552 2 ii,5; 4Q553 6 ii,4*) sowie ein Text, der das „Neue Jerusalem" genannt wird (*4Q554 3 iii,19*). Überall ist Babel die große Feindmacht, das erste von vier Reichen, das Land des Exils, das von fremden Göttern und bösen Geistern (Dämonen) bewohnt wird und in dem es gilt, am jüdischen Gesetz festzuhalten, bis am Ende der Zeit auch dieser Feind vernichtet wird.

Von sämtlichen Texten stehen die letztgenannten dem *Gebet des Nabonid* am nächsten, da es derselben Gruppe der nichtbiblischen, apokryphen Texte angehört, die nicht zu den Qumran-Texten im engeren Sinne zählen. Und doch unterscheidet sich die Sicht des Königs von Babel im *Gebet des Nabonid* erheblich von dem Bild Babels als Feindmacht in diesen Apokryphen und allen anderen Belegen (CD, Pseudo-Daniel, Pesharim). Hier ist der Name Babel überall negativ besetzt, während das *Gebet des Nabonid* ein positives Bild des babylonischen Königs zeichnet. Hinzu kommt, dass das Verhältnis zur biblischen Überlieferung ein anderes zu sein scheint. Während die meisten anderen Belege eindeutig von der biblischen Tradition abhängig sind, ist das Verhältnis im Falle des *Gebets des Nabonid* nicht ganz so eindeutig.

## 4. Nabonid im Danielbuch

Wie schon gesagt, weist das *Gebet des Nabonid* verblüffende Ähnlichkeiten zu den Erzählungen des Danielbuches und hier besonders zu Dan 4–5 auf: Wie Nebukadnezar in *Dan 4–5* auf „Beschluss" (פתגם) der „Wächter" (Engel) des Höchsten (*4,14*) „sieben Zeiten" fern

32 *4Q163 6–7 ii,4* (in der Auslegung von *Jes 10,12–19*); *8–10,1* (in der Auslegung von *Jes 14*); *25,1* (in der Auslegung von *Jes 30–31*); *4Q165 8,1* (Referenztext unklar, vielleicht *Jes 14 oder 39*). Alle Angaben nach M.P. Horgan, Pesharim: Qumran Interpretations of Biblical Books (= Catholic Biblical Quarterly Monograph Series 8), Washington 1979.

von den Menschen zusammen mit den wilden Tieren in der Wildnis verbringen musste (*Dan 4,22.29.30; 5,21*), so hält sich in *4Q242* Nabonid „auf Beschluss" Gottes für „sieben Jahre" in der Wüstenoase Tema auf.[33] Wie Daniel in *Dan 4–5 (4,5–6; 5,11–16, bes. 5,13)* bringt auch im *Gebet des Nabonid* ein „Judäer von den Exulanten" dem König von Babylon die rettende Lösung seines Problems;[34] in *Dan 4–5* steht der Judäer zu den babylonischen „Wahrsagern" (גזרין) in Konkurrenz (*4,4–5; 5,7*) oder ist über sie gesetzt (*Dan 5,11*), in *4Q242* wird er selbst als „Wahrsager" (גזר) bezeichnet. Und schließlich: Wie in Dan 4–5 geht es auch im *Gebet des Nabonid* um die Frage der Anbetung des wahren Gottes.[35] Die Bekehrung zu diesem Gott schließt die Befreiung von Sünden (*Dan 4,24*) und die Überwindung des Götzendienstes, der Verehrung der Götter aus Silber und Gold, Erz und Eisen, Holz und Stein und Ton (*Dan 5,4.23*), mit ein.[36] Auffallend ist auch, dass sowohl Dan 4 (genauer *3,31–4,34*) als auch das *Gebet des Nabonid* auf weite Strecken als Bekenntnis des Königs in Ich-Form formuliert ist, der König also selbst von seinem Schicksal und seiner Bekehrung zum jüdischen Gott berichtet. Wie es scheint, besteht ein irgendwie gearteter traditionsgeschichtlicher Zusammenhang.[37]

Nun ist schon lange vor der Entdeckung der Qumran-Texte vermutet worden, dass in den beiden Erzählungen von Nebukadnezar und seinem Sohn Belsazar in *Dan 4–5* Stoffe verarbeitet sind, die ursprünglich von Nabonid handeln, der ja in Wahrheit der Vater Belsazars war. Die Vermutung wurde zuerst 1924 von Sydney Smith aufgrund des von ihm herausgegebenen „Schmähgedichts" (Verse Account) auf Nabonid geäußert und durch Wolfram von Soden bestätigt.[38] Die Bestätigung basierte auf den inzwischen aufgefunde-

33 Auf die Wortparallele von *4Q242 1–3,3* mit *Dan 5,21* שוי) und darauf, dass mit ihr leider wenig anzufangen ist, wurde bereits oben unter 1 z.St. hingewiesen.

34 Die Terminologie ist nicht identisch: Der Wahrsager in 4Q242 wird – wie Daniels Freunde in *Dan 3,8.12* – „Judäer" genannt und zugleich als einer von den Exulanten bezeichnet, während Daniel in *Dan 2,25; 5,13; 6,14* immer nur „einer von den Exulanten Judas" heißt.

35 „Ehre und Größe" (יקר ורבו), die dem Namen Gottes erwiesen werden sollen (*4Q242 1–3,5*), haben in *Dan 5,18* eine (um zwei Glieder erweiterte) sprachliche und – die Machtverhältnisse umkehrende – sachliche Parallele (vgl. auch *4,33; 5,19–20*).

36 Wie in *4Q242* begegnet auch in *Dan 4,24* das Wort חטי (חטא) für die „Sünden" des Königs, die hier durch Almosen „gelöst" (פרק) werden sollen. Der Diskussion scheint bisher entgangen zu sein, dass der Ratschlag Daniels in Dan 4,24 eine Auslegung des Traumbildes von dem stehengelassenen Baumstumpf ist (mit שבק „stehen lassen" in *Dan 4,12.20.23*); eine Zwischenstellung nimmt die Septuaginta (*Dan 4,27*) ein, in der Daniel dem König empfiehlt, sich wegen seiner Sünden bittend an Gott zu wenden und sein Unrecht durch Almosen „zu lösen". Über die Reihung der Materialien, aus denen die Götzenbilder gemacht sind, s.o. Anm. 21; ein Argument gegen eine Beziehung zu Dan 4–5 (A. Lange/M. Sieker, Gattung und Quellenwert des Gebets des Nabonid, Göttingen 1996, 5–6.24–25.26–27) ergibt sich daraus gerade nicht.

37 Er wird vehement und auf recht gewaltsame Weise von Lange/Sieker (s.o. Anm. 36) bestritten, doch überwiegen – bei allen Unterschieden (auch und gerade gegenüber der historischen Realität!), die eine Rezeptionsgeschichte üblicherweise nun einmal mit sich bringt – die signifikanten Gemeinsamkeiten, die sich im erhaltenen Text von *4Q242* finden.

38 W. von Soden, Eine babylonische Volksüberlieferung von Nabonid in den Danielerzählungen, Zeitschrift für die alttestamentliche Wissenschaft 53 (1935), 81–89, wiederabgedruckt in: ders., Bibel und Alter Orient, Berlin 1985, 1–9; ders., Kyrus und Nabonid: Propaganda und Gegenpropaganda, in: H. Koch/D.N. MacKenzie (Hgg.), Kunst, Kultur und Geschichte der Achämenidenzeit und ihr Fortleben (= Archäologische Mitteilungen aus Iran Ergänzungsband 10), Berlin 1983, 61–68. Vgl. zum Folgenden ausführlich R. Meyer, Das Gebet des Nabonid, Berlin 1962 sowie – in

nen Harran-Inschriften (*H1* und *H2*) und der gesamten Propaganda und Gegenpropaganda der späten Nabonid- und frühen Kyros-Zeit, die jüngst von Hanspeter Schaudig neu ediert wurde.[39] Auf dem Hintergrund dieses Materials ist evident, dass man es sowohl in *Dan 4–5* als auch im *Gebet des Nabonid* mit einer judaisierten Version der babylonischen Nabonidlegende zu tun hat.[40] Dass es sich nicht um direkte (literarische) Abhängigkeit, sondern um einen traditionsgeschichtlichen Zusammenhang handelt, bei dem noch mancherlei Zwischenstufen anzunehmen sind, versteht sich eigentlich von selbst, sei aber, um Missverständnissen vorzubeugen, eigens gesagt.

Aus den babylonischen Quellen erklärt sich eine Reihe von Eigenheiten der beiden jüdischen Fassungen der Nabonidlegende: der Ich-Stil, der Aufenthalt Nabonids in Tema bzw. Nebukadnezars in der Wildnis (in den babylonischen Quellen für zehn Jahre, im *Gebet des Nabonid* und Dan 4 für sieben Jahre), das Motiv der Krankheit als Grund für diesen Aufenthalt (in H 2: Krankheiten in Babylon, im *Gebet des Nabonid* die Krankheit des Königs selbst) und sachlich die Frage der Anbetung des höchsten Gottes, der Fremdgötter-Vorwurf sowie die Rolle von Träumen und Wahrsagern – um nur die wichtigsten und offensichtlichsten Parallelen zu nennen. Im Zuge der jüdischen Rezeption des Stoffs ist aus der innerbabylonischen Auseinandersetzung um Nabonids eigenwillige Religionspolitik, die den Mondgott Sin favorisierte, eine Belehrungsgeschichte geworden, in der Nabonid bzw. Nebukadnezar positiv gezeichnet, der Sohn Nabonids, Belsazar, hingegen als Negativbeispiel hingestellt wird. Die Rolle des höchsten Gottes (Sin von Harran oder Marduk von Babylon) nimmt hier der jüdische Gott ein. Ihm allein wird die Fähigkeit zugesprochen, den König in sein Herrscheramt ein- oder abzusetzen – wie nach Auffassung der babylonischen Priester der Gott Marduk oder nach der Auffassung Nabonids und seiner Mutter auch der Gott Sin – und von (Haut-)Krankheiten zu heilen – wie in der akkadischen Überlieferung der Gott Sin.[41] Die Rolle der Weisen und Traumdeuter, auf die sich Nabonid

manchem davon abweichend – R.G. Kratz, Translatio imperii, Neukirchen-Vluyn 1991, 99–111; ders., From Nabonidus to Cyrus, in: A. Pananino/G. Pettinato (Hgg.), Ideologies as Intercultural Phenomena, Melammu Symposia III, Milano 2002, 143–156 (im Folgenden R.G. Kratz, From Nabonidus to Cyrus, Milano 2002) und die hier genannte ältere Literatur. Mit dem Vergleich haben sich unterdessen auch andere eingehender befasst, z.B. F. Garciá Martínez, The Prayer of Nabonidus, Leiden 1992, 129–133; J.J. Collins, Daniel: A Commentary on the Book of Daniel, Hermeneia, Minneapolis 1993, 216–219; ders., DJD XXII, Oxford 1996, 85–87; ders., New Light on the Book of Daniel from the Dead Sea Scrolls, in: F. García Martínez/E. Noort (Hgg.), Perspectives in the Study of the Old Testament and Early Judaism (= Vetus Testamentum Supplementum 73), Leiden 1998, 180–196, hier 181–187; K. Koch, Gottes Herrschaft über das Reich des Menschen, Louvain 1993 sowie ders., Daniel 1–4, Neukirchen-Vluyn 2005, 408–415.

39 H. Schaudig, Die Inschriften Nabonids von Babylon und Kyros' des Großen samt den in ihrem Umfeld entstandenen Tendenzschriften: Textausgabe und Grammatik (= Alter Orient und Altes Testament 256), Münster 2001; vgl. dazu P.-A. Beaulieu, The Reign of Nabonidus King of Babylon 556–539 B.C. (= Yale Near Eastern Researches 10), New Haven/London 1989; ders., An Episode in the Fall of Babylonia to the Persians, Journal of Near Eastern Studies 52 (1993), 241–261.

40 Ein analoger Fall ist die Judaisierung des assyrisch-aramäischen Achiqar im Tobitbuch (*Tob 1,21–22; 2,10; 11,18; 14,10*).

41 S. Ackermann, The Prayer of Nabonidus, Atlanta 1997, 58–60 mit Verweis auf die akkadischen Äquivalente zu aramäisch *šḥn*, nämlich *epqum, išrubum, šaḫaršubbum/-šuppu*. Eine andere Ableitung haben J.V. Kinnier Wil-

selbst in seinen Inschriften beruft, um sein religionspolitisches Programm zu rechtfertigen, übernimmt der anonyme judäische Wahrsager bzw. Daniel, der dem König die Erkenntnis des höchsten Gottes vermittelt.

Wie es scheint, lässt sich anhand der beiden jüdischen Versionen sogar der Weg des babylonischen Stoffs in die jüdische Tradition verfolgen. Zwar ist das Verhältnis zwischen *Dan 4–5* und dem *Gebet des Nabonid* von Qumran in der Forschung umstritten. Die einen meinen, dass der Qumran-Text, wie wir ihn vor uns haben, zwar nicht unbedingt älter sein müsse als das Danielbuch, aber eine ältere Fassung des Stoffs und damit das Bindeglied zwischen den babylonischen Quellen und dem Danielbuch dokumentiere.[42] Andere meinen, dass der Qumran-Text von der kanonischen Fassung des Danielbuchs abhängig sei.[43] Wieder andere behaupten, dass beide Fassungen gar nichts miteinander zu tun hätten.[44] Doch spricht vieles für die erste Möglichkeit.

Sowohl die namentliche Erwähnung Nabonids und seines Aufenthaltsortes Tema als auch die Anonymität des judäischen Wahrsagers im *Gebet des Nabonid* sind Motive, die sich nicht aus *Dan 4–5* ableiten lassen und somit m. E. eindeutige Indizien für die Priorität der qumranischen Fassung des Stoffs sind. Wäre das *Gebet des Nabonid* von *Dan 4–5* abhängig wie etwa die – in der Septuaginta und der Vulgata überlieferten – Gebete der drei Freunde von *Dan 3*, die Gebete des griechischen vom hebräischen *Esterbuch* oder das apokryphe *Gebet des Manasse* von dem Manassekapitel in *Könige* und *Chronik*, würde man ein Gebet des Nebukadnezar oder des Belsazar oder des Dareios erwarten. Dass es sich aber auch

son/I.L. Finkel 2007 vorgeschlagen (On *būšānu* and *di'u* or why Nabonidus went to Tema, Le Journal des Médecines Cuneiformes 9 [2007], 16–22). In בשחנא von *4Q242 1–3,2.6* (transkribiert als *bushāna*!) finden sie das akkadische Wort *būšānu* und deuten es als „the evil-smelling disease" und „in modern terms" als „scurvy and diptheria" (a.a.O. 20). In dieser Krankheit sehen Kinnier Wilson und Finkel den wahren Grund für den Tema-Aufenthalt und kombinieren das Zeugnis von *4Q242* mit der Angabe in *H 2 I, 21*, wonach der König die Stadt Babylon wegen „Kopfkrankeit und Hunger" (*di'u u sugû*) verlassen habe. Die Hypothese ist jedoch aus zweierlei Gründen problematisch: Zum einen berücksichtigt sie nicht den Kontext in *H 2*, der eindeutig apologetischen Charakter hat. Zum anderen ist die sprachliche Ableitung schwerlich richtig: *būšānu* (vom Verbum *ba'āšu*) ist das akkadische Äquivalent zu dem Adjektiv באישא „übel riechend", während בשחנא die mit Präposition ב versehene, determinierte Form des Nomens שחן (hebräisch שחין) „Geschwür, Aussatz" ist, das von „heiß" abzuleiten und daher eher mit dem von S. Ackermann (The Prayer of Nabonidus, Atlanta 1997) angeführten *šaḫaršubbum/-šuppu* zu vergleichen ist (so auch Dr. Nils P. Heeßel, Heidelberg, brieflich). In *Dan 4* ist aus der Krankheit von *H 2*, die von den Babyloniern in *4Q242* auf Nabonid selbst übergegangen ist und seine soziale Ausgrenzung zur Folge hatte, die Verwandlung des Königs in ein wildes Tier geworden.

42 So seit J.T. Milik, „Prière de Nabonide" (1956) und dem ausführlichen Nachweis von R. Meyer, Das Gebet des Nabonid, Berlin 1962 die meisten; vgl. R.G. Kratz, Translatio imperii, Neukirchen-Vluyn 1991, 101; F. García Martínez, The Prayer of Nabonidus, Leiden 1992, 129–132; J.J. Collins, DJD XXII, Oxford 1996, 86. Die Handschrift *4Q242* wird in die Zeit zwischen 75–50 v. Chr. (F.M. Cross, Fragments [1984]) oder etwas später in hasmonäisch-herodianische Zeit, d.h. nach 50 v. Chr. (J.T. Milik, „Prière de Nabonide" [1956], E. Puech, La Prière de Nabonide [1996]) datiert. Doch damit ist für das Werk selbst lediglich ein *terminus ad quem* gewonnen; der Stoff kann erheblich älter sein.

43 H.M.L. Gevaryahu, Qumran Fragment, Jerusalem 1957; A. Dupont-Sommer, Die essenischen Schriften, Tübingen 1960, 350–351; W. Dommershausen, Nabonid im Buche Daniel, Mainz 1964, 84–85; A. Steinmann, The Chicken and the Egg (2002).

44 A. Lange/M. Sieker, Gattung und Quellenwert des Gebets des Nabonid, Göttingen 1996 (s.o. Anm. 37).

nicht um zwei unabhängige Versionen handelt, geht daraus hervor, dass das Motiv des siebenjährigen Aufenthaltes in Tema bzw. der Wildnis sowie der Vorwurf der Fremdgötterverehrung, die beide aus den babylonischen Quellen abgeleitet sind, auch in der qumranischen Fassung (noch) in der Geschichte des Nabonid miteinander verbunden sind, während sie in *Dan 4–5* auf die beiden Erzählungen über Nebukadnezar und den Sohn Nabonids, Belsazar, aufgeteilt wurden. An die Stelle des Vorwurfs der Fremdgötterverehrung als der „Sünde" Nabonids in der Fassung von *4Q242* ist in *Dan 4* das Motiv der Hybris des Bauherrn Nebukadnezar getreten.[45] Die „Sünde" wird auch in Dan 4 noch erwähnt, ist hier aber – über den Weg des Bittgebetes in *Dan 4,27 LXX* (?) – zum Nebenmotiv geworden (*Dan 4,24*), während der Vorwurf der Fremdgötterverehrung auf Belsazar in *Dan 5* übergegangen ist.

Somit lässt sich eine nachgerade lückenlose Rezeptionsgeschichte rekonstruieren, in die im Übrigen auch die griechische Überlieferung (griechische Historiker, Septuaginta) hineingehört.[46] Ihren Ausgang hat diese Rezeptionsgeschichte bei den unterschiedlichen Sichtweisen Nabonids in den babylonischen Quellen (Nabonid-Inschriften, Harran-Stelen, Kyros-Zylinder, „Schmähgedicht") genommen. Auch diese Quellen sind bereits Teil der Rezeptionsgeschichte und nicht etwa historische Zeugen, an denen die spätere Überlieferung in *4Q242* oder *Dan 4–5* zu messen wäre. In der Fassung des Qumran-Textes hat der Nabonid-Stoff sodann eine judäische Vereinnahmung erfahren, die noch von Nabonid und einem anonymen judäischen Wahrsager handelt. Über diese judaisierte Fassung ist der Stoff schließlich in das Danielbuch gelangt, wo aus Nabonid – aus naheliegenden Gründen – Nebukadnezar und „sein Sohn" Belsazar und aus dem anonymen judäischen Seher der Held des Buches Daniel wurden. Das *Gebet des Nabonid* von Qumran kann somit als überlieferungsgeschichtliche Vorlage für *Dan 4–5* gelten. Doch wie ist diese Vorlage nach Qumran gelangt und welche Bedeutung kommt ihr hier zu?

45 Hierfür hat vielleicht die „Volksüberlieferung der Chaldäer" Pate gestanden, die durch Euseb (Praeparatio Evangelica IX, 41) in seinem Auszug des Historikers Abydenos (2. Jh. n. Chr.), der sich seinerseits auf Megasthenes (um 300 v. Chr.) beruft, auf uns gekommen ist. Vgl. dazu R.G. Kratz, Translatio imperii, Neukirchen-Vluyn 1991, 101–104; K. Koch, Daniel 1–4, Neukirchen-Vluyn 2005, 406–408.415.

46 Vgl. R.G. Kratz, From Nabonidus to Cyrus, Milano 2002; zur alten Septuagintafassung von Dan 4–5 vgl. R. Albertz, Der Gott des Daniel. Untersuchungen zu Dan 4–6 in der Septuagintafassung sowie zu Komposition und Theologie des aramäischen Danielbuches (= Stuttgarter Bibelstudien 131), Stuttgart 1988; K. Koch, Daniel 1–4, Neukirchen-Vluyn 2005, 377–380 und passim. Die Septuaginta steht in manchem der qumranischen Fassung, in anderem der masoretischen Fassung näher und scheint ein weiteres Zwischenglied zu repräsentieren. Während der Drucklegung erschien der Beitrag von C.A. Newsom, Why Nabonidus? Excavating Traditions from Qumran, the Hebrew Bible, and Neo-Babylonian Sources, in: S. Metso u.a. (Hgg.), The Dead Sea Scrolls: Transmisson of Traditions and Production of Texts (= Studies on the Texts of the Desert of Judah 92), Leiden 2010, 57–79, der sich weniger mit der Überlieferung selbst als mit Fragen der „oral composition and performance" (ebd. 73) beschäftigt und daher „necessarily speculative" (ebd. 76) bleibt.

## 5. Bekehrungsgeschichten in Qumran

Aufgrund der Überlieferungsgeschichte könnte man meinen, das *Gebet des Nabonid* sei als Teil der Danielüberlieferung in Qumran tradiert worden. Hierfür ließe sich auf das umfängliche Danielmaterial in Qumran verweisen: die Handschriften des biblischen Danielbuches in Qumran, die Zitate in den Qumranschriften sowie die übrigen Werke, die in den Umkreis des Danielbuchs gehören und in Qumran überliefert wurden, wie Pseudo-Daniel und der Text über die vier Reiche.[47] Doch diese Möglichkeit hat wenig Wahrscheinlichkeit für sich. Die Rezeption in Qumran zeigt durchweg ein Interesse an den eschatologischen (apokalyptischen) Partien des Danielbuches, nicht so sehr an den Erzählungen, und zudem ist im Unterschied zu den anderen Texten das *Gebet des Nabonid*, soweit wir sehen, eben auch keine Danielliteratur.

Als Alternative hat man an das Motiv von Krankheit und Heilung gedacht, das den Text für die Gemeinschaft von Qumran so anziehend gemacht habe. So wurde vorgeschlagen, dass mit dem „Wahrsager" (גזר) im *Gebet des Nabonid* ein „Exorzist" (Beschwörer) gemeint sei und der Text in Qumran somit als Beispiel für einen gelungenen jüdischen Exorzismus gedient habe oder sogar in der Gemeinschaft von Qumran selbst entstanden sei.[48] Doch auch wenn das Motiv von Krankheit und Heilung in dem *Gebet des Nabonid* zweifellos eine wichtige Rolle spielt und auch wenn Magie und Mantik nicht weit voneinander entfernt sind, trifft diese Erklärung nicht den Punkt. Die primäre Bedeutung des aramäischen Wortes גזר ist nun einmal nicht „Exorzist", und von einem Exorzismus ist im erhaltenen Text auch nirgends die Rede. Die Semantik von גזר und שחנא באישא „bösartige Krankheit" allein vermag diese Erklärung nicht zu tragen.

Und doch führt das Motiv von Krankheit und Heilung vielleicht auf die richtige Spur, sieht man es im Zusammenhang mit den anderen Motiven, die in dem Text begegnen, der Sündenvergebung und der Götzenverehrung, und mit der Gattung „Gebet", in der sich der Text selbst präsentiert. Aufschlussreich ist der sprachliche Hintergrund in der biblischen Tradition. Die bösartige Krankheit, an der Nabonid leidet, gilt hier als Unreinheit, für die es

47 Vgl. A. Mertens, Das Buch Daniel, Stuttgart 1971; P.W. Flint, The Daniel Tradition at Qumran, Michigan/ Cambridge 1997 und ders., The Daniel Tradition at Qumran, Leiden 2001.

48 A. Dupont-Sommer, Die essenischen Schriften, Tübingen 1960, 348–349.351–352; ders., Remarques linguistiques sur un fragment arameén de Qoumrân („Prière de Nabonid"), Comptes Rendus du Groupe Linguistique d'Études Chamito-Sémitique 8 (1958–1960), 48–56; ders., Exorcism, Leiden 1960 unter Berufung auf G. Furlani, Aram. GĀZRĪN = scongiuratori, Atti della Academia Nazionale dei Lincei, Classe di scienze morali, storiche e filologiche, serie ottava 4 (1948) 177–196. Ihm folgen J. Carmignac/É. Cothenet/H. Lignée, Les textes de Qumran: Traduits et annotés II, Paris 1963; J.A. Fitzmyer/D.J. Harrington, A Manual of Palestinian Aramaic Texts (= Biblica et Orientalia 34) Rom 1978; F. García Martínez, The Prayer of Nabonidus, Leiden 1992; F. García Martínez/E.J.C. Tigchelaar, Study Edition I, Leiden 1997; vgl. dagegen E. Puech, La Prière de Nabonide (1996), 218–219; K. Beyer, Die aramäischen Texte vom Toten Meer, Göttingen 1984, 542 und die Wörterbücher (Gesenius-Buhl, Köhler-Baumgartner). Zum Thema „Exorzismus in Qumran" vgl. W. Kirchschläger, Exorzismus in Qumran?, Kairos 18 (1976), 135–153, hier 144–148 auch über das Gebet des Nabonid.

in *Lev 13* genaue Verfahrensregeln gibt; die einschlägige Stelle aus der Tora ist auch in einer Handschrift von Qumran (*4Q365 18–19*) belegt. Es ist dieselbe Krankheit, mit der auch der König Hiskia in *2Kön 20* (*Jes 38*) und Hiob (in *Hi 2,7* sowie im Targum Hiob *11Q10 XVI,2* zu *Hi 30,13–20*) geschlagen sind, die beide aufgrund ihrer Frömmigkeit auf wundersame Weise wieder geheilt werden. Und es ist nach *Ex 9 (V. 9–11 = 4Q365 3)* eine der Plagen Ägyptens, die in dem Fluch von *Dtn 28 (V. 27.35)* auch Israel angedroht werden für den Fall, dass es das Gesetz des Mose nicht hält.[49]

Im Blick auf das *Gebet des Nabonid* ist es von besonderem Interesse, dass sich der Ausdruck „bösartige Krankheit" schon in den biblischen Parallelen teilweise mit anderen Motiven verbindet. Der König Hiskia wird aufgrund eines Gebets von seiner Krankheit durch den Propheten Jesaja geheilt. Hiob wird am Ende seiner langen Klagen wiederhergestellt, und seinen Freunden, die ihm Vorwürfe gemacht haben, werden auf Bitten Hiobs ihre Sünden vergeben. Die Fürbitte für die Freunde steht schon im hebräischen Text von *Hi 42,8–10*, die Sündenvergebung fügt der Targum zum Hiobbuch hinzu, der sich in Qumran gefunden hat (*11Q10 XXXVIII,2–3*).[50] Schließlich ist zu beachten, dass die Krankheit in *Ex 9* ganz Ägypten, also eine Fremdmacht, trifft und – liest man etwas weiter im Text – die ägyptischen Plagen insgesamt auf die Erkenntnis des Pharao abzielen, „dass es meinesgleichen nicht gibt in allen Landen" (*Ex 9,14*). Dem pädagogischen Zweck der Plagen entspricht, dass die „bösartige Krankheit" im Fluch von *Dtn 28 (V. 35)* damit verbunden wird, dass Israel als Strafe für seine Sünden in das babylonische Exil gehen und dort Göttern aus Holz und Stein dienen muss.

Kurz: Von den biblischen Belegen her ergibt sich ein innerer Zusammenhang von Krankheit, Sünde und (babylonischer) Götzenverehrung bzw. umgekehrt von Heilung, Sündenvergebung und Bekenntnis zum einen und einzigen Gott. Die Wende markiert fast immer das Gebet. Und genau dieser innere Zusammenhang prägt auch das *Gebet des Nabonid*. In ihm ist es allerdings – anders als bei Hiskia in *2Kön 20/Jes 38* (oder bei Manasse im apokryphen *Gebet des Manasse*) – nicht der judäische, sondern der fremde, babylonische König, der, vermittelt durch den judäischen Wahrsager, die Wende vollzieht und sich von den Götzen abwendet und zum einen Gott bekehrt.[51] Wenn man nach dem Rezeptionsinteresse für das *Gebet des Nabonid* in der Gemeinschaft von Qumran fragt, dann muss man also fragen, ob dieser innere Zusammenhang der Motive, die Bekehrungsge-

49 Die Qumran-Texte nach den Editionen von E. Tov/S.A. White, Reworked Pentateuch$^{c}$, in: Qumran Cave 4.VIII, DJD XIII, Oxford 1994, 255–318 + Pl. XXII–XXXII (4Q365) und F. García Martínez, 11Qtargum Job, in: Qumran Cave 11.2: 11Q2–18, 11Q20–31, DJD XXXIII, Oxford 1998, 79–180 + Pl. IX–XXI 1998 (11Q10), im Folgenden: F. García Martínez, 11Qtargum Job, Oxford 1998.

50 Text F. García Martínez, 11Qtargum Job, Oxford 1998; zur Parallele vgl. G. Fohrer, 4 Q Or Nab, 11 QtgJob und die Hioblegende, Zeitschrift für die alttestamentliche Wissenschaft 75 (1963), 93–97; F. García Martínez, The Prayer of Nabonidus, Leiden 1992, 133–135.

51 Das Gegenbild findet sich in *1 Henoch 67,8*: Hier werden die Könige im Endgericht mit (heißen, brennenden) Wassern gereinigt und geheilt und zugleich an Leib und Geist dafür bestraft, dass sie den Namen Gottes nicht erkannt haben.

schichte des heidnischen Königs, in der Gemeinschaft von irgendeinem besonderen Interesse war.

Die Frage ist nicht leicht zu beantworten. Zweifellos dürften der Gemeinschaft von Qumran die biblischen Beispiele solcher Bekehrungsgeschichten bekannt gewesen sein: die Bekehrung des aramäischen Feldhauptmanns Naaman in *2Kön 5*, der ebenfalls an einer Hautkrankheit litt, und natürlich die Danielerzählungen in Dan 1–6 (samt Zusatzstücken der Septuaginta). Einschlägig ist auch das aramäische Genesis-Apokryphon, eine Art Targum zur Genesis, der sich in Qumran gefunden hat. Hier ist es der Abraham von Gen 12 und Gen 20, der für den mit einem „bösen Geist" (רוח ביאשא) geschlagenen ägyptischen Pharao und sein Haus betet und ihn so von seiner Krankheit befreit, nachdem die ägyptischen Weisen (חכימיא) und Beschwörer (אשפיא) nichts haben ausrichten können. Ebenfalls einschlägig dürfte ein Werk sein, das an das Buch Ester erinnert und darum in der Forschung *Proto-Ester* genannt wird (4Q550). In diesem Text wird erzählt, wie ein judäischer Mann dem persischen König (Xerxes, Sohn des Dareios) gute Dienste leistet und ihn wahrscheinlich von der Macht des jüdischen Gottes überzeugt hat.

Doch das alles sind Überlieferungen, die – wie das *Gebet des Nabonid* – zwar in Qumran vorhanden waren und wohl auch gelesen und studiert wurden, aber nicht dort entstanden sind. In den Schriften, die mit einiger Sicherheit von der Gemeinschaft selbst stammen, begegnet das Motiv der Bekehrung eines heidnischen Königs nirgends. Insofern lässt sich nur mutmaßen, warum die Bekehrungsgeschichten überliefert wurden und welches Interesse die Gemeinschaft von Qumran daran hatte, eine Gemeinschaft, die so großen Wert auf Abgrenzung gegen alles Unreine und Fremde legte.

Ich vermute, dass die Bekehrung des Heiden zum jüdischen Gott als Beleg für die eigene Position aufgefasst wurde. Der judäische Protagonist in den Texten wurde natürlich mit einem Vertreter derjenigen Richtung des Judentums in hellenistischer Zeit gleichgesetzt, der die Gemeinschaft von Qumran angehörte. Wenn auf sein Wirken hin selbst ein heidnischer König das Bekenntnis zum jüdischen Gott annahm, so musste dies bedeuten, dass alle anderen Richtungen des Judentums auf dem Holzweg waren, so, wie es die Schriften der Gemeinschaft von Qumran ja auch ausdrücklich sagen. Nun haben die Bekehrungsgeschichten eine werbende, apologetische Tendenz, die nicht zuletzt den hellenisierten oder der hellenistischen Kultur nicht völlig abgeneigten Teilen des Judentums entgegenkam und von ihnen für ihre Zwecke genutzt wurde.[52] Diese Tendenz dürfte für die Gemeinschaft von Qumran in ihr Gegenteil, eine nach innen gerichtete, polemische Tendenz, umgeschlagen sein. Inwieweit die ursprüngliche apologetische Tendenz hier noch wirksam war und man in der Gemeinschaft von Qumran tatsächlich mit einer Be-

52 Als Beispiel sei hier nur der pseudonyme Aristeasbrief genannt, die Entstehungslegende der Septuaginta, die die griechische Übersetzung der Tora des Mose legitimiert und im Zuge dessen sowohl dem ptolemäischen König den wahren Gott nahebringt als auch die Abgrenzung des Judentums durch die Gesetze der Tora begründet.

kehrung der zeitgenössischen, hellenistischen Herrscher gerechnet hat, ist schwer zu sagen, aber auch nicht auszuschließen.[53] Doch damit ist ein anderes, zentrales und noch wenig erforschtes Kapitel der Qumranforschung, nämlich Qumran und der Hellenismus, aufgeschlagen, das hier nicht hingehört und darum auch nicht mehr behandelt werden soll.

53 Oder ist an potentielle Konvertiten, speziell aus dem Bereich der aramäischsprachigen und arabischstämmigen Nachbarn, gedacht, für die Nabonids Bekehrung in Tema als Vorbild dienen sollte (Annette Steudel, Göttingen, mündlich)?

*Willem F. Smelik*

# Myth and Truth: Babylon in the Babylonian Talmud

Ever since Israel's ancestors were 'wandering Aramaeans' who hailed from Mesopotamia and migrated to the Levant, but vividly recalled their Eastern origins, the cradle of Israel was remembered to have stood in Babylonia. In the heyday of Jewish Babylonia under Sasanian rule, when its leaders outshone the remainder of the Jewish world, this connection was not lost on the rabbis. Yet Babylonia also represents the nadir of Jewish history in the Ancient Near East, when the cream of Judaean society was deported to Mesopotamia in several waves in the early 6th c. CE – the Northern Kingdom already having been lost one and a half century previously. Later still, when the Romans had destroyed the Second Temple, and in the 2nd c. CE had declared Jerusalem a no-go area for Jews, Babylonia would become a major seat of diaspora Jewry. H<ow then did Babylonia fare in Jewish reception?

The second diaspora was not of Babylonian making, but the first was, and it was to be remembered as part of Jewish liturgy. The famous opening verse of Ps. 137 reflects the misery of the Judaeans who had been exiled to Babylonia in the early sixth c. BCE: 'By the rivers of Babylon, there we sat, sat and wept, as we thought of Zion'. Among those whose feelings of loss and alienation this psalm represents, their stay 'on alien soil' did not produce a sense of attachment to their new environment. The psalmist fosters the memory of Jerusalem while praying for the punishment of her conqueror (Ps. 137.8–9):

> Fair Babylon, you predator![1]
> Happy shall be whoever repays you in kind
> what you have inflicted on us!
> Happy shall be whoever seizes your babies
> and dashes them against the rocks!

This bitter psalm[2] posits Babylon as the 'implacable antithesis' to Zion.[3] Originally functioning within a cultic setting during the Second Temple period, the psalm was to be in-

1 The translation 'predator' assumes that שדודה has an active sense in the present co-text. Yet it is equally possible to retain the Masoretic vowels and read it as a passive participle, 'devastated one'. Since the reference to the grief in Babylon is in the past, this psalm was presumably composed upon return to Jerusalem. Cf. D.N. Freedman, 'The Structure of Psalm 137', in: H. Goedicke (ed.), Near Eastern Studies: In Honor of William Foxwell Albright, Baltimore 1971, 187–205.

2 Sometimes called a scandal from a Christian perspective of forgiveness; see E.H. Peterson, Answering God: The Psalms as Tools for Prayer, San Francisco 1989, 96, and contrast J. Eaton, The Psalms, London 2005), 454–56.

3 L.C. Allen, Psalms 101–150, Word Biblical Commentary 21 (2002).

cluded in the standard liturgy for the Ninth of Av, the day to commemorate the destruction of the First and Second Temples in Jerusalem. Still, grief for the loss of land and the shattering of the holiness did not only bring about a desire for an avenger, but also reflection on the reasons for God's apparent decision to have Jerusalem destroyed and its inhabitants deported. God authorised the evil that came down upon Jerusalem, as the Hebrew Bible itself already suggests.[4] The late Targum Psalms puts the desire of retribution in the mouth of the angels Michael and Gabriel, thus dissociating the biblical text from a 'human seeking vengeance' and turning it into a 'judgement on a higher plane'.[5]

As Jewish history in Babylonia was to last until the mid-twentieth century CE these feelings did not remain. Before long many Jews acculturated to their new surroundings to the extent that they chose not to return to Jerusalem when the opportunity presented itself but stayed put by those very same rivers of erstwhile longing. Eventually, Babylonian Judaism enjoyed its golden age during the first millennium CE, when it turned into the envy of the Jewish world, including Palestine, for its learning and relatively peaceful conditions. Babylonian Judaism was to recognise Persian rule in Shmuel's famous statement that דינא דמלכותא דינא 'the law of the kingdom is law'.[6] Babylonian Jews even expressed a sense of superiority over against their Palestinian peers in terms of learning and practice.

This article surveys the reception of Babylon among Jewish Babylonian rabbis from three points of view: (1) What were Babylonian Jewry's views of Ancient Babylonia? (2) How did Babylonian Judaism construct its own history on Babylonian soil? (3) To what extent did Babylonian Judaism acknowledge any acculturation to Sasanian Babylonia?

## The Memory of Ancient Babylonia

The name Babylon brims with history in the Talmud for both historical and literary-theological reasons. The Jews of the Sasanian period could still witness the historical signs of Babylonia's forlorn age in the form of ancient monuments, ruins, and sites identified with famous literary events. The memory of Ancient Babylonia was so visible that Rav Hamnuna formulated specific blessings for their perception (b. Ber. 57b):

> R. Hamnuna said in a discourse: If one sees the wicked Babylon, he should say five benedictions: On seeing [the city] Babylon itself he says, Blessed be He who has destroyed the wicked Babylon. On seeing the palace of Nebuchadnezzar, he says, blessed be He who destroyed the palace of the wicked Nebuchadnezzar ...

4 So too Flavius Josephus, Antiquitates Judaicae 10.3.2.40.

5 So M.J. Bernstein, 'Translation Technique in the Targum to Psalms. Two test Cases: Psalms 2 and 137', in: E.H. Lovering (ed.), Society of Biblical Literature 1994 Seminar Papers (Atlanta 1994), 326–45. For the connection with Tosefta Targum Isa. 21.5, see T. Edwards, Exegesis in the Targum of Psalms: The Old, the New, and the Rewritten, Piscataway 2007, 80–84.

6 b. Ned. 28a; b. Giṭ. 10b; b. B. Qam. 113a (3x); 113b (2x); b. B. Bat. 54b; 55a.

Other blessings include those for seeing the furnace, into which Abraham was thrown, or the lion's den of Daniel and his companions. Whether *all* such historical sites were correctly identified, which seems questionable in the case of Abraham's furnace and Daniel's den, is beside the point: these traditions expressed a strong awareness of living in the ancient cradle of civilization between the two rivers that flow from paradise, where ancient monuments were still visibly present in the geographical landscape.[7]

The memory could and would be developed in different directions, without being contradictory. There is the sense of antiquity that is rooted in narratives of Babylonia's cruel interference with Jewish history, but there is also a prouder sense of shared antiquity. Certain Talmudic traditions give expression to a Jewish bond with Mesopotamia, where Jews had been living from Abraham to Daniel. Whether or not the den was Daniel's or the furnace Abraham's, the rabbis point to a historical connection with the Mesopotamian environment which would be further developed in the rabbinic valuation of *Jewish* Babylonia as we will see below ('Jewish Babylonia'). And finally, in a third sense ancient Babylonia becomes the pawn of current theological reflection, a literary record replayed on the chessboard of history to answer certain contemporary questions and gain new insights.

Towering over the memory of Babylonia are two narratives of destruction: that of the Tower (and city) of Babylon and that of the First Temple of Jerusalem. In the rabbinic construction of Ancient Babylonia, these episodes were intrinsically linked to one another through their Babylonian protagonists, Nimrod, Sennacherib, and Nebuchadnezzar. Nimrod features in connection with the Tower of Babel-narrative. Although his name has not been found in cuneiform sources, the biblical and post-biblical tradition unanimously marks him out as the legendary founder of Mesopotamian cities.[8] In the so-called Table of Nations (Gen. 10) we first hear of Babylon as the beginning of Nimrod's kingdom, together with Erekh and Akkad (Gen. 10.10). All of these cities are located in the valley of Shinar, that is southern Mesopotamia, where the earliest generations of humankind dwelled (Gen. 11.2).[9] Thus Shinar and Babylon link Gen. 10 and 11.

Because Babel was his turf, it seemed natural enough to hold Nimrod responsible for the hubris of constructing the Tower of Babel. This allows for an important hermeneutic strategy, namely to read Gen. 10 as the outcome of Gen. 11 despite the fact that it comes before the latter chapter. Such a reading solves a discrepancy between these chapters. When the Table of Nations (Gen. 10) describes how Noah's descendants multiplied and spread out over the world, it appears to describe their dispersion and multiple languages as the out-

7 I. M. Gafni, 'Expressions and Types of "Local Patriotism" among the Jews of Sasanian Babylonia', in: S. Shaked and A. Netzer (eds.), Irano-Judaica, II, Jerusalem 1990, 63–71.

8 For a proposed identification with Ninurta, see K. van der Toorn and P. W. van der Horst, Nimrod before and after the Bible, Harvard Theological Review 83 (1990), 6–16.

9 The name Shinar denotes the area of southern Mesopotamia. For a discussion of this name, see K. van der Toorn and P. W. van der Horst, 'Nimrod Before and After the Bible', Harvard Theological Review 83 (1990), 1–29 (2–4).

come of an entirely natural process of human reproduction. Yet Gen. 11 sees the dispersion of a formerly unified people by the fragmentation of their languages as the punishment God meted out to the Babylonians. If Nimrod is the culprit of Gen. 11, the Table of Nations does not represent the antecedents of Gen. II but its aftermath.

Thus Nimrod is thought to have initiated the building of the Tower of Babylon, also called the 'temple of Nimrod' (b. A. Zar. 53b), although the biblical account itself does not illuminate the exact nature of his and his generations' sin. The rabbis admitted as much, as they compare the sins of the generation of Noah, which is stated clearly, to those of the generation of the Big Dispersion, which is not.[10] But most interpreters conceived of Nimrod's hubris and the failure of the people to value peace the major factor contributing to God's dismay. Yet the biblical account of Nimrod is not necessarily negative, as born out by earlier Jewish interpretations which preserve a more positive appreciation of Nimrod's life. Within the brief lists of the Table of Nations, Nimrod is unusual because four verses are allocated to him (Gen. 10.8–12). Brief as this description may be for historical purposes, it is relatively long for this chapter, indicating that Nimrod was a ruler of extraordinary reputation. Interestingly, his reception is somewhat mixed. The biblical description was almost universally taken to reflect Nimrod's enmity towards God: 'He was a mighty hunter before the Lord' (10.9) implied, as far as the rabbis were concerned, that Nimrod trapped people in his evil plans. The etymology of his very name (מרד 'to rebel') augured his rebellion against God,[11] reading 'before the Lord' as '*against* the Lord'.[12]

Typical for rabbinic lore is the association of Babylonian tyrants with one another. Despots like Nimrod, Sennacherib, and Nebuchadnezzar did not just share a cruel streak and sinful behaviour, but were in fact family. Nimrod was Nebuchadnezzar's grandfather (b. Hag. 13a), just as Nebuchadnezzar in turn was Sennacherib's son-in-law (TosTg 2 Kgs 19:35; Isa. 10:32–33).[13] Both Sennacherib and Nebuchadnezzar were assumed to have participated in the siege against Jerusalem (b. San. 95b); eventually they find themselves together in the deplorable conditions of the Netherworld (TosTg Ezek. 1.1). The apple never falls far from the tree. But all this is not just about stereotyping, but also to create associative patterns. To link historical figures is a common mnemonic technique. Thus b. Meg. 11a teaches that 'three [potentates] ruled over the whole world, that is Aḥab, Ahasuerus and Nebuchadnezzar'. In similar ways, dating the destructions of the first and second Temple and Bar Kochva's stronghold to the ninth of Av serves to foster the memory of these events.

10 Gen. R. 38.6; TanB xn 24.

11 C. Uehlinger, Weltreich und 'eine Rede': Eine neue Deutung der Sogennanten Turmbauerzählung (Gen 11, 1–9) (= Orbis Biblicus et Orientalis 101), Göttingen 1990, 66.

12 b. Eruv. 53a; see also b. Pes. 94b, b. Hag. 13a. Philo, Quaestiones in Genesim 2.81–82, already exploits this possibility, suggested by the Septuagint's translation of לפני by ἐναντίον.

13 See P. Grelot, Le Targoem d'Isaïe, X, 32–34 dans ses diverses recensions, Revue Biblique 90 (1983), 202–28.

The Biblical narratives soon became a vehicle for inner-rabbinic reflection on the course of history. The contemporary perspective transforms the memory of Ancient Babylonia into something that is neither myth nor history but theological truth. In a series of expositions about Sennacherib we find that he infers, as a true Jewish scholar, from Isa. 8.6–7 that God set him up to campaign against the kingdom of Judah (b. San. 94b). The text cited in his support raises the question of why he was punished if God Himself willed Sennacherib to embark upon his military campaign. The answer is that he overstepped his remit, as he should not have extended his hand against Jerusalem but only against the Northern Kingdom. Another tradition has Nebuchadnezzar quoting the scriptural translation of Aquila in defense of his own idolatry (Lev. R. 33.6).[14] Playful as such claims to scriptural support for their actions are on the part of Israel's notorious enemies, at a deeper level they address the theological questions of divine intervention and human responsibility. Sennacherib and Nebuchadnezzar are pawns in the Jewish reflection on God's providence.

Similar reflections are discernible in another widespread motif. Seemingly at variance with the portrayal of the deep-rooted evil of Sennacherib and his ilk are stories which claim that Sennacherib converted to Judaism. Thus b. Giṭ. 57b states:

> A Tanna taught, Naaman was a resident alien; Nebuzaradan was a righteous proselyte; descendants of Haman learnt the Torah in Benai Berak; descendants of Sisera taught children in Jerusalem; descendants of Sennacherib gave public expositions of the Torah. Who were these? Shemaya and Avṭalion.

The narrative setting evokes this tannaitic tradition, because the context is all about destruction and massive slaughtering, brought together under the header of the Fall of Bethar, the stronghold of Bar Kokhva. The setting includes a reference to the rivers of blood which Nebuzaradan caused to flow. However, trying to understand the miracle of Zekharya's blood, he repents as he realises: 'If such is the penalty for slaying one soul, what will happen to me who has slain such multitudes? So he fled away, and sent a deed to his house disposing of his effects and became a convert.' The disposal of his possessions is necessary, because they are tainted by idolatry. In the tradition quoted above, the anonymous Tanna lists several notorious antagonists of Israel's past: (1) the Aramaean Naaman, the courtier of the King of Aram (2 Kgs 5); (2) Nebuzaradan, who took orders from Nebuchadnezzar, burnt the Temple of Jerusalem; (3) Haman tried to destroy the entire Jewish population of the Persian empire; (4) Sisera took up battle against the tribes of Israel, to be defeated by the non-Israelite Yael;[15] (5) Sennacherib's offspring, Shemaya and Avtalion, became legendary leaders of the pre-rabbinic movement in the Second Commonwealth.

14 Cf. Lieberman, Greek in Jewish Palestine, New York 1942, 44–46 for parallels (Lieberman does not discuss Lev. R. 33.6 in this connection).

15 We do not hear of Sisera's children at all in biblical or rabbinic literature. However, there is a rabbinic tradition that interprets Yael's method of defeating Sisera as sexual seduction. Yael herself has been admitted to the ranks of Israel in post-biblical tradition. Thus we may speculate whether his converted offspring were in fact the children he fathered with her?

Leaving any illusory historical events aside, the motif of prominent non-Jews converting to Judaism serves several purposes. At one level, any such conversion represents a coup and source of pride for a people without political independence, frequently finding themselves living in the margins of other societies. As with the famous Roman proselyte Aquila, they bolster the pride of Israel and highlight the truth of Judaism. Yet another purpose is sociohistorical commentary. Although the motif of a ruler who converts from one faith to another is widespread and its purport would seem obvious enough, its embedding in the narrative context, in particular the rabbinic story-cycles, requires attention. The story of Sennacherib's conversion and his offspring which became legendary leaders of the pre-rabbinic movement in the Second Commonwealth, Shemaya and Avtalion, forms part of a long narrative cycle, which determines its narrative function and scope (b. Giṭ. 55b-58a).[16] The importance of proper literary contextualization in combination with a diachronic analysis becomes apparent in this cycle, which must be viewed from the perspective of the legal peg on which it enters the folios of the Talmud, the structure of the narrative, and its development of earlier motifs in rabbinic literature. On the course of history by way of reflection on Jewish culpability for the catastrophic events of history.

There are in fact several stories with an inordinate number of conspicuous conversions to Judaism, which include Nero and Sennacherib: b. San. 95b, b. A. Zar. 8a-11b and b. Giṭ. 55b-58a. The latter two involve Onqelos (the nasalised pronunciation of Aquila).[17] Inversions of appropriate behaviour within the matrix of Roman–Jewish relationships are paramount in these stories. The first of the cycles mentioned is linked to a discussion of pagan festivals and Roman-Jewish relationships. In order not to benefit from or contribute to pagan practices, Jews should avoid any financial deal with idolaters from three days before the beginning of a pagan festival until its end (m. A. Zar. 1.3). The conversions demonstrate not just Israel's superiority, but also the limited leeway of Roman–Jewish interaction: a very restricted model of coexistence that highlights enmity and warfare, separation and isolation, as well as conversion. And even the conversions have a touch of the inappropriate about them.[18] The second cycle is related to legal discussions about rabbinic amendments 'for the sake of good order' and 'in the interests of peace'. The narratives themselves, however, spin a grand carpet of destruction: the Temple, the King's Mountain, and the stronghold of the second revolt against the Romans, Bethar. These destructions, it is suggested,

16 A. Weiss, על היצירה הספרותית של האמוראים, New York 1962, 261–63; E. Yassif, The Cycle of Tales in Rabbinic Literature, Jerusalem Studies in Hebrew Literature 12 (1990), 103–46; R.L. Kalmin, Sages, Stories, Authors, and Editors in Rabbinic Babylonia, Atlanta 1994, 43–50; I.J. Yuval, 'Two Nations in Your Womb': Perceptions of Jews and Christians in Late Antiquity and the Middle Ages, Berkeley 2006, 49–56; J.L. Rubenstein, Talmudic Stories: Narrative Art, Composition, and Culture, Baltimore 2003, 139–75; P. Schäfer, Jesus in the Talmud, Princeton 2007, 82–94.

17 For a fuller discussion of both passages, see my forthcoming book *Rabbis, Language and Translation in Late Antiquity*.

18 For example, the circumcision of Qetia bar Shalom (b. A. Zar. 10b) or the necromancy of Onqelos (b. Giṭ. 56b-57a).

came about by the combined lack of legal innovation and historical spine. There is no need to go into further detail in the present connection. For the moment, these examples demonstrate how ancient and modern times are blurred as history becomes the theatre of current concerns. Indeed the name 'Babylon' comes as a guise for Rome, contemptuously referred to as an 'old whore'.[19] The name Babylon now assumes the quality of a chiffre, one which Babylonian sages inherited from Palestinian Judaism. The equation of Rome with Babylon, odd as it may seem to be from a geographic and political perspective, although it has a parallel in the NT, presumably derives from the practice of actualising the four kingdoms in the visions of Daniel (Dan. 2 and 7).[20] These kingdoms are symbolical entities, their identities may vary, just as historical memory may blur with contemporary society.

## Jewish Babylonia

Although not the promised land, in the eyes of Babylonian Jews Babylonia came as close as possible. They could reasonably lay claim to the longest pedigree in comparison with other Diaspora communities, since their connection to Babylonia reached all the way back to Abraham whose home it had been to begin with. As God's unfaithful partner, the Judaeans were sent back to 'their father's home' just as unfaithful wife would have been sent to hers (b. Pes. 87b). Apart from the Israelites who had been exiled to the Assyrian empire after the fall of Samaria, the true beginnings of Babylonian Judaism came about with the waves of deportation in the early sixth century (597, 587 and 582 BCE). We know little about the early Jewish Babylonian history up to the third century CE, and even during the talmudic period our data remain sparse. The Murashu archives, which belonged to a wealthy family of merchants, are one of the exceptions. They were discovered at Nippur and date from the fifth century BCE.[21]

Despite the lack of tangible data there is no doubt about the continuous Jewish presence in Mesopotamia and Babylonia ever since 597 BCE. Isolated references point to the existence of a considerable Jewish minority. One of our sources is the historian Flavius Josephus,[22] who reports about a deportation of Jews by Antiochus the Great from Mesopotamia and Babylonia to Lydia and Phrygia by the end of the third century BCE. Even though the report has the for Josephus typical bias of Jewish loyalty to their overlords, which serves his purposes all too well, the presence of Jewish 'communities' accords well with his other references to Babylonian Judaism (Ant. 12.147–150). For example, according to Josephus

19 S. Lieberman, Greek in Jewish Palestine, New York 1942, 45–46.

20 Cf. U. Gleßmer, Die "Vier Reiche" aus Daniel in der Targumischen Literatur, in: J.J. Collins and P.W. Flint (eds.), The Book of Daniel: Composition and Reception, Leiden 2001, II, 468–89.

21 G. Cardascia, Les Archives des Murašû, Paris 1951.

22 D. Goodblatt, Josephus on Parthian Babylonia (Antiquities XVIII, 310–379), Journal of the American Oriental Society 107 (1987), 605–22.

Herod the Great claimed to originate from a Babylonian-Jewish family, and during his rule the famous Hillel came from Babylonia to Judea. The communities there would send their temple tax and donations to Jerusalem. In the first century CE, the kingdom of Adiabene converted to Judaism, and the New Testament speaks of pilgrims who came to Jerusalem from Parthia, Media, Elam and Mesopotamia. The history may be patchy but Jewish presence is not in doubt.

In the third century CE the Babylonian rabbis began to rival the learning of their Palestinian counterparts, culminating in the composition of the Babylonian Talmud between the third and sixth-seventh centuries CE. This collection of 37 tractates[23] is the apex of classical rabbinic literature, towering over all subsequent sources of Jewish learning and teaching. By way of contrast, its Palestinian counterpart, the Talmud Yerushalmi or Palestinian Talmud, never commanded the same level of assent as the Bavli, which soon outshone its sibling in terms of influence and dissemination.

As the site of an increasingly important rabbinic culture, closely connected with their Palestinian peers, Babylonian sources sometimes express their own unique contributions which reverse the pecking order to some degree (b. Suk. 20a):

> For in ancient times when the Torah was forgotten from Israel, Ezra came up from Babylon and established it. [Some of] it was again forgotten and Hillel the Babylonian came up and established it. Yet again was [some of] it forgotten, and R. Hiyya and his sons came up [from Babylonia] and established it [once more].

Although Israel is still the acknowledged seat of Torah-learning, it has only been able to maintain its position because their learning was propped up by pillars of Babylonian learning. Universally recognised for their halakhic rigour and learning, Babylonian rabbis became increasingly assertive about their own status vis à vis Palestine. The rivalry found expression in two distinct constructions of patriotism. Whereas Palestinian rabbis cultivated a theology of 'the Land', which underpinned their own claims to legal hegemony, the Babylonian rabbis articulated their own form of local patriotism which is unique in the history of the Diaspora. They emphasized a connection with pre-destruction Jerusalem to the extent that they almost stood on a par with Palestine.[24] Their high pedigree was of both a genealogical and institutional character. On the basis of 2 Kgs 24.12–16 they argued that the 'craftsmen' and 'smiths' deported to Babylonia before the destruction of the First Temple were Torah-scholars (b. Giṭ. 88a). In contrast to Palestine, their learning was never contaminated by foreign suppression, hence purer and better. Their genealogy was purer than that of their Palestinian peers, since Ezra had only taken men of doubtful provenance back to Palestine, and intermarriage had further compromised their purity (b. Qid. 69b, 71b). Jehoiachin, the last king of Davidic lineage had been deported to Babylonia, relating even the Davidic line to Babylonia rather than Palestine.

23 Or 36½, since tractate Tamid did not receive a full commentary.
24 See I. Gafni, Land, Center, and Diaspora: Jewish Constructs in Late Antiquity, Sheffield 1997, 113–17.

## Acculturation & Acknowledgement: The Assyrian Script

When the Sasanids took control, Babylonian Jewry had already lived under Persian rule for many centuries. While explicit references to the Zoroastrians in Jewish sources remain of limited scope, there is every reason to believe that there were many cross-cultural influences between the Jews and the Persians.[25] Such influences range from language contact and loanwords to interpretative techniques shared with mantic practitioners, legal principles and literary motifs.[26] Rabbinic exegesis may have found its expression in the Graeco-Roman world, but employed techniques first developed in the Ancient Near east 'where', as Tigay writes, 'texts were studied intensively in scribal academies'.[27] The preservation of mantic literature expressed the belief that 'canonical' portents retained their value and could be applied anew. Tigay identified the actualization of names as one of the earliests Mesopotamian techniques indispensable to the diviners who employed handbooks of ancient divination as guides for the present day.

During the Amoraic period, the Sasanian authorities appear to have left their Jewish subjects in peace.[28] They expressed no interest in making converts among the Jews, while actions taken against Jewish practices are invariably related to Zoroastrian religious sensibilities. These sensibilities were taken on board by the rabbis. When Gen. R. 38.13 has Nimrod urge Abraham to 'worship the fire', the tradition appears to be based on the identification of Nimrod with Zoroaster.[29] This equation of Nimrod with Zoraster may have Jewish antecedents, although it is most commonly encountered in Christian sources. Be this as it may, Jewish Babylonian sources remain prudently silent on this topic.

In many other aspects, Persian influences are discernible. Kalmin has argued that Babylonian Rabbis were more Persian than Roman, and that their Persian acculturation distinguished them from their Palestinian peers.[30] As observed above, the absence of persecution relative to their Palestinian peers was reason to argue that Babylonian learning was

25 G. Herman, Ahasuerus, the former Stable-Master of Belshazzar, and the Wicked Alexander of Macedon: Two Parallels between the Babylonian Talmud and Persian Sources, AJS Review/The Journal of the Association for Jewish Studies 29 (2005), 283–297.

26 See, e.g., J. Tigay, An Early Technique of Aggadic Exegesis, in: H. Tadmor and M. Weinfeld (eds.), History, Historiography and Interpretation: Studies in Biblical and Cuneiform Literatures, Jerusalem 1984, 169–89; M. Macuch, Iranian Legal Terminology in the Babylonian Talmud in the Light of Sasanian Jurispudence, Irano-Judaica 4 (1999), 91–101; G. Herman, Priests and Amoraic Leadership in Sasanian Babylonia, in: M. Altshuler *a.o.* (eds.), Twelfth World Congress of Jewish Studies. Division B: History of the Jewish People, Jerusalem 2000, 59–68.

27 J. Tigay, An Early technique of Aggadic Exegesis, in: H. Tadmor and M. Weinfeld (eds.), History, Historiography and Interpretation: Studies in Biblical and Cuneiform Literatures, Jerusalem 1984, 170 n. 4.

28 R. Brody, Judaism in the Sasanian Empire: A Case Study in Religious Coexistence, Irano-Judaica 2 (1990), 52–62.

29 K. van der Toorn and P. W. van der Horst, Nimrod before and after the Bible, Harvard Theological Review 83 (1990), 26–27.

30 See R. Kalmin, The Sage in Jewish Society of Late Antiquity, London 1999; Idem, Jewish Babylonia between Persia and Roman Palestine: Decoding the Literary Record, New York 2006.

superior and of highest purity. Such claims, on the other hand, of higher purity in learning and genealogy would not favour any acknowledgement of cultural cross-influences. Here I will focus on one example of a borrowing which, after initial acknowledgement, was increasingly denied to have taken place and which, conveniently for our topic, focuses on the use of *Ashurit* ('Assyrian').

Between the Babylonian Exile and the end of the Bar Kokhba period, Hebrew texts were written in two different scripts: a late form of the old Hebrew script, itself a descendant of the proto-Canaanite script, and the square or 'Jewish' script, which rabbinic sources call *Ashurit*. The latter evolved from the Aramaic script and reflects the influence of international administration and trade during the Babylonian and Persian periods, and would eventually supersede the old Hebrew script. In the Palestinian Talmud, some of these historical circumstances are reflected in the recognition that the 'square' script originated in the Aramaic-speaking environment and had become the script of choice for Hebrew texts:[31]

> Said R. Yonathan of Beth Guvrin, Four languages are appropriately used in the world, and these are, every day speech [Greek] for song, Latin for war, Sursi [Aramaic] for wailing, Hebrew for discourse. And some say, also *Ashurit* [Assyrian] for writing. *Ashurit* has a script but no language, whereas Hebrew has a language but no script. They chose for themselves the *Ashurit* script and the Hebrew language.
> Why is [this script] called '*Ashurit*'? Because it is straight in [its mode of writing]. Said R. Levi, Because it came up with them from Ashur.

The four languages which the passage figures were prevalent in the Tannaitic period, and, although the tradition clearly stylized it taxonomy, it probably echoes the dominant usage of each: Latin as the language of the military superpower, Rome; Greek as the language of culture; Aramaic as the vernacular of the indigenous people; and Hebrew as the medium of the [rabbinic] scholars. But the claim that Hebrew has no script of its own finds its place a little too easily in the account of languages and script. Despite the fact that every Hebrew text was probably scripted in *Ashurit* by the second half of the second century CE, at least some of the Sages still knew that there once existed a (paleo-)Hebrew script called 'Hebrew' (עברי), leaving little doubt about its original status as the script in which every Hebrew text was written.[32] And not without reason: up to the Bar Kokhba period the עברי script was used on coins, in epitaphs and, for writing the Tetragrammaton, in Hebrew and Greek manuscripts, not to mention the Samaritan tradition which continues to use (a form of) this script down to today. Long after the failed revolt this script must have remained visible in Palestinian epigraphical artefacts, if not in manuscript evidence.

31 y. Meg. 1.11(8), 71b (par. y. Sot. 7.2, 21c). Assyria or Syria. Both *Assyrian* or *Syrian* were common designations for Aramaic in Greek and Hebrew; see, e.g., A. Wasserstein and D.J. Wasserstein, The Legend of the Septuagint: From Classical Antiquity to Today, Cambridge 2006), 55 n. 12.
32 Rashi's explanation (b. San. 21b), כתב עברי. של בני עבר הנהר, is etymologically possible but wholly implausible.

Why, then, does this passage deny a script to Hebrew, and by the same token the Aramaic language to *Ashurit*? The pragmatic acknowledgement that the Jewish script had its provenance in Assyria agrees with other ideologically charged, and historically false, rabbinic views on the origin and authenticity of *Ashurit*. These traditions claim that *Ashurit* – if not the authentic script all along which had only been abandoned in the face of Israel's sins – was authorized by no one less than Ezra, whose status balanced out that of Moses. There is little historical merit in these traditions as far as the script is concerned, and they are mutually exclusive. Yet all of them agree at one point: *Ashurit* is the script to be used for any ritually valid document, unless stated otherwise. Consequently, the Yerushalmi's narrative at the beginning of this chapter is ostensibly true and sensible, but in fact steel-faced, as it conceals the challenge to the authenticity and legitimacy of *Ashurit* for writing the Torah and the Book of Esther, let alone for phylacteries and the texts attached to door-posts (*mezuzot*). The urge to validate the innovative square script is the prime mover behind all these traditions.

Rabbinic literature relates different views on the origin of *Ashurit*, which can be divided into three main positions: (1) after the Exile the script changed from palaeo-Hebrew to square; (2) the script changed from square to palaeo-Hebrew, but was restored to its original form after the Exile; (3) the script had always been square, it was never changed. Each of these positions occurs in our three main sources, t. San. 4.7–8, y. Meg. 1.11(8), 71b/c and b. San. 21b/22a.[33] Key in the different appreciations of the alphabet change is Deut. 17.18. In the Tosefta two different interpretation of this text are provided. After noting Ezra's suitability for receiving the Torah, as evidenced by several prooftexts, t. San. 4.7 claims that Ezra introduced the Aramaic script and language.[34] The Tosefta first lists the geographical explanation, ascribed to R. Levi in the Palestinian Talmud quoted above, that the script is called 'Ashurit' because it came up with the Jewish exiles that returned from Ashur. Thus the Torah had been written in palaeo-Hebrew until the generations of Ezra and Nehemiah introduced the square script, which derived its name from their exilic provenance in Assyria. This assumption was challenged in the Tosefta by the assumption of a twofold change in script, from *Ashurit* to palaeo-Hebrew and back:

Rabbi says: 'In Assyrian writing the Torah was given to Israel, and when they sinned, it was changed to palaeo-Hebrew. But when they attained merit in the time of Ezra, Ashurit returned to them, as it is said, "Turn you to the stronghold, you prisoners of hope, even today do I declare that I will bring back – the change unto you"' (Zech. 9.12).

This is followed by the even stronger claim that the script really never changed, citing two prooftexts and explaining the name 'Ashurit' because the consonants 'are upright in

33 See further Sifre Deut. § 160; y. Soṭ. 7.2.21c; b. Zeb. 62a.

34 t. San. 4.7: 'Also through him were given a form of writing and language, as it is said, "and the writing of the letter was written the Aramaic character and interpreted in the Aramaic tongue" (Ezra 4.7). Just as its interpretation was in Aramaic, so is writing was in Aramaic. And it says, "but they could not read the writing, not make known to the king the interpretation thereof" (Daniel 5.8); this teaches that on that very day it was given.'

their manner of shaping letters'. Each element of the first tradition has been converted into a piece of testimony for the authenticity of *Ashurit*: the identity of the original script at Mt Sinai, the change of script, its provenance and name and the Deuteronomic prooftext.

Despite these latter views, the rabbis were still aware of the origins of the script – at least in Palestine. They knew of other formal characteristics such as the shape of the letters. This is evident from a discussion about the miracle of the letters which triggered a miracle when the Torah was written on stone: as the letters burned through the stone,[35] the inner part of either the samekh (in the square script) or the ayin (in the old script) stood as by a miracle. The rabbis discuss those letters featuring an enclosed space which are implicated in this miracle and correctly identify the two similar shapes which are closed forms without hooks between the two Hebrew scripts: *samekh* and *'ayin* (y. Meg. 1.11[8], 71b/c). Inevitably and typically, there is an element of stylizing here, since in the palaeo-Hebrew script far more letters would have resulted in a miracle, but these two forms are the most obvious ones to compare, particularly from the viewpoint of a reader used to *Ashurit*. Moreover, R. Yirmeya and R. Simon also supply correct paleographical information about the form of the letters *he* and *mem*. The distinction between medial and final forms of the *manzapek* (the letters *mem*, *nun*, *sade*, *peh* and *kaph*) first appears in Aramaic and early Jewish script towards the end of the Persian period. But in the Dead Sea scrolls they still occur inconsistently within a single manuscript, or not at all. The 1st c. CE Severus scroll may be another example of a manuscript without final forms.[36] The *he*, meanwhile, was often written as a closed letter akin to the *ḥeth*, from which it was distinguished only by an overhanging horizontal stroke (which stopped short in the *ḥeth*) but not by the disconnection of the left vertical bar.

The early rabbis thus appear to have been fully aware of the paleographical particularities of the square script, as is only natural. But this knowledge was compromised by dogmatic concerns in the Amoraic period. Likewise in later traditions, the ornamental crown of the letters שעטנ"ז ג"ץ – which, unlike the medieval form of this crown, would appear to have consisted of a little angular form on the top of the left vertical downstroke[37] – was also recognized; here, tradition would eventually relate these *tagin* to Moses at Mt Sinai, but not in early sources, and later sources would, moreover, no longer understand the exact object of reference of the כתר. And so we understand the change in the claims of the original script in which the Torah had been handed down to Moses. Distant in time and space from

35 See b. Meg. 3a; y. Meg. 1.11[8], 71b/c. For more discussion, see W.F. Smelik, 'Translation as Innovation in b. Meg. 3a', in: L. Teugels and R. Ulmer (eds.), New Developments in Midrash Research, Piscataway 2004, 25–49.
36 See J. Naveh, Early History of the Alphabet: An Introduction to West Semitic Epigraphy and Palaeography, Jerusalem/Leiden 1982), 170–72; A. Yardeni, The Book of Hebrew Script: History, Palaeography, Script Styles, Calligraphy & Design, Jerusalem 1997, 164–72; E. Tov, Textual Criticism of the Hebrew Bible, Assen 1992, 120, 210.
37 A. Yardeni, The Book of Hebrew Script: History, Palaeography, Script Styles, Calligraphy & Design, Jerusalem 1997, 210; cf. S.A. Birnbaum, The Hebrew Scripts, repr. Leiden 1971, 175. See b. Men. 29b.

epigraphic and manuscript evidence and spurred on to highlight the claims of the square script by dogmatic concerns, the at once realistic and polemic notion that the square script had been *selected* lost out to two alternative, and equally unlikely traditions, which claimed that *Ashurit* had been the one and only script all along. Even the notion of paleographic change was further compromised by the assertion that Ezra even changed the very *language* of Torah. All these later ideas are not only highly implausible but obviously political.

In general, the suppression of cultural borrowing is in keeping with the rabbinic strategy to relate all that carries halakhic importance – such as the script of the Torah – to the Torah and Mt Sinai itself. Historical explanations gave way to a form of reasoning that enshrines the current standard in Mosaic authorship. Even Ezra's role had to be rewritten. Prominent he still is, but from the authority behind a historical innovation he became the pious agent of restoration: Ezra is the one who restores what had been forgotten.

## Conclusion

In sum, Babylon and Babylonia represent the high and low of *biblical* history more than Jewish history, except for the heyday of Babylonian Judaism under Sasanian rule. Even then the reception of Babylonia in the Jewish tradition would be bound up with the Scriptures. However, the almost insular perspective of Scripture did not impair the possibility of using the historical protagonists to voice contemporary concerns. The memory of cruel Babylonians could be profitably used to mouthpiece current interests. We have also seen how Babylonian Jews were so proud of their own heritage that they turned Sasanian Babylonia into a true descendant of historical Israel.[38] Indeed, in Babylonia the rabbis created the true monument of Jewish learning, the Babylonian Talmud, because the relatively peaceful conditions and leeway to conduct their own affairs made this literary creation possible. Yet the enormous influence of the first Babylonian exile was eventually played down, and the cross-cultural influences of the Persians on Judaism – which existed in the script and the sciences – were not to be acknowledged explicitly.

38 See esp. I. Gafni, Land, center and Diaspora: Jewish Constructs in Late Antiquity, Sheffield 1997, 113–17.

*Christoph Markschies*[1]

# Babylon bei den Kirchenvätern

„Babylon bei den Kirchenvätern" – da denkt der halbwegs gebildete Zeitgenosse an die Johannesoffenbarung und den nordafrikanischen Bischof Augustinus und der Fachmann sieht Stapel von Literatur allein zu diesen beiden Komplexen vor seinem geistigen Auge. Beginnen wir unseren Erkundungsgang aber einmal nicht beim imaginierten Babylon der christlichen Antike, sondern bei dem realen Ort. Was wussten antike Christen über jenes reale Babylon?

„Geschichtliche u(nd) landeskundliche Mitteilungen über das alte B(abylon) sind bei kirchl(ichen) Schriftstellern sehr selten anzutreffen; eine längere zusammenfassende Übersicht ist nirgends zu finden"[2] – mit dieser ernüchternden Feststellung leitet Berthold Altaner seinen Überblicksartikel „Babylon Christlich" im renommierten „Reallexikon für Antike und Christentum" ein. Wenn die Stadt bereits im zweiten Jahrhundert, wie wir in der Fragmentenüberlieferung zum römischen Historiker Cassius Dio lesen, dem Kaiser Trajan nur „Schutthalden, Steine und Ruinen" und vereinzelte Sehenswürdigkeiten wie den Sterberaum Alexanders des Großen darbot,[3] dagegen etwas über sechzig Kilometer Luftlinie nördlich von ihr die prosperierende Hauptstadt Seleukia-Ktesiphon am Tigris bewundert werden konnte, verwundert das von Altaner beschriebene Schweigen eigentlich nicht. Bei Strabo heißt es mit dem Zitat eines Komikers: ἐρημία μεγάλη ἐστὶν ἡ μεγάλη πόλις („Doch große Wüstenei ist jetzt die große Stadt"),[4] und bei Theodoret von Cyrrhus wird die Stadt ἔρημος θάλασσα, „wüstes Meer", genannt.[5] Pausanias behauptet, es gebe von Babylon „fast nichts mehr außer der Mauer".[6] Und auch der Kirchenvater Hieronymus berichtet in seinem zu Beginn des fünften Jahrhunderts in Bethlehem verfassten Kom-

1 Praktisch alle Assistenten des Lehrstuhls halfen bei den Textrecherchen, also geht ein herzlicher Dank an Jan Bobbe, Marie-Theres Döhler, Barbara Hanus, Andreas Heiser, Henrik Hildebrandt und Anna Rack-Teuteberg.

2 B. Altaner, Art. „Babylon D. Christlich", Reallexikon für Antike und Christentum I, hg. von Th. Klauser, Stuttgart 1950, (1130–1134) 1130. Vgl. zum Thema auch: T. Boiy, Late Achaemenid and Hellenistic Babylon (Orientalia Lovaniensia Analecta 136), Leuven u.a. 2004, 186–192. Die einschlägigen literarischen, papyrologischen und epigraphischen Quellen sind (etwas summarisch) a.a.O. 49–54 bibliographiert.

3 Cassius Dio, Epitome LXVIII 30 (III, 218,7–10 Boissevain = Tusculum Bücherei V, 219 Veh); Text und Übersetzung auch bei A. Oppenheimer [in collaboration with B. Isaac and M. Lecker], Babylonia Judaica in the Talmudic Period, Tübinger Atlas des Vorderen Orients, Beihefte Reihe B Bd. 47, Wiesbaden 1983, (44–62) Nr. 20 S. 51.

4 Strabo XVI 1,5 (Text und Übersetzung bei Oppenheimer, Babylonia Judaica in the Talmudic Period, Nr. 14 S. 47–49); vgl. auch Zonaras XI 22 (Bibliotheca Teubneriana III, 69,18f. Dindorf) und XII 9 (ebd. 102,9–11).

5 Theodoret, Commentarius in Isaiam 20,2–21,1 (Sources chrétiennes 295, 150,491f. Guinot).

6 Pausanias, I 16,3 (Bibliotheca Teubneriana I, 35,34–37 Rocha Pereira).

mentar zum Propheten Jesaja, er habe von einem persischen Mönch erfahren, dass in den erhaltenen Stadtmauern Babylons wilde Tiere gehalten würden.[7]

Die in der Sekundärliteratur immer wieder begegnende These von einem raschen Niedergang der Stadt ist freilich in den letzten Jahren problematisiert worden; ein wirklich klares Bild fehlt aber bisher: Was in der Kaiserzeit von den Lehmziegelbauten der hellenistischen Stadt, beispielsweise vom griechischen Theater mit der Palaestra, noch vorhanden war und benutzt wurde, wissen wir leider nicht genau.[8] Vielleicht war in Babylon in der Spätantike aber doch noch mehr zu sehen als nur eine verwüstete Stadt, in der zwischen den erhaltenen Mauern wilde Tiere herumstreunten. Dafür spricht beispielsweise die Bemerkung des syrischen Mönchsbischofs Theodoret von Cyrrhus in seinem Jesajakommentar aus dem frühen fünften Jahrhundert, dass in der ruinierten Stadt weder Assyrer noch Chaldäer wohnen würden, sondern vor allem Juden.[9] Das müsste man mit entsprechenden Nachrichten bei Flavius Josephus zusammen lesen, die von einer zeitweiligen Austreibung der Juden im ersten nachchristlichen Jahrhundert berichten; im Bereich der von Babylon rund sechzig Kilometer entfernten Stadt Sippar soll es nach Angaben desselben Autors sogar zu einer mehrjährigen Herrschaft zweier jüdischer Brüder gekommen sein.[10] Dazu treten Kleinfunde wie einige der bekannten Zauberschalen (magic bowls), die in den Ausgrabungen gefunden wurden, oder Inschriften, die allerdings mit wenigen Ausnahmen nicht aus Kaiserzeit oder Spätantike stammen;[11] die sechs Zauberschalen, die Austen

7 Hieronymus, Commentarii in Isaiam V 14 in Jes. 13,20c (Vetus Latina 27, 559,6–560,27 Gryson/Coulie), eine englische Übersetzung bei Oppenheimer, Babylonia Judaica in the Talmudic Period, Nr. 22 S. 51f.

8 R. Koldewey, Das wieder erstehende Babylon, 5. überarb. u. erweiterte Aufl., hg. v. B. Hrouda, München 1990, 290–296; neuere Untersuchungen durch M.S.M. Ali, The Greek Theatre, Sumer 35 (1979), 111–116 und ders., The Southern Place, ebd. 93–110; insgesamt F. Wetzel, Babylon in der Spätzeit, Mitteilungen der Deutschen Orient-Gesellschaft 79 (1942), 3–7; ders., Babylon bei den klassischen Schriftstellern außer Herodot, Mitteilungen der Deutschen Orient-Gesellschaft 82 (1950), 47–53 und A. Mallwitz, C. Das Theater von Babylon, in: ders./E. Schmidt/A. Mallwitz, Das Babylon der Spätzeit, WVDOG 62 = Ausgrabungen der Deutschen Orient-Gesellschaft in Babylon 8 (1957), 3–22, bes. 19–22, (da der Versuch einer Datierung) sowie H. Klengel, Babylon zur Zeit der Perser, Griechen und Parther (=Forschungen und Berichte. Staatliche Museen zu Berlin 5), Berlin 1962, 40–53 (a.a.O. 50: „Im Wohnviertel verödeten und zerfielen [sc. während der Seleukidenzeit, C.M.] eine Reihe von Häusern und wurden teilweise mit Gräbern belegt"; a.a.O. 52: „Babylon nahm [sc. in parthischer Zeit] mehr und mehr einen dörfischen [sic] Charakter an") und Boiy, Late Achaemenid and Hellenistic Babylon, 190f.: „Since there are clear indications of habitation in Babylon (...), there is no doubt that Babylon was still inhabited" (a.a.O. 190).

9 Theodoret, Commentarius in Isaiam 13,20f. (Sources chrétiennes 295, 78,158f.); vgl. auch S. Shaked, Popular Religion in Sasanian Babylonia, Jerusalem Studies in Arabic and Islam 21 (1997), 103–117.

10 Flavius Josephus, Antiquitates Judaicae XVIII 310 (196,17–20 Niese) und 371–379 (207,19–209,17) zur schweren Lage der Juden und ebd. XVIII 314–371 (197,9–207,372) zu den Brüdern Anilaios und Asinaios; ausführlich A.M. Schwemer, Studien zu den frühjüdischen Prophetenlegenden Vitae prophetarum, Tübingen 1995 (= Texts and Studies in Ancient Judaism 1), 254f. sowie Oppenheimer, Babylonia Judaica in the Talmud Period, Wiesbaden 1983, 219, 285f., 360; vgl. auch A. Oppenheimer, Babylonian Synagogues with Historical Associations, in: Ancient Synagogues, hg. v. D. Urman/P.V.M. Flesher, Vol. 1 (= Studia post biblica 47/1), Leiden u.a. 1995, (40–48) 47f. = ders., Between Rome and Babylon. Studies in Jewish Leadership and Society (= Texts and Studies in Ancient Judaism 108), Tübingen 2005, (394–401) 400f. und ders., Babylonia Judaica in the Talmudic Period, 44–62, insbes. 54f.

11 Oppenheimer, Babylonia Judaica in the Talmudic Period, 57 (Nr. 9) zitiert eine Inschrift, die die Reparatur des Theaters für die Zeit 130–200 n.Chr. belegen könnte; vgl. auch Wetzel/Schmidt/Mallwitz, Das Babylon der Spät-

Henry Layard (1817–1894) 1853 publizierte und die seiner Aussage nach auf dem „Amran genannten Hügel" gefunden wurden, sind – wie solche Zauberschalen allgemein – nur schwer einer bestimmten Nutzergruppe zuzuordnen.[12] Gleiches gilt für die im Besitz der Berliner Staatlichen Museen befindlichen einschlägigen Ausgrabungsfunde, darunter drei Zauberschalen, eine in manichäisch-syrischer Schrift, eine zweite in der syrischen Estrangelo und eine dritte in mandäischer Schrift. Eine dieser Schalen stammt aus dem zwanzig Kilometer südwestlich von Babylon gelegenen Borsippa, die beiden anderen sind Geschenke mit unbekannter Provenienz.[13] Dazu kommen fünf Schalen mit jüdisch-aramäischen Aufschriften, gleichfalls unbekannter Provenienz.[14] Aufgrund heutiger Erkennt-

zeit, 20f. und insbesondere E. Schmidt, a.a.O. 49f. (Zeichnung, Publikation, Übersetzung und Datierung der Inschrift). – Bei Wetzel/Schmidt/Mallwitz ist die Inschrift W. Dittenberger, Orientis Graeci Inscriptiones Selectae 254 (Vol. 1, S. 414f. = U. Köhler, Zwei Inschriften aus der Zeit Antiochos' IV. Epiphanes, Sitzungsberichte der Preussischen Akademie der Wissenschaften zu Berlin – Philosophisch-Historische Klasse [SPAW.PH] 51 [1900], [1100–1108] 1107f.) im Berliner Vorderasiatischen Museum (= Oppenheimer, Babylonia Judaica in the Talmudic Period, 56f. (Nr. 5), nicht auf das zweite vorchristliche, sondern wegen „der Form der Buchstaben" auf das zweite Jahrhundert *nach* Christi Geburt datiert; wegen des Inhalts der Inschrift, einer Dedikation für Demokrates, den Sohn des Byttakos, Kommandanten der Garnison der Stadt (sc. Antiochia), scheint mir die Spätdatierung deutlich weniger wahrscheinlich als die gewöhnliche frühere Datierung (so auch Supplementum Epigraphicum Graecum VII, 39 und VII, 40): Der seltene Name ist nämlich bei Polybius (V 82,10 [Bibliotheca Teubneriana II, 206,11 Büttner/Wobst]) in seleukidischen Kontexten belegt (Köhler, a.a.O. 1107). Die Inschrift ist inzwischen auch publiziert in IK 65, Iscrizioni dello estremo oriente greco, Bonn 2004, Nr. 102 (mit Photo). Eine ähnliche Titulatur findet sich auch in einer Inschrift aus Ninive (ebd. Nr. 64), die sich leider auch nicht sicher datieren lässt. Mein Kollege Klaus Hallof von den *Inscriptiones Graecae* schreibt mir (20. 1. 2010): „Die Autorität von M. Holleaux (Études III 218), un monument de la domination parthe' hat sicher die Spätdatierung verursacht; mir scheint das 1. Jh. (oder Ende 2. Jh.) v. Chr. unbedingt richtig. Siehe noch Supplementum Epigraphicum Graecum 32,1609".

12 A.H. Layard, Discoveries in the Ruins of Niniveh and Babylon with Travels in Armenia, Kurdistan and the Desert. Being the Result of a Second Expedition, Undertaken for the Trustees of the British Museum, London 1853, 509–526 (hebräische bzw. syrische Transliterationen, englische Übersetzungen und Abbildungen auf S. 512–523). Die Funde von Layard werden im Britischen Museum aufbewahrt: M.A. Levy, Über die von Layard aufgefundenen chaldäischen Inschriften auf Topfgefäßen. Ein Beitrag zur hebräischen Paläographie und zur Religionsgeschichte, Zeitschrift der Deutschen Morgenländischen Gesellschaft 9 (1855), 465–491 sowie J. Naveh/Sh. Shaked, Amulets and Magic Bowls. Aramaic Incantations of Late Antiquity, Jerusalem [3]1998, 19.

13 K. Kessler, Das wahre Ende Babylons – die Tradition der Aramäer, Mandäer, Juden und Manichäer, in: Babylon. Mythos & Wahrheit, Bd. 2 Babylon. Wahrheit. Eine Ausstellung des Vorderasiatischen Museums. Staatliche Museen Berlin mit Unterstützung der Staatsbibliothek zu Berlin, hg. v. J. Marzahn u. G. Schauerte, München 2008, (467–486) 486 (Katalognummern 520–522, die letzten beiden Nummern Geschenke von Theodor Wiegand und James Simon), vgl. auch M. Lidzbarski, Ephemeris für Semitische Epigraphik, Bd. 1, Gießen 1902, 90–95, (Katalog Nr. 522 mit mandäischer Schrift); ders., Die Herkunft der manichäischen Schrift, SPAW.PH 1916, Berlin 1916, (1213–1222) 1213f. (Katalog Nr. 521 mit Estrangelo-Schrift); M. Moriggi, La lingua delle coppe magiche siriache (= Quaderni di semitistica 21), Florenz 2004, 263f., nr. 22 (Katalog Nr. 521); Ch. Müller-Kessler, Syrische Zauberschalen – Korrekturen und Nachträge, Welt des Orients 36 (2006), (116–130) 124–126, (Katalog Nr. 521) sowie L. Jakob-Rost, Borsippa, Forschungen und Berichte 27 (1989), (65–89) 73f. mit Abb. 17, p. 74 (Katalog Nr. 520).

14 D. Levene, Jüdisch-aramäische Zauberschalen aus Mesopotamien, in: Babylon. Wahrheit, 549–552 (Katalog Nr. 548–553). Der Oberkustos des Vorderasiatischen Museums, Dr. Joachim Marzahn, schreibt mir (22. 2. 2010), dass leider sämtliche Erwerbungsakten für die von James Simon geschenkten Katalognummern 548–552 „unter Verlust" seien, aber wahrscheinlich während der Vorexpedition 1887 erworben wurden. Im Inventar findet sich zu Nr. 548 (= VA 2180) die Angabe „Schêschfar". Einen solchen Ort kann Marzahn aber nicht lokalisieren. Zu den

nisse ist nahezu sicher, dass diese Zauberschalen erst aus der Spätantike (4.–7. Jh.) stammen können und im Fußboden von Häusern zur Abwehr von Attacken unterirdischer Dämonen vergraben waren. Man darf also vermuten, dass mindestens die Londoner Stücke als Spuren spätantiker Besiedlung der Ortslage Babylon interpretiert werden können, weniger als Hinweise auf schriftgelehrte Personen am Ort, die solche Schalen herstellen konnten – nichts spricht nämlich dagegen, daß man solche Schalen anderswo kaufte. Babylon war also in der Kaiserzeit und in der Spätantike eher kein unbewohnter, verwüsteter Ruinenhügel, sondern eine mehr oder weniger große Siedlung. Die Bemerkung des Hieronymus darf also offenkundig weniger als eine Nachricht über das historische Babylon genommen werden, sondern vielmehr als eine Bemerkung eines klugen Exegeten, der die eschatologische Verwüstung und Unbewohnbarkeit des Ortes, wovon die Bibel berichtet (Jes 13,20), schon in der Gegenwart realisiert sieht.

Mir geht es in meinem Beitrag weniger um die berühmten theologischen Konstrukte von der *civitas, quae mystice Babylonia dicitur*, wie es beispielsweise bei dem spätantiken Kirchenvater Augustinus heißt,[15] sondern um die Stadt, die im historischen Sinne Babel oder Babylon genannt wird, und die Frage, welche Kenntnisse davon spätantike christliche Theologen wirklich besaßen. Gibt es im christianisierten *Imperium*, anders als Altaner meint, nicht doch interessante geschichtliche und landeskundliche Mitteilungen über das alte Babylon? Sicher ist jedenfalls, daß der historische Ort Babylon in die christliche Erinnerungskultur der nachkonstantinischen Zeit einbezogen wurde: In einer christianisierten Fassung der frühjüdischen „Prophetenviten" aus dem sechsten Jahrhundert heißt es, dass „der heilige Daniel", also der gleichnamige Prophet, in Babylon starb und in der dortigen Grabhöhle der persischen Könige beigesetzt wurde. Und weiter wörtlich: „Und sein Grab ist bis auf den heutigen Tag allen in Babylon bekannt".[16] Die Aussage bezieht sich zwar auf das historische Babylon, gibt aber gleichzeitig für unser historisches Bild des Ortes in der Spätantike kaum etwas her: Joachim Jeremias hat schon vor fünfzig Jahren darauf hingewiesen, dass eine große Zahl von Autoren und darunter beispielsweise der Pilger Theodosius im nämlichen sechsten Jahrhundert eben dieses Grab nicht in Babylon, sondern in Susa lokalisierten; die geographischen Kenntnisse von der Region waren offenbar bei vielen christlichen Autoren nicht sehr präzise und beruhten weder im ersten Fall der Prophetenleben noch im zweiten Fall des möglicherweise aus Nordafrika stammenden Pilgers Theodosius auf Autopsie. Das wird beispielsweise klar, wenn der genannte Theodosius

Schalen vgl. J. Wohlstein, Ueber einige aramäische Inschriften auf Thongefässen des Königlichen Museums zu Berlin [II], Zeitschrift für Assyriologie 9 (1894), (11–41) 27–30 (Nr. 2426 = 550).

15 Augustinus, Contra Epistulam Parmeniani 2,9 (Corpus scriptorum ecclesiasticorum latinorum 51, 54,23 Petschenig); vgl. E. Lamirande, Art. „Babylon(ia)", Augustinus-Lexikon I, Basel 1986–1994, 566–569.

16 Ἐκεῖ οὖν ἐκοιμήθη ὁ ὅσιος Δανιὴλ ἐν Βαβυλῶνι ἐν εἰρήνῃ καὶ ἐτάφη ἐν τῷ σπηλαίῳ τῷ βασιλικῷ ἐνδόξος· καὶ ἔστιν ὁ τάφος αὐτοῦ ἕως τῆς σήμερον ἡμέρας πᾶσι γνωστὸς ἐν Βαβυλῶνι. Nach der ersten Epiphanius-Rezension im Paris. Graec. 1115, zitiert nach Schwemer, Synopse zu den *Vitae Prophetarum* = dies., Studien zu den frühjüdischen Prophetenlegenden, Bd. 2, Texte und Studien zum antiken Judentum 50, Tübingen 1996, 31*.

Susa *XXX milia de Babylonia* lokalisiert, obwohl es in Wahrheit rund 400 Kilometer von Babylon entfernt liegt.[17]

Auch auf den zweiten Blick scheint die Diagnose Altaners also zuzutreffen: Man scheint im christianisierten *Imperium* nichts Genaueres über die Stadt am Euphrat gewusst zu haben. So findet sich beispielsweise bei dem spätantiken Grammatiker Stephan von Byzanz in einem knappen Lexikoneintrag die Information, dass die Stadt auch Seleukia genannt werde[18] – mithin wurden in einem wichtigen geographischen Handbuch aus dem sechsten Jahrhundert Babylon und Seleukia-Ktesiphon einfach verwechselt. Aber dieser in der Sekundärliteratur verbreitete Eindruck trügt: Macarius (von Magnesia?) verwendet im fünften Jahrhundert in seiner apologetischen Argumentation gegen den neuplatonischen Philosophen Porphyrius präzisere Informationen: „Die so genannte Stadt Babylon war die Hauptstadt der Assyrer, sie hatte einen Umfang von achthundertfünfzig Stadien, die Breite der Mauer dehnte sich auf beinahe fünfzig Ellen aus und in der Höhe erhob sie sich auf zweihundertfünfzig Ellen, unbesiegbar und unauflöslich".[19] Die Zahlen bei Macarius weichen von denen anderer Autoren teils erheblich ab; Herodot spricht von einem Stadtumfang von vierhundertfünfzig Stadien, nach seinen Angaben ist die Mauer ebenfalls fünfzig Ellen breit, aber nur zweihundert Ellen hoch, spätere Autoren steigern die Zahlen ganz gewaltig;[20] der kaiserzeitliche christliche Autor Macarius, möglicherweise Bischof von Magnesia am Mäander, bemüht sich immerhin, nicht allzu phantastischen Angaben zu folgen, sondern wählt ebenso maßvolle wie präzise Größen der Stadt, die er ansatzweise beschreibt. Zu etwa derselben Zeit verfasste Prokop von Gaza seine bekannte Ekphrasis einer besonders bemerkenswerten Uhr in seiner Heimatstadt und verglich die Uhr mit anderen berühmten Bauwerken, darunter mit dem Tempel des Gottes Bel und den achtstöckigen Türmen der Stadt Babylon;[21] auch bei ihm wird vergleichsweise präzise vom historischen Babel gesprochen, von den acht Tortürmen ist, wenn ich recht sehe, schon in keilschriftlichen Texten die Rede. Eusebius von Caesarea, Enkelschüler des Origenes und Metropolit der Kirche von Palaestina, bietet ein feines Beispiel für die Kenntnisse eines hochgelehrten christlichen Theologen an der Wende vom dritten ins vierte Jahrhundert: Er zitiert in seiner apologetischen Sammelschrift *Praeparatio Evangelica* beispielsweise einen bei Alexander Polyhistor bewahrten samaritanischen Anonymus, den er fälschlich mit dem jüdisch-

17 Theodosius, p. 149,15f. Geyer; vgl. auch J. Jeremias, Heiligengräber in Jesu Umwelt (Mt. 23,29; Lk. 11,47). Eine Untersuchung zur Volksreligion der Zeit Jesu, Göttingen 1958, 113f.

18 Βαβυλών, Περσικὴ πόλις, μητρόπολις, Σελεύκεια καλουμένη (Ethnika 154,9f. Meineke).

19 Macarius, Apocriticus IV 11 (170 Blondel).

20 Eine Übersicht bei K. Brodersen, Die sieben Weltwunder. Legendäre Kunst- und Bauwerke der Antike (= Beck'sche Reihe 2029), München $^{6}$2004, 35–57.

21 Text: H. Diels, Über die von Prokop beschriebene Kunstuhr von Gaza. Mit einem Anhang enthaltend Text und Übersetzung der ΕΚΦΡΑΣΙΣ ΩΡΟΛΟΓΙΟΥ des Prokopius von Gaza, Abhandlungen der Preußischen Akademie der Wissenschaften – Philosophisch-Historische Klasse (APAW.PH) 26/7, Berlin 1917, (27–39) 28: „ein anderer mag bei seinem Aufenthalt in Babylon den Tempel des Bal Zeus und einen Turm über den anderen acht Stockwerk hoch geschichtet in Augenschein nehmen".

hellenistischen Historiker Eupolemus identifiziert. Der Anonymus trägt Theorien über die Gründung der Stadt Babylon durch Riesen vor, die den dortigen Turm gebaut hätten, und entmythologisiert so klassische babylonische Mythen.[22]

Alles in allem genommen, gibt es zwar – anders als Altaner meinte – geschichtliche und landeskundliche Mitteilungen über das alte Babylon bei kirchlichen Schriftstellern, aber eben doch nur mit relativ geringer Tiefenschärfe. Entsprechend haben wir eigentlich auch keine wirklich authentischen Nachrichten über die Christianisierung der Stadt Babylon. Der spätantike Bischof Theodoret von Cyrrhus schreibt im fünften Jahrhundert in einer seiner Predigten auf Johannes Chrysostomus (Clavis Patrum Graecorum III, 6225 = Bibliotheca hagiographica graeca 878t-x), Babylon sei nicht fern vom wahren Glauben.[23] Ob er damit wirklich die Stadt meint und nicht eine Landschaft, bleibt sehr unsicher; präzise Aussagen über den Grad der Christianisierung der Stadt sollte man diesem Satz nicht entnehmen. Die frühen Konzilsakten der persischen Kirche erwähnen Babylon nicht,[24] es gehörte seit 410 zur Metropolie Babylonien (Beth Armaje) mit dem Sitz in Seleukia-Ktesiphon.[25] In späten Aposteltraditionen wird dem Apostel Thaddäus die Christianisierung der Stadt Babylon zugewiesen.[26] Besonders interessant ist das kaum bekannte lateinische Martyrium der Apostel Simon Kananites und Judas Thaddäus (*Passio Simonis et Iudae* (*Thaddaei*), Bibliotheca hagiographica latina antiquae et medii aetatis [BHL] 7749–7751 = Clavis apocryphorum Novi Testamenti [CANT] 284) aus der sogenannten Abdias-Sammlung, deren lateinische Fassung letztmalig von Fabricius im Jahre 1703 kritisch herausgegeben wurde und letztmalig von Lipsius 1883/1884 ausführlicher analysiert wurde. Simon und Judas werden als jüdisches Brüderpaar aus Kana in Galiläa vorgestellt (*Si genus quaeris, Hebraei sumus; si conditionem, servi sumus Iesu Christi*)[27], die in Persien in allerlei Abenteuer verwickelt werden und gleich zu Beginn ihrer Tätigkeit in dieser *regio* mit zwei Magiern in einen regelrechten Zauberwettkampf geraten. Schließlich werden sie vom persischen König und einem seiner Generäle namens Varardach gebeten, in Babylon zu bleiben, „und sie

22 Eusebius, Praeparatio Evangelica IX 17,2–9 (Die griechischen christlichen Schriftsteller der ersten drei Jahrhunderte [GCS] Eusebius VIII/1, 502,17–504,9 Mras/Des Places), insbesondere 17,2 (502,19–503,1); deutsche Übersetzung mit Einleitung und Literaturhinweisen von N. Walter, Fragmente jüdisch-hellenistischer Historiker (= Jüdische Schriften aus hellenistisch-römischer Zeit I/2), Gütersloh 1976, 137–140. 141–143; ausführlich zu Autor und Text M. Hengel, Judentum und Hellenismus. Studien zu ihrer Begegnung unter besonderer Berücksichtigung Palästinas bis zur Mitte des 2. Jh.s v. Chr. (= Wissenschaftliche Untersuchungen zum Neuen Testament 10), Tübingen ³1988, 162–169.

23 Theodoret bei Photios, Bibliotheca cod. 273 508b (Collection des universités de France VIII, 109,29–31): Οὐκέτι τῆς εὐσεβοῦς θρησκείας τὰ Βαβυλῶνος ἀλλότρια.

24 E. Sachau, Vom Christentum in der Persis, SPAW.PH 39, 1916, Berlin 1916, 12–23.

25 E. Sachau, Zur Ausbreitung des Christentums in Asien, APAW.PH 1, 1919, Berlin 1919, 13f.

26 Cosmas Indicopleustes, Topographia Christiana II 77 (Sources chrétiennes 141, 393,1–5 Wolska-Conus).

27 Act. Sim. et Iud. 8 (Codex Apocryphus, 611 Fabricius = Codex Apocryphus, Abdiae Historia Apostolicae, 377,20f. Giles). Siehe auch: Passio Simonis et Judae/Passion de Simon et Jude, in: Écrits apocryphes chrétiens II, Édition publiée sous la direction de P. Geoltrain et J.-D. Kaestli, Paris 2005, 843–864, dort in französischer Übersetzung von D. Alibert u. a., hier S. 845 (Kap. 5).

verrichteten tagtäglich große Wunder: Sie machten Blinde sehend, Tauben gaben sie das Gehör wieder, Lahmen den Gang, machten Aussätzige rein, trieben auch aus besessenen Körpern jede Art von bösen Geistern".[28] Ob diese frühmittelalterlichen Erzählungen Lokalkolorit enthalten und insofern auf eine ursprünglich persische Tradition zurückgehen, wäre höchst interessant und müsste nach Vorarbeiten Gutschmids im vorletzten Jahrhundert einmal ausführlicher untersucht werden – allein die Tatsache, dass die Gegner der beiden Apostel einen schroffen, möglicherweise aus dem Manichäismus geschöpften Dualismus vertreten, spricht für Lokalkolorit.[29] Für unsere Fragestellung ist dagegen besonders interessant, dass der Text beschreibt, dass die beiden Apostel am Ende ihres Aufenthalts in Babylon einen gewissen Abdias als ersten Bischof von Babylon einsetzen (*Ordinavere autem Apostoli in civitate Babylonis episcoporum nomine Abdiam*)[30] und erst dann, als genügend Schüler geworben und Kirchen gebaut sind, durch die persischen Provinzen ziehen. Weitere für Babylon spezifische Traditionen finden sich in dem spannenden Text nicht, allerdings werden die Einwohner der Stadt implizit auch ganz offensichtlich als bekehrungswillig und kooperationsbereit vorgestellt, ein deutlicher Kontrast zu dem allgemeinen Babelbild vieler Kirchenväter (was für Lokalkolorit einer Grundschicht der *Passio* sprechen könnte).

Solchen eher – ich betone: eher – am historischen Babylon orientierten Passagen steht eine übergroße Fülle von antiken christlichen Texten gegenüber, in denen Babylon eine Chiffre für ungeheuere Schlechtigkeit ist – die Christen belegten ebenso wie die Juden das Schlechte mit der Chiffre Babylon. Das gilt nicht nur, wie von der Johannesapokalypse zu erwarten, für den römischen Staat, der das Christentum offiziell nicht zu tolerieren bereit ist, es gilt sogar für vergleichsweise alltagssprachliche Zusammenhänge: Den Ausdruck Βαβυλώνια πράγματα verwendet Palladius, der Biograph des Johannes Chrysostomus, um ungeheuere Schlechtigkeit zu bezeichnen; Βαβυλώνια κακά bedeutet bei Gregor von Nyssa „ungeheueres Leid".[31] Der Gotteszorn geht nach den sibyllinischen Orakeln von „des assyrischen Babylons hohen Mauern" aus.[32] Eine Fülle von Belegen für die Verwendung des Ortsnamens Babylon als Metapher in diesem Sinne bietet natürlich der nordafrikani-

28 Act. Sim. et Iud. 18 (Codex Apocryphus, 625 Fabricius = Codex Apocryphus, 387,19–23 Giles; Geoltrain/ Kaestli, S. 856 [Kap. 23]).

29 A.v. Gutschmid, Die Königsnamen in den apokryphen Apostelgeschichten, Revue d'histoire des missions 19 (1864), (161–183. 380–401) 380–385; so auch R.A. Lipsius, Die apokryphen Apostelgeschichten und Apostellegenden. Ein Beitrag zur altchristlichen Literaturgeschichte Bd. I, Braunschweig 1883, 176f.; Bd. II, 1884, 164–172; deutlich kritischer W. Speyer, Angebliche Übersetzungen des heidnischen und christlichen Altertums, in: ders., Frühes Christentum im antiken Strahlungsfeld, 81.

30 Act. Sim. et Iud. 20 (Codex Apocryphus, 628 Fabricius = Codex Apocryphus, 389,22f. Giles; Geoltrain/ Kaestli, S. 858 [Kap. 28]).

31 Palladius, Dialogus de vita Ioannis Chrysostomi 4 (Sources chrétiennes 341, 86,15 Malingrey/Leclercq); Gregor von Nyssa, Epistulae 19,17 (Gregorii Nysseni Opera VII/2, 67,7 Pasquali); weitere Belege bei Altaner, Art. „Babylon", Reallexikon für Antike und Christentum I, 1131.

32 Oracula Sibyllina III 809 (Die griechischen christlichen Schriftsteller der ersten drei Jahrhunderte 8, 89,809 Geffcken).

sche Kirchenvater Augustinus, obwohl sie ihm vermutlich durch die gnostische Gruppe der Manichäer vermittelt wurde[33] und schon beim ersten christlichen Universalgelehrten, Origenes, breit entfaltet ist, der Babylon als Stadt des Teufels und Sitz der gottfeindlichen Macht interpretiert. Auch der Prediger Ambrosius von Mailand nennt Babylon eine Region der Schlechtigkeit, freilich ist es wichtiger, babylonische Sitten zu meiden als schon aus dieser Region der Schlechtigkeit zu fliehen – dies ist offenkundig, solange der Mensch lebt, nicht möglich.[34] Augustinus sieht in den *Confessiones*, den Bekenntnissen, seine eigene Vergangenheit im Rahmen dieses Sprachspiels: „Welche Gesellschaft, in der ich mich auf den Straßen Babylons herumtrieb, in seinem Schmutz mich wälzte, als wäre er Balsam und köstliche Narde!". Seine Mutter ist allerdings schon „aus dem Inneren Babylons" geflohen;[35] er folgt ihr erst nach seiner Mailänder Bekehrung. Freilich ist Augustinus nicht nur in solchen vergleichsweise konventionellen Bahnen geblieben, sondern hat einen eigenständigen Entwurf einer christlichen Geschichtstheologie vorgelegt, in der Babylon eine zentrale Rolle spielt. Die umfangreichen Passagen, die Augustinus in *De civitate Dei* über die *civitas terrena seu Babylonica* schreibt, möchte ich hier nicht ausführlicher behandeln, zumal Augustinus das Werk selbst ein *opus magnum et arduum* nennt. Nur so viel: Babel steht beim nordafrikanischen Bischof wie für die ganze etymologisch orientierte antike christliche Exegese in den Bahnen von Origenes und seinem Enkelschüler Eusebius für „Verwirrung". Aber die besondere Pointe der Theologie des Augustinus ist, dass er die Vorstellung von zwei *civitates* von der irdischen Geschichte weitgehend ablöst und so in gewisser Weise befreit (Therese Fuhrer)[36], weil er die *civitates* ausschließlich von ihrer Stellung zum lebendigen Christus und seiner Lebensordnung her definiert. Zwei miteinander verbundene Gruppen von Bürgern (*civitas* steht gelegentlich synonym mit *societas*)[37] stehen einander gegenüber wie Licht und Finsternis. Wenn Augustinus die metaphernreiche biblische Sprache verwendet, werden diese beiden Gesellschaften als Städte repräsentiert, eben als Jerusalem und Babylon, er spricht aber auch an anderen Stellen ganz unumwunden von „zwei Arten menschlicher Gesellschaft"[38]. Rom ist „gleichsam das zweite, abendländische Babylon geworden"[39], aber als *Imperium* ebensowenig mit der *civitas diaboli* einfach identisch, wie Kirche und *civitas Dei* identisch sind. Man muss sich beim Verständnis

33 J. van Oort, Jerusalem and Babylon. A Study of Augustine's *City of God* and the Sources of his Doctrine of the Two Cities (= Supplements to Vigiliae Christianae 14), Leiden u.a. 1991; ders., *Civitas dei – terrena civitas:* The Concept of the Two Antithetical Cities and Its Sources (Books XI–XIV), in: Ch. Horn (Hg.), Augustinus, De civitate dei, Klassiker Auslegen 11, Berlin 1997, (157–171) 164f.

34 Ambrosius, De Isaac et Anima 6,54 (Fontes Christiani 48, 114,15–24 Dassmann).

35 Augustinus, Confessiones II 3,8 (Corpus Christianorum. Series Latina 27, 21,52–56 Verheijen).

36 Th. Fuhrer, Augustinus, Klassische Philologie kompakt, Darmstadt 2004, 138 sowie J. van Oort, De civitate dei (Über die Gottesstadt), in: V.H. Drecoll (Hg.), Augustin Handbuch, Tübingen 2007, (347–363) 362f.

37 Augustinus, De civitate dei XV 8 (Corpus Christianorum. Series Latina 48, 463,17f. Dombart/Kalb) und XIV 1 (ebd. 414,15–21); dazu van Oort, Civitas dei – terrena civitas, 161.

38 Augustinus, De civitate dei XIV 1 (Corpus Christianorum. Series Latina 48, 414,15–21).

39 Augustinus, De civitate dei XVI 17 (Corpus Christianorum. Series Latina 48, 522,33f.).

des Werkes immer deutlich machen, dass Augustinus seine zweiundzwanzig Bücher über den Gottesstaat, *De civitate Dei*, sukzessive von 413 bis 426 schrieb und versuchte, den Fall Roms von 410 theologisch zu bewältigen: Im Hochsommer des Jahres 410, Augustinus ging auf die Sechzig zu und amtierte seit etlichen Jahren als Bischof in der nordafrikanischen Kleinstadt Hippo Regius, war ein westgotisches Heer unter Alarich mit anderen Germanen, Hunnen und entlaufenen römischen Sklaven im Gefolge in die Stadt Rom selbst eingedrungen. In den Bahnen des Augustinus denken seitdem die allermeisten lateinischen Autoren: Von der *diaboli civitas, quae Babylonia dicitur*[40], spricht beispielsweise im sechsten Jahrhundert Cassiodor; er nennt Babylon *mater fornicationum totius que terrae pollutio*.[41]

Unser Überblick unter dem Thema „Babylon bei den Kirchenvätern“ führt auf einen merkwürdig disparaten Befund: Da sind die paar Bezüge auf den historischen Ort, hinter der höchst unsicheren Reichsgrenze gelegen; sie sind wohl notwendigerweise verschwommen. Da ist die große Geschichtstheorie des Augustinus, die die Tendenz zur Enthistorisierung des verschwundenen Ortes zur Vollendung bringt und doch mit dem klassischen Modell einer geschichtlich greifbaren Macht operierte – auf diese Weise natürlich die Verschwommenheit noch zusätzlich förderte. Alles das ist aber im Grunde doch kein Ausnahmefall, trotz der Besonderheiten Babylons. Es entspricht nämlich haargenau einer Tendenz zur Enthistorisierung und theologischen Überhöhung, die sich auch im Blick auf die Stadt Jerusalem feststellen lässt – also einen ungleich bedeutsameren Ort der antiken Christentumsgeschichte, der dann doch genauso behandelt wird wie Babylon.[42] Eine solche Beobachtung sollte allerdings davor warnen, die Befunde bei unserem Thema für allzu besonders zu halten.

40 Cassiodor, Expositio psalmorum 136 (Corpus Christianorum. Series Latina 98, 1231,34 Adriaen).

41 Cassiodor, Complexiones in epistolis apostolorum et actibus apostolorum et apocalypsi XXIV 17,1 (Corpus Christianorum. Series Latina 107, 124,8 Gryson = Patrologia Latina 70, 1413).

42 Details bei Ch. Markschies, Himmlisches und irdisches Jerusalem im antiken Christentum, in: M. Hengel, S. Mittmann u. A. M. Schwemer (Hgg.), La Cité de Dieu/Die Stadt Gottes. 3. Symposium Strasbourg, Tübingen, Uppsala 19.–23. September 1998 (= Wissenschaftliche Untersuchungen zum Neuen Testament 129), Tübingen 2000, 303–350.

*Jean-Jacques Glassner*

## Noch einmal: Der Turm zu Babel und die vielen Sprachen

Im Laufe der zweiten Hälfte des dritten Jahrhunderts vor Christus wollte Attalus I., König von Pergamon, die Galater, jene Gallier, die in Kleinasien siedelten, angreifen. Vor Eröffnung der Kriegshandlungen fragte er einen wohlbekannten, in Mathematik, Astronomie und Opferschau bewanderten babylonischen Wahrsager um Rat. Es galt, die Zustimmung der Götter zu erhalten. Dieser Wahrsager hieß Sudines, eine hellenisierte Form eines babylonischen Namens der auf *-iddin* endete. War diese Begegnung zwischen Kelten und Babyloniern nur ein Zufall? Allem Anschein nach nicht! Als man während der Renaissance in Europa die Gottheiten der keltischen Panthea wiederentdeckte, brachten die Philologen, Archäologen und Historiker einige sonderbare und unvermutete Konstrukte über das Thema des Turms zu Babel und die babylonische Sprachverwirrung zwischen Kelten und Babyloniern-Assyrern hervor.

Wir wollen den Blick auf Europa im 16. und 17. Jahrhundert richten. Es ist die Zeit, in der Bilder der antiken Gottheiten der keltischen Panthea wieder aufgefunden wurden. Sie tragen Namen wie Belenus, Teutates, Nehalennia, Abellio, Esus oder Taranis. Ihre Entdeckung war manchmal die Frucht eines Zufalls: am 5. Januar 1647 wurden nach einem Gewitter mehrere Stelen, die der Göttin Nehalennia gewidmet waren, in Zeeland am Ufer des Sees gefunden.

In dieser Zeit ließ die geringe Kenntnis der keltischen Religion und Mythologie der Phantasie der Gelehrten breiten Raum; es waren Philologen, Grammatikspezialisten, Archäologen oder Sprachforscher, die hier im Sinne ihrer Disziplin spekulierten. Auf dem Gebiet der Etymologie (das heißt der sprachlichen Herleitung) der Götternamen hat diese Imagination allerhand Wunderwerke geboren. Derlei etymologischen Abstammungen sollten die keltischen Gottheiten an Ur-Gestalten anschließen und auf diese Weise die Kelten in dem langen Zeitlauf der Geschichte verankern.

Im Folgenden seien einige Beispiele solcher Etymologien angeführt:

Vor allen anderen wurden Herleitungen aus dem Hebräischen bevorzugt. Hebräisch galt als die heilige Sprache, die Gott den ersten Menschen, Adam, erkennen ließ. Es war – so glaubte man – die einzige Sprache der Menschheit vor dem Bau des Turmes zu Babel.

Nach Louis Thomassin (1690) können alle Wörter, die mit der Religion der Kelten zu tun haben, von einem einzigem hebräischen Wort abgeleitet werden, dem Wort *halal*, das er mit „glänzend" übersetzt.[1]

1 Louis Thomassin, La méthode d'étudier et d'enseigner chrestiennement et utilement la grammaire ou les langues par rapport à l'écriture sainte en les réduisant toutes à l'hébreu, Paris 1690.

Mit Bezug auf die Göttin Nehalennia, Schutzgöttin der Händler und der Schiffer (die Herleitung ihres Namens ist bis heute unbekannt, offenbar weder keltisch noch germanisch), machte Thomas Reinesius (1682) einen Vorschlag, dem alle Gelehrten sofort folgten: Der Name sei von einem passiven Partizip eines hebräischen Verbums abgeleitet, ein Verbum das „sich hinlegen" bedeutete und von welchem sich das Wort „Fluss" ableite: nämlich *nahar* in Hebräisch oder *nehar* in chaldäisch (in dieser Zeit galt Chaldäisch als eine Variante des Hebräischen, eine Variante die man zum Beispiel im Buch des Esdras finden könne). Um diese Ableitung zu erreichen, war eine linguistisch mögliche Lautverschiebung der Konsonanten R und L vorauszusetzen: man gelangte also von *nahar* oder *nehar* zu *nahal* oder *nehal*.[2]

Was Belenus betrifft – nach heutiger Kenntnis wahrscheinlich soviel wie „der eine, der glänzt" in keltischer Sprache – hierfür hat derselbe Reinesius in seinem Namen die hebräische Wurzel *bel* „König, Herr" erkannt und um die Nachsilbe *-enus* erweitert.

Auch Ägypten war allgegenwärtig. Einstimmig erkannten die Gelehrten in dem ägyptischen Gottesnamen Toth die ursprüngliche Form des keltischen Namens Teutates, eine Abwandlung des römischen Gottes Merkur. Es galt als ausgemacht, dass die keltische Religion unter dem doppelten Zeichen von Rom und von Ägypten begründet worden sei. Wir wissen heute, dass Teutates eine Ableitung mit dem Suffix -ti vom keltischen teuta ist und soviel bedeutet wie, „der eine des Volkstammes".

Auch phoenikische und punische Etymologien boten sich an. Wiederum war Reinesius ein eifriger Verfertiger solcher Etymologien. Samuel Bochart (1646), der über eine gute Kenntnis des Keltischen verfügte, wusste, dass der Gottesname Taranis („der mit dem Donner zu tun hat") eine Ableitung von dem keltischen Wort *taran* („Blitz") war; dennoch versuchte er eine phoenikische Ethymologie herauszubilden und schuf somit einen phoeniko-punischen Präzedenzfall *tarem*.[3]

Nach Peter Borel (1655), einem Anhänger von Bochart, hätten die Phoeniker, namentlich Hannibal, auch nördlich der Alpen die Spuren ihrer Wanderungen hinterlassen. Demnach musste es viele Ähnlichkeiten zwischen phoenikischen und keltischen Gottheiten geben; zum Beispiel wurde der gallische *Herkules* auch in der Stadt *Tyr*, im heutigen Libanon, verehrt. Der keltische Gott *Esus*, dessen Namen man mit „Stolz" übersetzte, wurde mit

Bibliographie:

D. Droixhe, L'Étymon des dieux. Mythologie gauloise, archéologie et linguistique à l'âge classique, Droz, Genève 2002.

U. Eco, La ricerca della lingua perfetta nella cultura europea, Laterza, Roma 1994.

J.-J. Glassner, La Tour de Babylone. Que reste-t-il de la Mésopotamie?, Seuil, Paris 2003.

Chr. Goudineau, Regard sur la Gaule, Errance, Paris 1998.

A. Grafton, Forgers and Critics. Creativity and Duplicity in Western Scholarship, Princeton 1990.

M. Olender, Les Langues du paradis, Gallimard-Seuil, Paris 1989.

2 Thomas Reinesius, Syntagma inscriptionum antiquarium cum primis Romae veteris, quarum omissa est recensio in vasto J. Gruteri opere, Leipzig 1682.

3 Samuel Bochart, Geographiae sacrae pars prior, Phaleg; pars altera, Chanaan, Caen 1646.

Mars gleichgesetzt, weil der phoenikische *Haziz* oder *Azizos* vom Wort *hazis* („Stolz") abgeleitet wurde.[4] Wir wissen heute, dass Esus mit „Herr" zu übersetzen ist.

Pierre-Daniel Huet (1679) bietet einen Umweg über Mexiko; für ihn ist *Teutates* vom Wort *teutl/totl* abzuleiten, das heißt „Gott" in Nahuatl; hätten nicht die Phoeniker Amerika besiedelt? Es gäbe einen guten Beweis dafür: die Phoeniker teilten mit den Azteken den Brauch des Menschenopfers![5]

Und schließlich sind auch Versuche der etymologischen Ableitung einzelner Namen bzw. Begriffe aus dem Assyrischen bekannt. Georg Pictor (1532) spricht von einem assyrischen Apollo, der einen umgekehrten Korb als Kopfbedeckung trägt: Ist damit etwa die Hörnertiara gemeint?[6]

Gerardus Joannes Vossius (1641), betrachtete *Belenus* als eine Art Ableitung von *Hel*, der dem *Bel* entsprochen habe: die Konsonanten B und H seien angeblich gleichwertig. Von dieser Urform seien auch das griechische Wort *helios* und seine lakonische Form *bela* abgeleitet, und sogar der Beiname *hekêbolos*, „der Pfeile abschießt" eine Bezeichnung für den Gott Apollo.[7] Zwei Annahmen wären dann möglich: Belenus ist von Bel abgeleitet, entweder *belis, belin* oder *bel* mit den Suffixen *-in* und *-us*. Die Verbindung des assyrisch-babylonischen Baal/Bel, das heißt „König" oder „Herr", mit dem Sonnengott gewann durch die linguistische Fundierung besonderes Gewicht.

Nur wenig später sah Armand Maichin (1671) eine Inschrift in Clermont-Ferrand, in der Belenus genannt wurde, „ein Name", schrieb er, „der von Belus/Belos, dem König von Assyrien, dem Erfinder der Astrologie, abgeleitet war".[8] Diese Information bezieht sich wahrscheinlich auf eine Stelle bei Plinius, wonach in Babylon der Tempel des Bel, des Erfinders der Astrologie, als einziges Monument noch steht.[9]

Auch die Aenaeis wurde als eine Quelle der religiösen Etymologie verwendet. Die Königin Dido begrüßt die Trojaner und ruft den Schutz der Götter an, indem sie einen Becher ausgießt, einen Becher, den der assyrische König Belus benutzte und nach ihm alle Könige von Assyrien. Diese Nachricht überliefert Honoratus Servius, ein Grammatiker des 4. nachchristlichen Jahrhunderts. Man verehrte also in der Person des Belus, des sagenhaften ersten assyrischen Königs, zugleich den Sonnengott und Saturnus, Götter, die unter dem Namen *Bel* in phoenikischer Sprache und *Hel/El* in assyrischer Sprache bekannt gewesen waren.

Seit dem 17. Jahrhundert waren auch einige Begriffe und Ausdrücke der keltischen und germanischen Sprachen bekannt. Pierre Lescalopier (1660) brachte den Namen Esus in Verbindung mit dem Wort *vessus*, „Stolz", einem alten keltischen Wort, das bei Servius belegt ist. Er hielt es außerdem für erforderlich, diese Ableitung durch Verbindung mit

4 Peter Borel, Thresor de recherches et antiquitez gauloises et françoises, Paris 1655.
5 Pierre-Daniel Huet, Demonstration evangelica, Paris 1679.
6 Georg Pictor, Theologica mythologica, Anvers 1532.
7 Gerardus Joannes Vossius, De theologia gentili et physiologia christiana, Amsterdam 1641.
8 Armand Maichin, Histoire de Saintonge, Poitou, Aunis et Angoumois, Saint-Jean d'Angely 1671.
9 Ich danke E. Cancik-Kirschbaum für diesen Beleg.

dem hebräischen Wort *izzuz*, das er mit „Stolz“ übersetzte, zu bestätigen.[10] Aus eben dieser Zeit stammen Versuche den Namen der Göttin Nehalennia mit den europäischen Sprachen zu verbinden. Marc Zuer Boxhorn (1662) schlug eine flämische Ethymologie *Ea-land* oder *Zee-land*, „das Wasser-Land“ vor.

Bereits Pierre Pithou (1565) verband *Belenus* mit *Helenus*, einer Hypostase Apollos, und leitet den Namen vom Griechischen *belos* „Bolzen“ ab, gemeint wären die Pfeile des Apollo.[11]

Im Folgenden wird der Versuch unternommen, eine Ordnung in die obigen Beispiele zu bringen. Zu diesem Zweck muss der Mythos des Turms zu Babel noch einmal näher in den Blick genommen werden – genauer gesagt: Was gab es Neues im Westen im 16. und 17. Jahrhundert?

Seit den Kirchenvätern nahm man an, dass die Menschheit vor dem Bau des Turms zu Babel eine einzige Sprache sprach, was die Bibel jedoch überhaupt nicht behauptet (jede Nation hatte ihre eigene Sprache)! Nach dem Fall des Turms zu Babel wurde aber jene ursprüngliche Sprache durch Gott manipuliert, von der die vielen Sprachen der Menschheit abhingen. Aufgrund der daraus resultierenden Verwirrung, weiterer Veränderungen und vieler nachfolgender Entstellungen entfernten sich diese Sprachen immer weiter von der Ursprache, der als primärer Sprache wiederum das Hebräische am nächsten stand.

Die Gelehrten erarbeiteten nun die Etymologien, weil seit Platon allgemein angenommen wurde, dass die Wörter zur Kenntnis der Realität, inklusive der Geschichte, beitrugen. Nun waren diese Etymologien des Öfteren ziemlich brüchig, wenn nicht sogar völlig aus der Luft gegriffen. Jan van Goropius Becanus selbst (1569), obwohl ein großer Grammatikexperte, postulierte immer wieder geradezu aberwitzige Etymologien. Zum Beispiel solche vom Typus des bekannten Experiments des ägyptischen Pharaos Psammetich: Dieser König ließ Kinder in einer geschlossenen Umgebung erziehen, um anhand deren ersten spontan gesprochenen Worten die Ursprache der Menschheit zu identifizieren. So stellte er also fest, dass das erste Wort *becos* war, das im Phrygischen „Brot“ bedeutet (*defacto* meinte *becos* das Meckern einer Ziege). Goropius wiederum weist nach, dass darin nicht ein phrygisches Wort, sondern das flämische Wort *Becker* zu sehen sei! Daraus folgerte er: Das Flämische, nicht das Phrygische oder das Hebräische, sei die erste Sprache der Menschheit gewesen.[12]

Eine andere recht leichtfertige Methode liefert der Archäologe Noel Taillepied (1585): Der Straßennamen „Hundegasse“ in der Stadt Autun ist ihm Beweis genug anzunehmen, dass hier einmal ein Tempel des ägyptischen Gottes Anubis existierte.[13]

10 Pierre Lescalopier, Theologia veterum Gallorum, Paris 1660.

11 Pierre Pithou, Adversariorum subsecivorum libri II, Paris 1565.

12 Jan van Goropius Becanus, Origines antwerpianae, Anvers 1569.

13 Noel Taillepied, Histoire de l'estat et republique des Druides, Eubages, Sarronides, Bardes, Vacies, Anciens François etc., Paris 1585.

Zweite Bemerkung: Seit dem 15. Jahrhundert haben die Gelehrten die europäischen Volkssprachen wiederentdeckt. Ihr Studium begann in Spanien mit dem Erscheinen der ersten Grammatik des Kastilischen. Diese Vielfalt von Sprachen galt es nun, von einer einzigen Ursprungs-Sprache abzuleiten. Und zumindest am Anfang wurde allgemein angenommen, dass dies eben die hebräische Sprache sei. Diese Theorie einer Monogenese der Sprachen ging wiederum von der Annahme aus, dass es irgendwo noch einen gemeinsamen Kern gäbe, eine allen Sprachen gemeinsame Grundlage, die immer noch kenntlich sei, obwohl die Ereignisse um den Turmbau zu Babel sie verwischt habe. Im Jahr 1606 versuchte Estienne Guichard zu zeigen, dass alle Sprachen von hebräischen Wurzeln abgeleitet waren.[14] Auf die groteske Theorie von Vater Thomassin wurde bereits hingewiesen, für den alle Wörter, die mit der keltischen Religion in Verbindung standen, von einer einzigen hebräische Wurzel, dem Wort *halal*, abgeleitet waren.

1569 formulierte Goropius Becanus von Antwerpen eine neue Theorie. Er behauptete, dass die niederländische Sprache, genauer das Flämische, die Muttersprache der Menschheit gewesen sein müsste. Weshalb? Weil die Flamen keinen Anteil an der Erbauung des Turms zu Babel hatten, sei demzufolge auch ihre Sprache durch Gott nicht verändert worden.[15] Diese und ähnliche Theorien wurden mit großem Scharfsinn dargelegt. Zwischen 1600 und 1647 haben z. B. Raphael Raphelengius und Boxhorn auf seltsame Ähnlichkeiten zwischen dem Flämischen und dem Persischen hingewiesen. Zum Beispiel brachten sie das flämische Wort *biezen*, „strampeln" einer Ziege, mit persisch *bes*, „der Wind", französisch *bise* und altlateinisch *bisa* in Verbindung. Sie dachten, dass alle diese Wörter von einer einzigen Urform abgeleitet waren. Boxhorn nannte diese Muttersprache „skythisch"; im 19. Jahrhundert wird die komparative Grammatik diese Ursprache indo-germanisch nennen. In der Tat errang das Flämische auf diese Art eine bevorzugte Stellung als linguistisches Archiv der europäischen Sprachen.

Daneben wurden seit dem 16. Jahrhundert aber auch zunehmend neue Wege beschritten, die vom Hebräischen und von der Monogenesis fortführten. 1555 verglich Konrad Gesner 55 Sprachen miteinander und stellte fest, dass sie in keiner Verbindung zur hebräischen Sprache standen.[16] Sebastian Münster (1530) ging davon aus, dass in Europa drei Sprachen gesprochen wurden: Hebräisch (mit Chaldäisch; Sebastian Münster edierte 1527 eine Grammatik dieser Sprache), Griechisch und Lateinisch, von denen aber nur die beiden letztgenannten die Grundlage der modernen Sprachen bildeten.[17]

Im 17. Jahrhundert identifizierte wiederum Joseph-Justus Scaliger (1610), der größte Sprach-, ja Universalgelehrte seiner Zeit und Sohn des Julius Caesar Scaliger, einem ebenso großen Grammatikexperten, zwölf Ursprachen in Europa, die nicht miteinan-

14 Estienne Guichard, L'Harmonie étymologique des langues, Paris 1606.

15 Vgl. Anm. 12.

16 Konrad Gesner, Catalogus omnium scriptorum locupletissimus, in tribus linguis, latina, graeca et hebraica, Paris 1555.

17 Sebastian Münster, Dictionarium trilinguus, Basel 1530.

der verwandt waren.[18] Die These war eine klare Kriegserklärung gegen die Theorie der Monogenese der europäischen Sprachen. Wenig später erstellten Claude Saumaise, ein Nachfolger des Justus Scaliger, Hugo Grotius, Isaac Casaubon und Leibniz ein Verzeichnis der phonetischen Veränderungen in den europäischen Sprachen, und verbanden damit die Idee, dass es überhaupt keine Spur einer sogenannten Ursprache der Menschheit gibt.

Verkürzt lässt sich sagen, als Christ sah man im Hebräischen die Ursprache der Menschheit. Aber als Chauvinist versuchte man zu beweisen, dass Adam und Eva ihre eigene Muttersprache sprachen. Zum Beispiel war für Paul Pezron (1703), da die Kelten von Gomer, dem ältesten Sohn Japhets, abstammten, ihre Sprache mit dem Hebräischen verbunden.[19] Für Andreas Kempe (1688) sprach Gott hingegen schwedisch, Adam dänisch und Eva französisch, die Sprache des Geschlechts![20]

Besonders im Hinblick auf die semitischen Sprachen wurde das Hebräische aus seiner hervorragenden Stellung verdrängt zu Gunsten des Phoenikischen und Assyrischen. Samuel Bocharts Überlegungen (1646) zum Phoenikischen gründeten auf Zeugnissen aus Vergils' Aenaeis und auf Sanchuniaton, einen nur in Fragmenten überlieferten phoenikischen Historiker, der durch Philo von Byblos ins Griechische übersetzt worden war und von Eusebius ausführlich zitiert wird.

Thomas Reinesius (1682) hat die zahlreichen Verbindungen, die angeblich zwischen dem assyrischen *Bel* und dem phrygischen *Ballen*, dem griechischen *Apollo*, dem kretischen *Abelion* und dem keltischen *Belenus* bestanden, erforscht.[21] Leibniz hat ihm zu dieser Entdeckung der Einheit der Familie der semitischen Sprachen gratuliert, mit anderen Worten zu seiner Unterminierung der Vormacht der hebräischen Sprache. Reinesius ist ein Pionier in der komparatistischen Forschung auf dem Felde der semitischen Sprachen.

Abschließend bleibt festzuhalten, dass sich die Situation des Hebräischen von einer Prinzipat-Stellung als Ursprache verändert hat: Man gab die alte Theorie der Monogenese der Sprachen auf, um sich der neuen Theorie der Polygenese, das heißt des multiplen Ursprungs von Sprachen, zu nähern. Am Horizont des 17. Jahrhunderts werden die Unterschiede zwischen den beiden Familien der semitischen – heute afroasiatischen – und der indo-germanischen – heute indoeuropäischen – Sprache in Umrissen sichtbar. In diesem Prozess, der unter anderem im Bereich der Etymologie der keltischen Götternamen zur Geltung kommt, spielte Assyrien und seine Sprache sowie die Namen einiger seiner Gottheiten eine nicht ganz unerhebliche Rolle.

18 Joseph-Justus Scaliger, Thesaurus temporum, Amsterdam 1610.
19 Paul Pezron, Antiquité de la nation et de la langue des Celtes, Paris 1703.
20 Andreas Kempe, Die Sprachen des Paradises, Hamburg 1688.
21 Vgl. Anm. 2.

Dritte Bemerkung: Der Aufbau der linguistischen Identität und Expertise Europas ging in enger Verbindung mit dem Versuch einer Erweiterung des Wissens über die Anfänge der Geschichte der Menschheit einher. Wie bereits erwähnt verstand man Begriffe, ausgehend von der Annahme, dass die Beherrschung der Worte zugleich deren vollkommene Aneignung, das heißt die Beherrschung der Dinge und Ereignisse einschloss, als spontane, mechanische, onomatopoetische Emanation der Gegenstände, für die sie stehen. Sie bleiben dabei nicht etwa reduziert auf platonische Emblemata einer höheren Realität. Um die Geschichte der Menschheit zu schreiben, haben die Gelehrten der Renaissance einerseits eigene etymologische Hypothesen aufgestellt, aber ebenso lasen sie die griechischen und lateinischen Autoren, die verdrängt und unterdrückt waren. In der Bibel sowie in den Werken dieser alten Historiker fanden sie die Namen sowohl der assyrischen als auch der babylonischen Könige. 1498 wurde die von dem Dominikaner Annius von Viterbo publizierte lateinische Sammlung aller bekannten antiken Werke („Antiquitatum Variarum") um die Ausgabe der Werke des Berossos erweitert. Berossos war ein babylonischer Priester des 3. Jahrhunderts vor Christus, der auf Griechisch eine Geschichte Babyloniens geschrieben hatte. In derselben Zeit wurde jedoch nach und nach auch eine These aufgegeben, die bis ins hohe Mittelalter von großer Bedeutung war: die Annahme eines trojanischen Urbeginns der europäischen Völker. Entsprechende Rekonstruktionen folgten dem Modell des Mythos von der Abstammung der Römer, wie es zum Beispiel bei Vergil zu finden ist. Im 7. Jahrhundert nach Christus schrieb *Fredegaire* in seiner Chronik, wie Francion, der Bruder Aeneas, ein Königtum zwischen Rhein und Donau begründet habe. Im 13. Jahrhundert behauptete Rigord, der Historiograph Königs Philipp Augustus von Frankreich, dass sich 25000 Trojaner in der Pariser Gegend niedergelassen hätten. Seit dem 15. Jahrhundert beharrten jedoch die italienischen Humanisten darauf, dass diese trojanische Abkunft nur für die Römer galt. Dann betraten die Kelten die Bühne. Lediglich Pierre Pithou (1565) und Jean Lemaire von Belgien (zwischen 1510 und 1513) hielten die trojanische These weiterhin aufrecht. Nach Pierre Pithou kam der Trojaner Antenor nach Venetien, wo er mit einem anderen Volk zusammentraf, den Venetern der Bretagne; zusammen schlossen sie sich der Verehrung des Gottes Belenus an. Jean Lemaire schließlich verkehrte den Mythos des trojanischen Ursprungs: die Trojaner wären demnach aus Gallien und Germanien nach Kleinasien gekommen und hätten dort Troja gegründet!

Wie dem auch sei, 1714 vertrat als erster Nicolas Freret, bekannt als Erforscher der keltischen Sprache und Mitglied der französischen königlichen Akademie, eine Zeit lang durch Louis XIV. in die Bastille verbannt, den gallischen Ursprung der französischen Nation und bestritt die Troja-Theorien.[22] Hatte nicht früher Kardinal von Richelieu in seinem sogenannten „Testament" geschrieben: „Das Ziel meines Ministeriums war, die Grenzen, die die Natur Gallien zubestimmt hatte, ihr zurückzugeben, den Galliern einen gallischen König wieder zu bringen, Gallien mit Frankreich ‚in eins zu setzen'."

22 Nicolas Freret, Mémoire sur l'origine des Français, Paris 1714.

Die auf diese Weise wiederentdeckten Kelten und die Gallier mussten noch ihren Platz in der Chronologie der Völker finden. Diese Chronologie war genau diejenige, die ihren Ausgang von der Genealogie der Söhne Noahs nahm. Damit aber wurde die Beziehung zwischen Genesis und Babel-Ereignis für diese Genealogie wiederum grundlegend.

Für Jean Lemaire von Belgien, demzufolge Gallier Troja gegründet hatten, waren diese die Abkömmlinge eines gewissen Samothes, des vierten Sohns des Japhet, Königs von Gallien. Das ist reine Phantasie, denn ein solcher Samothes existierte nicht. Für Jacques Boulduc (1626) waren die Gallier Boten einer Kirche des einzigen Gottes, deren höchster Priester einst Sem war; Gomer, der erste Sohn des Japhet, war der erste König der Gallier, der ihnen die Enthüllung gebracht hatte.[23]

Dieser familiengeschichtliche Strang verbindet sich nun mit der Sprachgeschichte. Wie bereits ausgeführt, nahm mit der Verdrängung des Hebräischen als Referenz-Sprache der Frühgeschichte das Babylonisch-Assyrische diesen Platz ein.

Um die Chronologie zu etablieren, bediente man sich der Chronik des Eusebius von Caesarea, die Hieronymus ergänzt hatte. Zwei Autoren sind an erster Stelle zu nennen: John Selden (1617), der die keltischen Gottheiten Belenus und Abellio mit der Figur des assyrischen Baals gekoppelt hat. Er rückte den historischen Vorrang der biblischen Welt nach Babel-Assyrien. Baal hätte demnach großen Ruhm und ein langes Leben in Europa gefunden. Um die verschiedenen Aussprachen zu erklären, legte er dar, dass „die wandernden Wörter, die von Asien nach Europa gingen oder umgekehrt, meistens verändert“ waren.

Joseph-Justus Scaliger (1658), der John Seldens Überlegungen aufgriff, hat diesen noch übertroffen, als er eine Identifizierung von Bel/Baal mit dem ersten assyrischen König postulierte, und zudem diese Ereignisse chronologisch in die Zeit des Krieges zwischen den Riesen und den Titanen einreihte.[24] Scaliger benennt Assyrien als Wiege der semitischen Welt, eine Alternative zur Dominanz der Hebraica. Als König hat *Bel* eine Dynastie begründet, er war der Vater des *Ninus*, der wiederum die Stadt *Ninive*, das heisst Assyrien gegründet hat und seinen Platz in der Weltgeschichte an der Seite des Prometheus und der großen biblischen Figuren fand. Diese euhemeristische Erklärung ist von Isidorus von Sevilla übernommen worden: Die paganen Götter waren einst menschliche Könige, die Städte gründeten und nach ihrem Tod verehrt wurden.

Es wurde immer deutlicher, dass Assyrien die Wiege der Zivilisation war. Diese Vorstellung war neu. Seit den Griechen galten Assyrien und Babylonien mit ihren weibischen Königen eigentlich als Antipoden der Zivilisation. *Belus*, der erste assyrische König, wurde mit *Saturnus* und mit dem Sonnengott gleichgesetzt. Er trat ebenfalls in verschiedenen Erscheinungsformen auf, wie *Belphegor*, der *Bel* des Berges Fegor, eine Gottheit von außergewöhnlicher Fruchtbarkeit wegen seines außerordentlichen mächtigen Glieds, der in Griechenland dem *Priapus* gleichgestellt war. Endlich sah man eine manichäische Theologie in

23 Jacques Boulduc, De Ecclesia ante legem libri tres, Lyon 1626.
24 Vgl. Anm. 18.

der altorientalischen Religion, in der jede positive Gottheit ein entsprechendes negatives Gegenstück hatte. Jacob Spon (1679) beschrieb ein aus Palmyra stammendes Flachbild, das man in Rom gefunden hatte: Es trägt eine Widmung an *Glibolus* und *Malachbelus*, beide Götternamen sind mit *Bel* konstruiert, ein Wort, das für Spon auf den assyrischen Gott *Bel* zurückweist. Er setzte Glibolus, eine positive Gottheit, mit der Sonne, Malachbelus, eine negative Gottheit, mit dem Mond gleich.[25]

Rückblickend lässt sich also zusammenfassen, dass die Gelehrten des 16. und 17. Jahrhunderts auf der Suche nach der eigentlichen Identität der Kelten die Babylonier und Assyrer mit dem Beginn der Zivilisation in Verbindung brachten.

Offenbar waren Gelehrte jener Zeit eher an Detailstudien denn an einer umfassenden historischen Synthese interessiert und dies ungeachtet der Wiederentdeckung der antiken historiographischen Überlieferung – an erster Stelle natürlich die griechischen Autoren.

Zwangsläufig stützten sich diese Kenntnisse überwiegend auf die griechischen Historiker, die Annius von Viterbo (1498) gesammelt hatte. Er hatte besonders Berossos ins Lateinische übertragen, inklusive einer Besprechung von Texten anderer Historiker, die er ebenfalls ins Lateinische übersetzte. Auf diese Weise formte er die erste Sammlung von Schriften antiker Historiker. Er legte darüber hinaus die Regeln fest, nach denen die Auswahl der Autoren und die Art der Besprechung erfolgten. Tatsache ist, dass ihm Griechisch unbekannt war, und dass er bei anderen Autoren abschrieb und sie manchmal verbessert hat! Kurzum, er war ein schlauer Fälscher. Angezeigt und entlarvt von Scaliger und Becanus, wurde er von anderen wie Jean Bodin hoch gelobt. Er hat die alte Theorie der vier Reiche, von der Abfolge des Assyrischen, des Persischen, des Griechischen und schließlich des Römischen Reiches, wieder publik gemacht. Seine Arbeit wird erst im 20. Jahrhundert durch Felix Jacobys Sammlung der Fragmente der antiken Historiker ersetzt.

25 Jacob Spon, Ignotorum atque obscurorum quorundam deorum arae, Lyon 1679.

*Gotthard Strohmaier*

# Ḥarrān – die Stadt des Sin in islamischer Zeit

Der König Nabonid (regierte 555–539 v. Chr.) war als Herrscher zwar gescheitert, aber dem von ihm geförderten Kult des Mondgottes Sin war in der Stadt Ḥarrān und darüber hinaus ein langes Nachleben beschieden, das bis weit in die islamische Ära reichte.[1] Der Artikel Ḥarrān im „Reallexikon der Assyriologie" geht darauf leider mit keiner Silbe ein.[2] Die Einwohner Ḥarrāns, die sich in der Mehrheit nicht zum Christentum oder zum Islam bekehrt hatten, wurden Sabier genannt. Die Etymologie des Namens *aṣ-ṣābiʾa* oder *aṣ-ṣābiʾūna* ist strittig, und selbst wenn man sie überzeugend deuten könnte, wäre nichts über die Identität ihrer Träger gewonnen. Gegenwärtig ist in der Forschung noch nicht entschieden, ob mit den im Koran neben den Juden und Christen genannten Sabiern (Sure 2,62; 5,69; 22,17) die paganen Ḥarrānier oder die Mandäer gemeint sind. In der Forschung herrscht zwar gegenwärtig die keineswegs von allen geteilte Theorie vor, dass damit die Mandäer gemeint seien und dass die Ḥarrānier den Sabiernamen nur angenommen hätten, um einer drohenden Verfolgung unter dem Kalifen al-Maʾmūn (regierte 813–833) zu entgehen.[3] Ein bislang noch nicht berücksichtigtes frühes Zeugnis aus der Schule des nestorianischen Übersetzers Ḥunain ibn Isḥāq (809–873) setzt die Gleichsetzung von Ḥarrāniern und Sabiern aber als gegeben voraus.[4] Übrigens dürften die zur Zeit Mohammeds im Sumpfgebiet des südlichen Irak sehr zurückgezogen lebenden Mandäer kaum in den Gesichtskreis Mohammeds getreten sein. Von Ḥarrān aus bestanden hingegen seit Urzeiten stabile Handelsverbindungen in die arabische Halbinsel, nicht umsonst hat sich wohl Nabonid hierher nach der Oase Taima und Umgebung zurückgezogen.

Übersehen wurde auch, dass sich im Koran eine Reminiszenz an eine spezifisch sabische Legende findet. Die Stelle Genesis 12,1–5, die knapp von der Auswanderung Abrahams aus

1 G. Féhervári, Art. „Ḥarrān", in: The Encyclopaedia of Islam, 2. Aufl., Leiden, London 1960–2004, Bd. 3, S. 227–230.

2 J. N. Postgate, Art. „Ḥarrān", in: Reallexikon der Assyriologie und Vorderasiatischen Archäologie, Bd. 4, Berlin/New York 1998–2001, 122b–125a.

3 Vgl. die Erörterung in G. Strohmaier, Die ḥarrānischen Sabier bei Ibn an-Nadīm und al-Bīrūnī, in: Ibn an-Nadīm und die mittelalterliche arabische Literatur (= Beiträge zum 1. Johann Wilhelm Fück-Kolloquium [Halle 1987]), Wiesbaden 1996, 55f. (Nachdruck in: ders., Hellas im Islam. Interdisziplinäre Studien zur Ikonographie, Wissenschaft und Religionsgeschichte [= Diskurse der Arabistik 6, hg. von H. Bobzin/A. Neuwirth], Wiesbaden 2003, 169).

4 Außer F. de Blois, der auch auf das Zeugnis aufmerksam geworden ist, es jedoch für das Resultat einer sekundären Änderung durch einen späteren Kopisten hält (The ‚Sabians' [Ṣābiʾūn] in Pre-Islamic Arabia, in: Acta Orientalia 1995, 39–61, hier 41, Anm. 8). Seine Theorie, dass es sich bei den Sabiern des Korans um die Manichäer handelt, stößt sich m. E. an dem Umstand, dass sich im Koran nicht die leiseste Reaktion auf die gnostisch-dualistische Weltsicht findet.

Haran berichtet, war in der jüdischen und christlichen Überlieferung in dramatisierender Weise dahingehend ergänzt worden, dass Abraham zuvor die Götzenbilder im Hause seines Vaters zerschlagen habe. Das wird im Koran an mehreren Stellen nacherzählt, in Sure 37,83–93, die zu den älteren gerechnet wird, ist der Vorgang durch eine rätselhafte Einzelheit bereichert: „Zu seiner Gefolgschaft (d.h. Noahs) gehörte Abraham, da er reinen Herzens zu seinem Herrn kam, da er zu seinem Vater und zu seinem Volke sprach: ‚Was verehrt ihr da? Wollt ihr aus Lüge neben Gott noch andere Götter? Und was meint ihr über den Herrn der Welten?' Er warf einen Blick zu den Sternen. Dann sagte er: ‚Ich bin krank.' Da wandten sie sich von ihm ab und kehrten ihm den Rücken. Er begab sich zu ihren Göttern und sagte: ‚Warum esst ihr nichts? Was ist mit euch? Ihr redet nichts.' Da machte er sich an sie heran und schlug sie mit seiner Rechten." Was ist das für eine Krankheit, derentwegen seine Leute sich von ihm abwenden, und welche Rolle spielt sie in der Abkehr vom babylonischen Götzendienst? Zwar dürfen wir der Kombinationsgabe des Propheten einiges zutrauen, mit dem er beispielsweise den Judenfeind Haman des im elamitischen Susa angesiedelten Estherbuches zu einem Wesir des Pharao ernannt hat, der diesem so etwas wie den babylonischen Turm bauen soll, um damit zu dem Gott des Moses emporzusteigen (Sure 28,38 u. 40,36f.). Aber eine Krankheit Abrahams, die dem Frevel an den Götterbildern vorausgehen soll, erscheint singulär und auf seltsame Weise unmotiviert. Schlüssig und zusammenhängend erscheint die Geschichte bei al-Bīrūnī, der aus dem Buch eines sonst unbekannten christlichen Polemikers namens Ibn Sinkillā folgendes zitiert: „So sollen sie behaupten, dass Abraham, über dem Friede sei, nur deswegen ihre Gemeinde verlassen habe, weil sich auf seiner Vorhaut ein Aussatz gezeigt habe. Wem dies aber zustößt, der ist unrein, und sie halten keine Gemeinschaft mit ihm. Aus diesem Grunde schnitt er sich die Vorhaut ab, das heißt, er vollzog an sich die Beschneidung. Als er in einen ihrer Götzentempel eintrat, hörte er eine Stimme von dem Götzenbild her, die zu ihm sprach: ‚O Abraham, du bist von uns hinweggegangen mit einem Gebrechen und bist zu uns zurückgekehrt mit zweien. Geh hinaus und komme nie wieder zu uns zurück!' Da ließ er sich vom Zorn dazu hinreißen, dass er die Götzen in Stücke schlug und ihre Gemeinde verließ. Später aber reute ihn seine Tat, und er wollte seinen Sohn dem Planeten Jupiter opfern, gemäß ihrer angeblichen Gewohnheit, ihre Kinder als Schlachtopfer darzubringen. Als der Planet Jupiter die Aufrichtigkeit seiner Reue erkannte, löste er den Sohn durch einen Widder aus."[5] Al-Bīrūnī hält das für „lügenhaftes und törichtes Geschwätz", aber ich glaube, dass er hier mit seiner sonst so lobenswerten historischen Kritik zu weit

5 Aus: Al-Bīrūnī, Al-āṯār al-bāqiya ʿan al-qurūn al-ḫāliya, hg. v. E. Sachau, Leipzig 1878, 204,21–205,9; Übersetzung in: G. Strohmaier, Al-Bīrūnī. In den Gärten der Wissenschaft. Ausgewählte Texte aus den Werken des muslimischen Universalgelehrten, übersetzt u. erläutert, 3. neubearb. Aufl., Leipzig 2002 (= Reclam-Bibliothek 20045), Nr. 39; dazu G. Strohmaier, Eine sabische Abrahamlegende und Sure 37, 83–93, in: P. Nagel (Hg.), Studien zum Menschenbild in Gnosis und Manichäismus, Halle 1979 (= Wissenschaftliche Beiträge der Martin-Luther-Universität Halle-Wittenberg 1979/39 [K 5]), 223–227 (Nachdr. in: ders., Von Demokrit bis Dante. Die Bewahrung antiken Erbes in der arabischen Kultur [= Olms Studien 43], Hildesheim/Zürich/New York 1996, 125–129).

geht. Hier haben vielmehr die Sabier auf die jüdisch-christliche Tendenzlegende mit einer eigenen Version der Geschichte geantwortet, die außerdem noch die Pointe aufweist, dass Abraham wieder reumütig in den Schoß seiner alten Gemeinde zurückgekehrt sei. In der Tat war Ibrāhīm auch ein geläufiger Name bei den Sabiern.[6] Mohammed, der auf dem Markt den fruchtlosen Streit von Juden und Christen anhören musste, war auf der Suche nach einem dritten Weg. Und da konnten auch nur wenige Kontaktpersonen genügen, die sich auf Abraham beriefen, der weder Jude noch Christ und doch ein Freund Gottes war. Julius Wellhausen hat auch auf die interessante Nachricht hingewiesen, dass die ersten Muslime in Mekka von den Gegnern als Sabier beschimpft wurden.[7]

Die ḥarrānischen Sabier hatten bei der arabischen Eroberung im Jahre 640 eine günstige Übergabe ausgehandelt, und im Übrigen konnten sie über die neue Herrschaft erst einmal froh sei, denn sie waren unter der byzantinischen Herrschaft grausam verfolgt worden.[8] Noch unter dem Kalifen al-Ma'mūn (regierte 813–833) sollen sie ihren Opferkult zelebriert und feierlich mit Musik und Gesang einen Stier mit Blumen bekränzt und mit Glöckchen an den Hörnern in der Stadt herumgeführt haben, bis sie ihn ihren Göttern opferten.[9] Etwa um das Jahr 870 traf der Hofbeamte Muḥammad ibn Mūsā[10], von einer Reise aus byzantinischem Territorium kommend, in der Gegend von Ḥarrān auf einen Sabier namens Ṯābit ibn Qurra (um 826–901)[11], der sich mit der geistlichen Obrigkeit der Stadt entzweit hatte und in dem Dorf Kafartutha lebte. Muḥammad ibn Mūsā war an der astronomischen Wissenschaft brennend interessiert, und als er merkte, dass Ṯābit dazu außergewöhnliche Fähigkeiten besaß, überredete er ihn, mit ihm nach Bagdad zu kommen und sein Mitarbeiter zu werden. Ṯābit gewann hier auch noch die Freundschaft des Kalifen al-Muʿtaḍid (892–902) und holte seinen alten Vater nach, der in der Hauptstadt durch seine graue Gewandung und sein lang herab fallendes Haar auffiel. Arabische Historiker beschreiben Ṯābits Einfluss am Hofe auch so, als sei damit eine länger dauernde Vorrangstellung (*riyāsa*) der Sabier begründet worden. Neuere Gelehrte haben den arabischen Ausdruck so interpretiert, als ob Ṯābit in Bagdad und noch dazu am Hofe eine Art sabischen Gegen-

6 M. Tardieu, La filiation ascendante de Ṯābit b. Qurra, in: Perspectives arabes et médiévales sur la tradition scientifique et philosophique grecque (Actes du colloque de la Société internationale d'histoire des sciences et de la philosophie arabes et islamiques. Paris, 31 mars–3 avril 1993), hg. v. A. Hasnawi/A. Elamrani-Jamal/M. Aouad (= Orientalia Lovaniensia Analecta 79), Leuven/Paris 1997, 265.

7 Reste arabischen Heidentums, 3. unv. Aufl., Berlin 1961, 236f.

8 I. Rochow, Zu einigen oppositionellen Strömungen, in: F. Winkelmann u.a. (Hgg.), Byzanz im 7. Jahrhundert. Untersuchungen zur Herausbildung des Feudalismus, Berlin 1978, 233–236 u. 239.

9 The Chronography of ... Bar Hebraeus, übers. v. E. A. Wallis Budge, Oxford 1932, 128.

10 Über ihn s. D. R. Hill, Art. „Banū Mūsā", in: The Encyclopaedia of Islam (s.o. Anm. 1), Bd. 7, 640f.; G. Strohmaier, Die bemalten Weingefäße aus Samarra, in: Klio 63 (1981), 123–130 (Nachdr. in: Hellas im Islam (s.o. Anm. 3), 149–154).

11 B. A. Rosenfeld/A.T. Grigorjan, Art. „Thabit ibn Qurra, al-Ṣābiʾ al-Ḥarrānī", in: Dictionary of Scientific Biography, hg. v. Ch. C. Gillispie, Bd. 13, New York 1976, 288–295; R. Rashed/R. Morelon, Art. „Thābit b. Ḳurra", in: The Encyclopaedia of Islam (s.o. Anm. 1), Bd. 10, 428f.

papst eingesetzt hätte.[12] Immerhin dachte Ṯābit nicht daran, vom Glauben seiner Väter abzufallen. Er lehnte den strikten Monotheismus ab, und als ihn Muslime von der unbeschränkten Macht Gottes überzeugen wollten, soll er zurückgefragt haben, ob Gott auch machen könne, dass fünf mal fünf nicht gleich fünfundzwanzig sei.[13] In werbenden Schriften, aus denen der syrische Bischof Barhebräus noch zitieren konnte, ließ er sich folgendermaßen vernehmen: „Wir sind die Erben des Heidentums, das ruhmreich auf dieser Welt verbreitet war. Selig ist derjenige, der seine Last mit fester Hoffnung trägt um des Heidentums willen. Wer hat die Erde kultiviert und die Städte gebaut, wenn nicht die Edlen und die Könige des Heidentums? Wer hat die Häfen und die Kanäle angelegt, wer hat die geheimen Wissenschaften kundgetan? Wem hat sich die Gottheit offenbart, wem hat sie Orakel gegeben und zukünftige Dinge gelehrt, wenn nicht den Weisen unter den Heiden? Sie haben all das studiert, sie haben die Heilung der Seelen erläutert und ihre Erlösung kundgetan, sie haben auch die Heilung der Körper erforscht, und sie haben die Welt mit der Stabilität ihrer Institutionen und mit der Weisheit, der wichtigsten Tugend erfüllt. Ohne das Heidentum wäre die Erde arm und leer und im Elend versunken."[14]

Dieses „Heidentum", so habe ich das syrische Wort *ḥanpūtā* versuchsweise übersetzt, konnte auch gegenüber Juden, Christen und Muslimen mit eigenen Propheten aufwarten, wie dem ägyptischen Hermes, Agathodaimon und griechischen Philosophen wie Thales, Pythagoras, dazu Solon, der ein Großvater Platons mütterlicherseits sein sollte. Ein einheimischer Prophet hieß Baba mit unsicherer Datierung, der mit konkreten politischen Weissagungen nach Art der alttestamentlichen Propheten zitiert wird.[15] Im Jahre 943 sah der Schriftsteller al-Masʿūdī an einem Türklopfer eines sabischen Versammlungshauses eine syrische Inschrift, die ihm folgendermaßen übersetzt wurde: „Platon sagte: Wenn man sich selbst erkennt, wird man göttlich (*Man ʿarafa ḏātahu taʾallahā*)", was man nicht so interpretieren darf als ob es in Ḥarrān eine platonische Akademie gegeben hätte.[16]

Die Wissenschaft, die Ṯābit vertrat und auch weiterentwickelte, war ganz und gar die griechische, und wenn dabei babylonische Traditionen auftauchen, wie etwa die Jahreszählung nach der Ära des Nabonassar, der im Arabischen Buḫtnaṣar I. hieß, so sind sie durch

12 J. Hjärpe, Analyse critique des traditions arabes sur les Sabéens Ḥarrâniens, Uppsala 1972, 35–43; ders., The holy year of the Harranians, in: Orientalia Suecana 23–24 (1974–75), 68f.; kritisch dazu G. Strohmaier, Die ḥarrānischen Sabier ... (s.o. Anm. 3), 52–54 (Nachdr. in: Hellas im Islam [s.o. Anm. 3], 167f.).

13 H. Ritter (Hg.), Pseudo-Maǧrīṭī, Das Ziel des Weisen, Leipzig/Berlin 1933, 335 (Übersetzung v. H. Ritter u. M. Plessner, „Picatrix". Das Ziel des Weisen von Pseudo-Maǧrīṭī, London 1962, 351), zitiert nach einem Kommentar zur Aristotelischen „Metaphysik" von dem bekannten syrischen Experten Abū Bišr Mattā ibn Yūnus, vgl. dazu F.E. Peters, Aristoteles Arabus, Leiden 1968, 50.

14 The Chronography ... (s.o. Anm. 9), 153.

15 F. Rosenthal, The Prophecies of Bâbâ the Ḥarrânian, in: W.B. Henning/E. Yarshater (Hgg.), A Locust's Leg. Studies in honour of S. H. Taqizadeh, London 1962, 220–232.

16 So zutreffend J. Lameer, From Alexandria to Baghdad: Reflections on the Genesis of a Problematical Tradition, in: G. Endreß/R. Kruk (Hgg.), The Ancient Tradition in Christian and Islamic Hellenism. Studies on the Transmission of Greek Philosophy and Sciences dedicated to H. J. Drossaart Lulofs, Leiden 1997, 185–191.

den „Almagest“ des Ptolemaios vermittelt.[17] Die griechische Wissenschaft war nun einmal der altorientalischen überlegen gewesen und verfehlte darum nicht ihre Wirkung auf die Gebildeten der umliegenden Länder. Denken wir an Thales von Milet, der übrigens kein Grieche, sondern phönizischer oder halb phönizisch, halb karischer Abstammung war, und der die Sonnenfinsternis damit erklärte, dass der Mond, der von erdartiger Substanz sei, vor die Sonne geht.[18] In seine Zeit fiel die Verfinsterung vom Jahre 585 v. Chr., und Herodot (I 74) berichtet, dass sie während einer Schlacht zwischen den Lydern und den Medern eingetreten sei, was man als einen Wink der Götter verstand und den Kampf abbrach, was schließlich, vielleicht sogar unter Vermittlung Nabonids, wenn der Name Labynetos so gedeutet werden darf, zu einem Friedensschluss und einer dynastischen Hochzeit führte.[19] Wäre Thales damals mit seiner Wissenschaft schon Gehör geschenkt worden, hätte man den Krieg getrost fortsetzen können. Im Weltbild des neuplatonisch interpretierten Aristotelismus, wie er dem Vorderen Orient durch die Hochschule von Alexandrien vermittelt wurde, war dem gestirnten Himmel gegenüber dem Materialismus der Vorsokratiker wieder eine quasi göttliche Qualität zuerkannt worden, was sicher eine Verschmelzung mit der Religiosität der Sabier erleichterte, die einem Hochgottglauben mit einer neuplatonisch konzipierten Emanation durch die von Planetengeistern bewegten Sphären huldigten. Sie prägten mit ihrem Planetenkult bei den Muslimen das Bild der vorchristlichen griechischen Religion, auch al-Bīrūnī weiß in seiner großen Monographie über Indien von Sokrates zu berichten, dass er sich weigerte, „die Planeten mit seinem Munde Götter zu nennen, und wie elf der athenischen Richter mit Ausnahme des zwölften darin übereinkamen, ihn zum Tode zu verurteilen, so dass er sterben musste, indem er die Wahrheit nicht widerrief“.[20]

Angesichts dessen kann man die Frage stellen, ob es bei den Sabiern überhaupt noch genuin babylonische Traditionen gab. Vor der Entzifferung der Keilschrift meinte man, dass die arabischen neben den griechischen Nachrichten über die Sabier die einzige Möglichkeit böten, in die Vergangenheit des Zweistromlandes einzudringen, und die monumentale, deutsch geschriebene und noch nicht ganz ersetzte Monographie des russischen Arabisten Daniel Chwolson[21] „Die Ssabier und der Ssabismus“ vom Jahre 1856 war auch als Versuch in dieser Richtung gemeint.[22] Chwolson hatte übrigens schon eine erste Kunde

17 Al-Bīrūnī, Al-āṯār al-bāqiya ... (s.o. Anm. 5), 27,4–7.

18 G. Wöhrle (Hg.), Die Milesier, Bd. 1: Thales (= Traditio Praesocratica 1), Berlin/New York 2009, 136f. (Nr. Th 158); die Nachricht war auch im Arabischen zugänglich, vgl. 384f. (Nr. Th 489).

19 Ebd., 30–33 (Nr. Th 10); die Nachricht war auch im Arabischen zugänglich, vgl. 382f. u. 404f. (Nr. Th 487 u. 507).

20 G. Strohmaier, Das Bild des Sokrates in der arabischen Literatur des Mittelalters, in: H. Kessler (Hg.), Sokrates. Bruchstücke zu einem Porträt (= Sokrates-Studien III), Kusterdingen 1997, 106 (Nachdr. in: Hellas im Islam [s.o. Anm. 3], 50).

21 Y. Slutsky/L. Nemoy, Art. „Daniel Chwolson“, in: Encyclopaedia Judaica, Bd. 4, 2. Aufl., 2007, 724f.

22 2 Bde., St.-Petersburg 1856 (Nachdruck Amsterdam 1965); vgl. dazu V. V. Bartol'd, K voprosu o sabijach, in: ders., Sočinenija, Bd. 6, Moskau 1966, 469–486; vgl. von Chwolson auch: Über die Überreste der altbabylonischen Literatur in arabischen Übersetzungen, in: Mémoires présentés à l'Académie impériale des sciences de St.-Petersbourg 8 (1859), 329–524.

von der Keilschrift aus den Untersuchungen von Henry Creswicke Rawlinson und anderen.[23] Er glaubte aber nicht, dass sich dahinter eine semitische Sprache verbirgt.[24] Mit der Entzifferung der keilschriftlichen Urkunden sind nun fundierte Untersuchungen möglich geworden, wie sie von Hildegard Lewy[25] und Tamara M. Green[26] vorgelegt wurden. An uraltem Gut sind, wie zu sehen ist, die Namen von Göttern lebendig geblieben, aber auch sie sind schon teilweise durch die griechischen ersetzt. So sind nach einer Quelle, die Ibn an-Nadīm wiedergibt, die Wochentagsgötter folgendermaßen verteilt. Sonntag gehört dem Helios, Montag dem Sin, Dienstag dem Ares, Mittwoch dem Nabu[27], Donnerstag dem Bel, Freitag der Balṯā und der Sonnabend wieder dem griechischen Kronos.[28]

Sabier waren als geschickte Kunsthandwerker geschätzt und gaben so ihre traditionellen Motive an Muslime weiter, die es mit dem Bilderverbot weniger genau nahmen. Aus dem nordmesopotamischen Raum des 12., 13. und 14. Jahrhunderts gibt es prachtvolle tauschierte Metallarbeiten mit astrologischer Thematik. Nicht zufällig wurden im gleichen Umfeld auch Münzen mit Bildern der Tierkreiszeichen und der alten Planetengötter geprägt.[29] Merkur ist auf den Gefäßen im Ring der sieben Planeten als ein sitzender Schreiber dargestellt (Abb. 1).[30] Er hat damit weder etwas mit dem griechischen Götterboten noch mit dem paganen Offenbarungsträger der so genannten Hermetik zu tun, wohl aber mit dem altmesopotamischen Nabu. Auch in der arabischen Literatur findet sich Merkur als *kātib* („Schreiber") bezeichnet.[31]

Eine Eigenheit der astrologischen Motive des Kunsthandwerks besteht darin, dass die figürlichen Planetendarstellungen dort, wo es ästhetisch befriedigend war, mit einem der „Häuser" des Tierkreises, in dem sie eine besondere Macht ausüben sollten, kombiniert sind. So sitzt Jupiter zwischen den beiden Fischen, Mars reitet auf dem Widder, Saturn auf dem Steinbock oder Venus, hier züchtig bekleidet, spielt die Laute auf dem Rücken des

23 Vgl. E. Cancik-Kirschbaum/G. Chambon, Les caractères en forme de coins: Le cas du cunéiforme, in: Revue d'Assyriologie 100 (2006), 25, 33, 37 u. 39.

24 „Die Ssabier" ... (s. o. Anm. 22), Bd. 2, 163 u. 165.

25 Points of Comparison between Zoroastrianism and the Moon-Cult of Ḥarrân, in: A Locust's Leg (s. o. Anm. 15), 139–161.

26 The City of the Moon God. Religious Traditions of Harran, Leiden/New York/Köln 1992, 144–161.

27 Die Lesung Nābuq wurde von J. Fück überzeugend zu Nābū korrigiert (Arabische Kultur und Islam im Mittelalter. Ausgewählte Schriften, Weimar 1981, 61).

28 Fihrist, Bd. 1, hg. v. G. Flügel, Leipzig 1871, 321,27–31; vgl. weiter Lewy (s. o. Anm. 25), 139–145 und Green (s. o. Anm. 26), 150–160.

29 Einige Abbildungen in: G. Strohmaier, Denker im Reich der Kalifen, Leipzig/Jena/Berlin 1979, 49.

30 Weitere Abbildungen bei G. Strohmaier, Die neunzehn Kugeln des Pantokrators. Ein orientalisches Motiv an der Sponsus-Sponsa-Plastik des Magdeburger Doms, in: J. Dummer/J. Irmscher (Hgg.), Byzanz in der europäischen Staatenwelt, Berlin 1983, Tf. 5 (im inneren Medaillonkranz links oben) u. Tf. 7 (im Medaillonkranz links) (Nachdr. in: Hellas im Islam (s. o. Anm. 3), Abb. 5 auf S. 203 u. Abb. 7 auf S. 205; ältere Belege s. in: F. Pomponio, Art. „Nabû A. Philologisch", in: Reallexikon der Assyriologie ... (s. o. Anm. 2), Bd. 9, Berlin/New York 1998–2001, 21b f. und 24a.

31 J. Samsó, Art. „ʿUṭārid", in: The Encyclopaedia of Islam ... (s. o. Anm. 1), Bd. 10, 940–942.

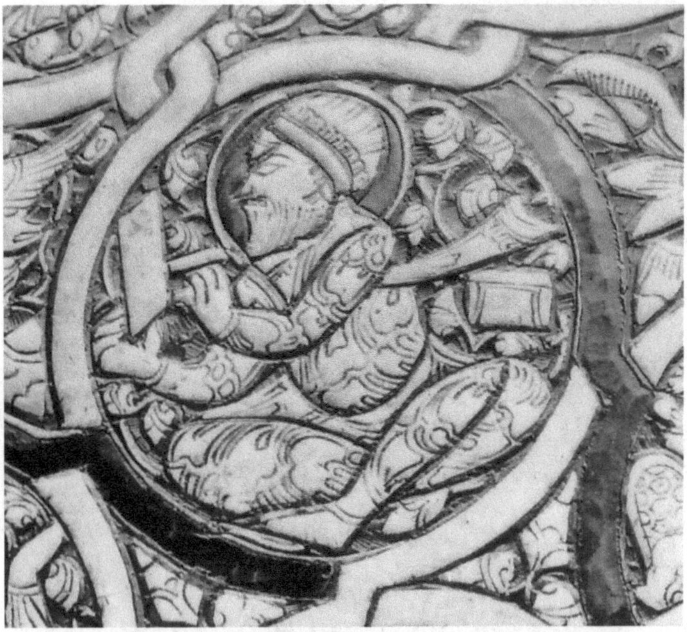

Abb. 1 | Merkur als Schreiber in einem Kreis der sieben Planeten auf einem tauschierten Bronzebecken aus dem 14. Jahrhundert (Florenz, R. Museo Nazionale, Collezione Carrand, Inv.-Nr. 364c).

Stiers. Zwischen den Zwillingen aber ließ sich der schreibend sitzende Merkur schwerlich unterbringen. Stattdessen halten die beiden jungen Männer ein zwischen ihnen stehendes langes gerades Gebilde fest, das mit einem Gesicht oder einer leeren Scheibe bekrönt ist (Abb. 2 u. 3). Willy Hartner hat das letztere als den Kopf des *ǧauzahar* gedeutet, eines unsichtbaren Himmelsdämons, der die Mond- und Sonnenfinsternisse verursacht, und im „Haus" der Zwillinge seine so genannte astrologische „Erhöhung" hat.[32] Wo aber ist der Planet, der den Zwillingen zugehört? Es kann nur Merkur, der Herr ihres „Hauses" sein. Hartner verfiel auf die Auskunft, dass er mit der Figur eines der Zwillinge verschmolzen wurde.[33] Aber wenn man bedenkt, dass der Name des Nabu noch bekannt war und er außerdem in der Funktion des Schreibers dargestellt wurde, sollte man nicht eher annehmen, dass auch sein altes Symbol des Griffels noch geläufig war, der hier ebenso mannshoch erscheint wie auf den kassitischen Kudurrus (Abb. 4).[34] Er ließ sich zwanglos zwischen den Zwillingen unterbringen.

Das kombinierte Motiv taucht sogar noch in der abendländischen Kunst des Mittelalters auf. Einen Beleg hat Theodor Dombart in einer Serie von Tierkreismedaillons in

32 W. Hartner, The Vaso Vescovali in the British Museum. A study on Islamic Astrological Iconography, in: Kunst des Orients 9 (1973/74), 105–114 u. 121, vgl. seinen Art. „al-djawzahar", in: The Encyclopaedia of Islam (s. o. Anm. 1), Bd. 2, 501 f. (für förderliche Hinweise danke ich Gisela Helmecke, Museum für Islamische Kunst, Berlin).

33 The Vaso Vescovali ..., 114.

34 U. Seidl, Art. „Nabû A. Archäologisch", in: Reallexikon der Assyriologie ... (s. o. Anm. 2), Bd. 9, Berlin/New York 1998–2001, 24b–27b.

Abb. 2 | Das Tierkreiszeichen der Zwillinge auf dem „Wade Cup" (Zeichnung bei D. S. Rice, The Wade Cup in the Cleveland Museum of Art, Paris 1955, 18, Fig. 13b).

Abb. 3 | Das Tierkreiszeichen der Zwillinge auf dem „Vaso Vescovali" (Zeichnung bei D. S. Rice, The Wade Cup in the Cleveland Museum of Art, Paris 1955, 20, Fig. 15a).

Abb. 4 | Der Griffel des Nabu auf dem Kudurru VA Bab 4375 im Vorderasiatischen Museum (Staatliche Museen zu Berlin), hier auf dem Rücken eines Schlangendrachens (für die Aufnahme danke ich Ralf-Bernhardt Wartke und Olaf Teßmer, Berlin).

einem hebräischen Gebetbuch fränkisch-bayrischer Provenienz gefunden, welches in der Sammlung Kaufmann der Bibliothek der Budapester Akademie der Wissenschaften aufbewahrt wird.[35] Hier haben die Zwillinge, die beide die Stange mit der Scheibe halten, Tierköpfe (Abb. 5), was den Verfasser veranlasst hat, dafür ebenfalls nach babylonischen Vorbildern zu suchen. Er kam in Unkenntnis der Parallelen auf den muslimischen Metallgefäßen auf die so genannte Doppellöwenkeule, bei der die Tierköpfe allerdings von der Stange, mit der sie verwachsen sind, abgekehrt sind. In Wirklichkeit handelt es sich bei den Tierköpfen und damit um die Vermeidung der reinen Menschengestalt lediglich um eine mittelalterliche jüdische Konzession an das alttestamentliche Bilderverbot, wie sie in dieser Form aus dem gleichen zeitlichen und geographischen Umfeld mehrfach belegt ist.[36] Christen kannten diese Hemmungen nicht. In den Filetstopfereien des Zehdenicker Altartuches, das zu den besonderen Schätzen des Berliner Märkischen Museums zählt, und wo sich inmitten der christlichen Heilsgeschichte drei Reihen von Planetentieren be-

35 Eine bedeutsame mittelalterliche Darstellung des Tierkreisbildes der Zwillinge, in: Festschrift für Georg Leidinger zum 60. Geburtstag, München 1930, 47–49 u. Tf. XII.

36 U. u. K. Schubert, Jüdische Buchkunst, Teil 1, Graz 1983, 72–75; U. Schubert, Die Vogelkopf-Haggada. Ein künstlerisches Zeugnis jüdischen Selbstbewusstseins am Ende des 13. Jahrhunderts, in: Anzeiger des Germanischen Nationalmuseums und Berichte aus dem Forschungsinstitut für Realienkunde 1988, 35–57.

Abb. 5 | Das Tierkreisbild der Zwillinge in einem hebräischen Machzor fränkisch-bayerischer Herkunft (Bibliothek der Ungarischen Akademie der Wissenschaften, Sammlung Kaufmann; für die Fotografie danke ich István Ormos, Budapest).

Abb. 6 | Das Tierkreisbild der Zwillinge in Filetstickerei auf dem „Zehdenicker Altartuch" vom Ende des 13. Jahrhunderts (Leihgabe der „Evangelischen Kirche Berlin-Brandenburg-schlesische-Oberlausitz" an die „Stiftung Stadtmuseum Berlin", Inv.-Nr. III.57.33.k; für die Aufnahme danke ich Albrecht Henkys, Berlin).

finden,[37] sitzen die Zwillinge, hier ein Junge und ein Mädchen,[38] einander thronend gegenüber und halten zwischen sich die Stange, die nicht nur mit der Scheibe an der Spitze sondern auch noch mit Andeutungen von Ästen verziert ist (Abb. 6). Dass das Motiv auch hier den Planeten Merkur verkörpern soll, ersieht man daraus, dass ein darunter befindliches Medaillon mit der Büste des Mondes gefüllt ist, der sich im ptolemäischen System in seiner Sphäre unterhalb des Merkur bewegt. Über den Zwillingen befindet sich der Stier, der hier als ihr „Haus" die Venus vertritt. Links von diesen astrologischen Figuren sind Reihen von Tieren, die ihrer speziellen Herrschaft unterstehen sollen. Auf die orientalische Herkunft von mutmaßlichen imitierten Vorbildern weist der Umstand, dass die Planeten rechts außen voran stehen, wie es der arabischen Schreibrichtung entspricht. So hat der Griffel des Nabu eine lange Spur hinterlassen, ausgehend von den babylonischen Kudurrus über die Astrologie und das Kunsthandwerk der Sabier bis hinein in die Textilkunst unseres Mittelalters.

37 G. Strohmaier, Arabische Astrologie auf dem Zehdenicker Altartuch, in: Jahrbuch des Märkischen Museums 4 (1978), 106–108 u. 203–208 (S. 204 die Abb. der Zwillinge; S. 206 u. 207 eine Miniatur mit dem analogen Motiv der Planetenkinder; die beiden Tafeln sind leider ungeschickt angeordnet; die Planeten befinden sich auch hier rechts außen, der arabischen Schreibrichtung entsprechend an der Spitze des Bereichs, dem sie vorstehen).

38 Bei einer Beschreibung des indischen Tierkreises erwähnt al-Bīrūnī (973–1048), dass in ihm die Zwillinge ein Junge und ein Mädchen sind (Al-Bīrūnī, India, hg. von E. Sachau, London 1887, 108,1f.; Übersetzung v. dems., Bd. 1, London 1888, 219), dies im Gegensatz zu der sonst üblichen Darstellung zweier Jünglinge. Ob ein Zusammenhang besteht, wird sich wohl nicht mehr aufklären lassen.

*Andrzej Reiche*

# Die altorientalischen Denkmäler in den Museen in Polen
## Geschichte und Bestand

Einen wichtigen Faktor in der Rezeption einer alten und entfernten Kultur durch eine moderne Bevölkerung bildet neben den literarischen Überlieferungen auch der materielle Bestand dieser Kultur, der in öffentlichen Sammlungen dem breiten Publikum zugänglich ist. Nachfolgend soll hier in aller Kürze der Bestand der altvorderasiatischen Sammlungen in den polnischen Museen mit einem Bezug auf deren Geschichte dargestellt werden. Es gibt heute zwei größere Sammlungen, die sich in Warschau und Krakau befinden.

## I.

In der zweiten Hälfte des 19. Jahrhunderts wurden in Europa, oft mit staatlicher Unterstützung, die großen Sammlungen vorderasiatischer Denkmäler gebildet. Damals existierte Polen als Staat nicht, sondern war unter den Nachbarstaaten Russland, Preußen und Österreich aufgeteilt. Dennoch wurden schon zu jener Zeit von polnischen aristokratischen Familien zahlreiche Sammlungen Antiker Kunst eingerichtet. Hauptsächlich betraf dies die klassisch-antike Kunst, daneben – in kleinerem Ausmaß – ägyptische Denkmäler und am Ende in sehr geringer Zahl auch die vorderasiatischen Objekte. Die in dieser Zeit in Polen verbreitete Bildung in antiker Geschichte und Kunst sowie die damalige Mode und nicht zuletzt die dekorative Kraft der Antiken Kunst spielten hierbei eine wichtige Rolle. Einige der Sammlungen entstanden im Ausland nur zum eigenen Vergnügen der Sammler, so zum Beispiel die berühmte Sammlung des Grafen Michał Tyszkiewicz (1828–1897),[1] und viele von ihren Denkmälern haben manche Museen außerhalb Polens bereichert.

1 W. Froehner, Collections du Château de Gołuchów. Antiquités, Paris 1899, 103–104, 1–4, 4a, 5; siehe auch A. Snitkuviené, The Epilogue of the Ancient Art Collection of Michał Tyszkiewicz. The Auction in Paris 1898, Archeologia 58 (2009), 197–209 (Pl. I–II) [im Druck]. Die Sammlungen wurden im Ausland zusammengetragen und einiges davon befindet sich heute in Polen (Czartoryski-Krakau, Czartoryski-Gołuchów, Potocki-Krzeszowice). In der Tyszkiewicz-Sammlung überwogen bei weitem ägyptische Objekte. Vieles davon befindet sich heute als Geschenk von Tyszkiewicz im Louvre. Weitere Einzelstücke kann man verstreut in anderen Museen finden, darunter z. B. ein hethitisches Stempelsiegel im Museum of Fine Arts, Boston. Allerdings waren die altorientalischen Fundstücke nicht zahlreich. Ein geringer Anteil vorwiegend ägyptischer Objekte war außerdem in seinem Palast in Łohojsk in Litauen deponiert, wovon sich heute immer noch ein Teil in Litauen befindet, ein weiterer Teil aber in Polen.

Es gab aber auch schon in dieser Zeit Sammlungen, die bewusst zum Zweck der Ausbildung des Volkes aufgebaut wurden. Hier muss man vor allem die Sammlung der Fürstin Izabella Czartoryska in Puławy vom Anfang des 19. Jahrhunderts hervorheben. Dieses älteste Museum in Polen wurde von der Fürstin unter dem Motto „um die Vergangenheit in Erinnerung zu halten, den aktuellen Bedürfnissen der Gesellschaft zu dienen und in die Zukunft zu führen" gestiftet. Dieser Idee treu bleibend, hat ihr Enkel Herzog Władysław Czartoryski (1828–1894) auf Grundlage dieser Sammlung und verschiedenen anderen verstreuten Sammlungen der Familie das Muzeum Książąt Czartoryskich (Museum der Herzöge Czartoryski) 1876 in Krakau der Öffentlichkeit zugänglich gemacht. Die Krakauer Gemeinde stiftete dieser Sammlung das Gebäude des ehemaligen Stadtarsenals.

Alle etwas später in dieses Museum aufgenommenen Altertümer wurden von Herzog Władysław in den Jahren 1875–1893 durch Vermittlung von Antikenhändlern vor allem in Italien aber auch in Frankreich und Ägypten gekauft.[2] Unter ihnen befindet sich auch ein kleiner Bestand altorientalischer Altertümer, der zum ersten Mal als ein Ganzes erst 1988 von J. Oelsner kurz besprochen und zum Teil veröffentlicht wurde.[3] Die größte Gruppe bilden 19 Roll- und Stempelsiegel, die zum Teil von S. Przeworski[4] und zum Teil von B. Kaim-Małecka[5] veröffentlicht wurden (Abb. 1–4).

Hinzu treten Fragmente von assyrischen Objekten: Reliefbruchstücke aus Kalchu (Nimrud)[6] und Ninive (Kujundschik)[7] (Abb. 5) sowie ein Fragment einer glasierten Knauf-

2 S. Komornicki, Muzeum Książąt Czartoryskich w Krakowie, Kraków 1929, 3.

3 J. Oelsner, Inschriften Mittelelamischer, Neuassyrischer und Neubabylonischer Herrscher auf Ziegeln in Kraków und Jena, in: J. Śliwa/E. Kluwe (Hgg.), From the Archaeological Collections of Cracow and Jena. Zeszyty Naukowe Uniwersytetu Jagiellońskiego, Kraków 1988,17–18.

4 S. Przeworski, Un cylindre hittite de Cracovie, Archiv für Orientforschungen 3 (1926), 172–74 (hethitisch); ders., Die Rennwagen-Darstellungen in der nordsyrischen Kunst, Archiv für Orientforschungen 5 (1928/29) Tf. XI,3 (syro-hethitisch?) und ders., Altorientalische Altertümer aus polnischen Sammlungen, Archiv für Orientforschungen 9 (1933/34), 123, Abb. 3–5 (akkadisch, zyprisch, altbabylonisch)

5 B. Kaim-Małecka, Catalogue des intalles orientales des Musées Polonais, Acta Archaeologica Lovaniensia 32 (1993), 59–95; Von den von Przeworski bearbeiteten Rollsiegeln hat die Verfasserin nur zwei gezeigt, das akkadische (Nr. 3) und das zyprische (Nr. 148). Dafür wurden zum ersten Mal veröffentlicht: ein zyprisches Rollsiegel (Nr. 136), zwei achämenidische Stempelsiegel (Nr. 43 und 44) und sieben sasanidische Stempelsiegel (Nr. 48, 51,67, 84, 92, 94 und 105). In der Czartoryski-Sammlung befinden sich noch vier Rollsiegel, die unveröffentlicht sind: Inv.-Nr. XI–789, XI–790 (Isin-Larsa Zeit) und XI–788, XI–1202 (nordsyrisch?).

6 Unveröffentlicht blieb ein Bruchstück (Inv.-Nr. XI–622a) eines Reliefs mit Darstellung einer Palmette des stilisierten „Lebensbaums" mit Fragmenten der „Standardinschrift" des Assurnasirpal II. (884–858 v. Chr.), was auf die Herkunft des Stückes aus dem Palast des Königs in Kalchu (Nimrud) hinweist. Der Text ist identisch mit RIMA II: A.0.101.23. (Die Identifizierung des Textfragmentes verdanke ich Dr. Małgorzata Sandowicz). Aus Kalchu (Nimrud) stammen vermutlich auch zwei Fragmente einer Marmor-Platte mit Keilschriftzeichen. Eine Identifizierung des Textes bleibt immer noch unbestimmt. Siehe auch J. Oelsner, Inschriften Mittelelamischer, Neuassyrischer und Neubabylonischer Herrscher auf Ziegeln in Kraków und Jena, in: J. Śliwa/E. Kluwe (Hgg.), From the Archaeological Collections of Cracow and Jena. Zeszyty Naukowe Uniwersytetu Jagiellońskiego, Kraków 1988, 18.

7 Es ist ein Fragment aus dem Nord-Palast Assurbanipals (669–631/629 v. Chr.) mit der Darstellung einer Schlacht gegen die Elamer, veröffentlicht in S. Przeworski, Ein assyrisches Relieffragment aus einer Krakauer Sammlung, Rocznik Orjentalistyczny 6 (1928), 84–88; siehe auch R.D. Barnett, Sculptures from the North Palace of Ashurbanipal at Niniveh, London 1976, 62, Pl. LXXI (u).

Abb. 1 | Rollsiegel mit Darstellung des thronenden Schamasch zwischen zwei, von Helden gehaltenen Türflügeln. Moderne Abrollung. Babylonien, Irak. Akkad-Zeit, um 2300 v. Chr. Onyx. H. 4,2 cm, D. 2,4 cm. Eigentum der Fundacja Książąt Czartoryskich; im Museum der Herzöge Czartoryski in Krakau, Inv.-Nr. MNK XI-390.

Abb. 2 | Rollsiegel mit Einführungsszene. Moderne Abrollung. Babylonien, Irak. Ur III-Zeit, 21. Jh. v. Chr. Serpentin. H. 2,4 cm, D. 1,2 cm. Eigentum der Fundacja Książąt Czartoryskich; im Museum der Herzöge Czartoryski in Krakau, Inv.-Nr. MNK XI-789.

Abb. 3 | Rollsiegel mit Anbetung der kriegerischen Ischtar. Die Inschrift: (1) a-we-il-li (2) mar al-la-ša-ri. Moderne Abrollung. Babylonien, Irak. Altbabylonische-Zeit, 18. Jh. v. Chr. Hämatit. H. 2,5 cm, D. 1,35 cm. Eigentum der Fundacja Książąt Czartoryskich; im Museum der Herzöge Czartoryski in Krakau, Inv.-Nr. MNK XI-214.

Abb. 4 | Rollsiegel mit einer Bankettszene. Moderne Abrollung. Anatolien, Türkei. Hethitisches Großreich, 2. Hälfte des 2. Jahrtausend v. Chr. Hämatit. H. 2,3 cm, D. 1,4 cm. Eigentum der Fundacja Książąt Czartoryskich; im Museum der Herzöge Czartoryski in Krakau, Inv.-Nr. MNK XI-1203.

Abb. 5 | Relieffragment mit einer Kampfszene der Assyrer mit den Elamitern. Nord-Palast des Assurbanipal in Ninive (Kujundschik). Assyrien, Irak. 7. Jh. v. Chr. Marmor. 17,9 × 20,5 cm. Eigentum der Fundacja Książąt Czartoryskich; im Museum der Herzöge Czartoryski in Krakau, Inv.-Nr. MNK XI-623.

fliese mit Inschriftresten[8] und ein Backstein mit einer Inschrift Salmanassars III. (858–824 v. Chr.).[9]

Dazu kommen noch einige andere Keilschriftdenkmäler. Es handelt sich dabei um drei Tonnägel mit Inschriften Gudeas von Lagasch (ca. 2143–2124 v. Chr.)[10] und eine unveröffentlichte Tontafel aus Umma der Ur-III Zeit, die auf das 2. Regierungsjahr Amar-Sins (ca. 2045–2037 v. Chr.) datiert ist[11]. Des Weiteren zählen dazu: zwei Dokumente, die aus den Archiven der altassyrischen Handelskolonie Kanesch (Kültepe) stammen.[12]

Die Sammlung der Herzöge Czartoryski wurde seit 1950 unter der Bezeichnung „Zbiory Czartoryskich" (Die Czartoryski-Sammlungen) vom Nationalmuseum in Krakau als eine eigene Abteilung verwaltet. Im Jahre 1991 wurde zudem im Nationalmuseum Kra-

8 Das Fragment stammt von einer glasierten, quadratischen Knauffliese, die mit einer Komposition von Palmetten und Granatäpfeln(?) verziert war. Für eine Typeneinordnung siehe P. Albenda, Decorated Assyrian Knobplates in the British Museum, Iraq 53 (1991), 50, Fig. 6 und Pl. VII und VIII. Das Inschriftfragment gehört zu Assurnasirpal II. Text RIMA II: A.0.101.109 (Die Lesung: $^{d}$NIN-[...„..] in Oelsner 1988: 18, Fußnote 5 ist mit: kid$^{9}$-mu-[ri šá URU kal-hi] zu ersetzen). Die Knauffliese gehörte also zum Besitz des Kidmuru Tempels in Kalchu, welcher der Göttin Ischtar gewidmet war (siehe D.J. Wiseman, A New Stela of Assur-Nasir-Pal II., Iraq 14 (1952), 34, Zeile 58).

9 Veröffentlicht in J. Oelsner, a.a.O., 18 und Anm. 6.

10 Die Inschriften wurden von Oelsner, a.a.O, 18, Anm. 3 identifiziert. Zwei von ihnen (XI–1206 u. XI–1207) sprechen über den Bau des Tempels für den Gott Nindara (RIME III/1.1.7.31), die dritte (XI–1208) über den Wiederaufbau des E-ninnu Tempels für den Stadtgott Ningirsu (RIME III/1.1.7.41).

11 Oelsner, a.a.O., 18, Anm. 3.

12 Veröffentlicht in L. Matouš, Deux tablettes cappadociennes de la collection des Czartoryski a Cracovie, Archeologia 8 (1956), 415–426 und ders., Zwei „kappadokische Tontafeln im Nationalmuseum zu Krakau", Journal of Juristic Papyrology 11/12 (1957/58), 111–118.

kau eine Stiftung (Fundacja Książąt Czartoryskich) gegründet, die – vermittels der Verwaltung des Nationalmuseums in Krakau – erneut die Disponierung über die Sammlungen übernommen hat.

Im Nationalmuseum in Krakau selbst wurde 1886 eine 97 Roll- und Stempelsiegel zählende Schenkung von Konstanty Schmidt-Ciążyński (1817–1889) deponiert. Schmidt-Ciążyński war ein Kunsthändler, der eine längere Zeit in St. Petersburg lebte und dort ein Antiquitätengeschäft besaß. Die Siegel umfassen eine Zeitspanne von der frühdynastischen bis in die sassanidische Periode und decken territorial die Regionen von Zypern bis nach Iran ab.[13] Die größte Gruppe bilden 76 sassanidische Siegel, die zweitgrößte Gruppe 12 altbabylonische Rollsiegel und 9 Siegel gehören zur neubabylonischen Periode (Abb. 6–11).

Parallel zur Krakauer Sammlung wurde in der zweiten Hälfte des 19. Jahrhunderts von Izabella Działyńska (1830–1899, geborene Czartoryska, Schwester des Herzogs Władysław Czartoryski) mit Hilfe ihres Gatten, des Grafen Jan Działyński (1829–1880), eine Sammlung aufgebaut, die ebenfalls zahlreiche Altertümer umfasste. Sie wurde anfangs in Paris – im Hôtel Lambert – aufbewahrt, jedoch 1883 in das private Museum im Schloss Gołuchów (Großpolen) überführt. Dieses Schloss wurde 1893 zum Majoratsgut der Herzöge Czartoryski, und nach dem Tode von Izabella erbte es ihr Neffe Witold Czartoryski (1864–1945), der die Sammlung weiter ausbaute. Den Kern dieser Sammlung von Altertümern bildet eine exzellente Kollektion griechischer Vasen. Vorderasiatische Denkmäler hingegen waren in der Sammlung in nur geringem Umfang vertreten. Immerhin sind aber in einem von W. Froehner 1899 publizierten Katalog sechs altorientalische Objekte verzeichnet (S. 103–104): (1–4) vier Tonnägel mit Bauinschriften Gudeas (ohne Beschreibung der Texte), (4a) ein Tonnagel mit Inschrift des Ba'u. Einer Übersetzung von Pierre Lacau zufolge berichtete die Inschrift über den Aufbau des Eninnu-Tempels für den Gott Ningirsu. Der Text könnte also RIME III/1.1.6.4 entsprechen,[14] (5) eine mit Keilschrift beschriebene Tontafel.

Die Objekte, die den Krieg überdauerten, befinden sich seit 1947 im Nationalmuseum in Warschau. Allerdings sind von den Tonnägeln mit den Gudeainschriften nur noch zwei und ein halber erhalten, einer spricht über den Bau des Tempels é-gír-su$^{ki}$ für den Gott Meslamta'ea[15]. Die übrigen (Inv.-Nr. 147058 und 147060 MNW) sind unveröffentlicht, beide sprechen von der Restaurierung des Eninnu-Tempels für den Gott Ningirsu[16].

13 B. Kaim-Małecka, Catalogue des intalles orientales des Musées Polonais, Acta Archaeologica Lovaniensia 32 (1993), 59–95: frühdynastisch (Nr. 1,2); akkadisch (3); neusumerisch (7, 8); Isin-Larsa (12,13); altbabylonisch (15, 16, 18–25, 28, 29), neubabylonisch (34–38, 40); achaemenidisch (41–44); syrisch (129, 131), kappadokisch (133), zyprisch (136, 138) und sasanidisch (48–59, 61–76, 78–92, 94–105, 107–128).

14 Dieser Tonnagel ist während des 2. Weltkrieges verschollen. Siehe G. Mizera (Hg.), Straty wojenne. Sztuka Starożytna. Obiekty utracone w Polsce w latach 1939–1945. Tom I (Wartime Losses. Ancient Art. Objects Lost in Poland in the Years 1939–1945) Vol. I, Poznań 2000, 329, Kat.1233.

15 A. Reiche, Katalognoten, in: W. Dobrowolski (Hg.), Galeria Sztuki Starożytnej. Egipt, Bliski Wschód. Przewodnik, Warszawa 2007, 129 (II.4); Der Text = RIME III/1.1.7.24.

16 Inv.-Nr. 147058 MNW und 147060 MNW (sehr fragmentarisch erhalten). Der Text = RIME III/1.1.7.41.

Abb. 6 | Rollsiegel mit einer Kampfszene zwischen Helden und Tieren. Moderne Abrollung. Babylonien, Irak. Frühdynastische Zeit, Mitte des 3. Jahrtausends v. Chr. Lapislazuli. H. 2,4 cm, D. 1,5 cm. Nationalmuseum Krakau, Inv.-Nr. MNK-IV G-1.

Abb. 7 | Rollsiegel mit einer Kampfszene zwischen Helden und Tieren. Moderne Abrollung. Babylonien, Irak. Frühdynastische Zeit, Mitte des 3. Jahrtausends v. Chr. Muschel. H. 2,1 cm, D. 0,9 cm. Nationalmuseum Krakau, Inv.-Nr. MNK-IV G-2.

Abb. 8 | Rollsiegel mit einer Darstellung des „Königs mit Keule" vor dem Gott Ea (?) stehend, hinter ihm ein kniender Held mit einer Vase mit fließendem Wasser. Moderne Abrollung. Babylonien, Irak. Altbabylonische Zeit, Erste Hälfte des 18. Jh. v. Chr. Hämatit. H. 3,5 cm. D. 1,3 cm. Nationalmuseum Krakau, Inv.-Nr. MNK-IV G-10.

Abb. 9 | Rollsiegel mit einer Anbetungsszene des Sonnengottes. Die Inschrift: (1) ìr $^{d}$MAR.TU (2) ìr $^{d}$*nè-eri*$_{11}$*-gal*. Moderne Abrollung. Babylonien, Irak. Altbabylonische Zeit, 18. Jh. v. Chr. Hämatit. H. 2,8 cm. D. 1,4 cm. Nationalmuseum Krakau, Inv.-Nr. MNK-IV G-11.

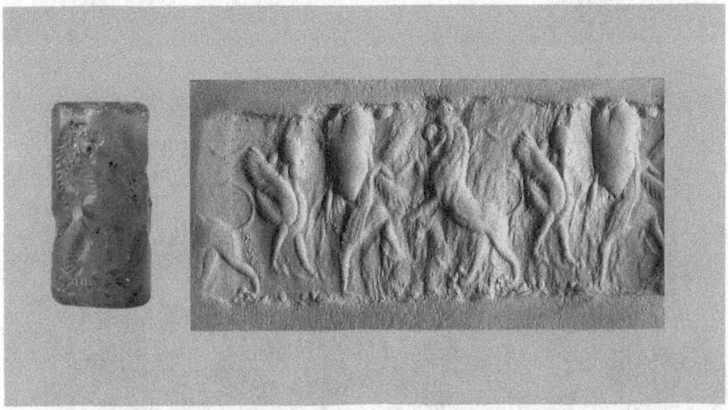

Abb. 10 | Rollsiegel mit Darstellung eines vierflügeligen Genius im Kampf mit einem Löwen. Ein Mischwesen später zugefügt. Moderne Abrollung. Babylonien, Irak. Neubabylonische Zeit, 9–8. Jh. v. Chr., und Spätbabylonische Zeit, 6.–5. Jh. v. Chr. Bergkristall. H. 2,5 cm, D. 1,2 cm. Nationalmuseum Krakau, Inv.-Nr. MNK-IV G-18.

Abb. 11 | Rollsiegel mit Darstellung eines Priesters(?), der ein Mischwesen anbetet. Moderne Abrollung. Babylonien, Irak. Spätbabylonische Zeit, 6.–5. Jh. v. Chr. Jaspis. H. 2,8 cm, D. 1,32 cm. Nationalmuseum Krakau, Inv.-Nr. MNK-IV G-20.

Abb. 12 | Rollsiegel mit Adorationsszene vor einem „Gottkönig". Moderne Abrollung. Babylonien, Irak. Altbabylonische Zeit, 19. Jh. v. Chr. Hämatit. H. 2,4 cm, D. 1,7 cm. Nationalmuseum Warschau, Inv.-Nr. MNW 148431.

Abb. 13 | Rollsiegel mit Adorationsszene mit dem „König mit Keule". Moderne Abrollung. Babylonien, Irak. Altbabylonische Zeit, 18. Jh. v. Chr. Hämatit. H. 2,1 cm, D. 1 cm. Nationalmuseum Warschau, Inv.-Nr. MNW 148432.

Die Ur III-zeitliche Tontafel stammt aus Umma und ist auf das 4. Regierungsjahr des Amar-Sin datiert. Es ist eine Quittung (nur die Zahl ist angegeben), die mit dem Siegelabdruck des A-kal-la dub-sar dumu Ur-e$_{11}$-e SAHAR versehen wurde.[17]

Zur Gołuchów Sammlung gehörte jedoch eine weitere, bei Froehner 1899 nicht erwähnte Sammlung von 22 Roll- und Stempelsiegeln, die möglicherweise erst von Witold Czartoryski gekauft wurde. Auch dieser Bestand befindet sich heute im Nationalmuseum in Warschau, aber erst seit 1954. Diese Siegelgruppe vertritt – ähnlich der vorher erwähnten Siegel-Sammlung aus Krakau – eine Zeitspanne vom Ende des 3. Jahrtausends vor Chr. bis in die sassanidische Zeit (6. Jahrhundert) und umfasst einen breiten geographischen Kreis von Zypern bis Iran.[18]

17 A. Reiche, Kolekcja sumeryjskich tabliczek klinowych w zbiorach Muzeum Narodowego w Warszawie, Rocznik Muzeum Narodowego w Warszawie 26 (1982)100–102, Nr. 7.
18 Die ganze Gruppe wurde von Kaim-Małecka 1993 veröffentlicht.

Unter den Bildmotiven dieser Siegelgruppe sind vor allem die so genannten Anbetungs-, Adorations- und Einführungsszenen von der Akkad-Zeit bis in die altbabylonische Zeit gut vertreten[19].

Darüber hinaus gibt es auch schöne, leider vereinzelte Beispiele assyrischer (Abb. 14), neubabylonischer (Abb. 15, Abb. 16) und achämenidischer Glyptik (Abb. 17), darunter auch zwei Beispiele der graeco-persischen Gemmen in Form eines Skarabäoids mit Darstellungen von zwei kämpfenden Kriegern (Abb. 18) und eines Bogenschützen in persischer Tracht (Abb. 19).[20]

Erwähnenswert ist hier auch ein hethitischer Hämatitstempel mit einer Hieroglypheninschrift[21] (Abb. 20). Eine separate Gruppe bilden syrische und zyprische Roll- und Stempelsiegel.[22] Zuletzt sind noch Beispiele der sassanidischen Glyptik zu nennen.[23]

Drei altvorderasiatische Altertümer befanden sich auch in der im 19. Jahrhundert entstandenen Antikensammlung der Grafenfamilie Potocki in Krzeszowice bei Krakau. Sie kamen als zufällige Schenkungen in die Sammlung. Ein neubabylonischer Tonzylinder und ein neubabylonisches Rollsiegel waren eine Schenkung der Herzogin Elisabeth Voroncov (Woronzow), geborene Branicka, für ihre Schwester die Gräfin Zofia Potocka.[24]

Der Tonzylinder (Abb. 21) trägt eine Inschrift Nebukadnezars II. (605–562 v. Chr.), die über den Bau des alten Palastes in Babylon berichtet.[25]

Das Rollsiegel (Abb. 22) zeigt eine Anbetungsszene, die von einer neubabylonischen Keilinschrift mit einem kurzen Gebet und dem Namen und Vatersnamen des Siegelbesitzers begleitet wird:

19 Insgesamt sind es sieben Rollsiegel; B. Kaim-Małecka, Catalogue des intalles orientales des Musées Polonais, Acta Archaeologica Lovaniensia 32 (1993), 59–95: Nr. 4 (akkadisch), 9 (neusumerisch), 11, 14 (Isin-Larsa), 17, 22, 27 (altbabylonisch).

20 B. Kaim-Małecka, a.a.O., Nr. 46,47; A. Mierzejewski, Katalognoten, in: W. Dobrowolski (Hg.), Galeria Sztuki Starożytnej. Egipt, Bliski Wschód. Przewodnik, Warszawa 2007, 149 (II.35; II.36).

21 P. Bieliński, Un cachet hittite du Musée National de Varsovie, Rocznik Muzeum Narodowego w Warszawie 22 (1978), 7–10.

22 B. Kaim-Małecka, a.a.O., Nr. 132, 135,137, 139,140.

23 Ebd., Nr. 60, 77, 106.

24 Die Kontakte der Fürstin Elisabeth mit ihrer Schwester Gräfin Zofia Potocka waren sehr intensiv. Die Objekte in der Potocki-Sammlung hatten Etiketten, die kurze Notizen – teilweise in Polnisch, teilweise in Französisch – enthielten. Auf dem Etikett des Tonzylinders kann man lesen: *„Ein Geschenk der Fürstin Woronzow geborene Branicka“* [*Dar Ksni Michałowej z Branickich Woronzow*] und *„cilendre [sic!] trouve a Ninive“*. Auf dem Etikett des Rollsiegels befinden sich folgende Notizen: *„Ein Zylinder von außerordentlicher Schönheit“* [Cylinder *bardzo* rzadkiej piękności] und „antiquité de Ninive“. Es ist interessant, dass in beiden Fällen Ninive als Fundort anstatt Babylon angegeben wurde.

25 Seit 1940 befindet er sich als Depot in der Czartoryski-Sammlung in Krakau (siehe K. Moczulska, O kolekcji starożytności rodziny Potockich z Krzeszowic, in: J. Śliwa (Hg.), MUNUS AMICITIAE. Studia archeologiczne poświęcone pamięci Olgi Hirsch-Dyczek, Kraków 1994, 191–204). Der Text wurde in J. Zabłocka/P.-R. Berger, Ein vollständiges Duplikat zur Nebukadnezar-Inschrift VAB 4 Nr. 46, Orientalia. Neue Serie 38 (1969), 122–125 publiziert.

Abb. 14 | Rollsiegel mit Darstellung von kultischen Handlungen am „Heiligen Baum". Moderne Abrollung. Assyrien, Irak. Neuassyrische Zeit, 9.–8. Jh. v. Chr. Serpentin. H. 3,5 cm, D. 1 cm. Nationalmuseum Warschau, Inv.-Nr. MNW 148429.

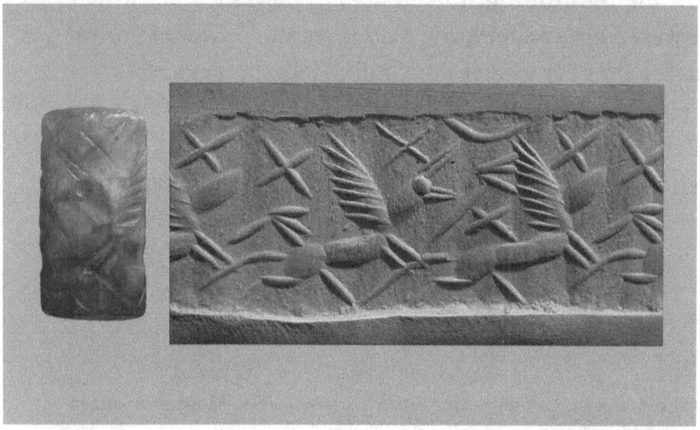

Abb. 15 | Rollsiegel mit laufenden Greifen. Moderne Abrollung. Babylonien, Irak. Neubabylonische Zeit, 6. Jh. v. Chr. Hämatit. H. 2 cm, D. 1 cm. Nationalmuseum Warschau, Inv.-Nr. 148437 MNW.

Abb. 16 | Stempelsiegel mit Anbetungsszene. Moderner Abdruck. Babylonien, Irak. Neubabylonische Zeit, 6.–5. Jh. v. Chr. Chalzedon. Stempelspiegel 1,5 × 2,5 cm, H. 3,1 cm. Nationalmuseum Warschau, Inv.-Nr. 148386 MNW.

Abb. 17 | Stempelsiegel mit Darstellung des Königs als Tierbezwinger. Moderner Abdruck. Babylonien, Irak. Achämenidisch, 6. Jh. v. Chr., Chalzedon. Stempelspiegel 1,5 × 2 cm, H. (abgebrochen) 1,4 cm. Nationalmuseum Warschau, Inv.-Nr. 148387 MNW.

Abb. 18 | Gemme mit Darstellung zweier kämpfender Krieger. Kleinasien (?). Graeco-persischer Stil, 5. Jh. v. Chr., Chalzedon. 2,2 × 2,7 cm, H. 1,5 cm. Nationalmuseum Warschau, Inv.-Nr. 148377 MNW.

Abb. 19 | Gemme mit Darstellung eines Bogenschützens in persischer Tracht. Kleinasien (?). Graeco-persischer Stil, 5. Jh. v. Chr., Chalzedon. 1,7 × 2 cm, H. 0,9 cm. Nationalmuseum Warschau, Inv.-Nr. 148388 MNW.

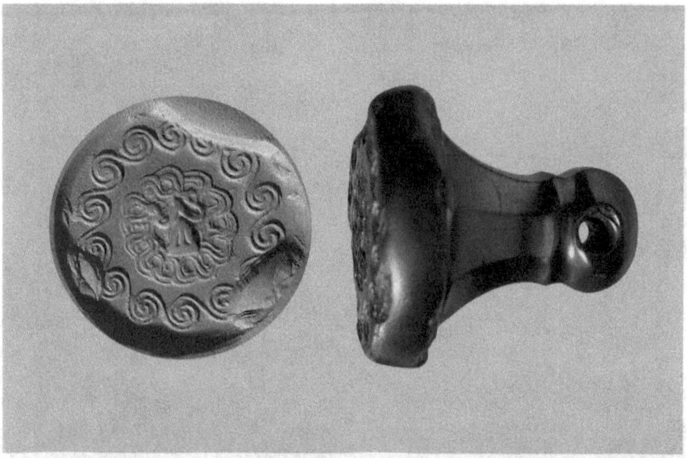

Abb. 20 | Stempelsiegel mit Flechtbändern und dem Namen *Tarhunwassu* oder *Tarhuntassu* in Hieroglyphenschrift. Moderner Abdruck. Anatolien, Türkei. Hethitisches Großreich, 15.–13. Jh. v. Chr. Hämatit. H. 3,1 cm, D. 3,3 cm. Nationalmuseum Warschau, Inv.-Nr. 148459 MNW.

Abb. 21 | Tonzylinder mit einer Inschrift Nebukadnezars II. (604–562 v. Chr.). Babylon, Irak. Ton. Leihgabe der Potocki-Familie aus Krzeszowice. L. 20 cm, D. 5,5 cm. Museum der Herzöge Czartoryski in Krakau, Inv.-Nr. DMNKCz 1652.

(1) $^{\mathrm{d}}$PA PAB ZI TIN BA-*šá* (2) NA$_4$.KIŠIB $^{\mathrm{md}}$PA.PAB.MU (3) A $^{\mathrm{md}}$PA.GIN.A (*Oh Nabû, schütze mein Leben, gib mir Gesundheit, Siegel des Nabû-ahu-iddin, Sohn des Nabû-mukīn-apli*).[26]

26 Seit Ende 1946 befindet es sich im Nationalmuseum in Warschau. Veröffentlicht in A. Mierzejewski, Katalognoten, in: D. Folga-Januszewska (Hg.), 111 Masterpieces of the National Museum in Warsaw, Warsaw 2000, 38–39 und ders., Katalognoten, in: D. Folga-Januszewska/K. Murawska-Muthesius (Hgg.), National Museum in Warsaw. Guide. Galleries and Study Collections, Warsaw 2001, 47, I.23; Das Rollsiegel wurde um 1816 von J. Raymond von Bagdad nach Paris gebracht und dort dem Grafen (ab 1845 Fürst) Michail Voroncov geschenkt oder verkauft (siehe F. Layard, Introduction à l'étude du culte public et des mystères du Mithra en Orient et en Occident. Vol. II, Paris 1847: Kommentar im Tafelverzeichnis zu Pl. LII, 4). Für eine ausführliche Beschreibung des Siegels und Verfolgung seiner Antiquargeschichte siehe A. Reiche/M. Sandowicz, Neo-Babylonian Seal from the Potocki Collection at the National Museum in Warsaw, in: O. Drewnowska (Hg.), Festschrift: Here and There, across the Ancient Near East. Studies in Honour of Krystyna Łyczkowska, Warszawa 2009, 195–200. (Die Lesung der Inschrift von M. Sandowicz).

Abb. 22 | Rollsiegel. Anbetungsszene mit neubabylonischer Inschrift und später hinzugefügten Darstellungen (eine Palme mit Hund und ein auf den Hinterbeinen stehendes Mischwesen). Babylonien, Irak. Neubabylonisch/-Achämenidisch(?), 6.–5. Jh. v. Chr. Chalzedon. H. 4 cm, D. 1,9 cm. Nationalmuseum Warschau, Inv.-Nr. 237444 MNW.

Was dieses Siegel besonders interessant macht, sind zwei später hinzugefügte Szenen, die Merkmale achämenidischer Kunst aufweisen. Das dritte Stück war ein sassanidisches Ringstempelsiegel, das dem begleitenden Etikett nach durch A. Humboldt gefunden wurde.[27]

## II.

In der Zeit zwischen den beiden Weltkriegen, als Polen als Staat wieder aufgebaut wurde, wurden die polnischen Museen mit vielen Schenkungen seitens Privatsammler bereichert. Leider enthielten nur drei von ihnen auch altvorderasiatische Denkmäler.

Im Jahr 1921 schenkte Władysław Semerau-Siemianowski, der als Arzt in der Türkei bei der „Société du Chemin de fer Ottoman d'Anatolie" angestellt war, seine Altertümersammlung dem Nationalmuseum in Warschau. Die Sammlung, die vor allem aus Münzen, Glasgefäßen und Terrakotten bestand, enthielt auch 28 Keilschrifttafeln. Achtzehn von ihnen[28] sind auf die Ur III-Zeit datierte Dokumente aus Umma und Dre-

27 Seit Ende 1946 im Nationalmuseum in Warschau; B. Kaim-Małecka, a.a.O.: Nr 93. Wie dieses Siegel in die Potocki-Sammlung geriet, bleibt eine offene Frage. Es könnte von A. von Humboldt selbst geschenkt worden sein, es ist aber auch möglich, dass es ein Geschenk der Fürstin Elisabeth Voroncov war. Für noch eine andere Möglichkeit siehe J. Żelazowski, Sassanidisches Siegel, Fundstück Alexander von Humboldts, Bulletin du Musée National de Varsovie 36 (1995), 1–6. Für die Beziehungen von A. von Humboldt zu der Familie des Grafen Michail Voroncov siehe K. Zielnica, Polonica bei Alexander von Humboldt. Ein Beitrag zu den deutsch-polnischen Wissenschaftsbeziehungen in der ersten Hälfte des 19. Jahrhunderts, Berlin 2004, 326–328.

28 Eine von ihnen, Inv. Nr. 42791= Bromski II, ist während des 2. Weltkrieges verschollen. Siehe G. Mizera (Hg.), Straty wojenne. Sztuka Starożytna. Obiekty utracone w Polsce w latach 1939–1945. Tom I (Wartime Losses. Ancient Art. Objects Lost in Poland in the Years 1939–1945) Vol. I, Poznań 2000 (im Folgenden Mizera 2000), 332, kat.1243.

Abb. 23 | Ur III-zeitliche Keilschrifttafel aus Umma, mit dem Text einer Quittung und Siegelabdruck des Ur-nungal. Sumer, Irak. Šusin 2. Regierungsjahr (um 2029 v. Chr.). Ton. 4 × 4,2 cm. Nationalmuseum Warschau, Inv.-Nr. 236417 MNW.

hem.[29] Die meisten von ihnen sind Quittungen, darunter einige mit Rollsiegel-Abrollungen und -Abdrücken (Abb. 23) versehen.

Eine Tafel gehört zur Gruppe der so genannten „messenger texts"[30] und eine gehört zu einer kleinen Gruppe, den so genannten „balanced accounts" der Kaufleute.[31] Dieses zuletzt erwähnte Dokument ist einzigartig, da es (wahrscheinlich) nach dem Tode des Kaufmanns hergestellt wurde (Rs. 10: nì-kas$_x$(ŠID)-ak Ur-TAR-luh dam-gàr / egir(LUM)-ba-úš-ta ni-kas$_x$(ŠID)-bi ba-aka).[32]

Von den 10 Tontafeln, die bisher nicht veröffentlicht wurden, sind vier während des 2. Weltkrieges verschollen[33] und die übrigen sind sehr schlecht erhalten. Vier von ihnen lassen sich als neubabylonisch datieren (Inv.-Nr. 199654 und 236426–428 MNW).

Eine Sammlung von 13 Keilschrifttafeln befindet sich seit 1932 auch im Muzeum Okręgowe (Bezirksmuseum) in Toruń (Thorn). Sie ist eine Leihgabe der Toruner Wissenschaftlichen Gesellschaft, in Folge einer Überschreibung von Walerian Amrogowicz

29 Veröffentlicht in J. Bromski, Les tablettes cunéiforms de la collection du Musée National à Varsovie, Rocznik Orientalistyczny 5 (1928), 1–20 (im Folgenden Bromski 1928); siehe auch A. Reiche, Dwa sumeryjskie dokumenty administracyjne z archiwum w Drehem, Rocznik Muzeum Narodowego w Warszawie 25 (1981), 55–64 und ders., Kolekcja sumeryjskich tabliczek klinowych w zbiorach Muzeum Narodowego w Warszawie, Rocznik Muzeum Narodowego w Warszawie 26 (1982), 91–115 (im Folgenden Reiche 1982).

30 Bromski 1928,16 = Reiche 1982:

31 Bromski 1928, 8 (42786) = Reiche 1982, 8; (Neue Inv. Nr. 236415); D.C. Snell, Ledgers and Prices. New Haven/London 1982, Pl. I.

32 Für eine andere Deutung dieser Zeile siehe Snell, a.a.O., 33–34 (und Tafel I).

33 Inv. Nr. 42806, 42807, 42811, 42813 siehe Mizera 2000, 332–33, kat.1244–1246, 1248.

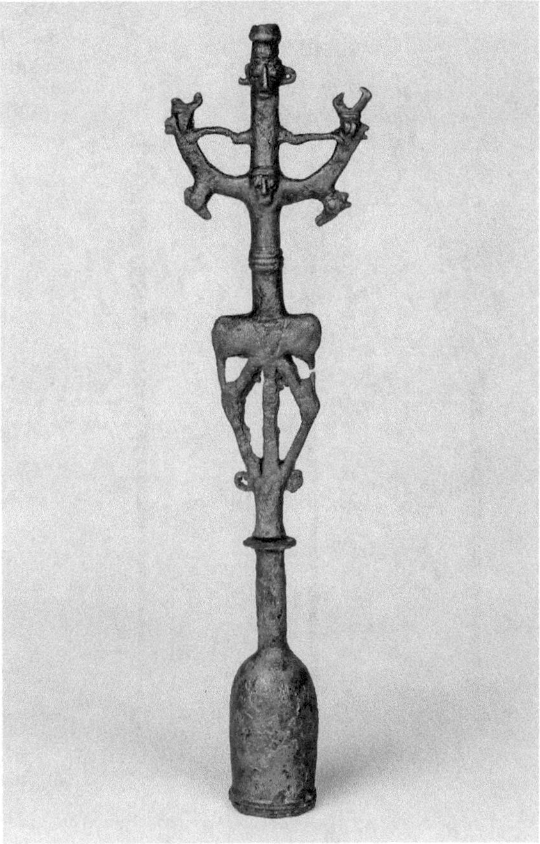

Abb. 24 | Standartenaufsatz in Form eines Tierbezwingers (sogenannter „Herr der Tiere") auf einem Ständer. Luristan. West-Iran. 9.–7. Jh. v. Chr. Bronze. H. (Standarte) 19,5 cm; (Ständer) 10 cm. Nationalmuseum Warschau, Inv.-Nr. 236736 und 236737 MNW.

(1863–1931) aus Sopot (Zopot). Amrogowicz war ein berühmter Numismatiker und Sammler. Zehn von diesen Tafeln sind Dokumente aus der Zeit der 3. Dynastie von Ur (einige von ihnen tragen ebenfalls Rollsiegelabdrücke) und stammen aus Umma und Drehem.[34] Eines der Dokumente gehört zur seltenen Gruppe von Ur III-zeitlichen Briefen, die mit der Formel lugal-gu$_{10}$ u-na-du$_{11}$ beginnen.[35]

Eine außerordentliche Schenkung von vierzehn so genannten „Luristanbronzen" hat das Nationalmuseum in Warschau, im Jahre 1932, seitens Solaiman Nassery, einem Antiquar aus Paris, bekommen.[36]

34 Erste Veröffentlichung: S. Szachno-Romanowicz, Tabliczki szumerskie ze zbiorów Towarzystwa Naukowego Toruńskiego, Rocznik Orjentalistyczny 11 (1935), 79–95. Siehe auch M. Stępień, Kolekcja neosumeryjskich tekstów klinowych w zbiorach Muzeum Okręgowego w Toruniu, Rocznik Muzeum w Toruniu 11 (1992), 63–98.

35 E. Sollberger, The Business and Administrative Correspondence under the Kings of Ur. Locust Valley 1966 (= Texts from Cuneiform Sources I): 41–2, No. 144.

36 S. Przeworski, Ceramika i bronzy staroirańskie w zbiorach polskich, Broń i Barwa. Biuletyn Stowarzyszenia Przyjaciół Muzeum Wojska 3/1 (1936), 1–26, Fig. 11, 12, 14, 16–26.

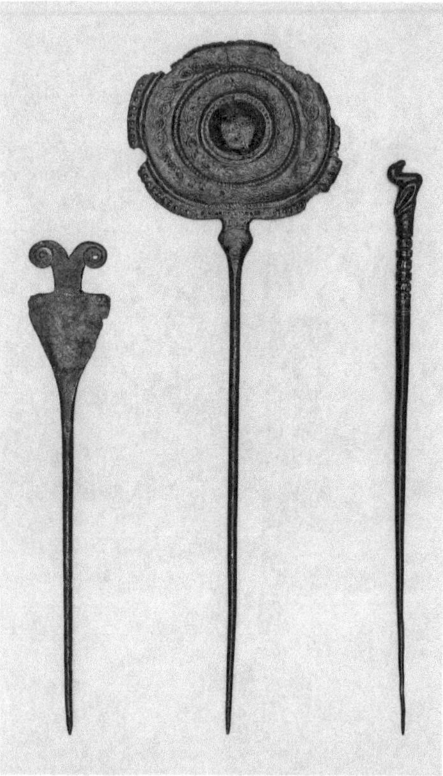

Abb. 25 | Drei verzierte Nadeln. Luristan. West-Iran. 8.–6. Jh. v. Chr. Bronze. L. 18,1; 25,7; 21,4 cm; Nationalmuseum Warschau, Inv.-Nr. 236191; 236193; 236192 MNW.

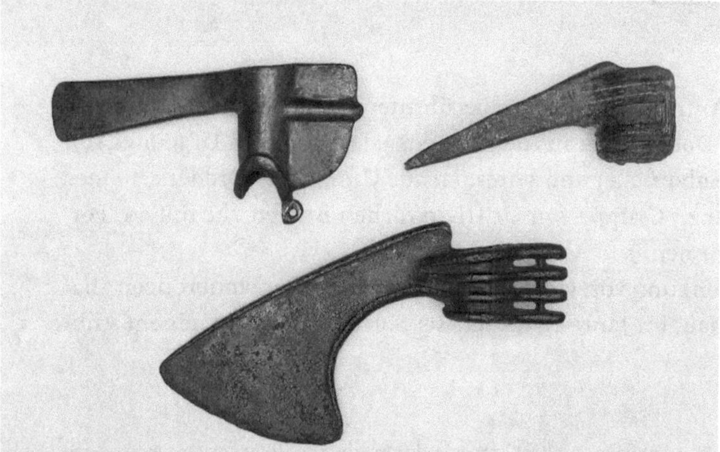

Abb. 26 | Streitäxte aus Luristan. West-Iran. Oben: 2. Hälfte des 3. Jahrtausends v. Chr., 1. Hälfte des 2. Jahrtausends v. Chr., unten: 2. Hälfte des 2. Jahrtausends v. Chr. Bronze. Länge: 14,4; 12,5; 19,3 cm. (b) Nationalmuseum Warschau, (a, c) Langfristige Ausleihe des polnischen Militärmuseums, Inv.-Nr.: 19 MWP*; 238670 MNW; 20 MWP*.

Sie enthielt u.a. einen Standartenaufsatz in Form eines Tierbezwingers[37] (so genannter „Herr der Tiere", Abb. 24), eine einfach gebrochene Trense[38], fünf auf verschiedene Art verzierte Nadeln[39] (Abb. 25) und Schmuck – ein Armband und zwei Ohrringe[40].

Kurz vor dem 2. Weltkrieg wurden von Władysław Dziewanowski zwei Äxte, die auf Luristan hinweisende Charakteristika zeigen,[41] dem Muzeum Wojska Polskiego (Polnisches Armee Museum) in Warschau geschenkt (Abb. 26).

## III.

Nach dem 2. Weltkrieg, als in Polen eine neue Gesellschaftsordnung errichtet wurde, deren Charakteristikum ein Hang zur Nationalisierung und Zentralisierung war, entschied der Staat, im Nationalmuseum in Warschau eine zentrale Sammlung antiker Kunst einzurichten. Aus diesem Grund wurden diesem Museum Altertümer u.a. der Czartoryski-Sammlung aus dem Schloss Gołuchów und der Potocki Sammlung aus Krzeszowice sowie Sammlungen aus den ehemaligen deutschen Gebieten im Westen und Norden Polens zugewiesen. Der Bestand wurde im geringeren Ausmaß auch durch Ankäufe und Schenkungen von Privatleuten bereichert. Meistens handelte es sich um einzelne Stücke oder kleine Gruppen von vorwiegend fragmentarisch erhaltenen Objekten, vor allem Tongefäßen (Abb. 27).

Eine wichtige Bereicherung der Sammlung mit originalen Objekten aus dem 3. Jahrtausend v. Chr., die aus Ausgrabungen stammen, erfolgte in der zweiten Hälfte der neunziger Jahre des vorigen Jahrhunderts.

Bereits 1948 gelangten in das Nationalmuseum drei Fragmente von assyrischen Reliefs, die einst vor dem 2. Weltkrieg im Palast der preußischen Aristokratenfamilie von Perponcher-Sedlnizky in Neudorf (heute Nowizna) in Niederschlesien bei Reichenbach (heute Dzierżoniów) im Eulengebirge in eine Wand eingemauert waren.

Des Weiteren ist ein Fragment mit einer Darstellung des adlerköpfigen Genius zu erwähnen, das aus Kalchu (Nimrud), aus dem Palast des Assurnasirpal II., stammt[42] (Abb. 28).

37 Ebd., 19, Fig.14; A. Mierzejewski Katalognoten, in: D. Folga-Januszewska/K. Murawska-Muthesius (Hgg.), National Museum in Warsaw. Guide. Galleries and Study Collections, Warsaw 2001, 48, I.24.

38 S. Przeworski, Ceramika i bronzy staroirańskie w zbiorach polskich, Broń i Barwa. Biuletyn Stowarzyszenia Przyjaciół Muzeum Wojska 3/1 (1936), 16, Ryc.11.

39 Ebd., 23, Fig.18–22; zwei Nadeln (Ryc. 19 und Ryc. 20 sind verschollen); A. Mierzejewski, Katalognoten, in: W. Dobrowolski (Hg.), Galeria Sztuki Starożytnej. Egipt, Bliski Wschód. Przewodnik, Warszawa 2007, 148, II.33.

40 S. Przeworski, Ceramika i bronzy staroirańskie w zbiorach polskich, Broń i Barwa. Biuletyn Stowarzyszenia Przyjaciół Muzeum Wojska 3/1 (1936), 24, Ryc. 23–25.

41 Ebd., 11, Ryc.5 und 15, Ryc. 10; A. Mierzejewski, Remarques sur les categories fonctionnelles des martels en bronze du Luristan, Etudes et Travaux 3 (1969), 17–35; ders., Katalognoten, in: W. Dobrowolski (Hg.), Galeria Sztuki Starożytnej. Egipt, Bliski Wschód. Przewodnik, Warszawa 2007, 145, II.29a,c.

42 I. Belke, Un fragment de bas-relief assyrien du Musée National de Varsovie, Bulletin du Musée National de Varsovie 10/1 (1969), 4–6; A. Mierzejewski, Katalognoten, in: D. Folga-Januszewska/K. Murawska-Muthesius (Hgg.), National Museum in Warsaw. Guide. Galleries and Study Collections, Warsaw 2001, 46, I.21.

Abb. 27 | Becher aus Nord-Mesopotamien und Babylonien. Irak. Kassitisch, 2. Hälfte des 2. Jahrtausends v. Chr.; mitannisch 14. Jh. v. Chr.; mittelassyrisch 13. Jh. v. Chr.; neuassyrisch 9.–8. Jh. v. Chr. Ton. H. 12,3; 12; 10,7; 10,1 cm. Nationalmuseum Warschau, Inv.-Nr. 238447; 237563; 237562; 237560 MNW.

Es gehört zu einer großen Wandplatte, deren Komposition aus zwei durch ein Keilschriftband getrennten Reliefstreifen besteht. Das Textband ist die so genannte „Standardinschrift" des Königs Assurnasirpal II. Das Fragment ist Teil einer im unteren Streifen dargestellten Szene von adlerköpfigen Genienfiguren, die einen „Heiligen Baum" befruchten.[43] An der Oberkante des Fragments sind auch einige Keilschriftzeichen der 21. Zeile der „Standardinschrift" (RIMA II: A.0.101.23) lesbar: KÙ.BABBAR.MEŠ KÙ.GI.MEŠ AN.[NA.MEŠ] *Silber, Gold, B[lei]*.

Zwei andere Relieffragmente stammen aus dem Südwestpalast des Sanherib in Ninive, jedoch sind beide auf die Zeit Assurbanipals (668–631) zu datieren[44] – eines stellt zwei elamische Bogenschützen dar (Abb. 29), die aus der Darstellung des Sieges Assurba-

43 So dekorierte Reliefplatten befanden sich im Raum I des Nordwestpalastes Assurnasirpals II. in Kalchu (Nimrud), siehe S.M. Paley/R.P. Sobolewski, The Reconstruction of the Relief Representations and their Positions in the Northwest-Palace at Kalchu (Nimrud) II. (Rooms: I, S, T, Z, West-Wing), Mainz a. Rhein 1987 (= Baghdader Forschungen 10), Pl. I.

44 I. Belke, Deux fragments de sculpture assyrienne du Musée National de Varsovie, Bulletin du Musée National de Varsovie 8/3 (1967), 57–63; R.D. Barnett/E. Bleibtreu/G. Turner, Sculptures from the Southwest Palace of Sennacherib at Niniveh (668–627), London 1998, 49, Pls. 27.14c und 28.15b; A. Mierzejewski, Katalognoten, in: D. Folga-Januszewska/K. Murawska-Muthesius (Hgg.), National Museum in Warsaw. Guide. Galleries and Study Collections, Warsaw 2001, 47, I.22.

Abb. 28 | Relieffragment mit Adlerkopf eines Genius. Aus dem Nordwestpalast des Assurnasirpal II. in Kalchu (Nimrud), Assyrien. Irak. 9. Jh. v. Chr. Alabaster. 27,5 × 33,7 cm. Nationalmuseum Warschau, Inv.-Nr. 199335 MNW.

nipals in der Schlacht gegen die Elamer am Ulai-Fluss stammen. Das zweite zeigt einen reitenden assyrischen Bogenschützen (Abb. 30). Die Spuren von beseitigten Keilschriftzeichen (oben rechts) verweisen auf eine sekundäre Verwendung der Platte. Alle drei Relieffragmente wurden 1847 von H. J. Ross dem Grafen Wilhelm von Perponcher-Sedlnitzky, der als preußischer Chargé d'affaires in Istanbul tätig war, verkauft. Ross, der zu dieser Zeit Ausgrabungen auf dem Hügel Kujundschik führte, wurde mit dieser Aufgabe brieflich von Sir Henry Layard beauftragt.[45]

Außer den schon oben erwähnten Keilschrifttafeln aus der Semerau-Siemianowski Sammlung befinden sich im Nationalmuseum in Warschau noch einige weitere Beispiele von Keilschrifttexten, die in den sechziger und siebziger Jahren des 20. Jahrhunderts als Ankäufe oder Schenkungen von privaten Leuten den Weg ins Museum fanden.

Zu ihnen gehören vier, der Form nach quadratische Backsteine mit Stempelabdrücken einer Inschrift des Königs Amar-Sin.[46] Die Inschrift wurde einmal zentral auf der quadratischen Grundfläche und zweimal auf den schmalen Seiten abgedrückt. Die 13-zeilige Bauinschrift des Königs nimmt Beziehung auf den Bau des Abzu für den Gott Enki (Z.12–13: Abzu ki-ág-a-ni mu-na-dù). Gemäß der Mitteilungen der Donatoren sollen die Backsteine aus Ur und aus Eridu stammen.

45 G. Turner, Sennacherib's Palace at Niniveh: the Drawings of H.A. Churchill and the Discoveries of H.J. Ross, Iraq 63 (2001), 124, Anm. 35.

46 Zwei Backsteine aus Ur (Inv.-Nr. 149787MNW, 149793 MNW) wurden dem Museum 1964 von Klemens Szatkiewicz und zwei weitere (Inv.-Nr. 237455 MNW, 237456 MNW) 1995 von Janusz Pawlak geschenkt. Veröffentlicht ist nur einer von ihnen, siehe A. Reiche, Katalognoten, in: W. Dobrowolski (Hg.), Galeria Sztuki Starożytnej. Egipt, Bliski Wschód. Przewodnik, Warszawa 2007,13, II.5.

Abb. 29 | Relieffragment mit elamitischen Bogenschützen, aus der Darstellung des Sieges Assurbanipals in der Schlacht gegen die Elamiter am Ulai-Fluss. Südwest-Palast des Sanherib in Ninive (Kujundschik), Assyrien, 7. Jh. v. Chr. Alabaster. 21 × 38 cm. Nationalmuseum Warschau, Inv.-Nr. 199333 MNW.

Abb. 30 | Relieffragment mit einem reitenden assyrischen Bogenschützen. Südwest-Palast des Sanherib (704–681 v. Chr.) in Ninive (Kujundschik). Assyrien, Irak. 7. Jh. v. Chr. Alabaster. 28,2 × 32 cm. Nationalmuseum Warschau, Inv.-Nr. 199334 MNW.

Vorhandene Beispiele sumerischer Herrscherinschriften wurden durch den Ankauf eines Backsteinfragments (Inv.-Nr. 237745 MNW) mit einer nur teilweise erhaltenen, handgeschriebenen Inschrift Gudeas aus Lagasch, die den Bau des Eninnu-Tempels erwähnt, ergänzt. Das unveröffentlichte Fragment gehört wohl zum Text RIME III, 1.1.6.4.

Weiterhin sind zwei Kaufmannsbriefe aus Kanesch (Kültepe) zu erwähnen, die aus der ehemaligen Rudolf Blankertz-Sammlung stammen.[47] Sie wurden nach dem Ende des

47 Veröffentlicht in: J. Lewy, Die Kültepetexte der Sammlung Rudolf Blankertz, Berlin 1929, 16, Text Nr. 2 (Inv.-Nr. 147872 MNW) und 34, Text Nr. 12 (Inv.-Nr. 147871 MNW).

Abb. 31 | Votiv-Tonobelisk mit Keilinschriften des Assurbanipal. Ninive (Kujundschik)? Assyrien, Irak. H. 13,2 cm. Nationalmuseum Warschau, Inv.-Nr. 200995 MNW.

2. Weltkrieges von sowjetischen Soldaten privat erworben und 1950 dem Nationalmuseum verkauft.[48]

Besonders hervorzuheben ist außerdem ein kleiner Tonobelisk mit Inschriften Assurbanipals, die über eine Löwenjagd und eine Opferung berichten[49] (Abb. 31). Das Objekt wurde während des 2. Weltkrieges in Beirut von einem polnischen Historiker, Professor Stanisław Kościałkowski (1881–1960), gekauft und von seinem Sohn Władysław dem Museum 1987 übereignet.

Im Jahr 1974 erhielt das Museum eine Schenkung seitens Herrn B. Barszcz, zu der ein Backsteinfragment mit einer elamitischen Keilinschrift des Untasch-Napirischa (ca. 1260–1235 v. Chr.) gehörte, die auf der Langseite anfängt und auf der Schmalseite

48 Wenigstens eine weitere Tontafel dieser Sammlung befindet sich in Polen in privaten Händen.

49 Abbildung und polnische Übersetzung in A. Mierzejewski, Katalognoten, in: W. Dobrowolski (Hg.), Galeria Sztuki Starożytnej. Egipt, Bliski Wschód. Przewodnik, Warszawa 2007, 141, II.22.

Abb. 32 | Glasierter Tonknauf mit dem Namen des elamitischen Königs Untasch-Napirischa. Dur-Untasch (Tschoga Zanbil), West-Iran. 13. Jh. v. Chr. H. 34 cm. Nationalmuseum Warschau, Inv.-Nr. 200398 MNW.

endet.[50] Hierzu gehörte weiter ein glasierter Tonknauf (Abb. 32) mit Spuren von Keilschriftzeichen, die Reste des Namens desselben Königs zeigen.[51] Im Text auf dem Backstein berichtet der elamische König über den Bau einer Zikkurrat und eines Tempels für den Gott Inschuschinak. Der Tempel soll aus glasierten Backsteinen hergestellt gewesen sein. Es ist sehr wahrscheinlich, dass die beiden Objekte aus Dur-Untasch (Tschoga Zanbil), dem elamischen Kultzentrum im südwestlichen Iran, stammen.

Im Nationalmuseum ist überdies ein Backstein mit einer siebenzeiligen handgeschriebenen Inschrift Salmanassars III. ausgestellt. Er ist eine langfristige Leihgabe des Muzeum Azji i Pacyfiku in Warschau, welches ihn als eine Schenkung in den Achtzigerjahren des 20. Jahrhunderts erhalten hat. Der Backstein stammt aus der Zikkurrat in Kal-

50 Inv.-Nr. 200399, veröffentlicht: A. Reiche, Zabytki pisma klinowego w Polsce (Abschnitt 8), in: C. B. F. Walker, Pismo klinowe, Warszawa 1998, 73–74, Fig.43. Der Text ist identisch mit der Untasch-Napirischa-Inschrift s. F. Malbran-Labat, Les inscriptions royales de Suse: briques de l'epoque paléo-elamite a l'empire néo-elamite. Paris 1995, 71–72, Text 6.

51 Hunderte von solchen glasierten Knäufen, die als architektonische Zierelemente dienten, wurden während der Ausgrabungen in Tchoga Zanbil (Dur Untasch) im südwestlichen Iran gefunden. Siehe R. Ghirshman et. al., Tchoga Zanbil (Dur-Untasch). I. La Ziggurat, Paris 1966 (= Mémoires de la Délégation Archéologique en Iran. Tome 39. Mission de Susiane), Taf. XVII, XIX, XCVIII (maßstabgerechte Zeichnungen).

Abb. 33 | Stierkopfprotome. Sumer. Irak. 3. Jahrtausend v. Chr. Bronze. Nationalmuseum Warschau, Inv.-Nr. 148570 MNW.

chu (Nimrud) was die zwei letzten Zeilen der Inschrift (6–7: *ri-<ṣip>-tú*$_7$ *šá* U$_6$.NIR *šá* URU Kal-ḫi) bestätigen.[52]
Es gibt daneben eine Anzahl von Backstein-Bruchstücken mit fragmentarisch erhaltenen Abdrücken von Stempelinschriften des neubabylonischen Königs Nebukadnezar II. Zumeist handelt es sich um Fragmente der sieben- und dreizeiligen Inschrift dieses Königs mit der Titulatur, die ihn als Wohltäter (za-nin) von Ezida und Esagila bezeichnet, ohne dass auf den Bau eines bestimmten Gebäudes Bezug genommen wird.[53]

Fünf vollständig erhaltene neubabylonische Backsteine mit Inschriften Nebukadnezars II. wurden zudem in den Nachkriegsjahren dem Archäologischen Museum in Krakau geschenkt. Die Ziegel und ihre Inschriften wurden von J. Oelsner veröffentlicht[54]. Es sind zwei 7-, zwei 6- und eine 3-zeilige Inschrift mit der Titulatur des Königs.

Die altvorderasiatische Rundplastik ist in der Warschauer Sammlung nur mit wenigen Beispielen vertreten. Drei von ihnen sind Darstellungen von Stierköpfen – zwei aus Bronze und einer aus Gold – und jeder stammt aus einer anderen Epoche und aus einem anderen geographischen Kontext.

52 A. Reiche, Katalognoten, in: W. Dobrowolski (Hg.), Galeria Sztuki Starożytnej. Egipt, Bliski Wschód. Przewodnik, Warszawa 2007, 132, II.5; Die Inschrift ist identisch mit RIMA III, A.0.102.111.

53 A(b'')$^7$ und A(b'')$^3$ der Klassifikation von Berger 1973 folgend. Nur zwei Fragmente sind veröffentlicht siehe I. Belke, Deux briques Néo-babyloniennes du Musée National de Varsovie, Bulletin du Musée National de Varsovie 2/2 (1961), 44–48.

54 Oelsner 1988: 21.

Abb. 34 | Stierkopfprotome. Urartu, Armenien. 8.–7. Jh. v. Chr. Bronze. Nationalmuseum Warschau, Inv.-Nr. 147077 MNW.

Der älteste von ihnen, ein 1956 gekaufter Bronze- oder Kupferguss, weist stilistische Merkmale der sumerischen Kunst auf und wird auf die 1. Hälfte des 3. Jahrtausends v. Chr. datiert.[55] Der Kopf (Abb. 33) könnte als Protome zur Verzierung des Resonanzkastens einer Leier gedient haben. Tierkopfprotome als Dekorationselemente waren zur dieser Zeit in Sumer verbreitet, worauf die Funde aus den Gräbern vom Königsfriedhof in Ur hinweisen.

Der zweite, aus Bronze gegossene Stierkopf könnte eine urartäische Kesselprotome aus dem 8.–7. Jahrhundert v. Chr. sein (Abb. 34). Seine Herkunft ist unbekannt, im Museum befindet er sich seit 1946.[56]

1965 wurde dem Nationalmuseum von Stanisław Hempel im Namen seiner verstorbenen Frau ein außergewöhnlicher Gegenstand überreicht.[57] Es ist eine aus getriebenem Goldblech hergestellte und durch Granulation dekorierte Stierkopfmaske (Abb. 35,

55 A. Mierzejewski, Protome sumérienne de bronze en forme de tête de taureau, Bulletin du Musee National de Varsovie 13/2–3 (1972), 29–34; ders., Katalognoten, in: W. Dobrowolski (Hg.), Galeria Sztuki Starożytnej. Egipt, Bliski Wschód. Przewodnik, Warszawa 2007,128, II.1.

56 Ebd. (2007), 129, II.3.

57 Hempel war ein polnischer Diplomat, der vor dem 2. Weltkrieg als polnischer Botschafter eine lange Zeit in Iran tätig war. Nach dem Kriegsende wohnte er mit seiner Frau in Paris.

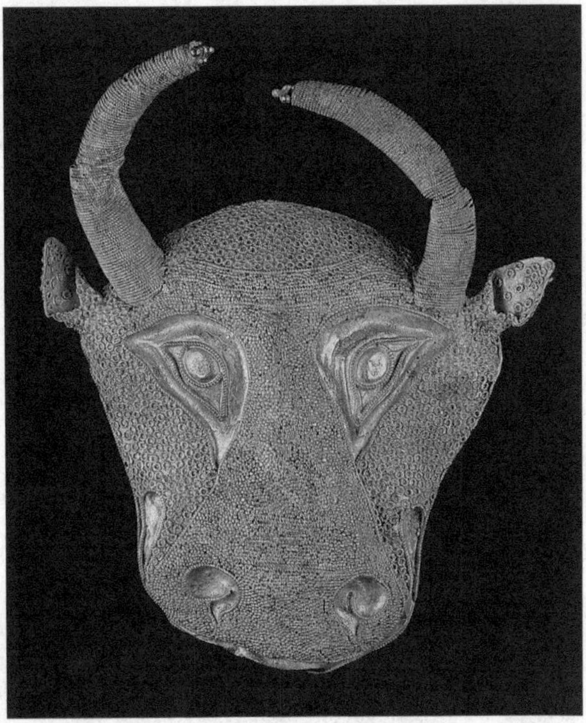

Abb. 35 | Stierkopfmaske. West-Iran (?). 2. Hälfte des 2. Jahrtausends (?). Goldblech. Nationalmuseum Warschau, Inv.-Nr. 149803 MNW.

Abb. 36). In der Mitte der flachen Hinterseite hatte sie einst eine Tülle gehabt[58], die zum Aufsetzen auf einen Bolzen dienen konnte. Es wird vermutet, dass man das Stück zur Verzierung von Prunkpferdegeschirr benutzte.[59] Hempels Informationen nach soll das Stück aus dem Iran stammen und auf die Achämenidenzeit (550–330 v. Chr.) zu datieren sein. Da man aber bis jetzt keine guten Vergleichsstücke für dieses Objekt finden kann, muss diese Datierung bis auf weiteres als unsicher angesehen werden. Aufgrund der Verbreitung des Motivs der Stierkopfmaske in der Kunst Vorderasiens in der zweiten Hälfte des 2. Jahrtausends v. Chr. wurde ein Vorschlag für eine frühere Datierung gemacht.[60] In solch einem Fall hätten wir es mit einem Erzeugnis des elamitischen Handwerks zu tun.

58 Eine derartige Tülle ist sichtbar auf einem Schwarz-Weiß-Foto, welches ein identisches Objekt zeigt. Die Aufnahmen wurde zusammen mit der Schenkung überlassen, leider ohne Informationen über das fotografierte Objekt [Abb. 24 a,b].

59 A. Mierzejewski, Katalognoten, in: D. Folga-Januszewska (Hg.), 111 Masterpieces of the National Museum in Warsaw, Warsaw 2000, 36–37; ders., Katalognoten, in: D. Folga-Januszewska/K. Murawska-Muthesius (Hgg.), National Museum in Warsaw. Guide. Galleries and Study Collections, Warsaw 2001, 49, I.26.

60 A. Reiche, Głowa byka (protoma). Katalognote, in: Wielkie Muzea: Polska, Muzeum Narodowe w Warszawie, Warszawa 2007, 20–21. Die Stierkopfmaske findet man beispielsweise als Füllmotiv in der zyprischen Glyptik (B. Kaim-Małecka, Catalogue des intalles orientales des Musées Polonais, Acta Archaeologica Lovaniensia 32 (1993), 59–95, Nr.137); in einem mitannischen Grab auf Tell Arbid wurde ein Karneolanhänger in Form einer

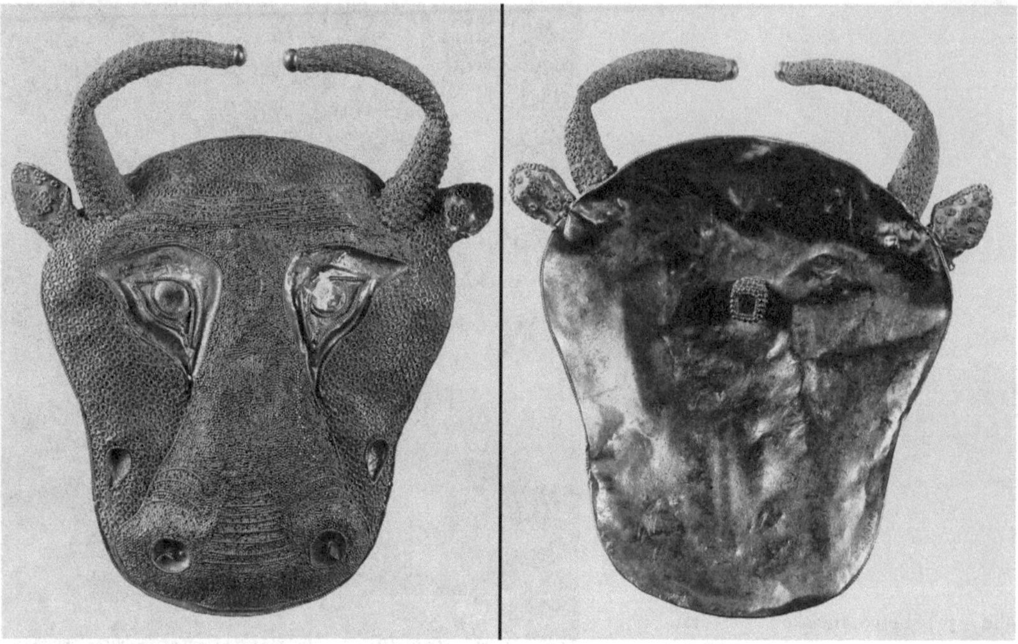

Abb. 36 | Archiv-Aufnahme einer Stierkopfmaske aus Goldblech, die ein Zwillingsexemplar zur Stierkopfmaske Abb. 35 darstellt. Herkunft und Besitzer unbekannt.

Dank der Ankäufe und Schenkungen sowie einer Leihgabe wurde der Bestand der so genannten „Luristanbronzen" im Museum wesentlich erweitert. Durch Schenkung gelangte 1959 seitens Edward Borowski eine Pferdetrense mit Mundstückstange und Wangenplatten in Form von Pferden (Abb. 37) in die Sammlung.[61]

Ankäufe von Dolchen und Kurzschwertern aus Luristan (Abb. 38), Pfeil- und Speerspitzen und einer Streitaxt in Form einer Hacke[62] (s. Abb. 26 oben), ergänzt durch zwei Äxte (Leihgaben aus dem Militärmuseum) und einem Kurzschwert (Leihgabe aus dem Muzeum Azji i Pacyfiku), erlaubten es, eine repräsentative Sammlung von Waffen aus Luristan aufzubauen.

Stierkopfmaske gefunden (A. Smogorzewska, Mittani Grave at Tell Arbid, Damaszener Mitteilungen 15 (2006), 86, Fig. 11,18; Tf. 6 b/3); vgl. die Stierkopfmaske in den Wandmalereien des Government House in Nuzi (R. F. S. Starr, Nuzi: Report on the Excavations at Yorgan Tepe near Kerkuk, Iraq, Conducted. 1927–1931, Vol. II (Plates), Cambridge/Mass. 1937, Pl. 128 H, '29 D).

61 A. Mierzejewski, Mors en bronze a traverse du Luristan, Bulletin du Musee National de Varsovie 10/4 (1969), 81–93.; ders., Katalognoten, in: W. Dobrowolski (Hg.), Galeria Sztuki Starożytnej. Egipt, Bliski Wschód. Przewodnik, Warszawa 2007, 146–7, II.31.

62 Ebd. (2007),145, II.25.

Abb. 37 | Pferdetrense mit Mundstückstange und Wangenplatten. Luristan. West-Iran. 8.–6. Jh. v. Chr. Bronze. B. 20 cm, H. der Platten 8,5 cm. Nationalmuseum Warschau, Inv.-Nr. 149510 MNW.

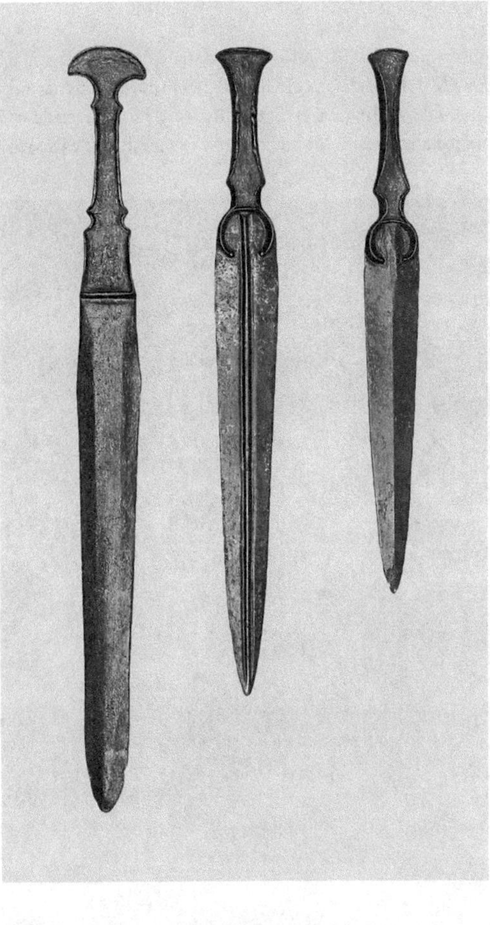

Abb. 38 | Dolche. Luristan. West-Iran. 9.–6. Jh. v. Chr. Bronze. L. 33,8; 28,5; 24,5 cm. Nationalmuseum Warschau, Inv.-Nr.: 200421; 200422; 200423 MNW.

Abb. 39 | Sechs Tongefäße aus dem 3. Jahrtausend v. Chr. aus Nord-Mesopotamien. Tell Abu Hafur (d), Tell Dschassa el-Gharbi (a, c) und Rad Schaqrah (b, e, f), Nordost-Syrien. Mitte des 3. Jahrtausends v. Chr. (a, c-f); 2. Hälfte des 3. Jahrtausends v. Chr. (b); Ton. H. 33; 14,5; 11,5; 29; 6,8; 9 cm. Nationalmuseum Warschau, Inv.-Nr.: 238143; 238190; 238158; 238132; 238174; 238185 MNW.

Abb. 40 | Handgemachter Pokal mit Ritzverzierung in Form von drei schematisch dargestellten Oranten (?). Tell Abu Hafur, Nordost-Syrien. Mitte des 3. Jahrtausends v. Chr. Ton. H. 8,3 cm. Nationalmuseum Warschau, Inv.-Nr. 238130 MNW.

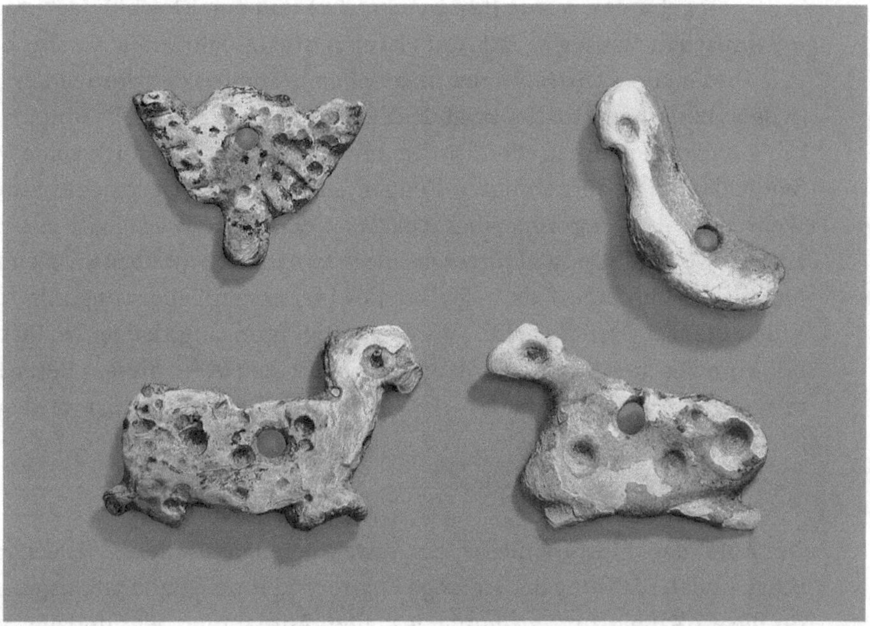

Abb. 41 | Anhänger in Form von Tiersilhouetten. Rad Schaqrah, Nordost-Syrien. Mitte des 3. Jahrtausends v. Chr. Perlmutt. H. 1,4; 2; 1,5; 1,6 cm. Nationalmuseum Warschau, Inv.-Nr.: 238211; 238208; 238210; 238212 MNW.

1995 und 1998 wurde die altorientalische Sammlung des Nationalmuseums in Warschau um Objekte aus den polnischen Rettungsgrabungen im Khaburgebiet im Nordosten Syriens bereichert.[63] Die Ausgrabungen fanden 1988 bis 1990 auf Tell Abu Hafur, Tell Dschassa el-Gharbi statt sowie 1991 bis 1995 in Rad Schaqrah. Die Hauptphase der Besiedlung dieser Hügel lässt sich in die erste Hälfte des 3. Jahrtausends v. Chr. datieren. Von einer späteren Besiedlung wurden nur geringe Reste gefunden (mitannische, neuassyrische und hellenistische auf T. Abu Hafur und T. Dschassa el-Gharbi; akkadische und spätassyrische/neubabylonische auf Rad Schaqrah). Den Grundstock dieser hundert Objekte zählenden Sammlung bilden verschiedene Arten von Tongefäßen, von denen die meisten als Grabbeigaben in Kindergräbern dienten (Abb. 39, 40).

Dieselbe Herkunft haben Schmuckelemente, wie verschiedene Anhänger aus Perlmutt (Abb. 41) und Perlen aus Fritte und Stein. Erwähnenswert sind auch zwei Gegenstände aus Kupfer: eine Sichel[64], die auf dem Fußboden eines akkadischen Hauses in Rad

63 A. Reiche, Polish Archaeological Research in North-Eastern Syria, Bulletin du Musée National a Varsovie XLII (2001, erschienen 2006), 95–106.

64 Ders., Katalognoten, in: W. Dobrowolski (Hg.), Galeria Sztuki Starożytnej. Egipt, Bliski Wschód. Przewodnik, Warszawa 2007, 142–3, II.25.

Schaqrah gefunden wurde (um 2350 v. Chr.) und eine Dolchklinge[65] vom Fußboden eines mitannischen Hauses in Tell Abu Hafur „Ost" (14. Jahrhundert v. Chr.).

Dieser kurze Überblick des materiellen Bestandes der altorientalischen Hochkulturen in den polnischen Museen zeigt deutlich, dass er als nicht sehr umfangreich und eher zufällig entstanden anzusehen ist. Dementsprechend muss auch seine Wirkung in der Öffentlichkeit als wahrnehmbare Zeugen einer mit der Geschichte Europas verbundenen Weltregion als eher gering eingeschätzt werden. Gegenwärtig gibt es keine Möglichkeiten, eine große Sammlung altorientalischer Denkmäler aufzubauen. Dennoch können diese Materialien zumindest als Gegenstände einer ersten Begegnung mit der altorientalischen Kultur dienen. Darüber hinaus wäre ein Programm langfristiger Leihgaben seitens der großen europäischen Museen zu diskutieren, mit dem die in diesem Bereich ärmeren Museen in verschiedenen Ländern des jetzt vereinigten Europas ergänzt werden könnten.

Der Autor dankt insbesondere den Verwaltungen der Nationalmuseen in Krakau und Warschau und der Stiftung der Herzöge Czartoryski für die Publikationsgenehmigungen sowie für die großzügig zur Verfügung gestellten Aufnahmen der Fundstücke.

65 Inv.-Nr. 238141, unveröffentlicht.

## Abstracts / ملخصات

**D. Charpin, Babylon in der altbabylonischen Zeit: eine Hauptstadt von vielen ... die als einzige übrig blieb**

Babylon owes its cultural importance, not only to the military victories and political maneuvers of Hammurabi, as is generally assumed, but also to the external circumstances prevailing in the Old Babylonian period. Only with the decline of the hugely significant empire of Eshnunna and Babylons takeover of its mediating role in the north as well as the acquisition and continuation of the Sumerian traditions of the south after the destruction of Isin and Larsa, did Babylon attain its important cultural position, a position from which it would influence all subsequent societies in the ancient Near East.

**J. Curtis, The Present Condition of Babylon**

The present condition of Babylon is influenced significantly by the exposure of the ruins by German excavations through 1917, the excavation and restoration of the site in the era of Saddam Hussein and the impact of military occupation since 2003. The extent of destruction caused by the stationing of allied troops at the site is documented in the reports of Polish archaeologists, the SBAH and in the author's own visits to the site. Not least, these reports have led to a joint committee consisting of UNESCO, SBAH, the World Monuments Fund and the Getty Conservation Institute, in order to stop any further deterioration of Babylon.

**M. Erler, Chaldeans in Platonism**

The Greeks are well known for taking over and perfecting a number of cultural traditions from earlier peoples. Alongside Egypt, Babylon was often referred to in the literary discourse as a place of general interest and as a city known for its wisdom. This paper illustrates various Babylonian elements in Greek philosophy and the role of Babylonian scholars in the intellectual and cultural history of Greece.

**F. M. Fales, Moving Babylon: On the Aramean and Chaldean presence in Southern Mesopotamia**

Babylon, as the center of the ancient world of Mesopotamia, was influenced and given its particular character by a number of very different factors in the course of its history. One important, but so far only partially recognized role in its development was played by the Aramaic and Chaldean tribes that inhabited the territory surrounding the city in the first half of the first millennium BC. This contribution investigates their importance for the region around Babylon, especially during the period of Assyrian hegemony.

**M. Geller, Die theoretische Grundlage der babylonischen Heilkunde**

Although not explicitly stated in the therapeutic texts, Babylonians possessed a set of theoretical foundations for their own therapeutical practice. The various aspects of Babylonian medicine provide a crucial insight into the epistemology in operation within Babylonian science (imagination, logic, observation and experience), which also demonstrates clear links with pre-hippocratic Greek medicine.

**J.-J. Glassner, Noch einmal: Der Turm zu Babel und die vielen Sprachen**

During the Renaissance the Celts, among others, and their various divine pantheons were identified as part of the early history of Europe. Since the Celtic language was unknown, however, and given the belief that the language of mankind prior to the construction of the Tower of Babel must have been Hebrew, well known scholars developed astounding Hebrew etymologies in order to explain the Celtic names. In searching for the identity of the Celts, the explanations grew so euhemeristic that it necessarily called for a rediscovery of the Assyrians and Babylonians as the cradle of the Semitic world.

**N.P. Heeßel, "Sieben Tafeln aus sieben Städten" – Überlegungen zum Prozeß der Serialisierung von Texten in Babylonien in der zweiten Hälfte des zweiten Jahrtausands v. Chr.**

The process through which various textual series were developed in the course of the first millennium BC (including the Hemerology for Nazimaruttaš, *šumma ālu*, the Uruk List of Kings and Sages from the Second Isin period, three fragmentary Middle Babylonian tablets from Babylon and representatives of the fourth tablet of the bārûtu series) is investigated on the basis of selected colophons. The texts clearly show that standardization was already quite advanced when the series format was first introduced in the Kassite period.

**F. Joannès, L'écriture du pouvoir à Babylone sous Nabuchodonosor II**

Various different textual sources (archives, libraries and monumental inscriptions) shed light on the image of Babylon and its social milieu. The inscriptions of the Chaldean rulers are of particular interest for the relationship between urban architecture and their religious and political background. This article describes the public display of literacy on the part of Nebuchadnezzar II in the broadest sense of the term. It is further demonstrated how the royal self-conception was largely formed through the construction of sacred architecture.

**D. Katz, Reconstructing Babylon: Recycling Mythological Traditions Toward a New Theology**

"Enūma eliš" represents an independent literary achievement of the post-Kassite period, which was formulated without direct dependence on a single literary model. In the myth, traditional motifs are altered and interwoven such that a generational shift among the gods, and in parallel a growing separation between the cities of Nippur and Babylon seems plausible.

**R.G. Kratz, Nabonid in Qumran**

The Prayer of Nabonidus, which is fragmentarily preserved in a Qumran manuscript, is analyzed in relation to the biblical book of Daniel as an example of the reception of Babylonian material in Qumran. The investigation of these and other texts from Qumran demonstrates the meaning that the last Neo-Babylonian king had for the inhabitants of Qumran.

**W.G. Lambert, Babylon: Origins**

The question of the origins of Babylon in the third and early second millennium BC is investigated on the basis of the name of the city and the position of the city god Marduk. The influence of Sippar, the city of the sun god Šamaš, on Babylon in the third millennium is demonstrated. Its importance waned in the course of this century, while in Sumer and Akkad a new triad of gods was established: An, Enlil and Enki. The possibility of a much earlier date than previously accepted for the "exaltation of Marduk" (prior to Hammurabi) on the basis of a newly constructed kinship relationship with Enki is also discussed.

**C. Markschies, Babylon bei den Kirchenvätern**

The question of what church fathers actually knew about the ancient city of Babylon, and the study of their historical and geographical investigations of the city as well as their efforts to synchronize various archaeological finds yields an ambivalent picture. The church fathers had only a very vague knowledge of Babylon. In contrast, there are ancient Christian texts in which Babylon is shown in a particularly negative light. Overall, this corresponds to the tendency of the time, cities were dehistoricized and represented in an exaggerated theological way.

**S.M. Maul, Die Wissenschaft von der Zukunft. Überlegungen zur Bedeutung der Divination im Alten Orient**

Ancient Near Eastern divination plays a prominent role in every period of Mesopotamian history, from the third millennium BC to the end of the fourth century AD. This contribution deals with the analytical techniques of divination, the world view associated with its practice, the significance of divination in the polity and its organizational apparatus. Not only are modern parallels to the predication of the future and a new look at the benefits of this ancient science proposed, but it is also demonstrated that ancient Near Eastern divination was an effective institution, acting as a control mechnaism in opposition to hierarchical structures.

**Umran M. Musa, The situation of the Babylon Archaeological Site until 2006**

From an archaeological point of view and in particular from the point of view of the State Board of Antiquities and Heritage, this paper offers a detailed description of the primary damage to structures both ancient and modern. Damage before 2003 and that which arose in 2003 and after will be carefully differentiated. The report concludes with a set of proposals, outlining necessary measures for the stabilization and preservation of the ruins as well as a brief report on activities that have been undertaken in collaboration with the World Monuments Fund in 2009.

**H. Neumann, Bemerkungen zu einigen Aspekten babylonischen Rechtsdenkens im Spannungsfeld von Theorie und Praxis**

The study of Babylonian law codes as well as legal documents and trial records shows that they make possible a description of both legal practice as well as the the socio-historical lifeworld of Mesopotamia from the third millennium BC to the end of the first Millennium BC. Of particular importance in this regard are the é-dub-ba-a-a (scribal school) materials and especially the literary trial records that were transmitted in the scribal schools. They prove that the scribes possessed not only a practical legal education, but also the capability of bringing theory and practice into a productive relationship with each other within the confines of the early history of law.

**M. Ossendrijver, Science in Action: Networks in Babylonian Astronomy**

This paper examines how the Babylonian astronomers of the Seleucid period cooperated in formation of a network of temple-based scholars in order to develop both astronomical observations and calculations. The colophons of scholarly texts from the library of the Reš-temple in Uruk serve as the main source for the reconstruction of this network. The role of the "scribe" and the "owner" of these tablets will be discussed as well as, for example, the curriculum vitae of the exorcist Nidinti-Anu/Anu-bēlšunu from the family of Ekur-zākir, who was active in Uruk ca. 210 BC.

**O. Pedersén, Excavated and Unexcavated Libraries in Babylon**

This contribution gives an overview of the known libraries of Babylon, including the current state of publication as well as future publications and planned research projects. One focus is the Neo-Babylonian libraries. Many libraries and archives may now be regarded as undetected, as only about 0.04% of the Old Babylonian, about 0.05% of Middle Babylonian and about 1.5% of the Neo-Babylonian city have been unearthed. This yields an incomplete and probably not representative picture of libraries as places in Babylon, where ancient knowledge was both preserved and perpetuated.

**D.T. Potts, The politai and the bīt tāmartu: The Seleucid and Parthian Theatres of the Greek Citizens of Babylon**

The re-analysis of both archaeological and especially written sources, including therein the not yet systematically incorporated data from the "Astronomical Diaries" leads to a reassessment of the dating of the Greek theater in Babylon. The inclusion of all relevant textual sources as well as recent studies the function of Greco-Roman theaters allows us to further specify the connection between the presence of Greek residents in Babylon and the use and political aims of the theatre in Babylon as a public gathering place.

**A. Reiche, Die altorientalischen Denkmäler in den Museen in Polen. Geschichte und Bestand**

This contribution presents a detailed overview of the current status of ancient Near Eastern collections in Polish museums, which have been collected since the nineteenth century through purchase, gifts as well as archeological excavation. The paper focuses on the two major collections of the National Museum in Warsaw and Krakow and concludes with a look at new ways cooperation between the museums of the ancient Near Eastern in Europe.

**W.F. Smelik, Myth and Truth: The Babylonian Talmud in Babylon**

This paper deals with the textual reception of Babylon from the point of view of Babylonian rabbis in terms of three different perspectives: (1) How did the Babylonian Jews see ancient Babylon? (2) How did Babylonian Jewry formulate its own history in Babylonia? And finally, (3) to what degree did Babylonian Jewry acknowledge or give in to acculturation in Sassanid Babylonia?

**G. Strohmaier, Ḥarrān – the City of Sin in the Islamic period**

The cult of the moon god Sin, sponsored by King Nabonidus, had a long historical aftermath in the city of Harran. The paper looks at the Sabians of Harran, who commited themselves to neither Christianity nor Islam, and had sporadic influence in the caliph's palace in Baghdad through the astronomer Ṯābit ibn Qurra (ca. 826–901), who was active in the palace. Sabian metalwork decorated with astronomical signs of the Zodiac, whose motifs were maintained down into the Western art of the Middle Ages, then serve as a primary source for the reconstruction of Babylonian traditions.

**W. van Soldt, The Role of Babylon in Western Peripheral Education**

The school texts, which have turned up as far afield as Anatolia, Syria, the Levant, Egypt and Assur, serve as some of the best sources in helping us to elucidate the kind of role that Babylon played in the transmission of Babylonian knowledge in the periphery. In looking at the babylonian lexical series ḪAR-ra, it can be shown that school materials from Babylonian, and perhaps even Babylon itself, were in use in the western periphery, and these materials may even have been brought by scholars who came from Babylon and settled in the periphery.

**د. شاربان، بابل في الفترة البابلية القديمة: عاصمة من بين العديد من العواصم.... الوحيدة التي بقيت**

لاتَدْينُ بابل بأهميتها الحضارية لانتصارات الملك حمورابي وسياسته فقط، كما يُعتقد بشكل عام، وانما أيضاً للظروف الخارجية المحيطة بها أثناء الفترة البابلية القديمة. فقد حازت
بابل على مكانتها التاريخية الحضارية المهمة، والتي أضاءت الدرب للأجيال القادمة، بعد أفول نجم مملكة أشنونا ذات الاهمية البالغة واستحواذ بابل على دورها كوسيط الى الشمال، ومن خلال تَبَنِّي واستمرارية التقاليد السومرية في الجنوب وذلك بعد تدمير إيسن ولارسا.

**ج. كورتس، الوضع الحالي في بابل**

أن وضع مدينة بابل اليوم قد تأثر بشكل كبير بسبب التنقيبات الآثرية الألمانية للموقع حتى عام ١٩١٧، وبسبب التنقيبات وأعمال الصيانة وأعادة الترميم في عهد صدام حسين، واخيراً بسبب الأحتلال العسكري منذ عام ٢٠٠٣.إن مدى الدمار الذي سببه تمركز قوات التحالف قد تم توثيقه في تقارير أعدها علماء آثار بولنديين والهيئة العامة للآثار والتراث في العراق وكاتب المقال نفسه. أخيرا وليس آخراً، فان هذه التقارير أدت إلى التزام مشترك بين اليونسكو، والهيئة العامة للآثار والتراث في العراق والصندوق العالمي للآثار ومعهد جيتي للصيانة، لأيقاف الانهيار المتزايد في بابل.

**م. أرلر، الكلدان في الفلسفة الأفلاطونية**

لقد أدرك اليونانيون، بانهم أقتبسوا الكثير من الشعوب التي سبقتهم وبأنهم اوصلوا هذه المعرفة لدرجة الكمال. فبالاضافة الى مصر يشار الى بابل بشكل خاص على أنها جزء من المناظرات الادبية، ومكان مثير للأهتمام العام ومدينة الحكمة. هذا المقال يوضح العناصر البابلية في الفلسفة اليونانية، والدور الذي لعبه العلماء البابليون في الثقافة والتاريخ الفكري لليونان.

**ف. م. فالس، بابل المؤثِرة: حول الوجود الآرامي والكلداني في جنوب بلاد الرافدين**

لقد تعرضت بابل وعلى مدى تاريخها كمركز لعالم بلاد الرافدين القديم لتأثيرات مختلفة كثيرة حددت شخصيتها وطابعها. لقد لعبت مجاميع القبائل الآرامية والكلدانية، والتي لم تلقَّ ألا القليل من الاهتمام حتى اليوم، دوراً مهماً في محيط المدينة في النصف الأول من الألف الأول ق.م. يهتم المقال بالسؤال عن أهمية الوجود الآرامي والكلداني في المنطقة المحيطة ببابل وبالاخص في زمن السيطرة الآشورية.

**م. جيلير، الأسس النظرية للطِبّ البابلي**

لقد أمتلك البابليون الأسس النظرية لممارساتهم العلاجية على الرغم من عدم وضوح ذلك في النصوص الطبية. من خلال الجوانب المتعددة للطب البابلي وضمن نطاق العلوم البابلية- الخيال، المنطق، المراقبة والتجربة- أمكن التعرف على نظرية المعرفة البابلية، والتي كانت لها أشارات واضحة إلى ما قبل طب أبقراط اليوناني.

**ج. ج. جلاسنر، مرة ثانية : برج بابل واللغات المتعددة**

خلال عصر النهضة تم الكشف عن صفحة من صفحات تاريخ أوروبا المبكر. من ضمن ما تم اكتشافه كان التعرف على الكلتيين ومجمع آلهتهم المختلفة. و بما أن اللغة الكلتية كانت هناك غير معروفة،و بما أنه كان هناك إيمان، بأن اللغة العبرية قد كانت اللغة الأصلية للأنسانية قبل بناء برج بابل، فقد طور علماء معروفون أشتقاقات عبرية مدهشة لأيضاح الأسماء الكلتية. وفي أثناء البحث عن هوية الكلتيين ظهرت تحليلات عقلانية (أوهيميرية) أدت بدورها الى أعادة أكتشاف الآشوريين والبابليين باعتبارهم مهداً لعالم الشعوب السامية.

**ن. ب، هيزل، "سبع الواح طينية من سبعة مدن". حول مسألة تطور تصنيف النصوص ووضعها في سلاسل في بابل في النصف الثاني من الألف الثاني ق. م.**

أن عملية نشوء وتطور سلسلة النصوص المختلفة من الالف الأول ق.م. قد تم تحليلها بالإعتماد على تذييلات نصوص مختارة، على سبيل المثال، النص الفلكي الخاص بالتقويم الذي إعتمده الملك "نازيماروتاش"، ونصوص الفأل الفلكية المعروفة باسم *شومّا آلو*، وايضاً "قائمة الحكماء والعلماء" من فترة أيسن الثانية، وثلاثة الواح طينية مكسرة من الفترة البابلية الوسيطة من بابل والتي تمثل اللوح الرابع من سلسلة- *باروتو* من الالف الأول ق.م. لقد أظهرت النصوص بوضوح وجود معايير نموذجية متقدمة جداً لتصنيف النصوص، حيث تم العمل في تراكيب سلسلة النصوص هذه بعد العصر الكشي.

**ف. جوانيس، "كتابات السلطة الملكية في بابل تحت حكم الملك نبوخذ نصر الثاني"**

ان صورة بابل ومحيطها الاجتماعي تتضح من خلال النصوص الكتابية المختلفة، الارشيفات، المكتبات والكتابات التذكارية. أن كتابات الحكام الكلدانيين ذات أهمية خاصة بالنسبة للعلاقة بين فن عمارة المدينة وبين خلفيتها الدينية والسياسية. في هذا المقال يتم أستعراض كتابات الملك نبوخذ نصر الثاني الرسمية على نطاق واسع. و سيتم توضيح كيف أن تشكيل مفهوم الذات الملكية كان يتم بشكل كبير عن طريق المباني الدينية.

**د. كاتس، إعادة بناء بابل وأحياء التقاليد الأسطورية نحو علم لاهوت جديد**

تمثل قصة الخليقة البابلية "*Enūma eliš*„ أنتاجاً أدبياً مستقلاً لما بعد الفترة الكاشية، دون الرجوع المباشر الى نصوص أدبية سابقة. لقد أصاب التغيير العناصر التقليدية في الأسطورة واصبحت متشابكة مع بعضها البعض، بحيث يبدو مفهوما التغيير الحاصل بين أجيال الآلهة و تزامنه مع الأنفصال الذي وقع بين مدينتي نفر وبابل.

**ر. ج. كراتس، الملك نبونائيد في لفائف قمران**

في هذا المقال سوف يتم تحليل نص "صلاة نبونائيد" المحفوظ بشكل جزئي بأحدى مخطوطات قمران و ذلك بالقياس الى كتاب دانيال التوراتي، حيث يمكن اعتبار هذا النص مثالاً على وصف بابل في كتابات قمران. أن البحث في هذا النص والنصوص الاخرى من قمران تظهر أية أهمية كانت لآخر ملك من ملوك الفترة البابلية الحديثة عند سكان قمران.

**و. ج. لامبرت، بابل : الأصول**

إن السؤال حول أصول بابل في الالف الثالث ومطلع الالف الثاني ق.م.، يتم عبر تحليل أسم المدينة وهيئة الاهها مردوخ. سوف يتم إظهار، أن بابل في الالف الثالث ق.م. كانت ترزح تحت نفوذ مدينة سبار، مدينة اله الشمس شمش. لقد تضاءلت أهميته (الاله شمش) في غضون هذا القرن، بينما نشأ في سومر واكد ثالوث آلهة جديد: آنو، أنليل وانكي. تتم هنا مناقشة أمكانية إرجاع زمن "رفع منزلة مردوخ" الى وقت أقدم بكثير مما كان مفترضاً في السابق من خلال التشكيل الجديد لعلاقة القرابة بأنكي/أيا وذلك قبل عهد الملك حمورابي.

**س. ماركشيس، بابل عند آباء الكنيسة**

أن السؤال حول معرفة آباء الكنيسة بمدينة بابل القديمة والبحث في ملاحظاتهم التاريخية والاقليمية عن المدينة ومطابقتها مع نتائج التنقيبات الآثرية المختلفة، قد أعطت صورة متناقضة. لقد أمتلك آباء الكنيسة معرفة محدودة جداً حول بابل. في المقابل هناك نصوص مسيحية قديمة قدمت صورة سلبية جداً عن بابل. في مجمل الأمر فان هذا يتطابق مع أتجاه العصر، بتقديم المدن بصورة لاتاريخية و برفعها لاهوتيا.

**ش. م. ماول، علم قراءة المستقبل، تأملات في أهمية العرافة في الشرق الأدنى القديم**

لقد لعبت العرافة دوراً بارزاً في كل الفترات الزمنية لبلاد الرافدين منذ الالف الثالث قبل الميلاد حتى نهاية القرن الرابع بعد الميلاد. تتحدث المقال عن تقنيات تحليل عرافة الأضاحي، و الرؤية الكونية المرتبطة بها، وأهمية العرافة في المجتمع وأجهزته التنظيمية. لا يقتصر الحال هنا فقط على أستعراض نظائر أو

مثائل حديثة لِتَنبؤات المستقبل، أو القاء نظرة جديدة على فوائد هذا العلم القديم، وانما يضاف الى ذلك أيضاً، الإقرار بان عرافة الشرق الادنى القديم كانت مؤسسة مؤثرة، بحيث شَكّلَتْ أنظمة مراقبة ضد الهيكليات المسيطرة.

**م. عمران موسى ، حالة الموقع الأثري في بابل حتى عام ٢٠٠٦**

من وجهة النظر الأثرية، وهنا بالتحديد من وجهة نظر الهيئة العامة للآثار والتراث في العراق، قد تم وضع قائمة مفصلة تصف حالة الاضرار الفادحة التي اصابت الهياكل القديمة والمباني الحديثة هناك. في هذا المضمار تم وبشكل دقيق إيضاح الاضرار التي نجمت ما قبل وتلك التي حصلت ما بعد عام ٢٠٠٣. لقد تم وضع سجل يحتوي على مقترحات بشأن التدابير اللازمة لتحقيق الاستقرار والحفاظ على الموقع الأثري. اضافة الى ذلك، وكجزء مكمل للتقرير، قد تم وضع ملخص عن الأنشطة التي تمت ، منذ عام ٢٠٠٩ بالتعاون مع الصندوق العالمي للآثار.

**هـ. نويمان، تعليقات على بعض جوانب الفكر القانوني البابلي بين أقطاب النظرية والتطبيق**

لقد أظهرت دراسة مجموعة القوانين البابلية و النصوص القضائية و نصوص المحاكم ونصوص الدعاوى، بأنها جميعها وبشكل متساوي تُشكل دليلاً على الممارسة القانونية. كما يظهر من خلالها أمكانية القاء نظرة على التاريخ الأجتماعي في فترة الالف الثالث ق.م. وصولاً لأواخر الألف الأول ق.م.
لقد حظيت (مدرسة الكَتَبَة) "é-dub-ba-a-a„ بأهمية خاصة وعلى وجه التحديد هنا ما عُثِرَ عليه من نصوص في أدب المحاكمات القضائية. لقد أثبتت هذه النصوص على أن الكَتَبَة كانوا يتلقوا مهارات في مبادئ علم القانون الى جانب تدريبهم القانوني العملي، بحيث يُشكل تناغم العناصر النظرية والعملية مع بعضها البعض علاقة مثمرة.

**م. أوسندريفر، العلم عند التطبيق : شبكة عمل لعلوم الفلك البابلية**

تبحث هذه المقالة في التعاون الذي كان قائماً بين الفلكيين البابليين من العصر السلوقي كمشاركين في شبكة من علماء المعبد لجمع المعلومات حول الأرصاد والحسابات الفلكية. لقد تم الاستعانة بتذييلات النصوص العلمية من مكتبة معبد ريش في الوركاء كمصدر رئيسي لأعادة تركيب هذه الشبكة. يتم أيضاً مناقشة دور "الكاتب" و"المالك" لهذه الالواح، وأيضاً تقديم مثالاً للسيرة الذاتية للمُعَوذ نيدينتُ- آنو/آنو- بيلشونو من بيت آيكور- زاكر، والذي مارس نشاطه في الوركاء في ٢١٠ ق.م.

**و. بيدرسون، المكتبات المكتشفة وغير المكتشفة في بابل**

تعطي المقالة لمحة عامة عن المكتبات المعروفة في بابل والحالة الراهنة للنشر وكذلك عن النشر المستقبلي والمشاريع البحثية. يتم هنا التركيز بشكل رئيسي على المكتبات في الفترة البابلية الحديثة. يمكن اعتبار الكثير من المكتبات والآرشيفات في عداد الغير مكتشفة لحد الان، حيث تم التنقيب في المدينة والكشف فقط عن حوالي 0,04% من الفترة البابلية القديمة وحوالي 0,05% من الفترة البابلية الوسيطة وحوالي1,5% من الفترة البابلية الحديثة. ولفهم وضع المكتبات في بابل كأماكن أحتوت وحفظت المعارف القديمة، فان الصورة تبدو حتى اليوم غير مكتملة، بل وربما غير نموذجية.

**د. ت. بوتس، المواطنين اليونانيين في بابل وبيت تامارتو (مكان الرؤية): المسارح السلوقية والبارثية للمواطنين اليونانيين في بابل**

أن إعادة تحليل المصادر الآثرية المعروفة لحد الآن والمصادر الكتابية بشكل خاص بما فيها هنا البيانات الخاصة بـ " اليوميات الفلكية"، والتي لم تُدرج حتى الان بشكل منهجي، قادت الى تقييم جديدة لتأريخ المسارح اليونانية في بابل. فمن خلال رَبْط جميع المصادر الكتابية ذات الصلة بالأبحاث التي أجريت مؤخراً حول وظيفة المسارح اليونانية- الرومانية، أمكن توضيح العلاقة بين وجود السكان اليونانيين في بابل وبين الأستخدام والهدف السياسي للمسرح هناك كمكان للتجمعات العامة.

**أ. رايشه، آثار الشرق الأدنى القديم في المتاحف في بولندا. التاريخ والموجودات**

في لمحة مفصلة حول محتويات المتاحف البولندية من مجموعات أثرية من الشرق الأدنى القديم، والتي تم جمعها منذ القرن التاسع عشر، سيتم أستعراض تاريخ شرائها، واهدائها واخيراً التنقيبات الآثرية التي كشفتها. يتم التركيز هنا على أثنين من أكبر مجموعات المتاحف الوطنية في مدينتي وارشو وكراكاو.
يختتم المقال بالقاء نظرة حول الفرص الجديدة للتعاون بين متاحف الشرق الأدنى القديم في عموم أوروبا.

**و. ف. سميلك، "الأسطورة والحقيقة : بابل في التلمود البابلي"**

بالاعتماد على المصادر الكتابية المتوفرة يتناول المقال صورة بابل من وجهة نظر الأحْبار البابليين، و ذلك من خلال ثلاث وجهات نظر مختلفة: (١) كيف تنظر الديانة اليهودية البابلية إلى بابل القديمة؟ (٢) كيف صاغت الديانة اليهودية البابلية تاريخها الخاص بها في بابل؟ واخيراً (٣) الى أي مدى تعترف اليهودية البابلية بالتأقلم مع ثقافة بلاد بابل الساسانية؟

**ج. شترومايز، حران، مدينة الآله سين في العصر الإسلامي**

كانت عبادة اله القمر سين، والتي رعاها الملك نبونائيد في حران، قد عاشت ذكراها لفترة طويلة من الزمن. أن مادة البحث هنا هم صابئة حران، الذين كانوا لا يدينون لا بالمسيحية ولا بالإسلام، و من خلال عالم البلاط الفلكي ثابت ابن قرة (٨٢٦-٩٠١) تمتعوا بتأثير مؤقت في بلاط الخلافة في بغداد. كمصدر رئيسي لإعادة تركيب التقاليد البابلية يتم الاعتماد على المصنوعات المعدنية الصابئية ذوات نقوش الأبراج الفلكية، والتي أستمرت أشكالها في الظهور في الفن الغربي للعصور الوسطى.

**و. فان سولت، دور بابل في التعليم في المناطق المحيطة به غرباً**

أن النصوص المدرسية المكتشفة في كل من بلاد الاناضول، سورية، بلاد شرق البحر الابيض المتوسط، مصر وآشور تخدم كمصدر رئيسي للإجابة على السؤال، حول الدور الذي لعبته بابل في نقل المعرفة البابلية الى تلك المناطق المحيطة بها من الغرب. فاستناداً الى السلسلة المعجمية البابلية خار-*را* = *خوبولو* أمكن تعزيز الدليل، بانه قد تم استخدام مواد تعليمية من بلاد بابل، وربما من بابل نفسها، في تلك المناطق المحيطة بها من الغرب، والتي من الممكن أنها جاءت عن طريق معلمين ذوي أصول ونشأة بابلية.

# Abkürzungsverzeichnis

| | |
|---|---|
| A | Museumssignatur der Asiatic Collection des Oriental Institute, Chicago |
| a.a.O. | am angegebenen Ort |
| aB | altbabylonisch |
| ACT | O. Neugebauer, Astronomical Cuneiform Texts: Babylonian Ephemerides of the Seleucid Period for the Motion of the Sun, the Moon, and the Planets, I–III, London 1955 |
| ADRT | A. J. Sachs/H. Hunger, Astronomical Diaries and Related Texts from Babylonia I–III. Diaries, Wien 1988–1996; H. Hunger/A. J. Sachs, Astronomical Diaries and Related Texts from Babylonia, Vol. V. Lunar and Planetary Texts, Wien 2001 (= Denkschriften der phil.-hist. Klasse 299); H. Hunger, Astronomical Diaries and Related Texts from Babylonia, Vol. VI. Goal-Year Texts, Wien 2006 (= Denkschriften der phil-hist. Klasse 346) |
| Anm. | Anmerkung |
| AO | Museumssignatur des Louvre (Antiquités orientales) |
| Art. | Artikel |
| Ash. | Tontafelsignatur des Ashmoleum Museum, Oxford |
| Aufl. | Auflage |
| av. | avant |
| B. | Breite |
| BA | Beiträge zur Assyriologie und semitischen Sprachwissenschaft, Leipzig 1889/90ff. |
| BAK | H. Hunger, Babylonische und assyrische Kolophone, Neukirchen-Vluyn 1968 (= Alter Orient und Altes Testament 2) |
| BAM | F. Köcher et al., Die babylonisch-assyrische Medizin in Texten und Untersuchungen, Berlin 1963ff. |
| BaM Beih | Baghdader Mitteilungen. Beihefte, Berlin/Mainz 1968ff. |
| BCHP | I. Finkel/R.J. van der Spek, Babylonian Chronicles of the Hellenistic Period |
| Bd. | Band |
| Beih. | Beiheft |
| BiMes | Bibliotheca Mesopotamica, Malibu 1975ff. |
| BM | Museumssignatur des British Museum in London |
| BRM | Babylonian Records in the Library of J. Pierpont Morgan, New Haven 1917ff. |
| CAD | The Assyrian Dictionary of he University of Chicago, Chicago 1956ff. |
| cit. | citatae |
| cf. | confer |
| CM | Cuneiform Monographs, Groningen/Leiden 1992ff. |
| CT | Cuneiform Texts from Babylonian Tablets in the British Museum, London 1896ff. |
| CTH | E. Laroche, Catalogue des textes hittites, Paris 1971 |
| D. | Durchmesser |
| DAI | Deutsches Archäologisches Institut |
| dem. | demselben |
| ders. | Derselbe |
| dies. | dieselbe |
| Diri | Lexikalische Serie DIRI = (*w*)*atru* |
| Diss. | Dissertation |

| | |
|---|---|
| DJD | Discoveries in the Judaean Desert, Publikationsreihe für die Erstheditionen der Schriftrollen vom Toten Meer |
| DOG | Deutsche Orient-Gesellschaft |
| Ea | Lexikalische Serie Ea A = *nâqu* |
| EA | J.A. Knudtzon, Die El-Amarna-Tafeln (= VAB II/1 [1908/15] – II/2 [1909/15]). |
| ebd. | Ebenda |
| ed./eds. | Editor |
| e.g. | for example |
| Fig. | Figure |
| Fn. | Fußnote |
| Fs. | Festschrift |
| H. | Höhe |
| Hg./Hgg. | Herausgeber |
| Hh. | Lexikalische Serie ḪAR-ra = *ḫubullu* |
| HSM | Museumssignatur des Harvard Semitic Museum |
| ibid. | ibidem |
| Id. | Idem |
| JANES | Journal of the Ancient Near Eastern Society, New York 1968/69 ff. |
| JAOS | Journal of the American Oriental Society |
| Jh. | Jahrhundert |
| K | Museumssignatur des British Museum in London (Kuyunjik) |
| KAR | E. Ebeling, Keilschrifttexte aus Assur religiösen Inhalts I/II, Leipzig/Berlin 1929/1934 (= WVDOG 28 und 34) |
| KTU | M. Dietrich/O. Loretz/J. Sanmartín, Die keil-alphabetischen Texte aus Ugarit, Münster 1976 (= Alter Orient und Altes Testament 24/1) |
| L. | Länge |
| Lu | Lexikalische Liste lú = *ša* |
| LXX | Septuaginta |
| mB | mittelbabylonisch |
| MC | Mesopotamian Civilizations, Winona Lake, Indiana |
| m.E. | meines Erachtens |
| MLC | Morgan Library Collection, Siglum der Yale Babylonian Collection, New Haven |
| MNW | Museumssignatur des Nationalmuseums Warschau |
| MSL | Materialien zum sumerischen Lexikon, Rom 1937 ff. |
| Nachdr. | Nachdruck |
| NCBT | Museumssignatur der Newell Collection of Babylonian Tablets (jetzt Yale University, New Haven) |
| neubearb. | neubearbeitet |
| O | Museumssignatur des Musée du Cinquantenaire (Brüssel), Museumssignatur Slg. Oheimir-Texte (Istanbul) |
| OECT | Oxford Editions of Cuneiform Texts, Oxford 1923 ff. |
| OGIS | Orientis graeci inscriptiones selectae (W. Dittenberger) |
| op. cit. | opere citato, opus citatum |
| OrNS | Orientalia, Nova Series, 1932 ff. |
| p. | pagino |
| PRU | Palais royal d'Ugarit. Mission de Ras Shamra, Paris 1955 ff. |
| RA | Revue d'Assyriologie Orientale, Paris 1886 ff. (neu: Revue d'Assyriologie et d'Archeologie) |
| Rev. | Reverse („Rückseite") |
| RGTC | Répertoire géographique des textes cunéiformes (= TAVO Beihefte, Reihe B Nr. 7, 1974 ff.) |

| | |
|---|---|
| RIMA | The Royal Inscriptions of Mesopotamia, Assyrian Periods, Toronto 1987ff. |
| RIME | The Royal Inscriptions of Mesopotamia, Early Periods, Toronto 1990ff. |
| Pl. | Plate |
| RlA | Reallexikon der Assyriologie, seit 1957: Reallexikon der Assyriologie und Vorderasiatischen Archäologie, Berlin/Leipzig 1928/32ff. |
| RIAA2 | T. Boiy, Hellenistic Legal Documents from Uruk in the „Royal Museums of Art and History“ (Brussels), Akkadica 124 (2003), 19–64 |
| Rs. | Rückseite |
| RS | Museumssignatur des Louvre und des Damaskus-Museums (Ras Shamra) |
| RTC | F. Thureau-Dangin, Recueil de tablettes chaldéennes, Paris 1903 |
| SEG | Supplementum epigaphicum graecum, 1923ff. |
| Sm | Museumssignatur des British Museum in London (Smith) |
| s. | siehe |
| s.o. | siehe oben |
| SpTU | Spätbabylonische Texte aus Uruk, I: H. Hunger (= Ausgrabungen der Deutschen Forschungsgemeinschaft in Uruk-Warka 9, 1976); II–III: E. von Weiher (=Ausgrabungen der Deutschen Forschungsgemeinschaft in Uruk-Warka 10, 1983; 12, 1988); IV–V: E. von Weiher (= Ausgrabungen in Uruk-Warka. Endberichte 12, 1993; 13, 1998) |
| TCL | Textes cunéiformes. Musée du Louvre, Département des Antiquités Orientales, Paris 1910ff. |
| Tf. | Tafel |
| TU | F. Thureau-Dangin, Tablettes d'Uruk: à l'usage des prêtres du Temple d'Anu au temps des Séleucides (= TCL 6, 1922) |
| u. | und |
| u.a. | und andere |
| UET | Ur Excavations. Texts, London 1928ff. |
| unpub. | unpubliziert |
| UVB | Vorläufiger Bericht über die Ausgrabungen in Uruk-Warka, 1930–1940 und 1956ff. |
| VAT | Museumssignatur des Vorderasiatischen Museums in Berlin (Vorderasiatische Abteilung. Tontafeln) |
| v. | von |
| viz. | videlicet |
| vgl. | vergleiche |
| Vol. | Volume |
| Vs. | Vorderseite |
| VS | Vorderasiatische Schriftdenkmäler der Königlichen Museen zu Berlin, Leipzig 1907–1917, NF Vorderasiatische Schriftdenkmäler der Staatlichen Museen zu Berlin, Berlin 1971ff. |
| YOS | Yale Oriental Series, Babylonian Texts, New Haven 1915ff. |
| W | Fundsignatur aus Uruk-Warka (Bagdad, Berlin) |
| WVDOG | Wissenschaftliche Veröffentlichungen der Deutschen Orient-Gesellschaft, Leipzig/Berlin 1936ff. |

# Indices

Die Schreibung der Eigennamen folgt einer (auch in sich uneinheitlichen) deutschsprachigen Konvention.

## 1. antike Orte/Regionen/Landschaften

## 1.1 antike Denkmäler



## 2. Sammlungen

## 3. Personen (Antike, Spätantike, Islam)

## 4. Personen der Zeitgeschichte

## 5. Völker (Altertum)



## 6. Institutionen der Zeitgeschichte



## 7. Quellen

### 7.1. Artefakte/ archäologische Objekte

## 7.2. Texte

### *7.2.1 Keilschriftliche Quellen (Sumerisch, Akkadisch, Hethitisch)*

Abkürzungen und Siglen folgen CAD bzw. Archiv für Orientforschung

### 7.2.2 *Hebräisch, Aramäisch*

#### 7.2.2.1 *Bibelstellen*

### *7.2.2.2 Qumran*

### 7.2.2.3 *Rabbinica*



### 7.2.3 *Griechisch*

### *7.2.4 Latein*

### *7.2.5 Arabisch*

## Abbildungsnachweise

J. Curtis, The Present Condition of Babylon, Fig. 1: J. E. Curtis
J. Curtis, The Present Condition of Babylon, Fig. 2. J. E. Curtis
J. Curtis, The Present Condition of Babylon, Fig. 3: J. E. Curtis
J. Curtis, The Present Condition of Babylon, Fig. 4: J. E. Curtis
J. Curtis, The Present Condition of Babylon, Fig. 5: J. E. Curtis
J. Curtis, The Present Condition of Babylon, Fig. 6: Trustees of the British Museum
J. Curtis, The Present Condition of Babylon, Fig. 7: J. E. Curtis
J. Curtis, The Present Condition of Babylon, Fig. 8: J. E. Curtis
J. Curtis, The Present Condition of Babylon, Fig. 9: J. E. Curtis
J. Curtis, The Present Condition of Babylon, Fig. 10: J. E. Curtis
J. Curtis, The Present Condition of Babylon, Fig. 11: J. E. Curtis
J. Curtis, The Present Condition of Babylon, Fig. 12: J. E. Curtis
J. Curtis, The Present Condition of Babylon, Fig. 13: J. E. Curtis
J. Curtis, The Present Condition of Babylon, Fig. 14: J. E. Curtis
J. Curtis, The Present Condition of Babylon, Fig. 15: J. E. Curtis
J. Curtis, The Present Condition of Babylon, Fig. 16: J. E. Curtis
M.U. Musa, The Situation of the Babylon Archaeological Site until 2006, Fig. 1: DAI-Orient-Abteilung
M.U. Musa, The Situation of the Babylon Archaeological Site until 2006, Fig. 2–31: State Board of Antiquities and Heritage, Iraq (Die Abbildungen Nr. 2; 3; 4; 5; 8; 19; 20; 21; 22; 23 und 32 sind erstmals für den "Report on Damage Assessment in Babylon" des "International Coordination Commitee for the Safeguarding of the Cultural Heritage of Iraq, Sub-commitee on Babylon (UNESCO 2009) freigegeben worden.)
M.U. Musa, The Situation of the Babylon Archaeological Site until 2006, Fig. 32: Satellite Imagery aquired by US State Department; interpretation by Elizabeth Stone
O. Pedersén, Excavated and Unexcavated Libraries in Babylon, Fig. 1: ADO25 Fig. 1, 23, and 50
O. Pedersén, Excavated and Unexcavated Libraries in Babylon, Fig. 2: ibid., Fig. 1
O. Pedersén, Excavated and Unexcavated Libraries in Babylon, Fig. 3: ibid., Fig. 2
O. Pedersén, Excavated and Unexcavated Libraries in Babylon, Fig. 4: VAM Berlin
O. Pedersén, Excavated and Unexcavated Libraries in Babylon, Fig. 5: op.cit., Fig. 23
O. Pedersén, Excavated and Unexcavated Libraries in Babylon, Fig. 6: op. cit., Fig. 25
O. Pedersén, Excavated and Unexcavated Libraries in Babylon, Fig. 7: VAM Berlin
O. Pedersén, Excavated and Unexcavated Libraries in Babylon, Fig. 8: op. cit., Fig. 50

O. Pedersén, Excavated and Unexcavated Libraries in Babylon, Fig. 9: op. cit., Fig. 73
O. Pedersén, Excavated and Unexcavated Libraries in Babylon, Fig. 10: op. cit., Fig. 98
O. Pedersén, Excavated and Unexcavated Libraries in Babylon, Fig. 11: VAM Berlin
O. Pedersén, Excavated and Unexcavated Libraries in Babylon, Fig. 12: op. cit., Fig. 112
S. Maul, Die Wissenschaft von der Zukunft – Überlegungen zur Bedeutung der Divination im Alten Orient, Abb. 1: S. M. Maul 2003, 78 (nach R. Leiderer, Anatomie der Schafsleber im babylonischen Leberorakel. Eine makroskopisch-analytische Studie, München/Bern/Wien/San Francisco 1990, 158, Abb. 2)
S. Maul, Die Wissenschaft von der Zukunft – Überlegungen zur Bedeutung der Divination im Alten Orient, Abb. 2: R. Leiderer, Anatomie der Schafsleber im babylonischen Leberorakel. Eine makroskopisch-analytische Studie, München/Bern/Wien/San Francisco 1990, 161, Abb. 7
S. Maul, Die Wissenschaft von der Zukunft – Überlegungen zur Bedeutung der Divination im Alten Orient, Abb. 5: Trustees of the British Museum
N. Heeßel, „Sieben Tafeln aus sieben Städten" – Überlegungen zum Prozeß der Serialisierung von Texten in Babylonien in der zweiten Hälfte des zweiten Jahrtausends v. Chr., Fig. 1: VAT 11609: Kopie von N. P. Heeßel
M. Ossendrijver, Science in Action:Networks in Babylonian Astronomy, Fig. 1: M. Ossendrijver
D. T. Potts, The politai and the bīt tāmartu: The Seleukid and Parthian Theatres of the Greek Citizens of Babylon, Fig. 1: nach A. Mallwitz, Das Theater von Babylon, in: F. Wetzel/E. Schmidt/A. Mallwitz, Das Babylon der Spätzeit (= WVDOG 62), Berlin 1957, 3–22, Taf. 8a
D. T. Potts, The politai and the bīt tāmartu: The Seleukid and Parthian Theatres of the Greek Citizens of Babylon, Fig. 2: ibid. Taf. 8b
D. T. Potts, The politai and the bīt tāmartu: The Seleukid and Parthian Theatres of the Greek Citizens of Babylon, Fig. 3: ibid. Taf. 9a
D. T. Potts, The politai and the bīt tāmartu: The Seleukid and Parthian Theatres of the Greek Citizens of Babylon, Fig. 4: ibid. Taf. 11a
D. T. Potts, The politai and the bīt tāmartu: The Seleukid and Parthian Theatres of the Greek Citizens of Babylon, Fig. 5: Fig. 5: gezeichnet nach R. Frederiksen, in: T.H. Nielsen (ed.), Even More Studies in the Ancient Greek Polis: Papers from the Copenhagen Polis Centre 6 (= Historia Einzelschriften 162), Stuttgart 2002, 65–124
G. Strohmaier, Ḥarrān – die Stadt des Sin in islamischer Zeit, Abb. 1: Florenz, R. Museo Nazionale, Collezione Carrand
G. Strohmaier, Ḥarrān – die Stadt des Sin in islamischer Zeit, Abb. 2: Zeichnung bei D. S. Rice, The Wade Cup in the Cleveland Museum of Art, Paris 1955, 18, Fig. 13b
G. Strohmaier, Ḥarrān – die Stadt des Sin in islamischer Zeit, Abb. 3: Zeichnung ibid., 20, Fig. 15a

G. Strohmaier, Ḥarrān – die Stadt des Sin in islamischer Zeit, Abb. 4: VAM, Aufnahme R.-B.Wartke / O. Teßmer

G. Strohmaier, Ḥarrān – die Stadt des Sin in islamischer Zeit, Abb. 5: Bibliothek der Ungarischen Akademie der Wissenschaften, Sammlung Kaufmann, Aufnahme I. Ormos

G. Strohmaier, Ḥarrān – die Stadt des Sin in islamischer Zeit, Abb. 6: Leihgabe der „Evangelischen Kirche Berlin-Brandenburg-schlesische-Oberlausitz" an die „Stiftung Stadtmuseum Berlin", Inv.-Nr. III.57.33.k, Aufnahme A. Henkys

A. Reiche, Die altvorderasiatischen Denkmäler in den Museen in Polen. Geschichte und Bestand, Abb. Abb. 1: Museum der Herzöge Czartoryski in Krakau

A. Reiche, Die altvorderasiatischen Denkmäler in den Museen in Polen. Geschichte und Bestand, Abb. 2: Museum der Herzöge Czartoryski in Krakau

A. Reiche, Die altvorderasiatischen Denkmäler in den Museen in Polen. Geschichte und Bestand, Abb. 3: Museum der Herzöge Czartoryski in Krakau

A. Reiche, Die altvorderasiatischen Denkmäler in den Museen in Polen. Geschichte und Bestand, Abb. 4: Museum der Herzogen Czartoryski in Krakau

A. Reiche, Die altvorderasiatischen Denkmäler in den Museen in Polen. Geschichte und Bestand, Abb. 5: Museum der Herzöge Czartoryski in Krakau

A. Reiche, Die altvorderasiatischen Denkmäler in den Museen in Polen. Geschichte und Bestand, Abb. 6: Nationalmuseum Krakau

A. Reiche, Die altvorderasiatischen Denkmäler in den Museen in Polen. Geschichte und Bestand, Abb. 7: Nationalmuseum Krakau

A. Reiche, Die altvorderasiatischen Denkmäler in den Museen in Polen. Geschichte und Bestand, Abb. 8: Nationalmuseum Krakau

A. Reiche, Die altvorderasiatischen Denkmäler in den Museen in Polen. Geschichte und Bestand, Abb. 9: Nationalmuseum Krakau

A. Reiche, Die altvorderasiatischen Denkmäler in den Museen in Polen. Geschichte und Bestand, Abb. 10: National Museum Krakau

A. Reiche, Die altvorderasiatischen Denkmäler in den Museen in Polen. Geschichte und Bestand, Abb. 11: National Museum Krakau

A. Reiche, Die altvorderasiatischen Denkmäler in den Museen in Polen. Geschichte und Bestand, Abb. 12: National Museum Warschau

A. Reiche, Die altvorderasiatischen Denkmäler in den Museen in Polen. Geschichte und Bestand, Abb. 13: National Museum Warschau

A. Reiche, Die altvorderasiatischen Denkmäler in den Museen in Polen. Geschichte und Bestand, Abb. 14: National Museum Warschau

A. Reiche, Die altvorderasiatischen Denkmäler in den Museen in Polen. Geschichte und Bestand, Abb. 15: Nationalmuseum Warschau

A. Reiche, Die altvorderasiatischen Denkmäler in den Museen in Polen. Geschichte und Bestand, Abb. 16: Nationalmuseum Warschau

A. Reiche, Die altvorderasiatischen Denkmäler in den Museen in Polen. Geschichte und Bestand, Abb. 17: Nationalmuseum Warschau

A. Reiche, Die altvorderasiatischen Denkmäler in den Museen in Polen. Geschichte und Bestand, Abb. 18: Nationalmuseum Warschau

A. Reiche, Die altvorderasiatischen Denkmäler in den Museen in Polen. Geschichte und Bestand, Abb. 19: Nationalmuseum Warschau

A. Reiche, Die altvorderasiatischen Denkmäler in den Museen in Polen. Geschichte und Bestand, Abb. 20: Nationalmuseum Warschau

A. Reiche, Die altvorderasiatischen Denkmäler in den Museen in Polen. Geschichte und Bestand, Abb. 21: Museum der Herzöge Czartoryski in Krakau

A. Reiche, Die altvorderasiatischen Denkmäler in den Museen in Polen. Geschichte und Bestand, Abb. 22: Nationalmuseum Warschau

A. Reiche, Die altvorderasiatischen Denkmäler in den Museen in Polen. Geschichte und Bestand, Abb. 23 : Nationalmuseum Warschau

A. Reiche, Die altvorderasiatischen Denkmäler in den Museen in Polen. Geschichte und Bestand, Abb. 24: Nationalmuseum Warschau

A. Reiche, Die altvorderasiatischen Denkmäler in den Museen in Polen. Geschichte und Bestand, Abb. 25: Nationalmuseum Warschau

A. Reiche, Die altvorderasiatischen Denkmäler in den Museen in Polen. Geschichte und Bestand, Abb. 26: Nationalmuseum Warschau (b); Langfristige Ausleihe des Polnischen Militär Museum in Warschau (a,c)

A. Reiche, Die altvorderasiatischen Denkmäler in den Museen in Polen. Geschichte und Bestand, Abb. 27: Nationalmuseum Warschau

A. Reiche, Die altvorderasiatischen Denkmäler in den Museen in Polen. Geschichte und Bestand, Abb. 28: Nationalmuseum Warschau

A. Reiche, Die altvorderasiatischen Denkmäler in den Museen in Polen. Geschichte und Bestand, Abb. 29: Nationalmuseum Warschau

A. Reiche, Die altvorderasiatischen Denkmäler in den Museen in Polen. Geschichte und Bestand, Abb. 30: Nationalmuseum Warschau

A. Reiche, Die altvorderasiatischen Denkmäler in den Museen in Polen. Geschichte und Bestand, Abb. 31: Nationalmuseum Warschau

A. Reiche, Die altvorderasiatischen Denkmäler in den Museen in Polen. Geschichte und Bestand, Abb. 32: Nationalmuseum Warschau

A. Reiche, Die altvorderasiatischen Denkmäler in den Museen in Polen. Geschichte und Bestand, Abb. 33: Nationalmuseum Warschau

A. Reiche, Die altvorderasiatischen Denkmäler in den Museen in Polen. Geschichte und Bestand, Abb. 34: Nationalmuseum Warschau

A. Reiche, Die altvorderasiatischen Denkmäler in den Museen in Polen. Geschichte und Bestand, Abb. 35: Nationalmuseum Warschau

A. Reiche, Die altvorderasiatischen Denkmäler in den Museen in Polen. Geschichte und Bestand, Abb. 36: Archivaufnahme Nationalmuseum Warschau

A. Reiche, Die altvorderasiatischen Denkmäler in den Museen in Polen. Geschichte und Bestand, Abb. 37: Nationalmuseum Warschau

A. Reiche, Die altvorderasiatischen Denkmäler in den Museen in Polen. Geschichte und Bestand, Abb. 38: Nationalmuseum Warschau

A. Reiche, Die altvorderasiatischen Denkmäler in den Museen in Polen. Geschichte und Bestand, Abb. 39: Nationalmuseum Warschau

A. Reiche, Die altvorderasiatischen Denkmäler in den Museen in Polen. Geschichte und Bestand, Abb. 40: Nationalmuseum Warschau

A. Reiche, Die altvorderasiatischen Denkmäler in den Museen in Polen. Geschichte und Bestand, Abb. 41: Nationalmuseum Warschau

FSC
www.fsc.org
MIX
Papier | Fördert gute Waldnutzung
FSC® C083411

Kolibri 360 GmbH
Ferdinand-Jühlke-Straße 7
99095 Erfurt, Deutschland
produktsicherheit@kolibri360.de